VOLUME FIVE HUNDRED AND EIGHTY EIGHT

METHODS IN ENZYMOLOGY

Molecular Characterization of Autophagic Responses, Part B

METHODS IN ENZYMOLOGY

Editors-in-Chief

ANNA MARIE PYLE
Departments of Molecular, Cellular and Developmental Biology and Department of Chemistry
Investigator, Howard Hughes Medical Institute
Yale University

DAVID W. CHRISTIANSON
Roy and Diana Vagelos Laboratories
Department of Chemistry
University of Pennsylvania
Philadelphia, PA

Founding Editors

SIDNEY P. COLOWICK and NATHAN O. KAPLAN

VOLUME FIVE HUNDRED AND EIGHTY EIGHT

METHODS IN ENZYMOLOGY

Molecular Characterization of Autophagic Responses, Part B

Edited by

LORENZO GALLUZZI

Department of Radiation Oncology, Weill Cornell Medical College, New York, NY, United States; Université Paris Descartes/Paris V, Paris, France

JOSÉ MANUEL BRAVO-SAN PEDRO

Université Paris Descartes/Paris V, Paris; Gustave Roussy Cancer Campus, Villejuif, France

GUIDO KROEMER

Université Paris Descartes/Paris V, Paris; Gustave Roussy Cancer Campus, Villejuif, France

ELSEVIER

ACADEMIC PRESS

An imprint of Elsevier

elsevier.com

Academic Press is an imprint of Elsevier
50 Hampshire Street, 5th Floor, Cambridge, MA 02139, United States
525 B Street, Suite 1800, San Diego, CA 92101–4495, United States
The Boulevard, Langford Lane, Kidlington, Oxford OX5 1GB, United Kingdom
125 London Wall, London, EC2Y 5AS, United Kingdom

First edition 2017

ISBN: 978-0-12-809674-1
ISSN: 0076-6879

For information on all Academic Press publications
visit our website at https://www.elsevier.com/books-and-journals

Publisher: Zoe Kruze
Acquisition Editor: Zoe Kruze
Editorial Project Manager: Helene Kabes
Production Project Manager: Magesh Kumar Mahalingam
Cover Designer: Mark Rogers

Typeset by SPi Global, India

CONTENTS

CONTRIBUTORS

M.O. Abiodun
Laboratory of Plant Nutrition, Faculty of Agriculture, Kyushu University, Fukuoka, Japan

Y. Araki
Center for Frontier Oral Science, Graduate School of Dentistry, Osaka University, Osaka, Japan

E. Ari
Eötvös Loránd University, Budapest, Hungary

E. Arias
Institute for Aging Studies, Albert Einstein College of Medicine, Bronx, NY, United States

A. Ballabio
Telethon Institute of Genetics and Medicine (TIGEM), Pozzuoli; Medical Genetics, Federico II University, Naples, Italy; Baylor College of Medicine; Jan and Dan Duncan Neurological Research Institute, Texas Children's Hospital, Houston, TX, United States

Y. Bao
Iowa State University, Ames, IA, United States

D.C. Bassham
Iowa State University, Ames, IA, United States

A.D. Baudot
Cancer Research UK Beatson Institute, Glasgow, United Kingdom

L. Bezu
Metabolomics and Cell Biology Platforms, Gustave Roussy Comprehensive Cancer Institute, Villejuif; Equipe 11 labellisée Ligue contre le Cancer, Centre de Recherche des Cordeliers; Université Paris Descartes, Sorbonne Paris Cité; Université Pierre et Marie Curie, Paris; University of Paris Sud XI, Kremlin Bicêtre, France

M. Bonora
Section of Pathology, Oncology and Experimental Biology, Laboratory for Technologies of Advanced Therapies (LTTA), University of Ferrara, Ferrara, Italy

P. Boya
Centro de Investigaciones Biológicas, CSIC, Madrid, Spain

J.M. Bravo-San Pedro
Gustave Roussy Cancer Campus, Villejuif; INSERM, U1138; Equipe 11 labellisée par la Ligue Nationale contre le Cancer, Centre de Recherche des Cordeliers; Université Paris Descartes/Paris V, Sorbonne Paris Cité; Université Pierre et Marie Curie/Paris VI, Paris, France

D. Carmona-Gutierrez
Institute of Molecular Biosciences, NAWI Graz, University of Graz, Graz, Austria

K. Castillo
Centro Interdisciplinario de Neurociencia de Valparaíso, Facultad de Ciencias, Universidad de Valparaíso, Valparaíso, Chile

H.-Y. Chang
Institute of Biological Chemistry, Academia Sinica; Genome and Systems Biology Degree Program, National Taiwan University, Taipei, Taiwan

L. Chen
State Key Laboratory of Medicinal Chemical Biology, Tianjin Key Laboratory of Protein Science, College of Life Sciences, Nankai University, Tianjin, China

Q. Chen
State Key Laboratory of Medicinal Chemical Biology, Tianjin Key Laboratory of Protein Science, College of Life Sciences, Nankai University, Tianjin; State Key Laboratory of Membrane Biology, Institute of Zoology, Chinese Academy of Sciences, Beijing, China

Y.-P. Chu
Institute of Biological Chemistry, Academia Sinica, Taipei, Taiwan

Y.-L. Chung
Cancer Research UK Cancer Imaging Centre, The Institute of Cancer Research, London, United Kingdom

P. Codogno
Institut Necker-Enfants Malades (INEM), INSERM U1151-CNRS UMR 8253, Université Paris Descartes-Sorbonne Paris Cité, Paris, France

A. Cristobal-Sarramian
Institute of Molecular Biosciences, University of Graz and BioTechMed-Graz, Graz, Austria

D. Denton
Centre for Cancer Biology, University of South Australia, Adelaide, SA, Australia

N. Dupont
Institut Necker-Enfants Malades (INEM), INSERM U1151-CNRS UMR 8253, Université Paris Descartes-Sorbonne Paris Cité, Paris, France

T. Eisenberg
Institute of Molecular Biosciences, NAWI Graz, University of Graz; BioTechMed-Graz, Graz, Austria

L. Esteban-Martinez
Centro de Investigaciones Biológicas, CSIC, Madrid, Spain

T.R. Eykyn
Cancer Research UK Cancer Imaging Centre, The Institute of Cancer Research; The Rayne Institute, St Thomas' Hospital, King's College London, London, United Kingdom

T. Farkas
Cell Death and Metabolism, Center for Autophagy, Recycling and Disease, Danish Cancer Society Research Center, Copenhagen, Denmark

E. Fodor
Eötvös Loránd University, Budapest, Hungary

S. Forveille
Metabolomics and Cell Biology Platforms, Gustave Roussy Comprehensive Cancer Institute, Villejuif; Equipe 11 labellisée Ligue contre le Cancer, Centre de Recherche des Cordeliers; Université Paris Descartes, Sorbonne Paris Cité; Université Pierre et Marie Curie, Paris, France

S.A. Freeman
Hospital for Sick Children, Toronto, ON, Canada

L. Galluzzi
Gustave Roussy Cancer Campus, Villejuif; INSERM, U1138; Equipe 11 labellisée par la Ligue Nationale contre le Cancer, Centre de Recherche des Cordeliers; Université Paris Descartes/Paris V, Sorbonne Paris Cité; Université Pierre et Marie Curie/Paris VI, Paris, France; Weill Cornell Medical College, New York, NY, United States

C. Giorgi
Section of Pathology, Oncology and Experimental Biology, Laboratory for Technologies of Advanced Therapies (LTTA), University of Ferrara, Ferrara, Italy

L.C. Gomes-da-Silva
Metabolomics and Cell Biology Platforms, Gustave Roussy Comprehensive Cancer Institute, Villejuif; Equipe 11 labellisée Ligue contre le Cancer, Centre de Recherche des Cordeliers; Université Paris Descartes, Sorbonne Paris Cité; Université Pierre et Marie Curie, Paris, France; University of Coimbra, Coimbra, Portugal

R. Gómez-Sánchez
University of Groningen, University Medical Center Groningen, Groningen, The Netherlands

S. Grinstein
Hospital for Sick Children; Keenan Research Centre, St. Michael's Hospital, Toronto, ON, Canada

A. Hamaï
Institut Necker-Enfants Malades (INEM), INSERM U1151-CNRS UMR 8253, Université Paris Descartes-Sorbonne Paris Cité, Paris, France

J. Han
State Key Laboratory of Medicinal Chemical Biology, Tianjin Key Laboratory of Protein Science, College of Life Sciences, Nankai University, Tianjin, China

C. Hetz
Biomedical Neuroscience Institute, Faculty of Medicine; Program of Cellular and Molecular Biology, Institute of Biomedical Sciences, Center for Molecular Studies of the Cell, University of Chile; Center for Geroscience, Brain Health and Metabolism, Santiago, Chile; Buck Institute for Research on Aging, Novato, CA; Harvard School of Public Health, Boston, MA, United States

Y.-H. Hung
Institute of Biological Chemistry, Academia Sinica; Institute of Biochemical Sciences, College of Life Sciences, National Taiwan University, Taipei, Taiwan

V. Izzo
Gustave Roussy Cancer Campus, Villejuif; INSERM, U1138; Equipe 11 labellisée par la Ligue Nationale contre le Cancer, Centre de Recherche des Cordeliers; Université Paris Descartes/Paris V, Sorbonne Paris Cité; Université Pierre et Marie Curie/Paris VI, Paris, France

M. Jäättelä
Cell Death and Metabolism, Center for Autophagy, Recycling and Disease, Danish Cancer Society Research Center, Copenhagen, Denmark

K. Kainz
Institute of Molecular Biosciences, NAWI Graz, University of Graz, Graz, Austria

O. Kepp
Metabolomics and Cell Biology Platforms, Gustave Roussy Comprehensive Cancer Institute, Villejuif; Equipe 11 labellisée Ligue contre le Cancer, Centre de Recherche des Cordeliers; Université Paris Descartes, Sorbonne Paris Cité; Université Pierre et Marie Curie, Paris, France

S. Kira
Center for Frontier Oral Science, Graduate School of Dentistry, Osaka University, Osaka, Japan

S.D. Kohlwein
Institute of Molecular Biosciences, University of Graz and BioTechMed-Graz, Graz, Austria

G. Kroemer
INSERM, U1138; Equipe 11 labellisée par la Ligue Nationale contre le Cancer, Centre de Recherche des Cordeliers; Université Paris Descartes/Paris V, Sorbonne Paris Cité; Université Pierre et Marie Curie/Paris VI; Pôle de Biologie, Hôpital Européen Georges Pompidou, AP-HP, Paris; Metabolomics and Cell Biology Platforms, Gustave Roussy Cancer Campus, Villejuif, France; Karolinska University Hospital, Stockholm, Sweden

S. Kumar
Centre for Cancer Biology, University of South Australia, Adelaide, SA, Australia

M.O. Leach
Cancer Research UK Cancer Imaging Centre, The Institute of Cancer Research, London, United Kingdom

M. Leduc
Metabolomics and Cell Biology Platforms, Gustave Roussy Comprehensive Cancer Institute, Villejuif; Equipe 11 labellisée Ligue contre le Cancer, Centre de Recherche des Cordeliers; Université Paris Descartes, Sorbonne Paris Cité; Université Pierre et Marie Curie, Paris, France

Y.-M. Lee
National University of Singapore, Singapore, Singapore

K. Lengyel
Eötvös Loránd University, Budapest, Hungary

C. Leroy
Institut Necker-Enfants Malades (INEM), INSERM U1151-CNRS UMR 8253, Université Paris Descartes-Sorbonne Paris Cité, Paris, France

T.-K. Lim
National University of Singapore, Singapore, Singapore

Q. Lin
National University of Singapore, Singapore, Singapore

P. Liu
Metabolomics and Cell Biology Platforms, Gustave Roussy Comprehensive Cancer Institute, Villejuif; Equipe 11 labellisée Ligue contre le Cancer, Centre de Recherche des Cordeliers; Université Paris Descartes, Sorbonne Paris Cité; Université Pierre et Marie Curie, Paris; University of Paris Sud XI, Kremlin Bicêtre, France

J.S. Long
Cancer Research UK Beatson Institute, Glasgow, United Kingdom

K. Ma
State Key Laboratory of Medicinal Chemical Biology, Tianjin Key Laboratory of Protein Science, College of Life Sciences, Nankai University, Tianjin, China

F. Madeo
Institute of Molecular Biosciences, NAWI Graz, University of Graz; BioTechMed-Graz, Graz, Austria

M.C. Maiuri
Gustave Roussy Cancer Campus, Villejuif; INSERM, U1138; Equipe 11 labellisée par la Ligue Nationale contre le Cancer, Centre de Recherche des Cordeliers; Université Paris Descartes/Paris V, Sorbonne Paris Cité; Université Pierre et Marie Curie/Paris VI, Paris, France

S. Marchi
Section of Pathology, Oncology and Experimental Biology, Laboratory for Technologies of Advanced Therapies (LTTA), University of Ferrara, Ferrara, Italy

K. Matsuoka
Laboratory of Plant Nutrition, Faculty of Agriculture; Biotron Application Center; Research Center for Organelle Homeostasis, Kyushu University, Fukuoka, Japan

D.L. Medina
Telethon Institute of Genetics and Medicine (TIGEM), Pozzuoli, Naples, Italy

Y. Mugume
Iowa State University, Ames, IA, United States

K. Müller
Metabolomics and Cell Biology Platforms, Gustave Roussy Comprehensive Cancer Institute, Villejuif; Equipe 11 labellisée Ligue contre le Cancer, Centre de Recherche des Cordeliers; Université Paris Descartes, Sorbonne Paris Cité; Université Pierre et Marie Curie, Paris, France

M. New
The Francis Crick Institute, London, United Kingdom

T. Noda
Center for Frontier Oral Science, Graduate School of Dentistry; Graduate School of Frontier BioSciences, Osaka University, Osaka, Japan

M. Oñate
Biomedical Neuroscience Institute, Faculty of Medicine; Program of Cellular and Molecular Biology, Institute of Biomedical Sciences, Center for Molecular Studies of the Cell, University of Chile; Center for Geroscience, Brain Health and Metabolism, Santiago, Chile

J. O'Prey
Cancer Research UK Beatson Institute, Glasgow, United Kingdom

M.E. Papandreou
Institute of Molecular Biology and Biotechnology, Foundation for Research and Technology—Hellas; Faculty of Medicine, University of Crete, Heraklion, Greece

S. Patergnani
Section of Pathology, Oncology and Experimental Biology, Laboratory for Technologies of Advanced Therapies (LTTA), University of Ferrara, Ferrara, Italy

T. Pendl
Institute of Molecular Biosciences, NAWI Graz, University of Graz, Graz, Austria

F. Pietrocola
Gustave Roussy Cancer Campus, Villejuif; INSERM, U1138; Equipe 11 labellisée par la Ligue Nationale contre le Cancer, Centre de Recherche des Cordeliers; Université Paris Descartes/Paris V, Sorbonne Paris Cité; Université Pierre et Marie Curie/Paris VI, Paris, France

P. Pinton
Section of Pathology, Oncology and Experimental Biology, Laboratory for Technologies of Advanced Therapies (LTTA), University of Ferrara, Ferrara, Italy

M. Radulovic
Institute of Molecular Biosciences, University of Graz and BioTechMed-Graz, Graz, Austria

F. Reggiori
University of Groningen, University Medical Center Groningen, Groningen, The Netherlands

C. Ruckenstuhl
Institute of Molecular Biosciences, NAWI Graz, University of Graz, Graz, Austria

K.M. Ryan
Cancer Research UK Beatson Institute, Glasgow, United Kingdom

J. Sakamaki
Cancer Research UK Beatson Institute, Glasgow, United Kingdom

A. Saric
Ryerson University, Toronto, ON, Canada

A. Sauvat
Metabolomics and Cell Biology Platforms, Gustave Roussy Comprehensive Cancer Institute, Villejuif; Equipe 11 labellisée Ligue contre le Cancer, Centre de Recherche des Cordeliers; Université Paris Descartes, Sorbonne Paris Cité; Université Pierre et Marie Curie, Paris, France

C. Settembre
Telethon Institute of Genetics and Medicine (TIGEM), Pozzuoli;
Dulbecco Telethon Institute (DTI); Medical Genetics, Federico II University, Naples, Italy

H.-M. Shen
Yong Loo Lin School of Medicine, National University of Singapore, Singapore; NUS Graduate School for Integrative Sciences and Engineering, National University of Singapore, Singapore

J. Sánchez-Wandelmer
University of Groningen, University Medical Center Groningen, Groningen, The Netherlands

V. Sica
Gustave Roussy Cancer Campus, Villejuif; INSERM, U1138; Equipe 11 labellisée par la Ligue Nationale contre le Cancer, Centre de Recherche des Cordeliers; Université Paris Descartes/Paris V, Sorbonne Paris Cité; Université Pierre et Marie Curie/Paris VI, Paris; Faculté de Medicine, Université Paris Saclay/Paris XI, Le Kremlin-Bicêtre, France

T. Sigmond
Eötvös Loránd University, Budapest, Hungary

S. Subramani
Section of Molecular Biology, Division of Biological Sciences, University of California, San Diego, CA, United States

J. Tadic
Institute of Molecular Biosciences, NAWI Graz, University of Graz, Graz, Austria

K. Takács-Vellai
Eötvös Loránd University, Budapest, Hungary

S. Tan
School of Biological Sciences, Nanyang Technological University, Singapore, Singapore

N. Tavernarakis
Institute of Molecular Biology and Biotechnology, Foundation for Research and Technology—Hellas; Faculty of Medicine, University of Crete, Heraklion, Greece

S.A. Tooze
The Francis Crick Institute, London, United Kingdom

V. Valenzuela
Biomedical Neuroscience Institute, Faculty of Medicine; Program of Cellular and Molecular Biology, Institute of Biomedical Sciences, Center for Molecular Studies of the Cell, University of Chile; Center for Geroscience, Brain Health and Metabolism, Santiago, Chile

T. Van Acker
The Francis Crick Institute, London, United Kingdom

M. Varga
Eötvös Loránd University, Budapest, Hungary

T. Vellai
Eötvös Loránd University, Budapest, Hungary

B. Villarejo-Zori
Centro de Investigaciones Biológicas, CSIC, Madrid, Spain

J. Wang
Yong Loo Lin School of Medicine, National University of Singapore, Singapore; Interdisciplinary Research Group in Infectious Diseases, Singapore-MIT Alliance for Research & Technology (SMART), Singapore

W. Wang
Section of Molecular Biology, Division of Biological Sciences, University of California, San Diego, CA, United States

E. Wong
School of Biological Sciences, Nanyang Technological University, Singapore, Singapore

T. Xu
Centre for Cancer Biology, University of South Australia, Adelaide, SA, Australia

W.Y. Yang
Institute of Biological Chemistry, Academia Sinica; Institute of Biochemical Sciences, College of Life Sciences; Genome and Systems Biology Degree Program, National Taiwan University, Taipei, Taiwan

J. Zhang
Yong Loo Lin School of Medicine, National University of Singapore, Singapore, Singapore

L. Zhao
Metabolomics and Cell Biology Platforms, Gustave Roussy Comprehensive Cancer Institute, Villejuif; Equipe 11 labellisée Ligue contre le Cancer, Centre de Recherche des Cordeliers; Université Paris Descartes, Sorbonne Paris Cité; Université Pierre et Marie Curie, Paris; University of Paris Sud XI, Kremlin Bicêtre, France

H. Zhou
Metabolomics and Cell Biology Platforms, Gustave Roussy Comprehensive Cancer Institute, Villejuif; Equipe 11 labellisée Ligue contre le Cancer, Centre de Recherche des Cordeliers; Université Paris Descartes, Sorbonne Paris Cité; Université Pierre et Marie Curie, Paris; University of Paris Sud XI, Kremlin Bicêtre, France

Y. Zhu
State Key Laboratory of Medicinal Chemical Biology, Tianjin Key Laboratory of Protein Science, College of Life Sciences, Nankai University, Tianjin, China

A. Zimmermann
Institute of Molecular Biosciences, NAWI Graz, University of Graz, Graz, Austria

PREFACE

L. Galluzzi*,†,‡,§,¶,||,1, J.M. Bravo-San Pedro†,‡,§,¶,||,1,
G. Kroemer†,‡,§,¶,#,**,††,1

*Department of Radiation Oncology, Weill Cornell Medical College, New York, NY, United States
†Equipe 11 labellisée Ligue contre le Cancer, Centre de Recherche des Cordeliers, Paris, France
‡INSERM, U1138, Paris, France
§Université Paris Descartes/Paris V, Sorbonne Paris Cité, Paris, France
¶Université Pierre et Marie Curie/Paris VI, Paris, France
||Gustave Roussy Comprehensive Cancer Institute, Villejuif, France
#Department of Women's and Children's Health, Karolinska Institute, Karolinska University Hospital Q2:07, Stockholm, Sweden
**Metabolomics and Cell Biology Platforms, Gustave Roussy Comprehensive Cancer Institute, Villejuif, France
††Pôle de Biologie, Hôpital Européen George Pompidou, AP-HP, Paris, France
[1]Corresponding authors: e-mail address: deadoc@vodafone.it; chemabsp@gmail.com; kroemer@orange.fr

1. INTRODUCTION

Autophagy is an evolutionary ancient biological process by which eukaryotic cells dispose of unwanted cytosolic entities through lysosomal degradation (Kaur & Debnath, 2015). Three distinct forms of autophagy have been characterized so far, differing from each other in the mechanisms through which autophagic substrates are delivered to lysosomes. In the course of *microautophagy*, cytosolic entities destined to disposal are taken up by lysosomes directly, upon invagination of the lysosomal membrane (Li, Li, & Bao, 2012). *Chaperone-mediated autophagy* relies on the recognition of cytosolic proteins bearing a KFERQ motif by members of the heat–shock protein (HSP) family, coupled to the lysosome-associated protein 2 (LAMP2)-dependent translocation of such cargo:chaperone complexes across the lysosomal membrane (Cuervo & Wong, 2014). *Macroautophagy* involves the progressive sequestration of autophagic cargoes by a double-membraned organelle (commonly known as autophagosome) that—upon sealing—becomes able to fuse with lysosomes (Lamb, Yoshimori, & Tooze, 2013). Although the pathophysiological relevance of microautophagy and chaperone-mediated autophagy is being increasingly recognized (Schneider & Cuervo, 2014; Uytterhoeven et al., 2015), macroautophagy arguably remains the best-characterized form of autophagy (and will be referred to as autophagy here below) (Galluzzi, Pietrocola, et al., 2015).

Autophagy continuously operates at low rates in virtually all eukaryotic cells, ensuring the degradation of normal by-products of life that may become dangerous upon accumulation, like damaged mitochondria and redox-active protein aggregates (Green, Galluzzi, & Kroemer, 2011). Such a baseline autophagic activity is critical for the maintenance of normal cellular functions, as demonstrated by the fact that defects in essential components of the autophagic machinery are associated with clinically relevant disorders including neurodegeneration, aging, and cancer (Choi, Ryter, & Levine, 2013; Galluzzi, Pietrocola, et al., 2015; Menzies, Fleming, & Rubinsztein, 2015). Moreover the autophagic flux, i.e., the actual degradation of autophagic substrates within lysosomes, increases in response to potentially lethal perturbations of intracellular and extracellular homeostasis, including nutritional, metabolic, physical, and chemical cues (Galluzzi, Pietrocola, Levine, & Kroemer, 2014; Green, Galluzzi, & Kroemer, 2014). In this setting, autophagy plays a key adaptive role as it supports the recovery of homeostasis and the survival of stressed cells. Accordingly, the pharmacological or genetic inhibition of essential constituents of the molecular machinery for autophagy generally (but not always) accelerates the demise of cells exposed to adverse microenvironmental conditions (Galluzzi, Bravo-San Pedro, et al., 2015). This said, autophagy can also contribute to cellular demise in a causal manner, both in developmental scenarios (e.g., in the maturation of salivary glands in *Drosophila melanogaster* larvae) (Berry & Baehrecke, 2007) and in cells responding to specific perturbations of homeostasis (e.g., neonatal neurons succumbing to hypoxia/ischemia) (Galluzzi, Bravo-San Pedro, Blomgren, & Kroemer, 2016). In both these cases, the pharmacological or genetic inhibition of autophagy de facto retards the cellular demise, warranting the use of the term "autophagic cell death" (Galluzzi, Bravo-San Pedro, et al., 2015). One particular instance of autophagic cell death that also impinges on the plasma membrane Na^+/K^+ ATPase is commonly known as autosis (Liu et al., 2013). In summary, autophagy mediates robust cytoprotective effects in multiple pathophysiologically relevant settings, but can also exert cytotoxic activity in some scenarios (Galluzzi, Bravo-San Pedro, et al., 2015).

Of note, autophagy can be relatively unselective or exquisitely specific (Sica et al., 2015). Rather unselective forms of autophagy dispose of various, nonessential cytoplasmic constituents and are generally activated by an increased demand of autophagic products (e.g., macromolecules for bioenergetic or anabolic purposes). Conversely, highly selective forms of autophagy target to lysosomal degradation-specific cytoplasmic entities and are often triggered by an increased availability in autophagic substrates (e.g., mitochondria or bacteria decorated with ubiquitin moieties) and/or functions (e.g., the removal of damaged mitochondria or pathogens) (Okamoto, 2014). In particular, "mitophagy" specifically degrades damaged mitochondria, "reticulophagy" nonfunctional portions of the endoplasmic reticulum, "nucleophagy" nuclear buds containing damaged chromatin, "ribophagy" supernumerary ribosomes, "aggrephagy" redox-active protein aggregates, "xenophagy" invading bacteria that escape endocytic vacuoles, "virophagy" constituents of the viral capsid, and "lipophagy" lipid droplets destined to lipolysis (Sica et al., 2015). The existence of all these autophagic programs further underscores the importance of autophagy in the maintenance of cellular homeostasis (Galluzzi, Pietrocola, et al., 2015). In addition, autophagy is critically required for the optimal secretion of ATP by malignant cells undergoing so-called immunogenic cell death (Kroemer, Galluzzi, Kepp, & Zitvogel, 2013), hence playing a key role in the elicitation of anticancer immune responses triggered by some forms of chemotherapy (Galluzzi, Buque, Kepp, Zitvogel, & Kroemer, 2015). Thus, autophagy is also involved in cell-extrinsic mechanisms that ensure the preservation of organismal homeostasis.

Autophagy is regulated by a complex network of interacting signal transduction cascades (Galluzzi et al., 2014; Sica et al., 2015). One of the major regulators of autophagic responses is mechanistic target of rapamycin (MTOR) complex 1 (MTORC1), a supramolecular complex with protein kinase activity that tonically suppresses autophagy in physiologically conditions (Laplante & Sabatini, 2012). Various inducers of autophagy, including nutrient deprivation, operate indeed by shutting down MTORC1 signaling (Mihaylova & Shaw, 2011). Thus, when intracellular ATP levels drop, ADP and AMP accumulate and become able to activated AMP-dependent protein kinase (AMPK), which catalyzes the inactivating phosphorylation of MTORC1 and the activating phosphorylation of unc51-like kinase 1 (ULK1) (Laplante & Sabatini, 2012; Mihaylova & Shaw, 2011). ULK1 initiates a beclin 1 (BECN1)-dependent signal transduction cascade that ultimately results in the formation of autophagosomes (Egan et al., 2011). This process is generally accompanied by the formation of phosphatidylinositol-3-phosphate by phosphatidylinositol 3-kinase catalytic subunit type 3 (PIK3C3, best known as VPS34), as well as by the lipidation of microtubule-associated protein 1 light chain 3 (MAP1LC3, best known as LC3), which accumulates in the autophagosomal membrane to operate as an autophagic adaptor (Lamb et al., 2013). Additional proteins that are required for canonical autophagic responses are autophagy-related 4 (ATG3), ATG4, ATG5, ATG7, ATG9, ATG10, ATG12, and ATG16L1

(Lamb et al., 2013). ATG3, ATG4, and ATG7 cooperate to catalyze the lipidation of LC3 and other members of the Atg8 protein family. ATG7 also cooperates with ATG10 to generate an ATG5–ATG12–ATG16L1 complex that contributes to the elongation of forming autophagosomes. Finally, ATG9 appears to play a critical role in the first steps of autophagosome formation (Lamb et al., 2013). Of note, noncanonical BECN1- and VPS34-independent variants of autophagy have been reported (Codogno, Mehrpour, & Proikas-Cezanne, 2012; Niso-Santano et al., 2015), suggesting the existence of at least some degree of redundancy in the molecular mechanisms that initiate autophagic responses.

Not surprisingly, autophagy has been the subject of an ever more intense wave of investigation throughout the past 2 decades, a progression that has been accompanied by a significant methodological evolution (Ohsumi, 2014). Indeed, dozens of techniques are nowadays available for the characterization of virtually all aspects of autophagic responses, in vitro and in vivo (Klionsky et al., 2016). However, various conceptual/methodological problems that have permeated autophagy research in the past persist, and should be taken into attentive consideration when experimental determinations are designed as well as when results are interpreted. Consensus guidelines recently published in *Autophagy* provide a very comprehensive and detailed analysis of most (if not all) of these problems (Klionsky et al., 2016). Here, we would like to briefly recall three of them, which we feel still have a major negative impact on current studies on autophagy. First, investigators have tended (and still tend) to misemploy end-point biomarkers of autophagy, like the accumulation of autophagosomes in the cytoplasm or the lipidation of LC3, as indicators of an ongoing autophagic response (Klionsky et al., 2016). This is particularly problematic as many of these biomarkers can also accumulate when the last steps of the autophagic program are inhibited. Several approaches are currently available to properly measure autophagic flux in vitro, whereas it remains complicated to perform such measurements in fixed samples (Klionsky et al., 2016). Second, many of the chemicals that have been (and still are) employed to modulate autophagy in experimental settings, including the MTORC1 inhibitor rapamycin, the PI3K inhibitors 3-methyladenine and wortmannin, as well as the lysosomal inhibitor chloroquine, are relatively unspecific and affect various cellular processes other than autophagy, in an on-target or off-target manner (Klionsky et al., 2016). Genetic tools (e.g., homologous recombination, RNA interference, the CRISP/Cas9 system) provide a comparatively more robust approach to ascribe a particular observation or phenotype to autophagy (Dominguez, Lim, & Qi, 2016; Ghildiyal & Zamore, 2009). However, (1) several components of the autophagic machinery exert autophagy-unrelated functions; and (2) the stable absence of a specific protein (as in knockout animals) may cause a considerable rewiring of signaling and metabolism, and/or the activation of compensatory processes. Genetically targeting at least two, if not three, distinct constituents of the autophagic machinery with RNA interference may be sufficient to circumvent the first of these issues (Ghildiyal & Zamore, 2009). Conversely, more refined strategies including the use of activatable Cre-coding constructs and the timely administration of (tissue-specific) Cre-encoding viral vectors are required to avoid cellular and organismal adaptation to the lifelong absence of one or more proteins (Masiero et al., 2009; Rao et al., 2014). We surmise that the development of novel, highly specific chemical inhibitors of autophagy and ever more refined strategies for the timed, tissue-specific genetic blockage of autophagic responses will shed new light on the implication of autophagy in several human disorders.

In *Molecular Characterization of Autophagic Responses* (an issue of the successful *Methods in Enzymology* series) world-leading experts in the field offer a comprehensive panel of techniques that can be used to measure virtually all aspects of autophagy, in vitro and in vivo. Thus, each of the 55 chapters of *Molecular Characterization of Autophagic Responses* provides a detailed description of one or a few protocols that are suitable for the characterization of autophagy in models as diverse as cultured human cancer cells, mice, fish, insects, worms, plants, and yeasts. The book (which consists of two parts, part A and part B) is equally addressed to expert investigators who may wish to expand their methodological competences, and to beginners in this exciting and rapidly expanding area of research.

ACKNOWLEDGMENTS

Authors are supported by the Ligue contre le Cancer (équipe labellisée); Agence National de la Recherche (ANR)—Projets blancs; ANR under the frame of E-Rare-2, the ERA-Net for Research on Rare Diseases; Association pour la recherche sur le cancer (ARC); Cancéropôle Ile-de-France; Institut National du Cancer (INCa); Fondation Bettencourt-Schueller; Fondation de France; Fondation pour la Recherche Médicale (FRM); the European Commission (ArtForce); the European Research Council (ERC); the LabEx Immuno-Oncology; the SIRIC Stratified Oncology Cell DNA Repair and Tumor Immune Elimination (SOCRATE); the SIRIC Cancer Research and Personalized Medicine (CARPEM); and the Paris Alliance of Cancer Research Institutes (PACRI).

Author disclosure: The authors have no conflicts of interest to disclose.

REFERENCES

Berry, D. L., & Baehrecke, E. H. (2007). Growth arrest and autophagy are required for salivary gland cell degradation in Drosophila. *Cell*, *131*, 1137–1148.

Choi, A. M., Ryter, S. W., & Levine, B. (2013). Autophagy in human health and disease. *The New England Journal of Medicine*, *368*, 651–662.

Codogno, P., Mehrpour, M., & Proikas-Cezanne, T. (2012). Canonical and non-canonical autophagy: Variations on a common theme of self-eating? *Nature Reviews. Molecular Cell Biology*, *13*, 7–12.

Cuervo, A. M., & Wong, E. (2014). Chaperone-mediated autophagy: Roles in disease and aging. *Cell Research*, *24*, 92–104.

Dominguez, A. A., Lim, W. A., & Qi, L. S. (2016). Beyond editing: Repurposing CRISPR-Cas9 for precision genome regulation and interrogation. *Nature Reviews. Molecular Cell Biology*, *17*, 5–15.

Egan, D. F., Shackelford, D. B., Mihaylova, M. M., Gelino, S., Kohnz, R. A., Mair, W., et al. (2011). Phosphorylation of ULK1 (hATG1) by AMP-activated protein kinase connects energy sensing to mitophagy. *Science*, *331*, 456–461.

Galluzzi, L., Bravo-San Pedro, J. M., Blomgren, K., & Kroemer, G. (2016). Autophagy in acute brain injury. *Nature Reviews. Neuroscience*, *17*, 467–484.

Galluzzi, L., Bravo-San Pedro, J. M., Vitale, I., Aaronson, S. A., Abrams, J. M., Adam, D., et al. (2015). Essential versus accessory aspects of cell death: Recommendations of the NCCD 2015. *Cell Death and Differentiation*, *22*, 58–73.

Galluzzi, L., Buque, A., Kepp, O., Zitvogel, L., & Kroemer, G. (2015). Immunological effects of conventional chemotherapy and targeted anticancer agents. *Cancer Cell*, *28*, 690–714.

Galluzzi, L., Pietrocola, F., Bravo-San Pedro, J. M., Amaravadi, R. K., Baehrecke, E. H., Cecconi, F., et al. (2015). Autophagy in malignant transformation and cancer progression. *The EMBO Journal*, *34*, 856–880.

Galluzzi, L., Pietrocola, F., Levine, B., & Kroemer, G. (2014). Metabolic control of autophagy. *Cell*, *159*, 1263–1276.

Ghildiyal, M., & Zamore, P. D. (2009). Small silencing RNAs: An expanding universe. *Nature Reviews. Genetics*, *10*, 94–108.

Green, D. R., Galluzzi, L., & Kroemer, G. (2011). Mitochondria and the autophagy-inflammation-cell death axis in organismal aging. *Science*, *333*, 1109–1112.

Green, D. R., Galluzzi, L., & Kroemer, G. (2014). Cell biology. Metabolic control of cell death. *Science*, *345*, 1250256.

Kaur, J., & Debnath, J. (2015). Autophagy at the crossroads of catabolism and anabolism. *Nature Reviews. Molecular Cell Biology*, *16*, 461–472.

Klionsky, D. J., Abdelmohsen, K., Abe, A., Abedin, M. J., Abeliovich, H., Acevedo Arozena, A., et al. (2016). Guidelines for the use and interpretation of assays for monitoring autophagy (3rd edition). *Autophagy*, *12*, 1–222.

Kroemer, G., Galluzzi, L., Kepp, O., & Zitvogel, L. (2013). Immunogenic cell death in cancer therapy. *Annual Review of Immunology*, *31*, 51–72.

Lamb, C. A., Yoshimori, T., & Tooze, S. A. (2013). The autophagosome: Origins unknown, biogenesis complex. *Nature Reviews. Molecular Cell Biology*, *14*, 759–774.

Laplante, M., & Sabatini, D. M. (2012). mTOR signaling in growth control and disease. *Cell*, *149*, 274–293.

Li, W. W., Li, J., & Bao, J. K. (2012). Microautophagy: Lesser-known self-eating. *Cellular and Molecular Life Sciences*, *69*, 1125–1136.

Liu, Y., Shoji-Kawata, S., Sumpter, R. M., Jr., Wei, Y., Ginet, V., Zhang, L., et al. (2013). Autosis is a Na+, K+-ATPase-regulated form of cell death triggered by autophagy-inducing peptides, starvation, and hypoxia-ischemia. *Proceedings of the National Academy of Sciences of the United States of America*, *110*, 20364–20371.

Masiero, E., Agatea, L., Mammucari, C., Blaauw, B., Loro, E., Komatsu, M., et al. (2009). Autophagy is required to maintain muscle mass. *Cell Metabolism*, *10*, 507–515.
Menzies, F. M., Fleming, A., & Rubinsztein, D. C. (2015). Compromised autophagy and neurodegenerative diseases. *Nature Reviews. Neuroscience*, *16*, 345–357.
Mihaylova, M. M., & Shaw, R. J. (2011). The AMPK signalling pathway coordinates cell growth, autophagy and metabolism. *Nature Cell Biology*, *13*, 1016–1023.
Niso-Santano, M., Malik, S. A., Pietrocola, F., Bravo-San Pedro, J. M., Marino, G., Cianfanelli, V., et al. (2015). Unsaturated fatty acids induce non-canonical autophagy. *The EMBO Journal*, *34*, 1025–1041.
Ohsumi, Y. (2014). Historical landmarks of autophagy research. *Cell Research*, *24*, 9–23.
Okamoto, K. (2014). Organellophagy: Eliminating cellular building blocks via selective autophagy. *The Journal of Cell Biology*, *205*, 435–445.
Rao, S., Tortola, L., Perlot, T., Wirnsberger, G., Novatchkova, M., Nitsch, R., et al. (2014). A dual role for autophagy in a murine model of lung cancer. *Nature Communications*, *5*, 3056.
Schneider, J. L., & Cuervo, A. M. (2014). Liver autophagy: Much more than just taking out the trash. *Nature Reviews. Gastroenterology & Hepatology*, *11*, 187–200.
Sica, V., Galluzzi, L., Bravo-San Pedro, J. M., Izzo, V., Maiuri, M. C., & Kroemer, G. (2015). Organelle-specific initiation of autophagy. *Molecular Cell*, *59*, 522–539.
Uytterhoeven, V., Lauwers, E., Maes, I., Miskiewicz, K., Melo, M. N., Swerts, J., et al. (2015). Hsc70-4 deforms membranes to promote synaptic protein turnover by endosomal microautophagy. *Neuron*, *88*, 735–748.

CHAPTER ONE

Renilla Luciferase-LC3 Based Reporter Assay for Measuring Autophagic Flux

T. Farkas, M. Jäättelä[1]
Cell Death and Metabolism, Center for Autophagy, Recycling and Disease, Danish Cancer Society Research Center, Copenhagen, Denmark
[1]Corresponding author: e-mail address: mj@cancer.dk

Contents

Abstract

Macroautophagy (autophagy) is a dynamic intracellular degradation pathway. Monitoring the flux through the autophagy pathway is experimentally challenging but obviously a prerequisite for the proper investigation of the process. Here, we present an indirect autophagy flux assay based on monitoring the degradation of an autophagosome-associated fusion protein Rluc-LC3 by luminescence detection. The

Methods in Enzymology, Volume 588
ISSN 0076-6879
http://dx.doi.org/10.1016/bs.mie.2016.09.072

method is suitable for screening purposes with a high number of parallel samples and can be used for measurements in cell lysates as well as in living cells. The Rluc-LC3 assay has proven useful for the identification of genes, miRNAs, and small molecules that regulate autophagy flux in mammalian cells.

1. INTRODUCTION

1.1 The Challenge of Measuring Autophagic Flux

An ideal assay for autophagic flux would quantitatively and continuously track intracellular cargo from its sequestration into autophagosomes to its breakdown in autophagolysomes and release of corresponding breakdown products back to the cytosol. The perhaps closest candidate to this ideal is the classical long-lived protein degradation assay (Bauvy, Meijer, & Codogno, 2009). The assay indeed measures breakdown products; this is achieved by monitoring the release of radioactive isotopes to the media after labeling of long-lived proteins. The definitive assessment of the contribution of autophagy to the readout however relies on the usage of genetic or chemical inhibition of the pathway. Such usage of inhibitors is problematic because of specificity issues and potential crosstalk between autophagy and other degradation systems like the proteasomal system (Korolchuk, Menzies, & Rubinsztein, 2010; Mizushima, Yoshimori, & Levine, 2010). Another more practical limitation to the long-lived protein degradation assay is that it is relatively laborious and not well suited to capture the dynamics of autophagy and from a practical point of view, not easily applicable for high-throughput analysis. The current experimental limitations in directly measuring bulk autophagic breakdown, combined with an increased understanding of autophagy, on the molecular level have prompted the development of simpler, more indirect, assays based on molecular markers. Among those, assays employing LC3 are the most widely used.

1.2 LC3 as a Marker of Autophagy

LC3 is a subfamily in the Atg8 family of ubiquitin-like proteins and is the only protein marker described to reliably bind the autophagosomal membrane and the phagophore (Klionsky et al., 2016). LC3 is synthesized as a proprotein, which is rapidly cleaved leaving a C-terminal glycine. The LC3-I molecule thus formed is diffusely located in the cytoplasm but upon autophagy induction, LC3-I is converted to LC3-II by conjugation of phosphatidylethanolamine to the C-terminus. This lipid serves as an anchor that mediates tight association of LC3-II to the phagophore as well as to the inner and outer autophagosomal membranes. After fusion with lysosomes, LC3-II on the inner membrane is degraded, whereas LC3-II on the outer membrane is delipidated and recycled (Tanida, Minematsu-Ikeguchi, Ueno, & Kominami, 2005).

Essentially all the described modifications of LC3 have been exploited in various assays to analyze autophagy (Klionsky et al., 2016). LC3 conjugation is thus easily observed by immunoblotting of the distinct LC3-II band (see chapter "Turnover of lipidated LC3 and autophagic cargoes in mammalian cells" by Rodríguez-Arribas et al. in Vol. 587). The consequent translocation of LC3 from diffuse into puncta has become a standard indicator of autophagosome formation and can be visualized either by antibody staining or by fluorescent microscopy of fusion proteins like enhanced green fluorescent protein (EGFP)-LC3. Due to the dynamics of autophagy, increased puncta formation can however either be caused by increased autophagosome formation or by inhibition of downstream processing, a situation in which autophagy is actually inhibited. This inherent ambiguity was ingeniously addressed by using a tandem fluorescent probe instead of only EGFP as LC3 fusion partner. Such a probe allows the distinction of autophagosomes before and after lysosome fusion, thus providing a visual indicator of the level of functional autophagy (Kimura, Noda, & Yoshimori, 2007). A more direct and often used indicator of flux is the degradation of LC3-II associated with the inner autophagosomal membrane, a phenomenon typically investigated by immunoblotting. This procedure is however not as straightforward as it may sound. In addition to the usual challenges of normalizing signal to house-keeping proteins and the limited dynamic range of LC3-immunoblots, LC3-II is subjected to similar dynamics as the autophagic puncta. To appreciate LC3-II degradation it is therefore often essential to block the turnover by including lysomal inhibitors in the assay (Klionsky et al., 2016; Mizushima & Yoshimori, 2007; Tanida et al., 2005). Another often used flux marker is sequestosome/p62; this protein binds members of the Atg8 family and functions as a cargo receptor for proteins destined for the autophagosome and is itself degraded via autophagy (Bjorkoy et al., 2009).

1.3 Cautions in Using LC3-II Turnover as a Marker of Autophagic Flux

The most basic concern in using LC3-II degradation as a marker of flux is that LC3-II is not a cargo but instead a marker of the autophagic membrane.

The exact role of LC3-II on the membrane is furthermore unclear. The correlation between LC3 turnover and flux was addressed in a recent paper; it was shown that it is possible to force an uncoupling of LC3-flux from flux of the cargo marker lactate dehydrogenase by the use of cycloheximide or LC3 siRNA (Szalai et al., 2015). The quantitative relevance of such uncoupling, when LC3 levels are not artificially lowered, is unclear. But the observations emphasizes the importance of corroborating LC3-II turnover data by independent means.

2. THE RLUC-LC3 ASSAY

2.1 Description of the Assay

The Rluc-LC3 assay is centered on the Rluc-LC3 fusion protein; this protein combines the high-dynamic range and sensitivity of the Renilla luciferase with the characteristic ability of LC3 to enter the lysosome in an autophagy-dependent manner. Basically, the Rluc-LC3 assay requires two cell populations, one expressing Rluc-LC3wt and another expressing the mutant, Rluc-LC3G120A, that cannot undergo lipidation and thus not bind the autophagosomal membranes. For a detailed validation of the Rluc-LC3 assay, see Farkas, Høyer-Hansen, and Jäättelä (2009).

As illustrated in the diagram in Fig. 1A and exemplified in the Western blot shown in Fig. 1B, Rluc-LC3wt is degraded when autophagy is induced, whereas the reference protein, Rluc-LC3G120A, is stable.

The changes in Rluc-LC3wt/G120A levels are mirrored by corresponding changes in luminescence (Fig. 1C). The increased autophagic flux induced by rapamycin, in the example, is thus displayed as a decrease in the ratio of luciferase activities in extracts from the two cell populations comparing treated to untreated conditions (see Fig. 1D).

In the example the increased Rluc-LC3 turnover is directly visible as a drop in the Rluc-LC3wt levels. But such simplicity cannot always be assumed. Some treatments might induce changes unrelated to autophagic degradation in the rate of synthesis or in lipidation-independent stability that affect the level of Rluc-LC3wt. Therefore, to considerably reduce such effects on the readout, it is essential to include the Rluc-LC3G120A data in the assay.

2.2 Applications of the Rluc-LC3 Assay

Essentially the Rluc-LC3 assay functions as a sensitive, internally controlled alternative to the classical procedure of estimating changes in endogenous

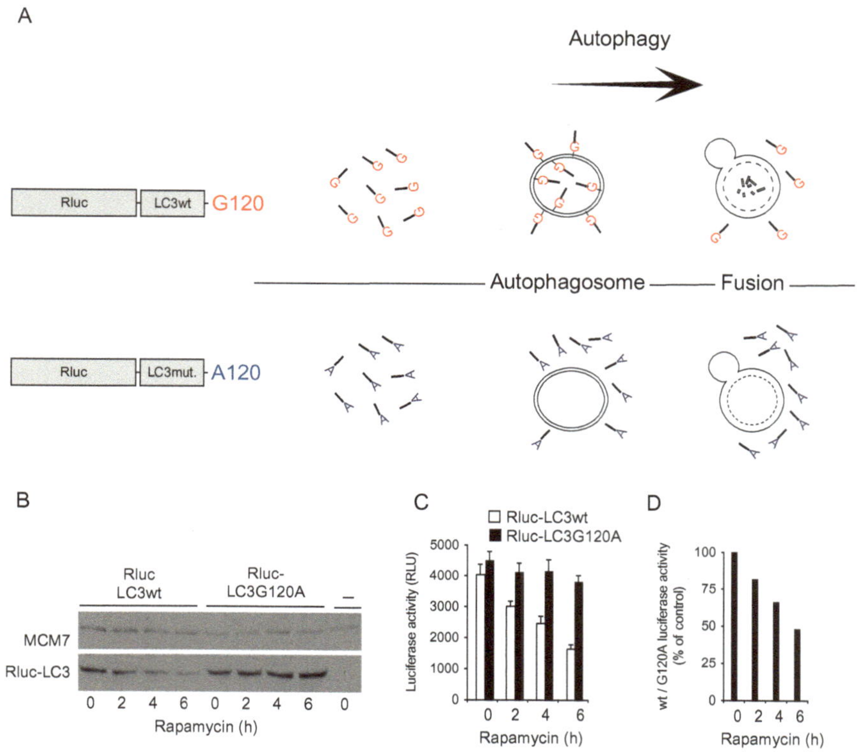

Fig. 1 Rapamycin induces autophagy-dependent decay of Rluc-LC3. (A) Illustration of the principle behind the Rluc-LC3 assay. In the control situation both Rluc-LC3wt (glycine in position 120 of LC3) and Rluc-LC3G120A are mainly present soluble in the cytoplasm. During autophagy RlucLC3wt binds both the inner and the outer membrane of the autophagosome, whereas Rluc-LC3G120A stays soluble. After fusion with lysosomes, Rluc-LC3wt on the inner membrane is degraded. (B) Immunoblotting with antibodies against Renilla luciferase (Chemicon, Mab 4410) and MCM7 (internal control) in total extracts from MCF7 cells stably expressing Rluc-LC3wt or Rluc-LC3-G120A and treated with 250 n*M* rapamycin for 0, 2, 4, or 6 h. (C) Rluc activity in extracts of MCF7 cells stably expressing Rluc-LC3wt or Rluc-LC3G120A and treated with 250 n*M* rapamycin for 0, 2, 4, or 6 h. The values represent mean ± standard deviation of a single experiment performed in sextuplicate. (D) The ratios of the luciferase measurements in (C) were calculated at each time point and expressed as percentages of the ratio obtained from untreated cells. *Fig. 1B–D is a partial reproduction of figure 1 from Farkas, T., Høyer-Hansen, M., & Jäättelä, M. (2009). Identification of novel autophagy regulators by a luciferase-based assay for the kinetics of autophagic flux. Autophagy, 5(7), 1018–1025. 9443 [pii] with permission from the publisher.*

LC3 by Western blotting. Notably, the assay, in addition, is readily applied to living cells, as well as extracts, in a microwell format. The Rluc-LC3 assay complements assays based on fluorescent-tagged LC3 such as EGFP-LC3. Such probes are optimized for microscopy analysis and are used to estimate the number of autophagosomes. They provide little information about the rate of autophagic processes. In support of a screen of siRNAs that induce EGFP-LC3 translocation the Rluc-LC3 assay was thus used to distinguish those hits that induced increased flux from those in which the increased translocation more likely was caused by other reasons such as reduced autophagosome turnover (Szyniarowski et al., 2011).

By adding luciferase substrate directly to the growth medium, the Rluc-LC3 assay offers the possibility of making multiple measurements on the same cells, thus detecting the propagation of autophagy in real time. This type of analysis can provide valuable information about the molecular events and putative causalities underlying autophagy. An example of a real-time analysis is shown in Fig. 2A. By comparing their time courses, the distinct effects of rapamycin and etoposide on Rluc-LC3 turnover are readily revealed. Rapamycin thus induces a fast response that persists with a nearly constant rate from 2 to 12 h. Etoposide on the other hand induces a slower response that reaches a maximum rate of Rluc-LC3 turnover after approximately 8 h. The distinct pattern of autophagy induction is paralleled by a distinct effect on phosphorylation of the mTorc1 substrate p70S6K. The Rluc-LC3 assay is suitable for multiwell plate analysis and has been used to identify novel autophagy regulators in small-molecule and microRNA libraries (Farkas, Daugaard, & Jäättelä, 2011; Farkas et al., 2009; Frankel et al., 2014, 2011).

3. DESIGNING THE RLUC-LC3wt AND RLUC-LC3G120A FUSION PROTEINS

The rat LC3B isoform with 22 amino acids downstream of glycine 120 was used in Rluc-LC3. This isoform is also widely used as a fusion partner in EGFP-LC3 to label autophagosomes (Kabeya et al., 2000). Among the several luciferases described in the scientific literature and in company catalogs, Rluc fulfills two criteria for being a suitable N-terminal fusion partner of LC3. First, protocols exist for assaying Rluc both in live cells and in extracts. Second, the protein has a long half-life; to elaborate on this, we use a variant of Rluc in Rluc-LC3 with a substitution (C124A) that increases the half-life of intracellular luciferase activity from 14 to 52 h (Loening, Fenn, Wu, & Gambhir, 2006). A high inherent resistance to inactivation of the luciferase moiety is preferable to support the notion that the

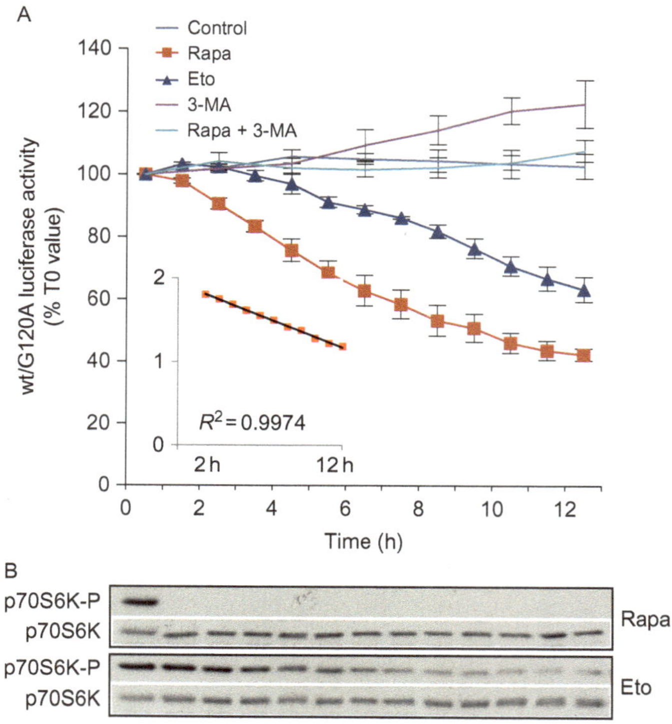

Fig. 2 Analysis of the kinetics of autophagic flux in living cells. (A) MCF-7 cells stably expressing RLuc-LC3wt or Rluc-LC3G120A were plated as described in step 1 in Section 6.2 and incubated with 50 n*M* of the live cell luciferase substrate EnduRen™ for 2 h prior to addition of either control medium ($n=3$) or drug treatments: 250 n*M* rapamycin (Rapa; $n=3$), 50 μ*M* etoposide (Eto; $n=3$), 10 m*M* 3-MA ($n=3$), or 250 n*M* rapamycin + 10 m*M* 3-MA (Rapa + 3-MA; $n=2$). Luciferase activity was measured with 1–2-h intervals as indicated. The values represent the mean ratio ± SEM of luciferase activities from the two cell lines expressed as percentages of the corresponding ratio in untreated cells at T0. The inset is a representation of the logarithm (base 10) of the rapamycin values after subtraction of 27%, as a function of time (from 2 to 12 h). The trendline and the corresponding R^2 were calculated with excel software. The value "27%" represents a hypothetical steady state of wt/G120A luciferase activity in the presence of rapamycin (established as the value giving a trendline with an R^2 closest to 1). (B) Immunoblotting with antibodies against p70S6K-P (Cell Signaling Technology, Danvers, MA, USA, 9206) and p70S6K (Cell Signaling Technology, 9202) of proteins from MCF-7 cells treated with rapamycin (250 n*M*) or etoposide (50 μ*M*) as above. *The figure is a partial reproduction of figure 4 from Farkas, T., Høyer-Hansen, M., & Jäättelä, M. (2009). Identification of novel autophagy regulators by a luciferase-based assay for the kinetics of autophagic flux.* Autophagy, *5(7), 1018–1025. 9443 [pii] with permission from the publisher.*

instability observed of Rluc-LC3 is mainly dictated by LC3, and thus dependent of autophagy. When creating the lipidation free reference protein, Rluc-LC3G120A, it is important to keep in mind that an alanine substitution in codon 120 not only prevents lipidation but also proper cleavage of pro-LC3 (Kabeya et al., 2000). Therefore, a stop codon next to codon 120 is also introduced in the expression construct.

4. ESTABLISHING RLUC-LC3wt- AND RLUC-LC3G120A-EXPRESSING CELLS

4.1 Expression Vectors and Cell Lines

We have mainly worked with MCF7 cells that express Rluc-LC3wt or Rluc-LC3G120A from stably transfected plasmids. The two expression plasmids pRluc-LC3wt and pRluc-LC3G120A are based on the CMV promoter-controlled pEGFP-C1 (Clontech) but with Rluc-LC3wt/G120A-open reading frames (ORFs) inserted instead of the EGFP ORF. We have also successfully used retrovirally transduced Hela-, U2OS-, and MCF7 cells in which the long terminal repeats of pBabehygro control the expression (Farkas et al., 2011).

4.1.1 Stable Transfection of MCF7 Cells (Transfections Are Performed Mainly According to the Fugene HD Protocol (Promega, Madison, WI, USA))

1. Distribute cells into three 10-cm plates with 10^6 cells in each. Two plates are needed for transfection of either pRluc-LC3wt or pRluc-LC3-G120A. One plate serves as control for G418 sensitivity. Grow cells in RPMI 1640+ GlutaMAX™ supplemented with 6% fetal calf serum (FCS), penicillin 100 U/mL, and streptomycin 100 μg/mL.
2. After 40 h, mix 5 μg pRluc-LC3wt or pRluc-LC3G120A with 250 μL DMEM (31966-021) and vortex. Add 15 μL Fugene HD, vortex again. Incubate for 15 min at room temperature.
3. Add 250 μL of Fugene HD/DNA mixture to cells. Apply by placing the pipette tip just below surface of 10 mL freshly shifted growth medium. Shake the plate gently. To reduce toxicity of the transfection mix, the growth medium is changed again after 12 h.
4. Start antibiotic selection 2 days after transfection. Use 400 μg/mL G418.
5. Follow cell death microscopically. Change G418-medium at least every third day. Usually, after 2–3 weeks of selection cells are ready to be

trypsinized, expanded in flasks, and frozen according to standard procedures. We use freezing medium containing 10% dimethylsulfoxide (DMSO) + 90% FCS.

6. The cells keep a stable expression of the fusion proteins for more than a month in culture and also keep responding to rapamycin. But it is advisable to take up fresh cells often to minimize drifting.

5. THE RLUC-LC3 ASSAY PERFORMED ON CELL LYSATES

5.1 Buffers and Solutions

1. Basic buffer: 100 m*M* Tris–HCl pH 7.4, 300 m*M* Na-ascorbate.
   We prepare the basic reaction buffer by mixing 200 mL pH-adjusted Tris–HCl with 11.9 g of Na-ascorbate.
2. Coelenterazine stock solution: Dissolve 1 mg coelenterazine (e.g., Synchem s053) in 950 μL acidified methanol (acidified methanol: 49 mL methanol + 1 mL HCl 1 *M*) to obtain a 2.5 m*M* stock solution. Keep at −20°C.
3. Reaction buffer: 10 mL basic buffer + 10 μL coelenterazine stock solution. If several plates are analyzed sequentially prepare fresh reaction buffer immediately before application to each plate.
4. Luciferase lysis buffer: Dilute 5 × passive lysis buffer (Promega E1941) in water to obtain a 1 × solution.

5.2 Protocol

The protocol shown here is adjusted for MCF7 cells in a 96-well plate format.

1. Plate Rluc-LC3wt- and Rluc-LC3G120A cells in the uneven and even numbered wells, respectively. Plating this way forms 48 pairs of neighboring wells, each pair establishing a unit inside which cells are treated identically and from which the Rluc-LC3wt/G120A activity ratio eventually is calculated. This plating strategy is intended to minimize putative location effects on the readout. We typically plate 8000 cells per well. But due to the high sensitivity of the luciferase assay it is feasible to plate less if needed.
2. Start the treatments the day after plating. Select a number of Rluc-LC3wt/G120A unit pairs to be used as untreated controls. Remaining pairs can be used for various treatments, for example, drug testing or starvation.

If starvation conditions are tested, it is important to rinse wells twice in starvation medium before final application of starvation medium.

3. At the end of treatment, remove medium from the wells. Add 40 μL of luciferase lysis buffer to each well. Freeze plate at −80°C.
4. Thaw plate on ice. Mix by vortexing 2500 rounds per minute for 20 s.
5. Transfer 6 μL of extract to corresponding wells of a white half volume 96-well dish.
6. Add 80 μL of freshly prepared reaction buffer.
7. Immediately after addition of reaction buffer measure luciferase activity in a plate reader set at room temperature.

6. THE RLUC-LC3 ASSAY PERFORMED ON LIVE CELLS

6.1 Buffers and Solutions

1. EnduRen™ stock solution: Dissolve 0.34 mg EnduRen™ Live Cell Substrate (Promega E6481) in 6 mL DMSO to obtain a 100 μ*M* stock solution. Heat to 37°C for 10 min and vortex to ensure complete resuspension. We make aliquots of 100 μL each and store them at −80°C.
2. Cell growth medium: RPMI-1640 without phenol red (cat. 11835, Thermo Fisher Scientific, Paisley, Scotland) supplemented with Glutamax™, 6% FCS, penicillin 100 U/mL, and streptomycin 100 μg/mL.

6.2 Protocol

Like above, the protocol is for MCF7 cells in a 96-well plate format. Here, it is described how to apply the method for drug analysis.

1. Plate cells 10,000 cells in 120 μL cell growth medium per well in white cell culture plates (Thermo Fisher Scientific, Roskilde, Denmark, 136101). Use the same plating strategy as described in step 1 in Section 5.2. Continue to step 2 the next day.
2. Take out an aliquot of EnduRen™ stock solution from the −80°C freezer. Incubate at 37°C for 10 min, vortex briefly. Make a 100 n*M* EnduRen™ solution in cell growth medium equilibrated to 37°C considering 10 mL per plate. Continue immediately to step 3.
3. Remove 60 μL of growth medium from each well of the cell culture plate. Add 60 μL of the 100 n*M* EnduRen™ solution made in step 2. The final concentration of EnduRen™ on cells is consequently

50 n*M*. Return plate to the cell incubator for a 2-h equilibration of the live cell substrate.

4. Read luminescence in a plate reader (we use a Varioskan Flash) preset at 37°C. Use this initial measurement to calculate the T0 values of the Rluc-LC3wt/G120A activity ratio for each pair of wells. These values can for simplicity be defined to 100%.
5. After measuring the T0 values, drugs are applied. We dilute drugs to a 5 × working concentration in a parallel 96-well plate, using cell growth medium with 50 n*M* EnduRen™ as diluent. 30 μL of 5 × drug solution is then transferred to the corresponding wells of the cell culture plate.
6. The luminescence is measured at appropriate intervals, for example, once per hour. For each measurement, the Rluc-LC3wt/G120A activity ratio of each pair of wells is calculated and related to the corresponding T0 values. The activity can be measured for at least 12 h.

6.3 Notes

1. In the initial testing of the live cell substrate EnduRen™ we used 60 μ*M* as suggested by the manufacturer. However, as described in Farkas et al. (2009), this concentration inhibits the turnover of Rluc-LC3 according to the assay. To eliminate this undesirable effect, we gradually reduced the concentration of EnduRen™ until no further inhibition of the response to rapamycin was evident. At this concentration of 50 n*M*, the quantitative response to rapamycin in the live cell and the extract settings of the assay was similar, suggesting that the concentration, albeit low, is appropriate to faithfully reflect the changes in the Rluc-LC3wt/G120A ratio occurring inside living cells.

   The low concentration of substrate might, however, pose a problem with detection in some cell lines. We speculate that the signal strength as well as the inhibition of Rluc-LC3 turnover is influenced by the intracellular concentration and conversion rate of EnduRen™, which could differ among cell lines. We therefore recommend that the concentration of EnduRen™ is optimized when establishing new reporter cell lines.

   In some cases, a cell line-specific optimization of expression vector might be beneficial for consistent detection in the live-cell setting.

   We do not know the exact steady-state amount of intracellular coelenterazine in cells grown with 50 n*M* EnduRen™, since it depends on the uptake/conversion rate, but we notice that 50 n*M* is far below the K_m value of Rluc-C124A, which is 2.7 ± 0.8 μ*M* (Loening et al., 2006).

We consider the option that, in rare cases, as described below, certain treatments could induce changes in the steady-state level of intracellular EnduRen™/coelenterazine that will affect the readout, even though such effects in principle should affect the two cell lines equally.

2. Usually the Rluc-LC3 assay shows good correspondence between data obtained in live cells and in lysates. We did however observe a discrepancy using the drug triciribine (Farkas et al., 2011). This protein kinase B/Akt inhibitor is known to induce autophagy (Evangelisti et al., 2011). And it does induce flux according to the Rluc-LC3 assay performed in lysates, but not in the live-cell setting. We do not know the reason for this difference, but we notice that the live-cell assay is more complex and possibly influenced by the uptake/conversion of EnduRen™ as described above and the intracellular localization of the luciferase and therefore, likely, more prone to generating artifacts. It suggests that it is advisable to make an initial comparison of new drugs in both settings, to avoid this putative artifact of the live-cell assay.

ACKNOWLEDGMENTS

This work was supported by a Center of Excellence grant from the Danish National Research Foundation and an Advanced Grant from the European Research Council (LYSOSOME; 340751).

REFERENCES

Bauvy, C., Meijer, A. J., & Codogno, P. (2009). Assaying of autophagic protein degradation. *Methods in Enzymology*, *452*, 47–61. http://dx.doi.org/10.1016/S0076-6879(08)03604-5.

Bjorkoy, G., Lamark, T., Pankiv, S., Overvatn, A., Brech, A., & Johansen, T. (2009). Monitoring autophagic degradation of p62/SQSTM1. *Methods in Enzymology*, *452*, 181–197. http://dx.doi.org/10.1016/S0076-6879(08)03612-4.

Evangelisti, C., Ricci, F., Tazzari, P., Chiarini, F., Battistelli, M., Falcieri, E., … Martelli, A. M. (2011). Preclinical testing of the Akt inhibitor triciribine in T-cell acute lymphoblastic leukemia. *Journal of Cellular Physiology*, *226*(3), 822–831. http://dx.doi.org/10.1002/jcp.22407.

Farkas, T., Daugaard, M., & Jäättelä, M. (2011). Identification of small molecule inhibitors of phosphatidylinositol 3-kinase and autophagy. *The Journal of Biological Chemistry*, *286*(45), 38904–38912. http://dx.doi.org/10.1074/jbc.M111.269134. M111.269134 [pii].

Farkas, T., Høyer-Hansen, M., & Jäättelä, M. (2009). Identification of novel autophagy regulators by a luciferase-based assay for the kinetics of autophagic flux. *Autophagy*, *5*(7), 1018–1025. 9443 [pii].

Frankel, L. B., Di Malta, C., Wen, J., Eskelinen, E. L., Ballabio, A., & Lund, A. H. (2014). A non-conserved miRNA regulates lysosomal function and impacts on a human lysosomal storage disorder. *Nature Communications*, *5*, 5840. http://dx.doi.org/10.1038/ncomms6840.

Frankel, L. B., Wen, J., Lees, M., Høyer-Hansen, M., Farkas, T., Krogh, A., ... Lund, A. H. (2011). microRNA-101 is a potent inhibitor of autophagy. *The EMBO Journal, 30*(22), 4628–4641. http://dx.doi.org/10.1038/emboj.2011.331. emboj2011331 [pii].

Kabeya, Y., Mizushima, N., Ueno, T., Yamamoto, A., Kirisako, T., Noda, T., ... Yoshimori, T. (2000). LC3, a mammalian homologue of yeast Apg8p, is localized in autophagosome membranes after processing. *The EMBO Journal, 19*(21), 5720–5728.

Kimura, S., Noda, T., & Yoshimori, T. (2007). Dissection of the autophagosome maturation process by a novel reporter protein, tandem fluorescent-tagged LC3. *Autophagy, 3*(5), 452–460.

Klionsky, D. J., Abdelmohsen, K., Abe, A., Abedin, M. J., Abeliovich, H., Acevedo Arozena, A., ... Zughaier, S. M. (2016). Guidelines for the use and interpretation of assays for monitoring autophagy (3rd edition). *Autophagy, 12*(1), 1–222. http://dx.doi.org/10.1080/15548627.2015.1100356.

Korolchuk, V. I., Menzies, F. M., & Rubinsztein, D. C. (2010). Mechanisms of cross-talk between the ubiquitin-proteasome and autophagy-lysosome systems. *FEBS Letters, 584*(7), 1393–1398. http://dx.doi.org/10.1016/j.febslet.2009.12.047.

Loening, A. M., Fenn, T. D., Wu, A. M., & Gambhir, S. S. (2006). Consensus guided mutagenesis of Renilla luciferase yields enhanced stability and light output. *Protein Engineering, Design & Selection, 19*(9), 391–400.

Mizushima, N., & Yoshimori, T. (2007). How to interpret LC3 immunoblotting. *Autophagy, 3*(6), 542–545.

Mizushima, N., Yoshimori, T., & Levine, B. (2010). Methods in mammalian autophagy research. *Cell, 140*(3), 313–326.

Szalai, P., Hagen, L. K., Saetre, F., Luhr, M., Sponheim, M., Overbye, A., ... Engedal, N. (2015). Autophagic bulk sequestration of cytosolic cargo is independent of LC3, but requires GABARAPs. *Experimental Cell Research, 333*(1), 21–38. http://dx.doi.org/10.1016/j.yexcr.2015.02.003.

Szyniarowski, P., Corcelle-Termeau, E., Farkas, T., Høyer-Hansen, M., Nylandsted, J., Kallunki, T., & Jäättelä, M. (2011). A comprehensive siRNA screen for kinases that suppress macroautophagy in optimal growth conditions. *Autophagy*, 7(8), 892–903. 15770 [pii].

Tanida, I., Minematsu-Ikeguchi, N., Ueno, T., & Kominami, E. (2005). Lysosomal turnover, but not a cellular level, of endogenous LC3 is a marker for autophagy. *Autophagy, 1*(2), 84–91.

CHAPTER TWO

Measurement of Autolysosomal pH by Dual-Wavelength Ratio Imaging

A. Saric*, S. Grinstein†,‡,1, S.A. Freeman†
*Ryerson University, Toronto, ON, Canada
†Hospital for Sick Children, Toronto, ON, Canada
‡Keenan Research Centre, St. Michael's Hospital, Toronto, ON, Canada
[1]Corresponding author: e-mail address: sergio.grinstein@sickkids.ca

Contents

Abstract

Cellular components sequestered by autophagosomes during macroautophagy must be degraded and their components recycled in order to maintain homeostasis. To this end cells orchestrate the fusion of autophagosomes with lysosomes, degradative organelles that are rich in hydrolases. Most of the lysosomal enzymes function optimally at low pH, and products of macromolecular catabolism are cotransported with protons across the autolysosomal membrane. These functions are facilitated by the ability of lysosomes to pump protons inward, acidifying their lumen. Clearly, proper homeostasis of the luminal pH is crucial for autolysosomal function. We describe a method for the measurement of the absolute pH of individual autolysosomes in live cells. This technique involves measurement of the fluorescence of a pH-sensitive probe initially delivered to lysosomes and subsequently determined to have reached autolysosomes. By measuring the fluorescence at two separate wavelengths and calculating their ratio, potential artifacts introduced by photobleaching or by changes in autolysosome size, shape, or positioning are minimized. Combining such ratio determinations with an in situ calibration procedure enables absolute measurements of pH, which are superior to the qualitative estimates obtained with fluorescent weak bases such as LysoTracker.

Methods in Enzymology, Volume 588
ISSN 0076-6879
http://dx.doi.org/10.1016/bs.mie.2016.09.073

1. INTRODUCTION

When cells are deprived of nutrients, incur organelle damage, or accumulate protein aggregates, the process of macroautophagy (hereafter referred to as "autophagy") is initiated. Autophagy is exquisitely controlled, ultimately leading to the production of autophagosomes that sequester cytoplasmic components—including proteins and entire organelles—within their lumen (Mizushima, 2007). Once trapped, the contents of the autophagic vacuoles must be broken down and recycled to provide the cell with endogenous nutrients. This requires delivery to the autophagosomes of degradative enzymes such as proteases, lipases, and nucleases. These are supplied to the autophagosomes by fusion with lysosomes; the resulting autolysosomes are hybrid structures endowed with properties of both precursor organelles (Baba, 1994; Dunn, 1990; Lawrence & Brown, 1992). Importantly, the lysosomal membrane contributes vacuolar-type ATPases (V-ATPases) that actively pump protons into the organelle lumen, rendering it acidic. This acidic environment is important for the activation and efficient function of the hydrolases, many of which function optimally at acidic pH. It follows that the perturbation of lysosomal or autolysosomal pH severely impairs autophagic flux and thereby can exacerbate the outcome of some diseases (Coffey, Beckel, Laties, & Mitchell, 2014; Kawai, Uchiyama, Takano, Nakamura, & Ohkuma, 2007; Lee et al., 2010; Lu et al., 2015). Understanding the processes whereby pH is regulated is therefore paramount to a complete understanding of autolysosome function. This requires precise means of measuring the autolysosomal pH.

Reliable measurements of organellar pH entail selection of a suitable probe and the ability to target it to the compartment of interest. For the most part autolysosomal pH has been estimated by qualitative or semiquantitative means using predominantly membrane-permeant weak bases that accumulate in acidic organelles. *N*-{3-[(2,4-Dinitrophenyl)amino]propyl}-*N*-(3-aminopropyl)methylamine, one such acidotropic compound, can be visualized by immunostaining with antidinitrophenyl antibodies and has been used in combination with electron microscopy (Eskelinen, 2005; Martinet, De Meyer, Andries, Herman, & Kockx, 2006). More commonly, however, the distribution of weak bases is monitored by fluorescence microscopy. Some probes, like LysoTracker, accumulate in acidic organelles without otherwise altering their fluorescence properties. Others, like LysoSensor, in addition undergo pH-dependent spectral shifts (Lee et al.,

2010; Neely, Green, & LaFerla, 2011; Zhang et al., 2012). Although these sensors are acceptable indicators of acidity, they are not without limitations. First, they are expected to accumulate not only in lysosomes/autolysosomes but also in any other compartments that are sufficiently acidic, e.g., in late endosomes. Second, when used at inappropriately high concentrations, membrane-permeant basic probes can themselves alter the pH of the compartments where they accumulate. Dextran-conjugated LysoSensor probes—that are taken up by endocytosis and are eventually delivered to lysosomes—have been used to assess the pH of lysosomal compartments during autophagy, though not that of autolysosomes specifically (Lee et al., 2010). pHrodo®, rhodamine-based dyes that exhibit strong fluorescence in acidic environments and can be visualized by microscopy, are increasingly used to assess the pH of organelles. Succinimidyl esters of pHrodo® have been conjugated to phagocytic targets like yeast and bacteria to assess phagosome formation and acidification (Bonilla et al., 2013; Deriy et al., 2009; Maneu et al., 2011). pHrodo-containing probes have been used to assess the pH of autophagic compartments, using an in vitro calibration procedure (Maulucci et al., 2015). Finally, and most recently, the coral-derived pH-sensitive fluorescent protein Keima has been utilized to estimate autolysosomal pH (Katayama, Kogure, Mizushima, Yoshimori, & Miyawaki, 2011; Kogure et al., 2006).

Here, we describe a quantitative method for the specific measurement of autolysosomal pH, based on the measurement of the fluorescence of a pH-sensitive probe at two separate wavelengths. Dual-wavelength ratio fluorescence imaging has been widely used to measure the pH of lysosomes (Canton & Grinstein, 2015; Ohkuma & Poole, 1978). This approach relies on the use of dyes that undergo spectral shifts as a function of pH. By selecting two wavelengths that are differentially affected by pH and calculating their ratio, it is possible to estimate the pH in a manner that is independent of the amount of the probe. This has distinct advantages: photobleaching of the probe as well as changes in the shape or position of the organelles, which can alter the intensity of the fluorescence recorded, have similar effects on both wavelengths and hence little net effect on their ratio. Moreover, for dyes with an isosbestic point, selection of wavelengths that change in opposite directions when the pH is altered improves the dynamic range of the measurements. Importantly, the technique is quantitative; when coupled to in situ calibration, the absolute pH of individual compartments in a given cell under a particular set of experimental conditions can be determined.

For measurement of lysosomal pH, ratiometric dyes are typically conjugated to nondegradable large biomolecules such as dextrans, which are taken up by endocytosis and delivered to lysosomes via vesicular traffic (Ohkuma & Poole, 1978). Targeting the autolysosome is somewhat more complex. In principle, one could target the probe to either of the precursor organelles that fuse to generate the autolysosome (i.e., the autophagosome or the lysosome). However, autophagosome biogenesis is dynamic, complex, and occurs minimally in resting (unstressed) cells, making specific targeting of the lumen difficult. By contrast, delivering a probe to the lysosomal lumen is a more agreeable option (Ohkuma & Poole, 1978). Probes that are soluble in the extracellular fluid phase will be taken up by endocytosis, and delivery to lysosomes can be ensured by following the pulse with a suitable chase period, usually $\geq$1 h (Vieira, Botelho, & Grinstein, 2002). While this approach ensures that some of the pH sensor will reach autolysosomes, it is insufficient to distinguish the latter from conventional lysosomes that have not undergone fusion with autophagosomes. Therefore, additional means are required to identify and selectively measure the pH of autolysosomes. In the method described in this chapter, *bona fide* autolysosomes are identified by the colocalization of the lysosomal pH probe with the conventional autophagy marker microtubule-associated protein light chain 3 (LC3). When autophagy is initiated, LC3 redistributes from a largely cytosolic pool to the membrane of autophagosomes, forming punctate structures (Fig. 1). As autolysosomes form, LC3 is shed from their inner surface into the organellar lumen (Tanida, Minematsu-Ikeguchi, Ueno, & Kominami, 2005). Its location in autophagosomes and autolysosomes can be visualized by fusing LC3 to a fluorescent protein such as mCherry. mCherry is convenient for several reasons: its red fluorescence is distinct from the green emission of the most commonly used pH-sensors, which are derivatives of fluorescein. Second, mCherry retains its fluorescence in the acidic environment of autolysosomes, due to the comparatively low pK_a of its fluorophore. Third, mCherry is relatively resistant to cleavage by lysosomal proteases, persisting within the autolysosomal lumen (Costantini et al., 2015; Pankiv et al., 2007; Shaner et al., 2004). Thus, the colocalization of the mCherry signal with that of the lysosomal-targeted pH probes is indicative of autolysosomes (Fig. 1).

Proper targeting is only one of the features that need to be fulfilled by a suitable autolysosomal pH sensor; its spectral properties and pH dependence are also critical. Since the pH of lysosomes has been estimated to range from 4.5 to 5.0 (Coffey & De Duve, 1968; Mellman, Fuchs, &

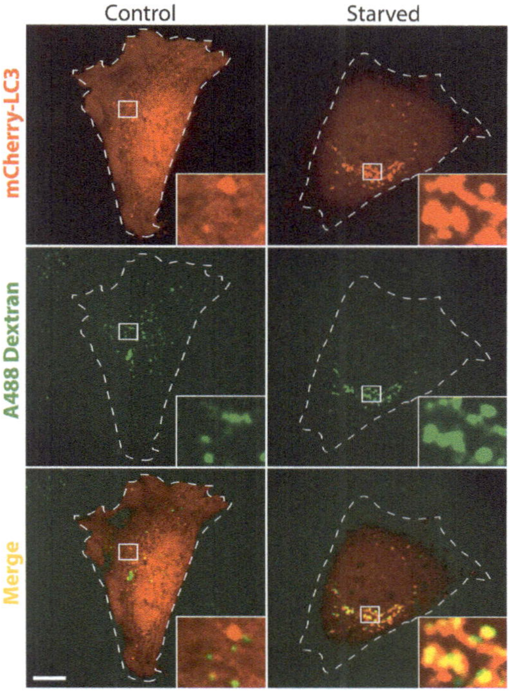

Fig. 1 Assessment of autolysosome formation in HeLa cells. Lysosomes were prelabeled by incubating HeLa cells with Alexa Fluor 488-dextran. The cells were then transfected with mCherry-LC3. In cells grown in DMEM + 5% FBS (control) LC3 is largely distributed diffusely in the cytosol, with a small number of LC3-positive puncta (see *red* channel, *top left*). A few puncta colocalize with the dextran, indicating that some basal amount of autophagy occurs in these cells, as expected. However, most red puncta do not colocalize with lysosomes and may be preautolysosomal structures (e.g., phagophores or amphisomes) or LC3 aggregates. Starvation was induced by switching the medium to HBSS for 4 h (starved). Note that the mCherry-LC3 signal is redistributed from the cytosol to perinuclear puncta that colocalize extensively with Alexa Fluor 488-dextran, indicating formation of autolysosomes (see inset at *bottom right*). *Dotted white lines* delimit the cells, as visualized by differential interference contrast microscopy. Scale bar = 10 μm.

Helenius, 1986), a similarly acidic value can be anticipated for the autolysosomal lumen. Based on this premise, the ideal probe should have a comparably acidic pK_a, to ensure an optimal dynamic range of the pH-induced fluorescence changes. The most widely used dextran-conjugated probes for measurement of lysosomal pH are fluorescein derivatives. Indeed, fluorescein and the fluorescein-derived Oregon Green dyes have been widely used for the ratiometric measurement of pH of a variety of organelles, including the Golgi complex and endolysosomes (Jiang, Maher, McCormick, & Schindler, 1990; Kim et al., 1996; Poët et al., 2006; Tabeta et al., 2006; Trombetta, Ebersold, Garrett, Pypaert, & Mellman, 2003; Wu et al., 2000). Fluorescein-dextran has a pK_a of 6.4, while its more photostable counterpart, Oregon Green 488-dextran, has a pK_a of 4.7. Both probes have an emission maximum near 520 nm that is exquisitely pH sensitive when excited at (or near) 490 nm. When excited at (or near) 440 nm, the changes in emission intensity caused by altering the pH are markedly smaller (e.g., Fig. 2A). For this reason, the ratios of fluorescence intensity produced by excitation at 490 nm (pH-sensitive wavelength) divided by that obtained at 440 nm (more pH-insensitive wavelength) are proportional to—and therefore indicative of—the pH of the environment (Fig. 2B). In the protocol detailed later Oregon Green 488-dextran is used due to its high photostability and low pK_a.

The proportionality between the fluorescence ratio and the pH of the medium can be validated by merely suspending Oregon Green 488-dextran (or any other probe of choice) in solutions of varying pH (Fig. 2B). In principle, this ratio vs pH curve obtained in vitro can be used to calibrate measurements made in live cells. However, the microenvironment of lysosomes or autolysosomes can affect the behavior of the probe. Furthermore, variation may occur between different cell types as well as different methods used to induce autophagy. A more accurate and highly recommended approach is therefore to conduct an in situ calibration. This necessitates manipulation of the organellar pH in a controlled and predictable manner. The method used most commonly to manipulate the pH is based on the ability of ionophores to increase the proton permeability of biological membranes. Specifically, alkali cation/H^+-exchanging ionophores such as nigericin or monensin are applied to cells suspended in media of suitable cation composition. Most simply, adding the K^+/H^+ ionophore nigericin to cells suspended in solutions containing a cytosolic $[K^+]$ approximating that of the cytosol ($\approx$140 m*M*) will ensure that the cytosolic pH will become (nearly) identical to the extracellular pH. Sequential incubation of nigericin-treated cells in $[K^+]$-rich buffers of varying pH will yield a calibration curve, as illustrated in Fig. 2C. Because nigericin will alter the pH of the lysosome or autolysosome, the calibration must be performed post hoc, and the pH of the organelle of interest is then calculated by interpolation of the ratio measured prior to the addition of the ionophore (Fig. 2C). For both in vitro and in situ calibrations it is important to select appropriate buffers to maintain the desired pH. For the pH 4–5 range acetate/acetic acid are recommended,

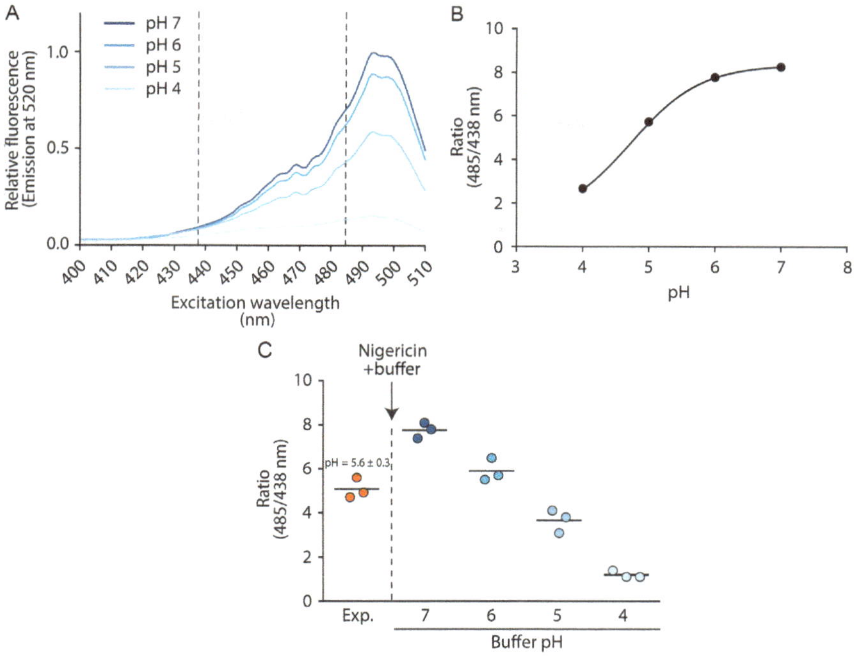

Fig. 2 Autolysosomal pH measurements of the fluorescence ratio of Oregon Green 488. (A and B) In vitro determinations of pH-dependent changes in the spectral properties of Oregon Green 488-dextran. (A) Fluorimetry was conducted to obtain the excitation spectrum of Oregon Green 488-dextran using calibration buffers of varying pH values (0.25 mg/mL dextran was used). Note the small changes in Oregon Green fluorescence in response to pH when using 438 nm excitation, compared to the larger changes at 485 nm excitation. *Dashed lines* indicate the peak excitation wavelength at which emission intensities are collected for generation of 485/438 ratios. (B) Background-subtracted 485/438 ratios from (A) were plotted against the corresponding buffer pH to generate a calibration curve. (C) In situ pH measurement of autolysosomes, made in live HeLa cells. An autolysosome was identified by colocalization of mCherry-LC3 and Oregon Green 488-dextran (as in Fig. 1 inset under "starved" condition). Background-subtracted 485/438 ratios were determined for the autolysosome in untreated cells (*red dots*). The pH was next clamped using nigericin and high [K^+] calibration buffers, and the 485/438 ratios were determined for the same organelle at various pH values (*blue dots*). A curve was fitted to the calibration data and used to interpolate the resting pH of the autolysosome (pH 5.6 ± 0.3 in this instance).

for pH 6–7 we suggest 2-(*N*-morpholino)ethanesulfonic acid (MES), and for pH 7–8 we use *N*-(2-hydroxyethyl)piperazine-*N*′-(2-ethanesulfonic acid) (HEPES).

Overall, this method relies on the identification of autolysosomes as mCherry-LC3- and Oregon Green 488-dextran-double positive structures.

Once the organelles of interest have been identified, regions of interest (ROIs) are drawn and the fluorescence of Oregon Green 488-dextran is recorded at the two chosen wavelengths, followed by in situ calibration. Interpolation of the experimental values in the respective calibration curves culminates the determination.

2. CELL PREPARATION

The following method is described for HeLa cells, though in principle it can be applied to any cell type. Similarly, while we use starvation as the means to initiate autophagy, the procedure should be applicable to other inducers of autophagy. Lysosomes are prelabeled with Oregon Green 488-dextran, the cells are then transfected with mCherry-LC3 and, once expression is apparent, starved to induce autophagy. After 4 h of starvation, HeLa cells exhibit maximal colocalization of the two fluorophores, indicating the formation of autolysosomes (Fig. 1). Researchers should predetermine the time frame of autolysosome formation in their particular system (cell type, autophagy inducer) by monitoring autolysosome formation as earlier.

1. On day 1, plate HeLa cells at ≈50% confluence on 18-mm sterile coverslips in a 12-well tissue culture dish containing Dulbecco's modified Eagle's medium (DMEM) (Wisent, St. Bruno, QC, Canada, 319-007-CL) supplemented with 5% heat-inactivated fetal bovine serum (FBS) (Sigma, St. Louis, MO, F4135).
2. Add Oregon Green 488-dextran (Invitrogen, Carlsbad, CA, D7171) to each well to a final concentration of 0.25 mg/mL and incubate the cells overnight at 37°C under 5% CO_2.
3. On day 2, chase the dextran by replacing the medium with dextran-free DMEM and incubate the cells at 37°C for at least 2 h to ensure progression of the probe to lysosomes.
4. Following the chase, transfect the cells with 0.5 μg mCherry-LC3 cDNA and 1.5 μL FuGENE 6 transfection reagent (Roche, Indianapolis, IN, E2692) per well, as per the manufacturer's instructions. Incubate the cells at 37°C overnight.
5. On day 3, induce starvation by replacing the medium with Hank's balanced salt solution (HBSS) (Wisent, 311-515-CL) and incubating the cells for 4 h at 37°C. Following this incubation, abundant autolysosomes can be observed by fluorescence microscopy as punctate structures with overlapping red (LC3) and green (dextran) fluorescence (Fig. 1).

3. EQUIPMENT SETUP AND SOFTWARE REQUIREMENTS

For live dual-wavelength ratio fluorescence imaging, an inverted light microscope (Leica, Wetzlar, Germany, DM4) fitted with a high-intensity arc lamp (EXFO Photonic Solutions Inc., Mississauga, ON, Canada, X-Cite 120), a self-cooling electron multiplier charge-coupled camera (Photometrics, Tucson, AZ, Cascade II EMCCD camera), and a heated stage were used. Image acquisition was controlled using the MetaFluor Fluorescence Ratio Imaging Software (Molecular Devices, Silicon Valley, CA); analysis of images was carried out using ImageJ software (National Institutes of Health, Bethesda, MD).

1. Set the stage temperature to 37°C.
2. For ratio fluorescence imaging of Oregon Green 488-dextran, use alternate excitation at 485 ± 10 and 438 ± 12 nm by selecting suitable filters.
3. Use a 505 nm dichroic mirror (Chroma Technology Corp., Bellows Falls, VT) to reflect excitation light through the objective and allow emitted light to reach the camera through a 535 ± 20 nm filter.
4. Similarly, ensure that acquisitions in the red channel can be made in order to visualize mCherry-LC3 and identify autolysosomes, using 587 nm excitation light and measuring the emission at 610 nm.
5. Allow several minutes for the camera to cool in order to reduce thermal noise.

4. IMAGE ACQUISITION

1. Transfer a coverslip of the prepared HeLa cells onto a Chamlide coverslip chamber (Quorum Technologies, Guelph, ON, Canada, CM-B-40) and fill with 1 mL HBSS, wiping dry the underside that will contact the objective.
2. Place the chamber onto the heated stage atop a $40\times$ objective.
3. Launch MetaFluor to begin a new protocol and configure the acquisition settings such as exposure time and gain.
4. Acquire sample images at the 485 and 438 nm excitation wavelengths in order to assess the fluorescence intensity of Oregon Green 488-dextran. If needed, further tune the acquisition settings to optimize the signal. Binning may be applied to enhance the signal-to-noise ratio; however, this will result in decreased spatial resolution. For example, data in Fig. 2C were collected from images with 2×2 binning which resulted

in a pixel size of 0.8 μm. Clusters of 2–5 pixels were traced as individual autolysosomes, which are typically 0.5–1.5 μm in size (Mizushima, Ohsumi, & Yoshimori, 2002).

5. In the red channel, locate mCherry-LC3 transfected cells and adjust the focus manually if necessary (devices that maintain the focus automatically are available and, if available in the experimenters laboratory, are recommended). It is recommended that manual focus adjustments be conducted throughout the experiment using this channel for two reasons: (1) mCherry is exceptionally photostable and (2) the red signal will be used solely for locating autolysosomes and will not contribute to the pH measurements; thus, one can afford to excite mCherry more frequently than Oregon Green 488, whose fluorescence integrity should be preserved to maximize the accuracy of the pH measurements.
6. Once the cells are in focus, acquire one image in the red channel and be mindful not to shift the stage from this point on.
7. Capture 3–5 image pairs, 1 min apart, of the Oregon Green 488 at 485 and 438 nm excitations such that two sequential images are generated per acquisition. These images (one red and three to five green) will be used to determine both the location of autolysosomes and the pH of these compartments.
8. In MetaFluor, draw ROIs around Oregon Green 488-containing compartments. Within each ROI, the 485/438 ratio is automatically calculated and plotted on a ratio vs time graph during live acquisitions. This feature allows for real-time monitoring of the ratio within individual cells or individual autolysosomes, depending on how the ROI is drawn. In the absence of stage disturbance and gross shifts in the focal plane, the ratio across the 3–5 acquisitions should be nearly identical within each ROI. It should be noted that the correct calculation of the fluorescence ratio requires separate and accurate subtraction of background fluorescence for each individual wavelength. This in turn requires establishment of separate ROIs in regions devoid of visible fluorescence (e.g., outside the cell).

5. IN SITU pH CALIBRATION

An in situ pH calibration is carried out to assign absolute pH values to the 485/438 ratios acquired. The intracellular pH can be "clamped" using high [K^+] buffers of known pH in combination with the K^+/H^+ ionophore nigericin. This forces the pH of cells and individual organelles to equilibrate with the pH of the surrounding environment (i.e., the calibration buffer).

1. Prepare pH calibration buffers that will span the range of pH 4–8. Each calibration buffer should contain 140 m*M* KCl, 1 m*M* $MgCl_2$, 1 m*M* $CaCl_2$, 5 m*M* glucose, and the optimal buffer for the desired pH (e.g., acetate/acetic acid for pH 4–5, MES for pH 6–7, and HEPES for pH 7–8).
2. Adjust the pH of each buffer using 1 *M* KOH or 1 *M* HCl, as appropriate.
3. Filter-sterilize each buffer and prewarm to 37°C.
4. Taking care not to disrupt the position of the coverslip on the heated stage, remove liquid from the coverslip, and wash cells with a calibration buffer of known pH (e.g., pH 8 buffer).
5. Add 1 mL of this calibration buffer containing 10 μg/mL nigericin and allow 3–5 min for the intraorganellar pH to equilibrate with the extracellular pH.
6. Acquire 2–3 sample images, 1 min apart, to ensure that the 485/438 ratio for each ROI is steady between time points. If large fluctuations are observed between ratios for a given ROI, wait an additional 2–3 min. When the ratios are stable and the intraorganellar pH has equilibrated with that of the surrounding medium, acquire and save 3–5 images.
7. Repeat steps 4–6 for all the calibration buffers ranging from pH 8 to 4 (e.g., pH 7 buffer, followed by pH 6 buffer, etc.) taking care to maintain the X/Y position of the coverslip and refocusing when needed.

6. IMAGE ANALYSIS AND DETERMINATION OF AUTOLYSOSOMAL pH

Once the raw image files are collected, they can be exported as TIFF files for analysis in ImageJ. In this method, definition of the ROIs using ImageJ is conducted manually. This software program allows the user to create multiple ROIs in a field of view that can be reapplied across multiple images, and average intensity measurements can be obtained for each ROI in each channel. Using this tool, data can be extracted from the images and further analyzed in a spreadsheet program such as Excel (Microsoft, Redmond, WA). In Excel, background subtraction is carried out individually for each image then the 485/438 ratios are calculated. ImageJ plug-ins such as Ratio Plus (https://imagej.nih.gov/ij/plugins/ratio-plus.html) can in principle be used to calculate ratios of two images, but caution must be applied to compensate for possible shift of the images. Since fluorescence is affected equally at all wavelengths, any effects from photobleaching are minimized when ratios are calculated. Having acquired 485/438 ratios at varying pH values, a calibration curve can be generated for each ROI to determine the absolute pH of that organelle.

1. Open all the image files in ImageJ. These will include one initial image of mCherry-LC3, 3–5 experimental images of the Oregon Green 488 at 485 nm excitation, 3–5 images of the Oregon Green 488 at 438 nm excitation, and all images from the calibration (3–5 images at 485 nm and 3–5 images at 438 nm per buffer used).
2. In the mCherry-LC3 micrograph, draw ROIs around all visible mCherry puncta. In this case puncta are traced manually using the freehand or elliptical selection tools; this process can be automated using thresholding to identify the brighter mCherry puncta over the weaker cytosolic mCherry signal.
3. Restore these ROIs onto the first experimental image acquired for Oregon Green 488 (485 nm excitation). If any of the ROIs appear to have no signal in the 485 nm channel, they can be ignored, as they are likely autophagosomes that have escaped fusion with lysosomes.
4. Draw at least three more ROIs or manually reposition unused ROIs from step 3 between cells to sample the background fluorescence intensity. Save these updated ROIs.
5. Collect the average fluorescence intensity for each ROI at 485 and 438 nm for each time point and across all calibration conditions and transfer this data to Excel. If the field of view between acquisitions varies slightly, reposition the ROIs such that they trace the original structures, save the ROIs, and continue with the analysis.
6. Determine the average background intensity for each image from the background ROIs and subtract this value from each ROI in that image. Repeat this step for each image individually.
7. Calculate the background-subtracted 485/438 ratio for each ROI for each time point. Average the time points and plot the calibration ratios against the pH of each buffer to generate a calibration curve.
8. Curve fitting is conducted by the least squares method and can be used to calculate the absolute pH of the organelle by interpolating the experimentally determined 485/438 ratio. In this case, a third order polynomial was fitted to the calibration data shown in Fig. 2C, and the pH of this autolysosome was determined to be 5.6 ± 0.3.

7. CONCLUSION

The use of dual-wavelength ratio fluorescence imaging allows for precise measurements of intraorganellar pH within living cells. This technique can be applied to estimate the absolute pH of autolysosomes. As with most fluorescence microscopy techniques, some caveats are pertinent. Like other fluorescein derivatives, the probes used for pH determination are photolabile. Excessive photobleaching of probes and photodamage to the cells caused by reactive oxygen species can diminish the accuracy of the results; for these reasons, care should be taken to minimize light exposure whenever possible. For Oregon Green 488-dextran, emission intensity at the 438 nm wavelength is particularly weak. If background subtraction is not conducted separately for each wavelength, the pH measurements will be affected adversely.

As an alternative to this single-probe dual-wavelength technique, the use of two dyes, one that is pH sensitive and the other pH insensitive, is also practical. However, when combining two separate probes one should be aware that their different chemistries endow different photostability; thus, their photobleaching rates may be significantly different and the calculation of fluorescence ratios will not correct for this disparity.

As mentioned briefly in Section 1, the fluorescent protein Keima has also been used to measure autolysosome pH. The bimodal excitation spectrum of this protein allows for the generation of fluorescence intensity ratios that vary according to pH. As used by Katayama et al. (2011), Keima has limitations. When expressed in the cytosol, some of the protein will indeed be trapped in autolysosomes. However, a substantial amount of Keima remains in the cytosol, where it reports a different (more alkaline) pH. Because fluorescence images are diffraction limited, it is virtually impossible to eliminate a contribution of cytosolic Keima to the autolysosomal determinations, particularly in the z-axis, where the point spread is greatest.

In addition, the pK_a of Keima is 6.5, which is not optimal for the determination of the more acidic pH values expected within autolysosomes.

Lastly, it is noteworthy that the use of high [K^+] plus nigericin for in situ calibration rests on the assumption that the [K^+] of autolysosomes is similar to that of the cytosol, an assumption that remains unproven. Deviations from this prediction will result in inaccurate pH determinations.

ACKNOWLEDGMENTS

We kindly thank Dr. Peter Kim (Hospital for Sick Children, Toronto) for providing the mCherry-LC3 construct and Dr. Roberto Botelho (Ryerson University, Toronto) for support. We thank Drs. Johnathan Canton and Danielle Johnson as well as Philip Ostrowski (Hospital for Sick Children, Toronto) for their advice. S.A.F. is supported by a fellowship from the Heart and Stroke Foundation of Canada. Supported by Grant FDN-143202 from the Canadian Institutes of Health Research to S.G.

REFERENCES

Baba, N. (1994). Ultrastructural analysis of the autophagic process in yeast: Detection of autophagosomes and their characterization. *The Journal of Cell Biology, 124*(6), 3–8.

Bonilla, D. L., Bhattacharya, A., Sha, Y., Xu, Y., Xiang, Q., Kan, A., et al. (2013). Autophagy regulates phagocytosis by modulating the expression of scavenger receptors. *Immunity, 39*(3), 537–547.

Canton, J., & Grinstein, S. (2015). Measuring lysosomal pH by fluorescence microscopy. In F. Platt & N. Platt (Eds.), *Methods in Cell Biology: 126.* (pp. 85–99). Amsterdam: Elsevier.

Coffey, E. E., Beckel, J. M., Laties, A. M., & Mitchell, C. H. (2014). Lysosomal alkalization and dysfunction in human fibroblasts with the Alzheimer's disease-linked presenilin 1 A246E mutation can be reversed with cAMP. *Neuroscience, 263*, 111–124.

Coffey, J. W., & De Duve, C. (1968). Digestive activity of lysosomes. *The Journal of Biological Chemistry, 243*, 3255–3263.

Costantini, L. M., Baloban, M., Markwardt, M. L., Rizzo, M., Guo, F., Verkhusha, V. V., et al. (2015). A palette of fluorescent proteins optimized for diverse cellular environments. *Nature Communications, 6*, 7670.

Deriy, L. V., Gomez, E. A., Zhang, G., Beacham, D. W., Hopson, J. A., Gallan, A. J., et al. (2009). Disease-causing mutations in the cystic fibrosis transmembrane conductance regulator determine the functional responses of alveolar macrophages. *The Journal of Biological Chemistry, 284*(51), 35926–35938.

Dunn, W. A. (1990). Studies on the mechanisms of autophagy: Maturation of the vacuole. *The Journal of Cell Biology, 110*, 1935–1945.

Eskelinen, E. L. (2005). Maturation of autophagic vacuoles in mammalian cells. *Autophagy, 1*(1), 1–10.

Jiang, L., Maher, V. M., McCormick, J. J., & Schindler, M. (1990). Alkalinization of the lysosomes is correlated with ras transformation of murine and human fibroblasts. *The Journal of Biological Chemistry, 265*(9), 4775–4777.

Katayama, H., Kogure, T., Mizushima, N., Yoshimori, T., & Miyawaki, A. (2011). A sensitive and quantitative technique for detecting autophagic events based on lysosomal delivery. *Chemistry and Biology, 18*(8), 1042–1052.

Kawai, A., Uchiyama, H., Takano, S., Nakamura, N., & Ohkuma, S. (2007). Autophagosome-lysosome fusion depends on the pH in acidic compartments in CHO cells. *Autophagy, 3*(2), 154–157.

Kim, J. H., Lingwood, C. A., Williams, D. B., Furuya, W., Manolson, M. F., & Grinstein, S. (1996). Dynamic measurement of the pH of the Golgi complex in living cells using retorograde transport of the verotoxin receptor. *The Journal of Cell Biology, 134*(6), 1387–1399.

Kogure, T., Karasawa, S., Araki, T., Saito, K., Kinjo, M., & Miyawaki, A. (2006). A fluorescent variant of a protein from the stony coral *Montipora* facilitates dual-color single-laser fluorescence cross-correlation spectroscopy. *Nature Biotechnology, 24*(5), 577–581.

Lawrence, B. P., & Brown, W. J. (1992). Autophagic vacuoles rapidly fuse with pre-existing lysosomes in cultured hepatocytes. *Journal of Cell Science, 102*(Pt. 3), 515–526.

Lee, J. H., Yu, W. H., Kumar, A., Lee, S., Mohan, P. S., Peterhoff, C. M., et al. (2010). Lysosomal proteolysis and autophagy require presenilin 1 and are disrupted by Alzheimer-related PS1 mutations. *Cell*, *141*(7), 1146–1158.

Lu, S., Kuo, C., Chen, H., Yang, Y., Liu, C., Anderson, R., et al. (2015). Insufficient acidification of autophagosomes facilitates Group A *Streptococcus* survival and growth in endothelial cells. *mBio*, *6*(5), 1–12.

Maneu, V., Yáñez, A., Murciano, C., Molina, A., Gil, M. L., & Gozalbo, D. (2011). Dectin-1 mediates *in vitro* phagocytosis of *Candida albicans* yeast cells by retinal microglia. *FEMS Immunology and Medical Microbiology*, *63*(1), 148–150.

Martinet, W., De Meyer, G. R. Y., Andries, L., Herman, A. G., & Kockx, M. M. (2006). *In situ* detection of starvation-induced autophagy. *The Journal of Histochemistry and Cytochemistry: Official Journal of the Histochemistry Society*, *54*(1), 85–96.

Maulucci, G., Chiarpotto, M., Papi, M., Samengo, D., Pani, G., & De Spirito, M. (2015). Quantitative analysis of autophagic flux by confocal pH-imaging of autophagic intermediates. *Autophagy*, *11*(10), 1905–1916.

Mellman, I., Fuchs, R., & Helenius, A. (1986). Acidification of the endocytic and exocytic pathways. *Annual Reviews of Biochemistry*, *55*, 663–700.

Mizushima, N. (2007). Autophagy: Process and function. *Genes and Development*, *21*(22), 2861–2873.

Mizushima, N., Ohsumi, Y., & Yoshimori, T. (2002). Autophagosome formation in mammalian cells. *Cell Structure and Function*, *27*, 421–429.

Neely, K. M., Green, K. N., & LaFerla, F. M. (2011). Presenilin is necessary for efficient proteolysis through the autophagy-lysosome system in a γ-secretase-independent manner. *The Journal of Neuroscience*, *31*(8), 2781–2791.

Ohkuma, S., & Poole, B. (1978). Fluorescence probe measurement of the intralysosomal pH in living cells and the perturbation of pH by various agents. *Proceedings of the National Academy of Sciences*, *75*(7), 3327–3331.

Pankiv, S., Clausen, T. H., Lamark, T., Brech, A., Bruun, J. A., Outzen, H., et al. (2007). p62/SQSTM1 binds directly to Atg8/LC3 to facilitate degradation of ubiquitinated protein aggregates by autophagy. *The Journal of Biological Chemistry*, *282*(33), 24131–24145.

Poët, M., Kornak, U., Schweizer, M., Zdebik, A. A., Scheel, O., Hoelter, S., et al. (2006). Lysosomal storage disease upon disruption of the neuronal chloride transport protein ClC-6. *Proceedings of the National Academy of Sciences*, *103*(37), 13854–13859.

Shaner, N. C., Campbell, R. E., Steinbach, P. A., Giepmans, B. N. G., Palmer, A. E., & Tsien, R. Y. (2004). Improved monomeric red, orange and yellow fluorescent proteins derived from *Discosoma* sp. red fluorescent protein. *Nature Biotechnology*, *22*(12), 1567–1572.

Tabeta, K., Hoebe, K., Janssen, E. M., Du, X., Georgel, P., Crozat, K., et al. (2006). The Unc93b1 mutation 3d disrupts exogenous antigen presentation and signaling via Toll-like receptors 3, 7 and 9. *Nature Immunology*, 7(2), 156–164.

Tanida, I., Minematsu-Ikeguchi, N., Ueno, T., & Kominami, E. (2005). Lysosomal turnover, but not a cellular level, of endogenous LC3 is a marker for autophagy. *Autophagy*, *1*(2), 84–91.

Trombetta, E. S., Ebersold, M., Garrett, W., Pypaert, M., & Mellman, I. (2003). Activation of lysosomal function during dendritic cell maturation. *Science*, *299*(5611), 1400–1403.

Vieira, O. V., Botelho, R. J., & Grinstein, S. (2002). Phagosome maturation: Aging gracefully. *The Biochemical Journal*, *366*(Pt. 3), 689–704.

Wu, M. M., Llopis, J., Adams, S., McCaffery, J. M., Kulomaa, M. S., Machen, T. E., et al. (2000). Organelle pH studies using targeted avidin and fluorescein-biotin. *Chemistry and Biology*, 7(3), 197–209.

Zhang, X., Garbett, K., Veeraraghavalu, K., Wilburn, B., Gilmore, R., Mirnics, K., et al. (2012). A role for presenilins in autophagy revisited: Normal acidification of lysosomes in cells lacking PSEN1 and PSEN2. *The Journal of Neuroscience*, *32*(25), 8633–8648.

CHAPTER THREE

Long-Lived Protein Degradation During Autophagy

N. Dupont, C. Leroy, A. Hamaï, P. Codogno[1]
Institut Necker-Enfants Malades (INEM), INSERM U1151-CNRS UMR 8253, Université Paris Descartes-Sorbonne Paris Cité, Paris, France
[1]Corresponding author: e-mail address: patrice.codogno@inserm.fr

Contents

Abstract

Macroautophagy, a major lysosomal degradative pathway for cytoplasmic components, is a process that can be stimulated in response to many stressful situations including cancer treatment. The central autophagic organelle is the autophagosome, a double-membrane-bound vacuole that sequesters cytoplasmic material. The ultimate destiny of the autophagosome is fusion with the lysosomal compartment, where cargo, including proteins, is degraded. Here, we report a method to measure the lysosomal degradation of long-lived proteins along the autophagic pathway.

1. INTRODUCTION

Macroautophagy (referred to later as autophagy) is a form of autophagy that degrades cellular macromolecules and cytoplasmic structures in eukaryotic cells (Boya, Reggiori, & Codogno, 2013; Yang & Klionsky, 2010). Autophagy begins with the multistep formation of a double-membrane-bound vacuole known as the autophagosome, which sequesters cytoplasmic material in bulk or in a selective manner via autophagy adapters such as SQSTM1/p62. The discovery of ATG (autophagy related) proteins

Methods in Enzymology, Volume 588
ISSN 0076-6879
http://dx.doi.org/10.1016/bs.mie.2016.09.074

(Mizushima, Yoshimori, & Ohsumi, 2011), which are involved in autophagosome formation was a major breakthrough in the understanding of autophagy and its importance in physiology and pathology (Mizushima & Komatsu, 2011; Ravikumar et al., 2010; Rubinsztein, Codogno, & Levine, 2012). In mammalian cells, once the autophagosome has been formed, it acquires acidic and degradative capacities by merging with endocytic compartments to form a compartment known as the amphisome. The final stage of autophagy is the fusion of autophagic vacuoles (amphisomes or autophagosomes) with lysosomes to form autolysosomes, where the autophagy cargo is totally degraded.

Two sets of information are of fundamental importance before any conclusion can be reached about the stimulation of autophagy: (1) determination of the increase in the number of autophagosomes formed (induction of autophagy) and (2) assay of the autophagic flux (i.e., the clearance of the autophagy cargo by the lysosomal compartment). An increase in the number of autophagosomes can be the consequence of an increase in the induction of autophagy and in autophagic flux, but it can also be the consequence of a blockade of the autophagic flux without any stimulation of the induction step (Klionsky et al., 2016; Mizushima, Yoshimori, & Levine, 2010). In this chapter, we will describe the method to investigate the autophagic flux based on the assay of the degradation of long-lived proteins. We will also briefly summarize other methods to assay the autophagic flux. Readers interested in these methods can consult other chapters in this series.

2. PROTEOLYSIS

2.1 Materials and Reagents

1. Dulbecco's modified Eagle's medium (DMEM, 4.5 g/L glucose, glutamax) (Invitrogen, Life Technologies, Saint-Aubin, France, ref 31966) supplemented with 10% fetal bovine serum (Invitrogen, Life Technologies, Saint-Aubin, France).
2. Phosphate-buffered saline (PBS): 137 m*M* NaCl, 2.7 m*M* KCl, 4.3 m*M* Na_2HPO_4, 1.4 m*M* KH_2PO_4, pH 7.3.
3. Hanks' balanced salt solution (HBSS) without sodium bicarbonate (Invitrogen, Life Technologies, ref 14025) or Earle's balanced salt solution (EBSS) (Invitrogen, Life Technologies, Saint-Aubin, France, ref 24010–043).
4. L-[U-^{14}C]-valine (266 mCi/mmol, 9.84 GBq/mmol, Amersham Biosciences, Buckinghamshire, UK, ref CFB75).

5. "Cold" valine (Merck Chemicals Limited, Nottingham, UK, ref 1.08495.0100).
6. Trichloroacetic acid (TCA) (Sigma-Aldrich, Saint Quentin Fallavier, France, ref T4885).
7. 0.2 *M* NaOH.
8. 3-Methyladenine (3-MA, Sigma-Aldrich, Saint Quentin Fallavier, France, ref M9281).
9. Ammonium chloride (NH_4Cl, Sigma-Aldrich, Saint Quentin Fallavier, France, ref A9434).
10. Leupeptin (Sigma-Aldrich, Saint Quentin Fallavier, France, ref L2884).

2.2 Protocol (see Fig. 1)

The protocol described later was originally validated for use with various cancer cells (see Bauvy, Meijer, & Codogno, 2009). Cells are seeded in the usual medium in 6-well plates and routinely used near confluence because protein degradation is enhanced in growing cells (Fuertes, Martin De Llano, Villarroya, Rivett, & Knecht, 2003). The standard used assay of radiolabeled long-lived protein is shown in Fig. 2. In a routine protocol, experiments are run in the presence or absence of 3-MA to block autophagic proteolysis (wortmannin another phosphatidylinositol 3-phosphate kinase can also be use to block autophagic proteolysis). However, wortmannin is less stable in culture medium than 3-MA). In addition, the total lysosomal proteolysis can be blocked by adding NH_4Cl to raise the lysosomal pH. Other lysosomotropic agents such as chloroquine or inhibitors of the vacuolar ATPase such as bafilomycin A1 can also be used. Leupeptin is added to block the residual cathepsin activity (Fuertes, Martin De Llano, Villarroya, Rivett, & Knecht, 2003). In addition, proteasome inhibitors such as MG132 can be used to enable evaluation of the contribution of the ubiquitin proteasome system not only to the degradation of short-lived proteins but also to some extent to the degradation of long-lived proteins (Fuertes et al., 2003; Zhao, Zhai, Gygi, & Goldberg, 2015). Herein, we will describe the standard protocol used to assay the degradation of radiolabeled protein.

(1) Intracellular proteins are labeled for 18 h at 37°C with 0.2 μCi/mL of L-[U-^{14}C] valine (specific activity 266 mCi/mmol) in complete medium (Pulse).

(2) Any unincorporated radioactivity is eliminated by rinsing the cells three times with PBS.

(3) The cells are then incubated with fresh complete medium containing 10 m*M* cold valine (see Note A in Section 2.3) for 1 h during which

Flowchart for proteolysis

Cells are seeded in the usual medium in 6-well plates and used near confluence.

Add 0.2 μCi/mL of L-[U-^{14}C] valine in complete medium and incubate cells for 18 h at 37°C (Pulse)

↓

3 × wash with PBS

↓

Add fresh complete medium containing 10 m*M* cold valine then incubate for 1 h

↓

Remove the medium and replace with fresh complete medium containing 10 m*M* cold valine or with EBSS or with EBSS with 10 m*M* 3-MA (Chase)

↓

Transfer supernatant to eppendorfs and add TCA to the medium at a final concentration of 10% (w/v)

↓

Spin-down the culture medium for 10 min at 470 × *g* at 4°C and read the acid-soluble radioactivity by liquid scintillation counting.

↓

Wash cells twice with cold TCA (w/v), plus 10 m*M* cold valine

↓

Dissolve the cell pellet at 37°C in 0.2 *M* NaOH and incubate for 2 h.

↓

Read the radioactivity from this cell pellet by liquid scintillation counting and calculate the ratio of the acid-soluble radioactivity in the medium to that in the acid-precipitable cell fraction.

Fig. 1 Flowchart for proteolysis.

time short-lived proteins are degraded (in some cell lines this period can be extended to 24 h).

(4) After this period, the medium is removed and replaced with the appropriate fresh complete medium supplemented with 10 m*M* cold valine, and cell are then incubated for a further 4 h (longer incubations in the chase medium are also possible, and in some cases it may be desirable to use multiple time points). Throughout the chase period, 3-MA

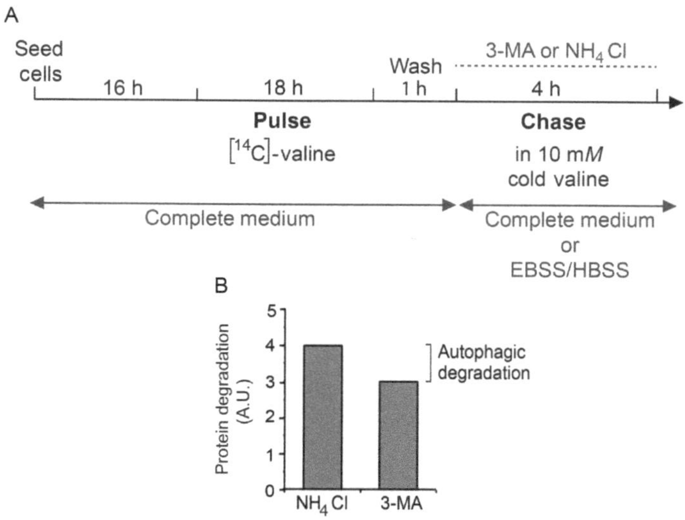

Fig. 2 Assay to measure long-lived protein degradation. (A) Experimental protocol to assay degradation of [^{14}C]-valine-labeled long-lived proteins. (B) Effect of inhibitors on the degradation of [^{14}C]-valine-labeled long-lived proteins. Keys: *3-MA*, 3-methyladenine; NH_4Cl, ammonium chloride. The concentrations shown are those that have a maximal effect in HeLa cells. A dose response with the inhibitors has to be performed to determine the saturation concentration at which they should be used in each cell line.

(see Note B in Section 2.3) can be added at a final concentration of 10 m*M* to inhibit the de novo formation of autophagic vacuoles (Seglen & Gordon, 1982). To stimulate autophagy HBSS (or EBSS) (see Note C in Section 2.3) containing 10 m*M* cold valine and 0.1% bovine serum albumin can be added.

(5) The medium is then precipitated overnight after adding TCA at a final concentration of 10% (w/v).

(6) After centrifuging the culture medium for 10 min at 470 × *g* at 4°C, the acid-soluble radioactivity is measured by liquid scintillation counting.

(7) The cells are washed twice with cold TCA (w/v), plus 10 m*M* cold valine to make sure that no radioactivity has remained adsorbed to the denatured proteins. The cell pellet is then dissolved at 37°C in 0.2 *M* NaOH and incubated for 2 h.

(8) The radioactivity is then measured by liquid scintillation counting. The rate of degradation of longer-lived proteins is calculated from the ratio of the acid-soluble radioactivity in the medium to that in the acid-precipitable cell fraction.

2.3 Notes and Potential Pitfalls

A. Valine is used because this amino acid does not interfere with autophagy. Since leucine is a physiological inhibitor of autophagy (Meijer & Dubbelhuis, 2004), the use of this amino acid at high concentration (10 m*M*) during the chase period can interfere with autophagic proteolysis. The choice of amino acid during labeling is also important to avoid a continued incorporation of the radioactive amino acid into proteins during the chase period because of a high pool of unincorporated amino acids (see Zhao, Garcia, & Goldberg, 2016).

B. 3-MA blocks autophagy by inhibiting class III phosphatidylinositol 3-kinase (Petiot, Ogier-Denis, Blommaart, Meijer, & Codogno, 2000). However, it should be kept in mind that 3-MA is also a phosphatidylinositol 3-kinase inhibitor (Blommaart, Krause, Schellens, Vreeling-Sindelarova, & Meijer, 1997) and interferes with other intracellular trafficking pathways dependent on the phosphatidylinositol 3-kinases. 3-MA also affects some other intracellular events (Tolkovsky, Xue, Fletcher, & Borutaite, 2002). 3-MA is routinely used at a final concentration of 10 m*M*. It is a good idea to prepare a stock solution at higher concentration (e.g., 100 m*M*) in water. It may be necessary to heat the solution under a warm-water faucet to dissolve such a high concentration of 3-MA. The prepared solution can be stored frozen, but the 3-MA may precipitate on thawing. It should be kept in mind that treatment of cells with 3-MA for longer than 24 h could stimulates autophagy (for discussion, see Wu et al., 2010). This is due to the relief of the inhibition of class III PI3K and to the inhibition of class I P3K (Petiot et al., 2000).

C. HBSS is used when the cells are incubated in a humidified chamber at 37°C in the absence of CO_2, otherwise EBSS is used. The reason is that EBSS, unlike HBSS, contains bicarbonate, and the pH of the medium is governed by the Henderson–Hasselbalch equation: $pH = pK_a + \log[HCO_3^-]/[CO_2] = 6.1 + \log[HCO_3^-]/[CO_2]$.

3. OTHER METHODS FOR MEASURING AUTOPHAGIC FLUX

Determination of long-lived protein degradation using radiolabeled amino acids is a classical method to measure the autophagic flux. Recently a method based on L-azidohomoalanine has been developed to quantify long-lived protein degradation (Zhang, Wang, Ng, Lin, & Shen, 2014). The azide-containing proteins obtained after a click reaction can be detected by an alkyline fluorescent dye coupled with flow cytometry. Other methods have been developed for measuring autophagic flux (Klionsky et al., 2016). Some of these methods are based on the turnover of specific autophagy cargoes. Many of the methods described are based on the detection of protein LC3 (a member of the ATG8 family) (Mizushima et al., 2011). During autophagy, LC3-I is converted into LC3-II by the formation of a covalent link between the C-terminus of LC3-I and the polar head of phosphatidylethanol amine. LC3-II and other members of the ATG8 family are involved in the elongation of the preautophagosomal membrane and in the sealing of the membrane (Mizushima & Komatsu, 2011; Mizushima et al., 2011; Weidberg et al., 2010). The fraction of LC3-II associated with the inner face of the autophagosome is transported into the lysosomal compartment where it is degraded. The turnover of LC3-II can be assayed by immunoblotting in the presence and absence of lysosomal inhibitors (such as inhibitors of cathepsins or by lysosomal acidification) (Sarkar et al., 2005; Tanida, Minematsu-Ikeguchi, Ueno, & Kominami, 2005). The lysosomal turnover of protein SQSTM1/p62, which is transported to the lysosome via the interaction between its LC3-interaction region (LIR) with LC3-II (Birgisdottir, Lamark, & Johansen, 2013) can be used to assay the autophagic flux. When this strategy is used, it is important to verify that under the condition used (stimuli, cell lines) the expression of *SQSTM1/p62* mRNA, which can fluctuate, does not influence the expression of the protein. As an example it has been shown that amino acids can upregulate the transcription of *SQSTM1/p62* mRNA in some cell lines, thus making the interpretation of the level of SQSTM1/p62 not a reliable tool for investigating the autophagic flux in this setting (Sahani, Itakura, & Mizushima, 2014).

The use of an mRFP-GFP tandem fluorescent-tagged LC3 has been proposed for visualizing the autophagic flux (Kimura, Noda, & Yoshimori, 2007). The method is based on the differential quenching of the fluorescence emitted by the red and green probes in the lysosomal environment. Briefly, autophagosomes appear yellow and lysosomes red, because of the disappearance of the GFP-LC3 signal at an acidic pH. When the autophagic flux is interrupted, the red fluorescence is no longer detectable, and only yellow dots are visible. Elazar and colleagues (Shvets, Faas, & Elazar, 2008) introduced a fluorescence-activated cell sorter (FACS)-based method to quantify the turnover of GFP-LC3 as a measure of the autophagic activity in mammalian cells. In this method, the decrease of the fluorescence reflects the lysosomal delivery of LC3 to the lysosomal compartment. FACS analysis presents several advantages. For example, this method is quantitative and allows one to perform large-scale screens for unknown modulators of autophagy.

Another interesting method for estimating autophagic flux is based on sequestration of the autophagic cargo (e.g., lactate dehydrogenase combined with an inhibitor of lysosomal function to prevent the intralysosomal degradation of lactate dehydrogenase). The assay measures the rate of transfer of the marker from the soluble (cytosol) to the particulate (sedimentable) cell fraction that contains the autophagic compartments. The method was originally developed by Seglen and coworkers for hepatocytes but can be applied to other cell types (Klionsky, Cuervo, & Seglen, 2007) and references therein).

The advantages and limitations of all these methods, including proteolysis, have been discussed in detail in a recent review of the guidelines to monitor autophagy (Klionsky et al., 2016). Readers can consult this review for more information about the various aspects of the autophagic flux.

4. CONCLUSION

The assays presented here have been widely used. However, other methods for measuring autophagic flux have also been developed, such FACS-based methods (Shvets et al., 2008) and the use of nonradioactive methods to assay the protein degradation (Zhang et al., 2014). Researchers should perform several autophagic assays before reaching any final conclusions about whether the initiation of autophagy or/and autophagy flux is affected under their experimental conditions.

ACKNOWLEDGMENTS

Work in Patrice Codogno laboratory is supported by funding from INSERM, CNRS, Université Paris-Descartes-Sorbonne Paris Cité, and grants from INCa and ANR. A.H. is supported by a fellowship from La Fondation de France, and N.D. is supported by a fellowship from the "Association pour la Recherche contre le Cancer" (ARC).

REFERENCES

Bauvy, C., Meijer, A. J., & Codogno, P. (2009). Assaying of autophagic protein degradation. *Methods in Enzymology*, *452*, 47–61.

Birgisdottir, A. B., Lamark, T., & Johansen, T. (2013). The LIR motif - crucial for selective autophagy. *Journal of Cell Science*, *126*(Pt 15), 3237–3247.

Blommaart, E. F., Krause, U., Schellens, J. P., Vreeling-Sindelarova, H., & Meijer, A. J. (1997). The phosphatidylinositol 3-kinase inhibitors wortmannin and LY294002 inhibit autophagy in isolated rat hepatocytes. *European Journal of Biochemistry*, *243*(1–2), 240–246.

Boya, P., Reggiori, F., & Codogno, P. (2013). Emerging regulation and functions of autophagy. *Nature Cell Biology*, *15*(7), 713–720.

Fuertes, G., Martin De Llano, J. J., Villarroya, A., Rivett, A. J., & Knecht, E. (2003). Changes in the proteolytic activities of proteasomes and lysosomes in human fibroblasts produced by serum withdrawal, amino-acid deprivation and confluent conditions. *The Biochemical Journal*, *375*(Pt 1), 75–86.

Kimura, S., Noda, T., & Yoshimori, T. (2007). Dissection of the autophagosome maturation process by a novel reporter protein, tandem fluorescent-tagged LC3. *Autophagy*, *3*(5), 452–460.

Klionsky, D. J., Abdelmohsen, K., Abe, A., Abedin, M. J., Abeliovich, H., Acevedo Arozena, A., et al. (2016). Guidelines for the use and interpretation of assays for monitoring autophagy (3rd edition). *Autophagy*, *12*(1), 1–222.

Klionsky, D. J., Cuervo, A. M., & Seglen, P. O. (2007). Methods for monitoring autophagy from yeast to human. *Autophagy*, *3*(3), 181–206.

Meijer, A. J., & Dubbelhuis, P. F. (2004). Amino acid signalling and the integration of metabolism. *Biochemical and Biophysical Research Communications*, *313*(2), 397–403.

Mizushima, N., & Komatsu, M. (2011). Autophagy: Renovation of cells and tissues. *Cell*, *147*(4), 728–741.

Mizushima, N., Yoshimori, T., & Levine, B. (2010). Methods in mammalian autophagy research. *Cell*, *140*(3), 313–326.

Mizushima, N., Yoshimori, T., & Ohsumi, Y. (2011). The role of Atg proteins in autophagosome formation. *Annual Review of Cell and Developmental Biology*, *27*, 107–132.

Petiot, A., Ogier-Denis, E., Blommaart, E. F., Meijer, A. J., & Codogno, P. (2000). Distinct classes of phosphatidylinositol 3′-kinases are involved in signaling pathways that control macroautophagy in HT-29 cells. *The Journal of Biological Chemistry*, *275*(2), 992–998.

Ravikumar, B., Sarkar, S., Davies, J. E., Futter, M., Garcia-Arencibia, M., Green-Thompson, Z. W., et al. (2010). Regulation of mammalian autophagy in physiology and pathophysiology. *Physiological Reviews*, *90*(4), 1383–1435.

Rubinsztein, D. C., Codogno, P., & Levine, B. (2012). Autophagy modulation as a potential therapeutic target for diverse diseases. *Nature Reviews. Drug Discovery*, *11*(9), 709–730.

Sahani, M. H., Itakura, E., & Mizushima, N. (2014). Expression of the autophagy substrate SQSTM1/p62 is restored during prolonged starvation depending on transcriptional upregulation and autophagy-derived amino acids. *Autophagy*, *10*(3), 431–441.

Sarkar, S., Floto, R. A., Berger, Z., Imarisio, S., Cordenier, A., Pasco, M., et al. (2005). Lithium induces autophagy by inhibiting inositol monophosphatase. *The Journal of Cell Biology*, *170*(7), 1101–1111.

Seglen, P. O., & Gordon, P. B. (1982). 3-Methyladenine: Specific inhibitor of autophagic/lysosomal protein degradation in isolated rat hepatocytes. *Proceedings of the National Academy of Sciences of the United States of America*, *79*(6), 1889–1892.

Shvets, E., Fass, E., & Elazar, Z. (2008). Utilizing flow cytometry to monitor autophagy in living mammalian cells. *Autophagy*, *4*(5), 621–628.

Tanida, I., Minematsu-Ikeguchi, N., Ueno, T., & Kominami, E. (2005). Lysosomal turnover, but not a cellular level, of endogenous LC3 is a marker for autophagy. *Autophagy*, *1*(2), 84–91.

Tolkovsky, A. M., Xue, L., Fletcher, G. C., & Borutaite, V. (2002). Mitochondrial disappearance from cells: A clue to the role of autophagy in programmed cell death and disease? *Biochimie*, *84*(2–3), 233–240.

Weidberg, H., Shvets, E., Shpilka, T., Shimron, F., Shinder, V., & Elazar, Z. (2010). LC3 and GATE-16/GABARAP subfamilies are both essential yet act differently in autophagosome biogenesis. *The EMBO Journal*, *29*(11), 1792–1802.

Wu, Y. T., Tan, H. L., Shui, G., Bauvy, C., Huang, Q., Wenk, M. R., et al. (2010). Dual role of 3-methyladenine in modulation of autophagy via different temporal patterns of inhibition on class I and III phosphoinositide 3-kinase. *The Journal of Biological Chemistry*, *285*(14), 10850–10861.

Yang, Z., & Klionsky, D. J. (2010). Eaten alive: A history of macroautophagy. *Nature Cell Biology*, *12*(9), 814–822.

Zhang, J., Wang, J., Ng, S., Lin, Q., & Shen, H. M. (2014). Development of a novel method for quantification of autophagic protein degradation by AHA labeling. *Autophagy*, *10*(5), 901–912.

Zhao, J., Garcia, G. A., & Goldberg, A. L. (2016). Control of proteasomal proteolysis by mTOR. *Nature*, *529*(7586), E1–E2.

Zhao, J., Zhai, B., Gygi, S. P., & Goldberg, A. L. (2015). mTOR inhibition activates overall protein degradation by the ubiquitin proteasome system as well as by autophagy. *Proceedings of the National Academy of Sciences of the United States of America*, *112*(52), 15790–15797.

CHAPTER FOUR

Proteomic Profiling of De Novo Protein Synthesis in Starvation-Induced Autophagy Using Bioorthogonal Noncanonical Amino Acid Tagging

J. Zhang*[,1], J. Wang*[,†,‡,1], Y.-M. Lee[†], T.-K. Lim[†], Q. Lin[†,2], H.-M. Shen*[,§,2]

*Yong Loo Lin School of Medicine, National University of Singapore, Singapore, Singapore
[†]National University of Singapore, Singapore, Singapore
[‡]Interdisciplinary Research Group in Infectious Diseases, Singapore-MIT Alliance for Research & Technology (SMART), Singapore
[§]NUS Graduate School for Integrative Sciences and Engineering, National University of Singapore, Singapore
[2]Corresponding authors: e-mail address: dbslinqs@nus.edu.sg; phsshm@nus.edu.sg

Contents

[1] These authors contributed equally to this work.

Methods in Enzymology, Volume 588
ISSN 0076-6879
http://dx.doi.org/10.1016/bs.mie.2016.09.075

Abstract

Autophagy is an intracellular degradation process activated by stress factors such as nutrient starvation to maintain cellular homeostasis. There is emerging evidence demonstrating that de novo protein synthesis is involved in the autophagic process. However, up-to-date characterizing of these de novo proteins is technically difficult. In this chapter, we describe a novel method to identify newly synthesized proteins during starvation-mediated autophagy by bioorthogonal noncanonical amino acid tagging (BONCAT), in conjunction with isobaric tagging for relative and absolute quantification (iTRAQ)-based quantitative proteomics. L-azidohomoalanine (AHA) is an analog of methionine, and it can be readily incorporated into the newly synthesized proteins. The AHA-containing proteins can be enriched with avidin beads after a "click" reaction between alkyne-bearing biotin and the azide moiety of AHA. The enriched proteins are then subjected to iTRAQ™ labeling for protein identification and quantification using liquid chromatography-tandem mass spectrometry (LC-MS/MS). By using this technique, we have successfully profiled more than 700 proteins that are synthesized during starvation-induced autophagy. We believe that this approach is effective in identification of newly synthesized proteins in the process of autophagy and provides useful insights to the molecular mechanisms and biological functions of autophagy.

1. INTRODUCTION

Macroautophagy, referred to as *autophagy* in this chapter, is an important intracellular degradation mechanism that is highly conserved in all eukaryotes (Bauvy, Meijer, & Codogno, 2009). The process of autophagy involves the formation of a double-membrane compartment, called the *phagophore*, to isolate and sequester cytoplasmic constituents such as protein aggregates and impaired organelles (Levine & Klionsky, 2004; Mizushima & Komatsu, 2011). The phagophore expands and matures into an autophagosome, which subsequently fuses with a lysosome, resulting in the formation of an autolysosome for the degradation of the sequestered materials (Levine & Klionsky, 2004; Mizushima & Komatsu, 2011; Shen & Mizushima, 2014). Under normal conditions, basal autophagy occurs constitutively, possibly reflecting its role in maintaining cellular homeostasis via the degradation of long-lived proteins and the removal of damaged cellular organelles (Mizushima & Klionsky, 2007). On the other hand, the presence of stresses such as starvation leads to autophagy induction and promotes dynamic recycling of the basic biomolecules such as amino acids to supplement vital cellular functions (Wirawan, Vanden Berghe, Lippens, Agostinis, & Vandenabeele, 2012). This is made possible by the inactivation of mTOR (mechanistic target of rapamycin) and subsequent activation of the ULK1 (unc-51-like autophagy-activating kinase 1) complex to initiate autophagosome formation (Chen & Klionsky, 2011; Mizushima & Komatsu, 2011).

It is well known that under starvation conditions, the global protein synthesis is significantly reduced, mainly by the suppression of mTOR activity. At present, it is controversial whether the autophagic process is regulated at the transcriptional level (Watanabe-Asano, Kuma, & Mizushima, 2014). Recently, there is emerging evidence demonstrating that the autophagic process involves a number of nuclear transcription factors and their transcriptional activity, such as TP53 (tumor protein p53), STAT3 (signal transducer and activator of transcription 3), NF-κB (nuclear factor of kappa light polypeptide gene enhancer in B-cells), HIF1 (hypoxia inducible factor 1), FOXO (forkhead box O), and TFEB (transcription factor EB) (Fullgrabe, Klionsky, & Joseph, 2014). Therefore, it is reasonable to speculate that de novo protein synthesis is implicated in both the basal and induced autophagic processes. Thus, identification and analysis of de novo proteins synthesized during autophagy will provide insight into understanding both the regulatory mechanisms and the biological functions of autophagy.

In recent years, several new techniques for labeling and detection of de novo proteins have been developed. Among them, bioorthogonal noncanonical amino acid tagging (BONCAT) has attracted much attention. Using the simple principle of bioorthogonal metabolic tagging, a noncanonical amino acid with a small bioorthogonal functional group, such as the azide-bearing L-azidohomoalanine (AHA), can be incorporated into the newly synthesized proteins (Dieterich et al., 2007; Dieterich, Link, Graumann, Tirrell, & Schuman, 2006; Ngo & Tirrell, 2011). As AHA is a methionine surrogate, it can be naturally and efficiently taken up by the cell's endogenous translational machinery (Dieterich et al., 2007; Kiick, Saxon, Tirrell, & Bertozzi, 2002). Importantly, the tagged de novo proteins can be subsequently enriched via an alkyne-bearing biotinylated tag, or labeled with a fluorophore via a Cu(I)-catalyzed azide alkyne cycloaddition (CuAAC) reaction (i.e., "click" chemistry) (Rostovtsev, Green, Fokin, & Sharpless, 2002; Tornoe, Christensen, & Meldal, 2002). Recently, we have developed a novel method to quantify autophagic protein degradation with AHA labeling (Zhang, Wang, Ng, Lin, & Shen, 2014), in which the newly synthesized proteins are pulse-labeled with AHA, and the reduced fluorescence intensity of the AHA-containing proteins induced by autophagy is measured by flow cytometry for quantification of autophagic protein degradation.

In this chapter, we present a detailed protocol to perform the AHA-tagging/iTRAQ-based quantitative proteomics analysis for the specific and unbiased identification of de novo protein synthesis in the course of autophagy, based on our recent work (Wang, Zhang, Lee, et al., 2016). Alternatively, other quantitative proteomics-based methods (such as SILAC, TMT, SWATH, etc.) coupled with AHA tagging can also be adopted to differentiate the specific de novo protein synthesis, such as pulsed azidohomoalanine labeling in mammals (PALM) (McClatchy et al., 2015), quantitating proteome dynamics in primary cells (QuaNCAT) (Howden et al., 2013), or BONCAT-pulse SILAC (BONCAT-pSILAC) (Bagert et al., 2014). In this protocol, we trigger starvation-induced autophagy in HeLa cells by culturing them in an amino acid-free medium. These methods, described in detail in the next sections, could be adaptable to other cell lines.

2. MATERIALS, EQUIPMENT, AND SOLUTIONS

2.1 Materials

- HeLa cells are commercially available from ATCC (ATCC® CCL-2™, Manassas, VA)
- Dulbecco's Modified Eagle's Medium: DMEM, containing 4500 mg/L D-glucose (Sigma-Aldrich, #D1152, St. Louis, MO)
- Heat-inactivated fetal bovine serum (FBS; HyClone, #SV30160.03, Logan, UT)
- Dialyzed FBS (Invitrogen, #26400044, Carlsbad, CA)
- Penicillin–streptomycin mixture (Life Technologies, #15140122, Carlsbad, CA)
- 1 × PBS, pH 7.2–7.4 (Life Technologies, #10010-023, Carlsbad, CA)
- Milli-Q water
- Molecular biology-grade water (Sigma-Aldrich, #W4502, St. Louis, MO)
- Dimethyl sulfoxide (DMSO; Sigma-Aldrich, #D2650, St. Louis, MO)
- AHA reagent (Invitrogen, #C10289; or Click Chemistry Tools, #1066-25, Carlsbad, CA)
- L-Methionine, $\geq$ 98% (Sigma-Aldrich, #M9625, St. Louis, MO)
- Trypsin–EDTA, 0.25% (Life Technologies, #25200, Carlsbad, CA)
- Biotin alkyne (Invitrogen, #B10185, Carlsbad, CA)

- Tris (2-carboxyethyl) phosphine hydrochloride, ≥ 98% purity (TCEP; Sigma-Aldrich, #C4706, St. Louis, MO)
- Tris [(1-benzyl-1*H*-1,2,3-triazol-4-yl) methyl] amine (TBTA; Sigma-Aldrich, #678937, St. Louis, MO)
- Copper sulfate, ≥ 99.99% purity ($CuSO_4$; Sigma-Aldrich, #451657, St. Louis, MO)
- Sodium dodecyl sulfate (SDS; Sigma-Aldrich, #862010, St. Louis, MO)
- Halt™ Protease Inhibitor Cocktail (Pierce, #88266, Waltham, MA)
- DNase (New England Biolabs, #M0303S, Ipswich, MA)
- RNase (Qiagen, #19101, Hilden, Germany)
- Streptavidin beads (Sigma-Aldrich, #S1638, St. Louis, MO)
- Urea (Sigma-Aldrich, #U5378, St. Louis, MO)
- Methyl methanethiosulfonate (MMTS; Pierce, #23011, Waltham, MA)
- Trypsin, 12.5 ng/μL (Promega, #V5111, Madison, WI)
- Triethylammonium hydrogen carbonate buffer, 1 *M* (TEAB; Sigma-Aldrich, #T7408, St. Louis, MO)
- iTRAQ Method Development Kit (SCIEX, #4352160, Framingham, MA)
- Acetonitrile (ACN; Sigma-Aldrich, #34967, St. Louis, MO)
- Formic acid (FA; Sigma-Aldrich, #V800192, St. Louis, MO)
- Phosphoric acid (Sigma-Aldrich, #V800287, St. Louis, MO)
- Acetone (Sigma-Aldrich, #34850, St. Louis, MO)
- Sodium hydroxide (NaOH; Sigma-Aldrich, #306576, St. Louis, MO)

2.2 Equipment

- Biological safety level 2 tissue culture hood (Thermo Fisher Scientific, Waltham, MA)
- Cell incubator at 37°C, 5% CO_2 (Thermo Fisher Scientific, Waltham, MA)
- Thermo Scientific Nunc Cell Culture/Petri Dishes 150 mm Dish (Thermo Scientific, #1256590, Waltham, MA)
- Centrifuge tubes with screw caps, 15 mL (BD Biosciences, #352196, Sparks, MD)
- Refrigerated centrifuge (Eppendorf, Model 5415R; Model 5810R, Hamburg, Germany)
- Microcentrifuge tubes, 1.5 mL (Axygen, #311-08-051, Union City, CA)
- Shaker for microcentrifuge tubes with temperature control (Eppendorf Thermomixer® C, Hamburg, Germany)
- Refrigerators, 4°C
- Freezer, −20°C

- Vortex mixer (LP Vortex Mixer from Thermo Scientific, #88880018, Waltham, MA)
- Minisart filters, 0.45-μm pore size (Sartorius, #51123103, Kyoto, Japan)
- Steritop-GP Filter Unit (Millipore, #SCGVT05RE, Billerica, MA)
- Water bath, 37°C
- Ultrasonic water bath cleaner (Elmasonic, SH250EL, Singen, Germany)
- Phase-contrast microscope (Nikon ECLIPSE, TE2000-S, Tokyo, Japan)
- A filter spin column (GE Healthcare, 27-3565-01, Chicago, IL)
- TripleTOF 5600 system (AB SCIEX, Framingham, MA)
- Parafilm
- C_{18} column Sep-Pak column (Waters, #WAT051910, Milford, MA)
- ChiPLC-nanoflex (Eksigent, Dublin, CA)
- Speedvac (Savant™ SPD131DDA SpeedVac™ Concentrator)
- Lyophilizer machine (American Lyophilizer, INC.)
- Eksigent nanoLC Ultra system, coupled to the ChiPLC-nanoflex system (Eksigent, Dublin, CA)
- ProteinPilot™ 4.5 (AB SCIEX, Framingham, MA)
- Probe sonicator (Hielscher—Ultrasound Technology, Teltow, Germany)
- Chemical fume hood (FLOW SCIENCES, INC.)
- Cation Exchange Buffer Pack (AB SCIEX,#4326747, Framingham, MA), containing individual 100-mL bottles of loading buffer, elution buffer, cleaning buffer, and storage buffer. The buffer pPack also includes one 0.2-mL cation exchange cartridge
- Cation Exchange column (AB SCIEX, #4326695, Framingham, MA)

2.3 Methionine-Free DMEM

Add 50 mL of dialyzed FBS to 450 mL of L-methionine-free DMEM containing 4500 mg/L D-glucose, without L-glutamine, sodium pyruvate, L-methionine, and L-cystine (Gibco, #21013-024). Stocks of L-glutamine (10 mL), L-cystine (10 mL), and sodium pyruvate (5 mL) are also added to the medium. This medium is freshly prepared per use. It should be warmed to room temperature immediately before use.

2.4 Amino Acid-Free Medium

For 500 mL, the following reagents (Table 1) are dissolved in 450 mL of Milli-Q water.

Adjust pH to 7.2, and then add 7.5 mg of phenol red (Sigma-Aldrich P3532). Top up to 500 mL with Milli-Q water.

Sterilize by filtration using Steritop-GP Filter Unit (Millipore, #SCGVT05RE).

Table 1 Components of Amino Acid-Free DMEM (500 mL)

Chemical	Quantity	Cat #
$NaHCO_3$	1850 mg	Sigma-Aldrich, S5761
NaCl	3030 mg	Sigma-Aldrich, S7653
KCl	200 mg	Sigma-Aldrich, P9541
$MgSO_4 \cdot 7H_2O$	100 mg	Sigma-Aldrich, 1374361
$CaCl_2 \cdot 2H_2O$	132 mg	Sigma-Aldrich, C8106
10 mg/mL $Fe(NO_3)_3$	5 μL	Sigma-Aldrich, 254223
D-Glucose	500 mg	Sigma-Aldrich, G8270
MEM vitamin solution (100 ×)	20 mL	Invitrogen, 11120-052
1 *M* HEPES (pH7.5)	7.5 mL	Sigma-Aldrich, H0887

2.5 C_{18} Buffers

C_{18} buffer A: 98% H_2O, 2% ACN, 0.05% FA
Elution buffer E1: 50% ACN, 50% H_2O; Elution buffer E2:75% ACN, 25% H_2O

2.6 Gradient Separation Buffer

Mobile phase A: 2% ACN, 0.1% FA
Mobile phase B: 98% ACN, 2% H_2O, 0.05% FA

3. FUNDAMENTALS: AHA LABELING COMBINED WITH THE iTRAQ APPROACH FOR IDENTIFICATION OF DE NOVO PROTEIN SYNTHESIS

3.1 General Principles

For comprehensive profiling of the de novo protein synthesis during autophagy under starvation conditions, it is critical to achieve specific and efficient enrichment of these low-abundant proteins. Later, we provide a detailed protocol for de novo protein analysis via metabolic labeling (Fig. 1). The method relies on depleting the cells' reserves of natural-occurring L-methionine, replacing it with the analog AHA during de novo protein synthesis. In addition, we used the iTRAQ-based quantitative proteomics in conjunction with BONCAT to differentiate the bona fide newly synthesized proteins from the nonspecific binding proteins during the affinity enrichment process (Fig. 2).

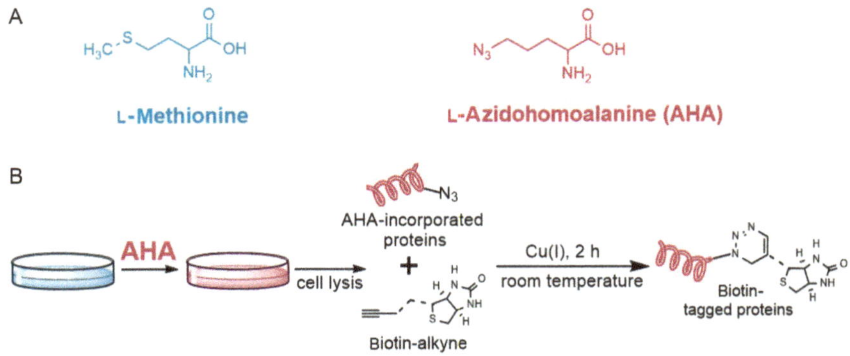

Fig. 1 De novo synthesized protein labeled by AHA: (A) Structure of natural L-methionine and L-azidohomoalanine (AHA). (B) AHA can be incorporated into de novo synthesized proteins and tagged with biotin by click chemistry.

3.2 Advantages of This Approach

First, AHA has a structural resemblance to the native L-methionine that can be readily taken up by the cell and subsequently incorporated into de novo synthesized proteins (Dieterich et al., 2007). In addition, this labeling does not affect the autophagic flux level (data not shown in this chapter). Second, AHA labeling can enrich low-abundant proteins, which eliminates the necessity of extensive peptide fractioning (Eichelbaum, Winter, Berriel Diaz, Herzig, & Krijgsveld, 2012). This enables one to study events that occur at early or intermediate stages of biological processes, such as autophagy, which require a short period of labeling. Third, an iTRAQ quantitative proteomics approach is used in tandem with BONCAT to differentiate any potential background contamination from selective enrichment (Wang, Zhang, Lee, et al., 2016; Wang, Zhang, Zhang, He, et al., 2015; Wang, Zhang, Zhang, Wong, 2016) to enhance the reliability of the results obtained. In autophagy studies, as the protein synthesis machinery is greatly suppressed, the background contamination may be more significant due to the low abundance of newly synthesized proteins. Thus, it is highly necessary to find a way to filter them out.

3.3 Applications of This Approach

This protocol is developed based on human cervical adenocarcinoma HeLa cells and could be optimized to work in other cell types from other organisms. Autophagy can also been induced by other stimuli for various treatment time. This would allow us to better understand the molecular

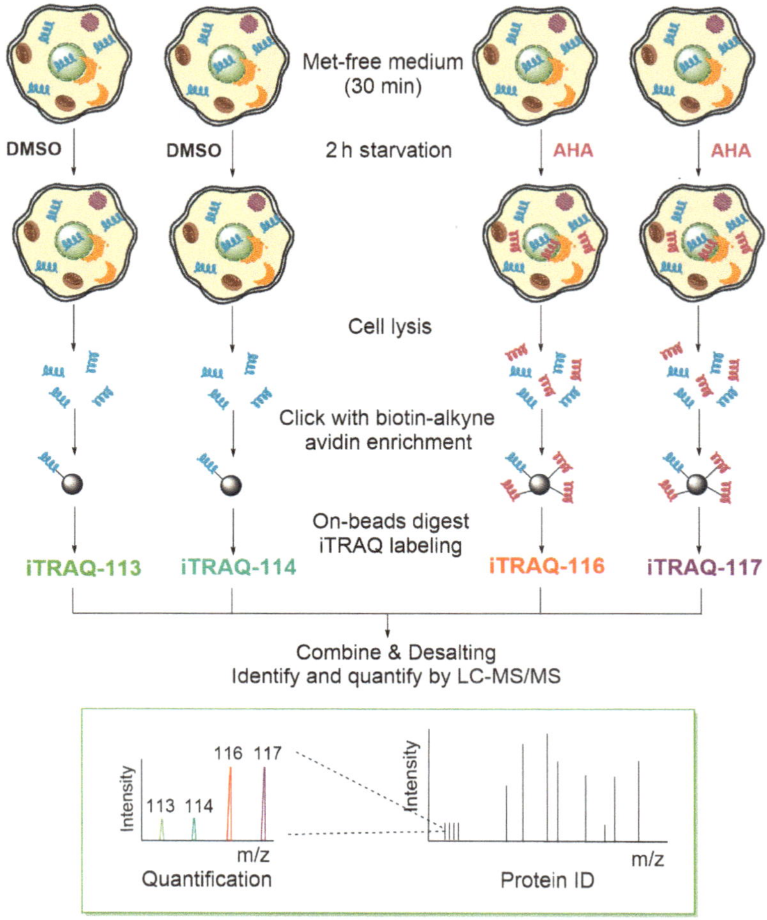

Fig. 2 Workflow for analysis of newly synthesized proteins in autophagy using AHA labeling combined with iTRAQ-based quantitative proteomics. Cells were first cultured in L-methionine-free medium to deplete the natural methionine reserves before labeling with AHA. After AHA incorporation into the de novo proteins, the cells were then harvested and lysed for a click reaction with a biotin-alkyne tag. Via avidin affinity isolation, AHA-containing proteins were enriched, digested on beads, and labeled with iTRAQ. Finally, the resulting peptides were pooled for identification and quantification via LC-MS/MS. The peptides of the two biological replicates of control pulled-down samples were labeled 113 and 114, while the two probe pulled-down samples were labeled 116 and 117. These labeled peptides were combined for identification and quantification via LC-MS/MS. For specific targets, the iTRAQ reporters should show highly differential intensities.

events that occur during the process of autophagy under different stimuli. The new knowledge obtained could allow us to better understand the relationship between autophagy and various diseases.

4. PROTOCOL

4.1 Cell Culture and Metabolic Labeling by AHA

1. Grow HeLa cells in a 150-mm cell culture dish in a regular growth medium and incubate them in a 5% CO_2 atmosphere at 37°C until cells reach 70–80% confluency.
2. Remove the medium and replace with L-methionine-free DMEM with 10% dialyzed FBS for 30 min to deplete the natural L-methionine reserves.
3. Label the cells with AHA (50 μ*M*) for up to 2 h in an amino acid-free medium. DMSO is used as a control (see Note 1).
4. Harvest the cells from each treatment in a 50-mL, round-bottom Corex tubes. Spin at 1200 × *g* for 3 min. Decant the medium and wash cells with 10 mL of 1 × PBS before spinning again to decant the PBS.
5. Lyse the cells with 2 mL 0.16% SDS in PBS with 1 × Halt™ Protease Inhibitor Cocktail, 50 μg/mL RNase and 50 μg/mL DNase. Sonicate the cell suspension with 0.5-s pulses for 60 s (see Note 2).
6. Centrifuge the sample at 14,000 × *g* for 5 min at 4°C. Keep the supernatant fraction and store at −80°C until use.

Pause Point: The supernatant can be stored in −80°C for several months without significantly interfering with the results.

Note 1: Biological replicates of AHA- or DMSO-treated samples were included to overcome experimental variations.

Note 2: For a particular cell line, concentration optimization experiments for AHA labeling may be needed to achieve satisfactory labeling efficiency. Higher concentrations of AHA may need to be used in certain cell lines.

Note 3: SDS will significantly decrease the efficiency of click chemistry. Therefore, it is recommended to keep its level below 0.2%.

4.2 Click Chemistry Tagging With Biotin Alkyne

7. Determine the protein concentration of each sample (two controls and two AHA-labeled samples). Start with an equal amount of proteins (i.e., 4 mg) to perform the following click chemistry tagging and quantitative proteomics.

8. Add 1 × PBS to each sample until the volume reaches 2 mL.
9. Prepare the stock solutions needed for the click reaction: 100 × biotin alkyne (10 m*M* in DMSO), 100 × TCEP (100 m*M* in water), 100 × TBTA (10 m*M* in DMSO), and 100 × $CuSO_4$ (100 m*M* in water) (see Note 3).
10. Add 20 μL of the 100 × biotin alkyne (from the 10-m*M* stock in DMSO from step 9) to each tube and vortex.
11. Add 20 μL of 100 × fresh TCEP solution (from step 9) and vortex.
12. Add 20 μL of 100 × TBTA (from step 9) and vortex.
13. Add 20 μL of 100 × $CuSO_4$ (from step 9) and vortex.
14. Incubate the samples in the dark for 2 h at room temperature under constant agitation (Eppendorf mixer, at 700 × *g*).
15. Transfer each sample to a 15-mL tube and add 10 mL of prechilled acetone (−20°C).
16. Leave the sample at −20°C for 4 h.
    Pause Point: The acetone-reconstituted samples can be stored in −20°C overnight or for a longer time.
17. Centrifuge the sample at 4000 × *g* at 4°C for 30 min to precipitate the proteins (see Note 4).
18. Remove the supernatant and air-dry the sample (see Note 5).
19. Add 700 μL of 1.2% SDS in PBS to each sample before sonicating the samples with 0.5-s pulses for 60 s. Heat the samples to 80–90°C for 5 min (see Note 6).
20. Dilute the sample with 10 mL of PBS (see Note 7).
21. Centrifuge the diluted sample at 4000 × *g* at room temperature. Keep the supernatant (see Note 8).

Note 4: Freshly prepare both TCEP and $CuSO_4$ prior to use. Stock solutions of 10 m*M* biotin alkyne and 10 m*M* TBTA can be stored at −20°C for several months. A slightly basic pH of the clicking buffer will be optimal for the click chemistry reaction.

Note 5: Acetone precipitation is used to concentrate the sample and remove the free biotin tag. The free biotin tag can compete with the labeled protein for binding with avidin in the following steps and significantly decrease the enrichment efficiency.

Note 6: Do not overdry the sample, as it will be difficult to redissolve in the following steps.

Note 7: Use high concentrations of SDS and heating to dissolve the protein fully.

Note 8: Dilute the samples to keep the concentration of SDS below 0.2%, as it will interfere with the following avidin beads binding steps.
Pause Point: Samples can be stored at −80°C for several days.

4.3 Avidin Affinity Purification

22. Add 100 μL of Streptavidin beads to a 15-mL falcon tube. Wash the beads with 5 mL of PBS and centrifuge the beads at 700 × *g* for 3 min at room temperature to remove the solvent from the beads (see Note 9).
23. Repeat the wash two more times.
24. Add the protein sample to the beads and incubate for 4 h at room temperature.
25. Centrifuge the beads at 700 × *g* for 3 min and remove the supernatant.
26. Wash the beads with 10 mL of 1% SDS in PBS. Place the beads on a rotator for 10 min before pelleting the beads at 700 × *g* for 5 min at room temperature.
27. Repeat the wash (step 26) for two more times.
28. Wash the beads with 10 mL of 0.1% SDS in PBS. Place the beads on a rotator for 10 min before pelleting the beads at 700 × *g* for 5 min at room temperature.
29. Wash the beads with 10 mL of 6 *M* urea. Place the beads on a rotator for 10 min before pelleting the beads at 700 × *g* for 5 min at room temperature.
30. Wash the beads with 10 mL of PBS. Place the beads on a rotator for 10 min before pelleting the beads at 700 × *g* for 5 min at room temperature.
31. Repeat the wash (step 30) for two more times.
32. Wash the beads with 10 mL of Milli-Q water. Place the beads on a rotator for 10 min before pelleting the beads at 700 × *g* for 5 min at room temperature.
33. Reconstitute the beads with 200 μL of 0.5 *M* TEAB.
34. Add 4 μL of reducing agent (TCEP, 50 m*M*) to each sample, vortex, and incubate the sample on the thermoshaker at 700 × *g* for 60 min at 60°C.
35. Cool the sample to room temperature. Add 2 μL of cysteine blocking reagent (MMTS, 200 m*M*) into each sample, vortex, and incubate the sample in the dark at room temperature for 15 min to block the cysteine residues (see Note 10).

36. Prepare the trypsin working solution. Add 50 μL of water to the 25-μg trypsin bottle to make the final trypsin concentration of 0.5 μg/μL (see Note 11).
37. Add 8 μL (4 μg) of trypsin into each sample and incubate at 37°C overnight (12–16 h) to digest the protein into peptides (see Notes 12 and 13).
38. Separate the digested peptides and the beads by a filter spin column, and collect the eluted sample solution.
39. Proceed to the next iTRAQ labeling step, or keep the sample at −80°C for further use.

Note 9: Centrifuge the beads at no more than 1000 × *g* as the beads might break.
Note 10: MMTS is a neurotoxin; handle it in the chemical hood.
Note 11: Use high-standard-sequencing grade-modified trypsin.
Note 12: The tube should be sealed with parafilm to avoid evaporation of the samples during the overnight incubation.
Note 13: The final concentration of trypsin should be at least 10 ng/μL.

4.4 Isobaric Tag for Relative and Absolute Quantification (iTRAQ) Labeling

40. Dry the peptide samples using a speed vacuum device (SpeedVac). It may take several hours to dry the sample totally (see Notes 14 and 15).
    Pause Point: The dried peptide samples can be stored at −20°C for several months.
41. Reconstitute the SpeedVac-dried peptide samples with 30 μL of 250 m*M* TEAB (dissolution buffer), vortex, and spin the sample (see Note 16).
42. Take 0.25–0.5 μL (smaller volumes recommended) of the sample and pipette it to a pH paper to check the sample's pH (see Note 17).
43. Take out the iTRAQ labeling reagent (4-/8-plex), warm to room temperature, and reconstitute it with 70 μL ethanol (4-plex) or 50 μL isopropanol (8-plex).
44. Mix the reconstituted iTRAQ reagent with the sample and react for 1 h (4-plex) or 2 h (8-plex). React the sample with iTRAQ reagent as follows: control 1, iTRAQ 113; control 2, iTRAQ 114; AHA-labeled group 1, iTRAQ 116; AHA-labeled group 2, iTRAQ 117 (see Notes 18 and 19).
45. Combine all iTRAQ-labeled samples in a new tube.

Note 14: Equal volume is important for the labeling process.

Note 15: The dried peptide sample can be stored at −80°C for several months.
Note 16: If the sample precipitates, sonicate to dissolve the substances.
Note 17: Optimally pH 8 (alkaline) for labeling. If not, add NaOH to increase its pH.
Note 18: Strictly transfer the iTRAQ reagent to the sample. The volume may be more than 100 μL. Each volume must be equal.
Note 19: Store sample at −20°C first before transferring it to a −80°C fridge. Do not label for more than 1 h. Add phosphoric acid to stop the reaction.

4.5 Sample Cleanup by Strong Cation Exchange Chromatography

46. Centrifuge the combined sample at 14,000 × *g* for 10 min and transfer the supernatant into a 50-mL tube. Reduce the concentrations of buffer salts and organics (dilute sample mixture around 10 times with cation exchange buffer-load) (see Notes 20 and 21).

47. Check the sample's pH using a pH paper (range: 3–4.5) and add phosphoric acid to adjust the pH to between 2.5 and 3.3 (see Note 22).

48. Inject 1 mL of the cation exchange buffer-clean to condition the cartridge and remove the buffer-storage solution already in the cartridge. Divert the flow-through to waste.

49. Inject 2 mL of the cation exchange buffer-load and divert the flow-through to waste.

50. Slowly inject (1 droplet/s) the diluted sample (removing the bubbles) into the strong cation exchange (SCX) cartridge and collect the flow-through in a 50-mL sample tube.

51. Inject 1 mL of buffer-load to wash the TCEP, SDS, and excess iTRAQ reagents from the cartridge and collect the flow-through in a 50-mL sample tube (see Note 23).

52. Slowly inject 500 μL of cation exchange buffer-elute to elute the peptides and capture the eluate in a new 1.5-mL tube.

53. Inject 1 mL of buffer-clean to wash the undigested proteins such as trypsin from the cartridge and divert the flow-through to waste.

54. Inject 2 mL of buffer-storage and divert the flow-through to waste. Disassemble the cartridge and store it at 4°C.

Note 20: Do not inject the precipitates into the column. The column will be clotted and the sample will be lost.
Note 21: All the SCX buffer should be stored at 4°C.
Note 22: Handle the concentrated phosphoric acid in a chemical hood.

Note 23: Keep the flow-through until you verify by MS/MS analysis that the sample loading onto the SCX cartridge succeeded. If loading fails, you can repeat loading using the flow-through after you troubleshoot the cause of the loading failure.

4.6 Desalting of Labeled Samples by the C_{18} Column

55. Dilute the sample with 3–4 mL of C_{18} buffer A (from Section 2.5) in a 15-mL falcon tube (remember to rinse the original 1.5-mL tube).
56. Rinse a new 5-mL syringe with 100% ACN and then connect the syringe to the short end of a Sep-Pak column. Condition the Sep-Pak column with 10 mL of 100% ACN (see Note 24).
57. Inject 10 mL of C_{18} buffer A (from Section 2.5) to condition the column.
58. Inject the sample into column slowly and collect flow-through (1 drop per second). Load the flow-through one more time and recollect the flow-through.
59. Load 3–5 mL of C_{18} buffer A (from Section 2.5) into the column to remover salts/contaminants.
60. Elute with 5 mL of Elution buffer E1 and 5 mL of Elution buffer E2 (from Section 2.5) and collect the flow-through in a 50-mL falcon tube.
61. Add 10 mL of Milli-Q water (flow-through volume: Milli-Q water is 1:1) to the flow-through, whirl to mix, and freeze at −80°C overnight.
62. Remove cap of sample tube, cover it tightly with parafilm by crossing the cap several times. Poke 5–10 holes on the parafilm with needle.
63. Lyophilize the sample with the lyophilizer machine overnight.
64. The lyophilized dried sample is reconstituted with 1.5 mL of mobile phase A (from Section 2.6).
65. Transfer the content into a 2-mL Eppendorf tube and dry the sample using SpeedVac (see Note 25).

Note 24: Be careful to avoid introducing air into the column, as air bubbles will significantly decrease the binding capacity of the column.

Pause Point: The dried sample can be stored at −80°C for several months.

4.7 Nano-LC Electrospray Ionization MS

66. Reconstitute the sample with 100 μL of C_{18} buffer A (from Section 2.5).
67. Separate the iTRAQ-labeled peptides via an Eksigent NanoLC-Ultra system coupled to the cHiPLC-Nanoflex system (Eksigent, Dublin, CA, USA), with a 75-μm × 150-mm analytical column packed with

Reprosil-Pur C-18 (Eksigent, 804-00011), in a Trap-Elute configuration. The sample (5 μL) is loaded into the LC system (see Note 26).

68. Separate peptides using gradient separation buffer (from Section 2.6): 5–12% of mobile phase B (20 min), 12–30% of mobile phase B (90 min), 30–90% of mobile phase B (2 min), 90% of mobile phase B (5 min), 90–95% of mobile phase B (3 min), and 5% of mobile phase B (13 min), at a flow rate of 300 nL/min.

69. Use a TripleTOF 5600 (SCIEX) mass spectrometer to acquire both the MS and MS/MS spectra (see Note 27). Each duty cycle contains one MS analysis of 250 ms, followed by MS/MS analyses of 20 precursors with at least 100-ms accumulation time per spectrum. MS spectra are acquired in high-resolution mode, with a mass range of 350–1250 *m/z*. MS/MS spectra are acquired in high-sensitivity mode, with a mass range of 100–1800 *m/z* and dynamic exclusion for 15 s. Both rolling collision energy and iTRAQ reagent collision adjustments setting are selected for the MS/MS analyses (Wang, Zhang, Chia, et al., 2015; Wang, Zhang, Zhang, et al., 2015).

Note 26: Other nano-LC systems and columns also can be used, depending on the availability and accessibility.

Note 27: Analysis of iTRAQ-labeled peptides has been optimized with SCIEX TripleTOF 5600 and 6600 series mass spectrometers. If other types of mass spectrometers are to be used, make sure that they are compatible with iTRAQ analysis. The low mass range should be able to detect the iTRAQ reporter ions (*m/z* 114–121) with sufficient resolution. The collision energy should be optimized to fragment the iTRAQ-labeled peptides efficiently and produce the reporter ions.

4.8 Protein Identification and Quantification Using ProteinPilot™ Software

70. Quantify and identify the peptide using ProteinPilot™ Software 4.5 (SCIEX) with the Paragon algorithm (4.5.0.0, 1654) (see Note 28). Use the SwissProt database (v2015.9) with a total of 40,406 *Homo sapiens* entries for the search, with the following parameter settings: Cysteine alkylation with MMTS; trypsin digestion; TripleTOF 5600; biological modifications (see Notes 29 and 30).

71. Export the Protein summary file (contains identification and quantification results) and import into Microsoft Excel for manual data analysis.

72. Use unused protein score ≥1.3 (corresponding to ≥95% confidence level) as the cutoff threshold to filter the protein list.

73. Calculate four iTRAQ ratios of the AHA-labeled and the DMSO control samples (116:113, 116:114, 117:113, and 117:114) for each identified protein. Convert them to log2 ratios.
74. Perform a 1-sample *t*-test to check whether the mean log2 ratio is truly different from 0 (see Note 31).
75. Retain only the proteins with $p \leq 0.05$ after the *t*-test analysis.
76. Delete the proteins with either the 114/113 or 117/116 ratio ≥ 1.3 or ≤ 0.77. This is to remove the results that are inconsistent in the two biological replicates (see Note 32).
77. Delete the proteins identified with a single peptide (see Note 33).
78. Retain only proteins with mean iTRAQ ratio (AHA-labeled vs DMSO control) ≥ 2 (see Note 34).

Note 28: Ideally, iTRAQ data are analyzed with the ProteinPilot software. If other software were to be used, make sure that only unique peptides are used to calculate the iTRAQ ratios. Peptides with incomplete iTRAQ labeling, trypsin miscleavage, or methionine oxidation also should be excluded.

Note 29: The false discovery rate for peptide identification is estimated through a decoy database search strategy. The Proteomics System Performance Evaluation Pipeline (PSPEP) feature of the ProteinPilot™ Software 4.5 is utilized to create a randomized database for this purpose.

Note 30: Do not normalize the iTRAQ ratio. For ProteinPilot, background correction is not recommended for this experiment.

Note 31: In Excel, for each set of four log2 ratios associated with each protein, create four 0 values and then perform a two-tailed paired *t*-test. This is equivalent to a 1-sample *t*-test.

Note 32: The iTRAQ ratio cutoff thresholds were set as 1.3 for upregulated proteins, and reciprocally 0.77 for downregulated proteins (Higuchi et al., 2013; Tan et al., 2008; Wang et al., 2014; Wang, Zhang, Lee, et al., 2016; Wang, Zhang, Zhang, et al., 2016). The standard deviation of all the ratios of the labeled peptides was computed as 0.15. Thus, with the use of a 1 ± 2 standard deviation formula, if the iTRAQ ratios between the replicates (114/113 or 117/116) exceed the cutoff thresholds, they are considered inconsistent.

Note 33: Doing this ensures the robustness of the data. Proteins identified with single peptides may still be true hits and may be retained if desired. However, additional justification would be needed for publication purposes.

Note 34: This is an arbitrary decision. A twofold cutoff is considered highly stringent due to the ratio compression commonly observed in iTRAQ experiments.

5. LIMITATIONS

While BONCAT is an attractive labeling technique to use in the autophagy field, there are still some technical disadvantages that need to be briefly addressed. First, BONCAT is limited to proteins that have at least one methionine residue, which excludes 1.02% of all known entries in the human protein database (Dieterich et al., 2006). During posttranslational processing, it has been observed that the sole N-terminal methionine in some proteins may be removed. The proteins that undergo such processing constitute about 5.08% of all human proteins (Dieterich et al., 2006). Thus, a small selected group of proteins may not be successfully identified. However, the majority (at least 94%) of the mammalian proteome remains suitable as candidates for protein identification via the BONCAT approach. Second, BONCAT has to be performed in a methionine-free medium with dialyzed FBS (to eliminate L-methionine from other sources) because the incorporation efficiency of AHA is much lower than natural methionine (Kiick et al., 2002). Therefore, theoretically the presence of endogenous methionine inside the cells may affect AHA labeling adversely.

REFERENCES

Bagert, J. D., Xie, Y. J., Sweredoski, M. J., Qi, Y., Hess, S., Schuman, E. M., et al. (2014). Quantitative, Time-Resolved Proteomic Analysis by Combining Bioorthogonal Noncanonical Amino Acid Tagging and Pulsed Stable Isotope Labeling by Amino Acids in Cell Culture. *Molecular & Cellular Proteomics*, *13*, 1352–1358.

Bauvy, C., Meijer, A. J., & Codogno, P. (2009). Assaying of autophagic protein degradation. *Methods in Enzymology*, *452*, 47–61.

Chen, Y., & Klionsky, D. J. (2011). The regulation of autophagy—Unanswered questions. *Journal of Cell Science*, *124*, 161–170.

Dieterich, D. C., Lee, J. J., Link, A. J., Graumann, J., Tirrell, D. A., & Schuman, E. M. (2007). Labeling, detection and identification of newly synthesized proteomes with bioorthogonal non-canonical amino-acid tagging. *Nature Protocols*, *2*, 532–540.

Dieterich, D. C., Link, A. J., Graumann, J., Tirrell, D. A., & Schuman, E. M. (2006). Selective identification of newly synthesized proteins in mammalian cells using bioorthogonal noncanonical amino acid tagging (BONCAT). *Proceedings of the National Academy of Sciences of the United States of America*, *103*, 9482–9487.

Eichelbaum, K., Winter, M., Berriel Diaz, M., Herzig, S., & Krijgsveld, J. (2012). Selective enrichment of newly synthesized proteins for quantitative secretome analysis. *Nature Biotechnology*, *30*, 984–990.

Fullgrabe, J., Klionsky, D. J., & Joseph, B. (2014). The return of the nucleus: Transcriptional and epigenetic control of autophagy. *Nature Reviews. Molecular Cell Biology*, *15*, 65–74.

Higuchi, S., Lin, Q., Wang, J., Lim, T. K., Joshi, S. B., Anand, G. S., et al. (2013). Heart extracellular matrix supports cardiomyocyte differentiation of mouse embryonic stem cells. *Journal of Bioscience and Bioengineering*, *115*, 320–325.

Howden, A. J. M., Geoghegan, V., Katsch, K., Efstathiou, G., Bhushan, B., Boutureira, O., et al. (2013). QuaNCAT: Quantitating proteome dynamics in primary cells. *Nature Methods, 10*, 343–346.
Kiick, K. L., Saxon, E., Tirrell, D. A., & Bertozzi, C. R. (2002). Incorporation of azides into recombinant proteins for chemoselective modification by the Staudinger ligation. *Proceedings of the National Academy of Sciences of the United States of America, 99*, 19–24.
Levine, B., & Klionsky, D. J. (2004). Development by self-digestion: Molecular mechanisms and biological functions of autophagy. *Developmental Cell, 6*, 463–477.
McClatchy, D. B., Ma, Y., Liu, C., Stein, B. D., Martínez-Bartolomé, S., Vasquez, D., et al. (2015). Pulsed azidohomoalanine labeling in mammals (PALM) detects changes in liver-specific LKB1 knockout mice. *Journal of Proteome Research, 14*, 4815–4822.
Mizushima, N., & Klionsky, D. J. (2007). Protein turnover via autophagy: Implications for metabolism. *Annual Review of Nutrition, 27*, 19–40.
Mizushima, N., & Komatsu, M. (2011). Autophagy: Renovation of cells and tissues. *Cell, 147*, 728–741.
Ngo, J. T., & Tirrell, D. A. (2011). Noncanonical amino acids in the interrogation of cellular protein synthesis. *Accounts of Chemical Research, 44*, 677–685.
Rostovtsev, V. V., Green, L. G., Fokin, V. V., & Sharpless, K. B. (2002). A stepwise huisgen cycloaddition process: Copper(I)-catalyzed regioselective "ligation" of azides and terminal alkynes. *Angewandte Chemie (International Ed in English), 41*, 2596–2599.
Shen, H. M., & Mizushima, N. (2014). At the end of the autophagic road: An emerging understanding of lysosomal functions in autophagy. *Trends in Biochemical Sciences, 39*, 61–71.
Tan, H. T., Tan, S., Lin, Q., Lim, T. K., Hew, C. L., & Chung, M. C. (2008). Quantitative and temporal proteome analysis of butyrate-treated colorectal cancer cells. *Molecular & Cellular Proteomics, 7*, 1174–1185.
Tornoe, C. W., Christensen, C., & Meldal, M. (2002). Peptidotriazoles on solid phase: [1,2,3]-triazoles by regiospecific copper(i)-catalyzed 1,3-dipolar cycloadditions of terminal alkynes to azides. *The Journal of Organic Chemistry, 67*, 3057–3064.
Wang, J., Tan, X. F., Nguyen, V. S., Yang, P., Zhou, J., Gao, M., et al. (2014). A quantitative chemical proteomics approach to profile the specific cellular targets of andrographolide, a promising anticancer agent that suppresses tumor metastasis. *Molecular & Cellular Proteomics, 13*, 876–886.
Wang, J., Zhang, C. J., Chia, W. N., Loh, C. C., Li, Z., Lee, Y. M., et al. (2015a). Haem-activated promiscuous targeting of artemisinin in Plasmodium falciparum. *Nature Communications, 6*, 10111.
Wang, J., Zhang, J., Lee, Y.-M., Koh, P.-L., Ng, S., Bao, F., et al. (2016a). Quantitative chemical proteomics profiling of de novo proteins synthesis during starvation-mediated autophagy. *Autophagy, 12*, 1931–1944.
Wang, J., Zhang, C. J., Zhang, J., He, Y., Lee, Y. M., Chen, S., et al. (2015b). Mapping sites of aspirin-induced acetylations in live cells by quantitative acid-cleavable activity-based protein profiling (QA-ABPP). *Scientific Reports, 5*, 7896.
Wang, J., Zhang, J., Zhang, C. J., Wong, Y. K., Lim, T. K., Hua, Z. C., et al. (2016b). In situ proteomic profiling of curcumin targets in HCT116 colon cancer cell line. *Scientific Reports, 6*, 22146.
Watanabe-Asano, T., Kuma, A., & Mizushima, N. (2014). Cycloheximide inhibits starvation-induced autophagy through mTORC1 activation. *Biochemical and Biophysical Research Communications, 445*, 334–339.
Wirawan, E., Vanden Berghe, T., Lippens, S., Agostinis, P., & Vandenabeele, P. (2012). Autophagy: For better or for worse. *Cell Research, 22*, 43–61.
Zhang, J., Wang, J., Ng, S., Lin, Q., & Shen, H. M. (2014). Development of a novel method for quantification of autophagic protein degradation by AHA labeling. *Autophagy, 10*, 901–912.

CHAPTER FIVE

Methods to Monitor and Manipulate TFEB Activity During Autophagy

D.L. Medina*, C. Settembre*,†,‡, A. Ballabio*,‡,§,¶,1

*Telethon Institute of Genetics and Medicine (TIGEM), Pozzuoli, Naples, Italy
†Dulbecco Telethon Institute (DTI), Naples, Italy
‡Medical Genetics, Federico II University, Naples, Italy
§Baylor College of Medicine, Houston, TX, United States
¶Jan and Dan Duncan Neurological Research Institute, Texas Children's Hospital, Houston, TX, United States
[1]Corresponding author: e-mail address: ballabio@tigem.it

Contents

Abstract

Macroautophagy is a catabolic process deputed to the turnover of intracellular components. Recent studies have revealed that transcriptional regulation is a major mechanism controlling autophagy. Currently, more than 20 transcription factors have been

Methods in Enzymology, Volume 588
ISSN 0076-6879
http://dx.doi.org/10.1016/bs.mie.2016.10.008

shown to modulate cellular autophagy levels. Among them, the transcription factor EB (TFEB) appears to have the broadest proautophagy role, given its capacity to control the biogenesis of lysosomes and autophagosomes, the two main organelles required for the autophagy pathway.

TFEB has attracted major attention owing to its ability to enhance cellular clearance of pathogenic substrates in a variety of animal models of disease, such as lysosomal storage disorders, Parkinson's, Alzheimer's, α1-antitrypsin, obesity as well as others, suggesting that the TFEB pathway represents an extraordinary possibility for future development of innovative therapies. Importantly, the subcellular localization and activity of TFEB are regulated by its phosphorylation status, suggesting that TFEB activity can be pharmacologically targeted.

Given the growing list of common and rare diseases in which manipulation of autophagy may be beneficial, in this chapter we describe a set of validated protocols developed to modulate and analyze TFEB-mediated enhancement of autophagy both in vitro and in vivo conditions.

1. INTRODUCTION

Autophagy is an evolutionary conserved catabolic process that provides energy and recycles cellular components through the targeting of intracytoplasmic cargo to lysosomes. Autophagy plays a crucial role in the cellular adaptation to environmental cues such as energy demanding conditions (i.e., nutrient deprivation). Autophagy relies on the biogenesis and cooperation of two organelles, the autophagosome and the lysosome.

Recent studies have shown that the transcription factor EB (TFEB) plays an important role in autophagy. TFEB belongs to the microphthalmia family of basic helix-loop-helix–leucine-zipper (bHLH-Zip) transcription factors (MiT family) (Steingrímsson, Copeland, & Jenkins, 2004). TFEB positively regulates lysosome biogenesis and function by enhancing the expression of lysosomal hydrolases and of lysosomal membrane proteins, such as the components of the v-ATPase proton pump complex (Medina et al., 2011; Palmieri et al., 2011; Sardiello et al., 2009). TFEB also promotes autophagosome biogenesis and autophagosome fusion with the lysosome by inducing the expression of autophagy-related genes (Settembre et al., 2011).

TFEB activity is primarily regulated by phosphorylation events, mainly occurring at serine residues. Notably, the mTORC1 kinase, a main negative regulator of autophagy, is a key modulator of TFEB function (Martina, Chen, Gucek, & Puertollano, 2012; Roczniak-Ferguson et al., 2012; Settembre et al., 2012).

In the presence of nutrients, mTORC1 phosphorylates TFEB at multiple serine residues on the lysosome surface and promotes the binding of TFEB to 14-3-3 proteins preventing its nuclear translocation and activation (Martina et al., 2012; Roczniak-Ferguson et al., 2012; Settembre et al., 2012).

During starvation mTORC1 dissociates from the lysosomal surface and its activity is inhibited. In parallel, upon starvation the activity of the phosphatase calcineurin is induced by lysosomal calcium release through the activation of the lysosomal cation channel MCOLN1 (Medina et al., 2015). Together, these events lead to TFEB dephosphorylation, nuclear translocation, and activation (Fig. 1). In the nucleus TFEB interacts with the transcriptional coactivator-associated arginine methyltransferase 1 (CARM1), and binds to the CLEAR sites in the promoters of many lysosomal and autophagy genes promoting their expression during starvation (Settembre et al., 2011; Shin et al., 2016).

Not surprisingly, the TFEB paralogue transcription factor E3 (TFE3) has also been shown to regulate lysosomal homeostasis by inducing the expression of genes encoding proteins involved in autophagy and lysosomal biogenesis in response to starvation and lysosomal stress. TFE3 regulation seems to share the same regulatory mechanisms as TFEB and therefore TFE3 is also regulated by mTORC1-dependent phosphorylation on the lysosomal surface as well as by the phosphatase calcineurin (Martina, Diab, Brady, & Puertollano, 2016; Martina, Diab, Li, & Puertollano, 2014; Martina, Diab, Lishu, et al., 2014).

The identification of a global transcriptional regulation of lysosomal function and autophagy has been exploited to boost cellular clearance in mouse models of a variety of disease conditions. TFEB overexpression results in the clearance of accumulating substrates in cells and tissues from mouse models of several types of LSDs, as well as of Parkinson's,

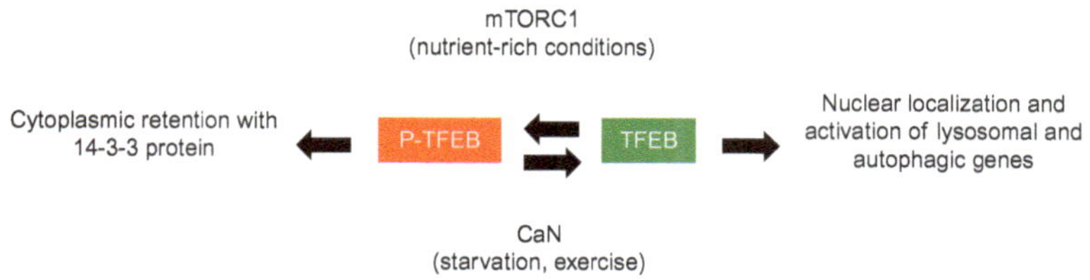

Fig. 1 TFEB regulation. TFEB is negatively regulated by nutrients through mTORC1-mediated phosphorylation and cytoplasmic retention by binding to 14-3-3 proteins. Conditions such as nutrient deprivation or exercise promote TFEB dephosphorylation by calcineurin (CaN). Dephosphorylated TFEB translocates into the nucleus activating lysosomal and autophagic genes.

Alzheimer's, Huntington's disease, α1-anti-trypsin deficiency, and spinal bulbar muscular atrophy (Cortes, 2014; Decressac et al., 2013; Medina et al., 2011; Pastore et al., 2013; Polito et al., 2014; Song et al., 2016; Spampanato et al., 2013; Tsumeni et al., 2012; Xiao et al., 2015). In the same way, overexpression of TFE3 promotes clearance in a cellular model of a lysosomal storage disorder, Pompe disease (Martina, Diab, Li, et al., 2014; Martina, Diab, Lishu, et al., 2014). These findings have opened a novel therapeutic strategy based on the modulation of cellular clearance, through the transcriptional activation of the lysosomal/autophagic pathway, with potential applicability to many diseases.

The mechanism by which TFEB promotes the clearance of stored materials needs to be further characterized. However, it is possible that TFEB-mediated cellular clearance results from a combination of the effects on lysosomal biogenesis, autophagy, and lysosomal exocytosis (Medina et al., 2011; Spampanato et al., 2013). Therefore, drug-screening approaches aimed at identifying molecules that promote TFEB nuclear translocation present an interesting path forward to treat several diseases (Moskot et al., 2014; Song et al., 2016) (Fig. 2). The important implications resulting from the study of

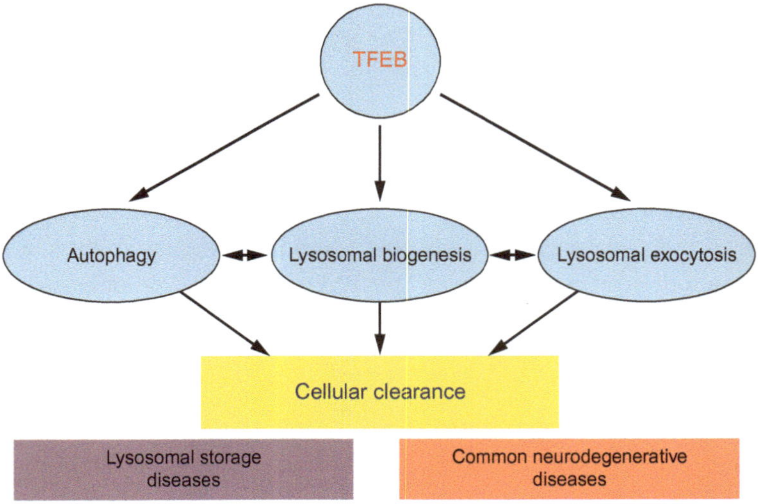

Fig. 2 Model of TFEB-mediated cellular clearance. TFEB overexpression modulates autophagy, lysosomal biogenesis, and induces lysosomal exocytosis. The activation of these biological pathways may contribute to the clearance of pathological accumulation in various disease conditions such as lysosomal storage disorders and common neurodegenerative disease.

TFEB-mediated function prompted us to present a comprehensive set of the major experimental procedures used to enhance TFEB/TFE3 signaling and examine the transcriptional induction of the autophagy pathway.

2. METHODS TO MONITOR TFEB/TFE3 ACTIVATION

Under basal conditions TFEB is mainly found in the cytoplasm in an inactive, phosphorylated state (Settembre et al., 2011). However, under specific conditions, such as nutrient deprivation, exercise, or lysosomal stress, TFEB is rapidly dephosphorylated and moves to the nucleus (Fig. 3). Although phosphoproteomic studies identified at least 10 different phosphorylation sites on the TFEB protein, mutation analysis indicated that two residues (Ser211 and Ser142) are particularly critical for TFEB subcellular localization (Martina et al., 2012; Roczniak-Ferguson et al., 2012; Settembre et al., 2012). In vitro studies suggest that Ser142 may either be phosphorylated by ERK2 or mTORC1, while only the latter phosphorylates Ser211. TFEB phosphorylation favors the interaction with 14-3-3 proteins and therefore its cytoplasmic retention (Roczniak-Ferguson et al., 2012). Interestingly, TFEB phosphorylation by the mTORC1 complex occurs on the lysosomal surface. Hence, TFEB subcellular localization may be used as a marker for TFEB activation. The key procedural steps to measure endogenous or overexpressed TFEB localization by western blot

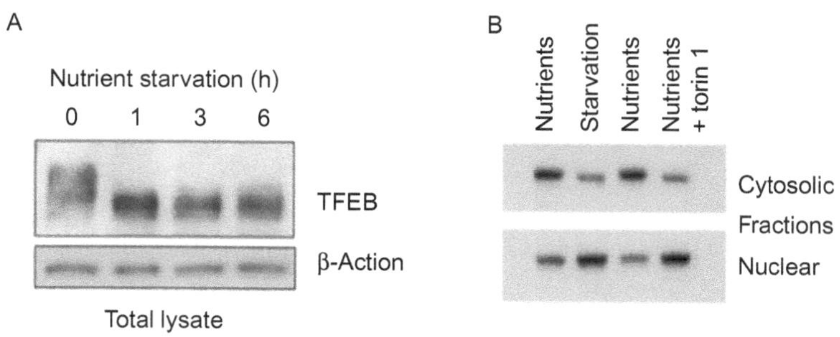

Fig. 3 Detection of TFEB phosphorylation and subcellular localization in HeLa cells by immunoblotting. (A) Total cellular lysate isolated from cells nutrient deprived (HBSS starvation) for the indicated time points. (B) TFEB subcellular localization. Cells were cultured in complete media (nutrients), in HBSS buffer for 1 h (starvation), or in complete media supplemented with torin 1 inhibitor. Endogenous TFEB was detected using Cell Signaling antibody (Cat n. 4240).

and immunofluorescence are summarized below. In addition, we have included a procedure to quantify TFEB subcellular localization by using high-content imaging analysis using a cell line stably overexpressing TFEB-GFP fusion protein (Medina et al., 2015).

2.1 Analysis of TFEB Subcellular Localization by Immunoblot

- TFEB antibodies: To detect endogenous TFEB in human cell extracts, the anti-TFEB antibody from Cell Signaling should be used (#4240). To detect endogenous TFEB in mouse cell extracts, the anti-TFEB antibody from Bethyl Laboratories (A311-347A) should be used.
- Prepare the Triton X-100 lysis buffer: 50 m*M* Tris–HCl (pH 7.5), 0.5% Triton X-100, 137.5 m*M* NaCl, 10% glycerol, 5 m*M* ethylenediaminetetraacetic acid (EDTA), PhosSTOP, and EDTA-free protease inhibitor tablets (Roche, Indianapolis, IN, USA). Keep on ice.
- Cell lysates: After three washes with cold phosphate-buffered saline (PBS), lyse 6 million cells with 0.5 mL of 0.5% Triton X-100 lysis buffer for 15 min on ice, shaking gently.
- Collect and centrifuge the cell lysate at 2000 rpm for 15 min at 4°C with a tabletop centrifuge and transfer 250 μL of the supernatant, which is composed of the cytosolic and membrane fractions, to a new Eppendorf tube. Discard the rest of the supernatant. Keep the pellet.
- Rinse the nuclear pellet three times with 0.5 mL of lysis buffer, and resuspend the pellet in 0.1 mL of lysis buffer, supplemented with 0.5% sodium dodecyl sulfate (SDS). Sonicate in cold room three times for 3 s at low output to shear genomic DNA.
- Preclear by centrifugation at 13,000 rpm for 15 min at 4°C with a tabletop centrifuge. Transfer the supernatant to a new tube.
- Determine the protein concentration in supernatant extracts using the colorimetric bicinchoninic acid protein assay kit (Pierce Chemical Co., Boston, MA, USA), following the manufacturer's instructions.
- Add an equal amount of (2 ×) SDS-PAGE sample buffer to each of the tubes containing the nuclear and cytoplasmic fractions and boil the samples for 10 min.
- Perform western blot, using 10% SDS-PAGE gels. Transfer protein onto PVDF (polyvinyl difluoride) membrane. Block the PVDF membrane with 5% nonfat milk in TBS-T buffer (TBS containing 0.05% Tween-20) for 1 h at room temperature (RT) under gentle shaking. Incubate with primary antibodies, anti-TFEB (1:1000 in 5% nonfat milk in

TBS-T buffer) overnight (ON) at 4°C, anti-TUBULIN (Sigma; 1:2000), and anti-H3 (Cell Signaling; 1:10,000) at RT for 2 h. Detect signals with horseradish peroxidase (HRP)-conjugated goat anti-mouse or anti-rabbit IgG antibody (1:2000, Vector Laboratories) and visualize with the Super Signal West Dura substrate (Thermo Scientific, Rockford, IL), according to the manufacturer's protocol.

- The purity of the fractions is confirmed by the lack of beta-tubulin in the nuclear fraction and of histone H3 in the cytosolic fraction, as detected by western blot.

2.2 Detection of TFEB Ser142 Phosphorylation by Immunoblot

Phosphorylation of TFEB at Ser142 is a key step in TFEB nuclear translocation during starvation. A custom phosphospecific-antibody, recognizing phosphorylated TFEB at the Ser142 residue, was generated by the GenScript Company. Recently, a similar antibody from EMD-MILLIPORE has become commercially available (ABE1971-Anti-phospho TFEB-Ser142 antibody). The TFEB P-Ser142 antibody is suitable for western blot; however, given the high homology between the amino acid sequences surrounding Ser142 and the other members of the MiT subfamily, it is possible that cross-reaction events may occur. This issue can be avoided by analyzing samples obtained from cells overexpressing TFEB.

- Follow standard western blot procedures and incubate primary antibody diluted 1:1000 in 5% milk ON at 4°C. Avoid many cycles freeze and thaw.

2.3 Detection of P-Ser211 TFEB by Using 14-3-3 Motif

The phosphorylation of the Ser211 site on TFEB protein favors the binding to the cytosolic chaperone 14-3-3, which keeps TFEB sequestered in the cytosol (Martina et al., 2012; Roczniak-Ferguson et al., 2012). Hence, the phosphorylation of TFEB on Ser211 can be determined by exploiting the selective interaction between Ser211-phosphorylated TFEB and 14-3-3 proteins. The following antibodies are required: anti-FLAG (1:1000) and anti-actin (1:4000) from Sigma-Aldrich; P-Ser-14-3-3 binding motif antibodies (1:1000) from Cell Signaling Technologies; anti-pan 14-3-3 antibodies (1:1000) from Santa Cruz Biotechnology.

- Rinse TFEB-3XFLAG-expressing cells twice with ice-cold PBS and lyse in ice-cold lysis buffer (400 m*M* NaCl, 25 m*M* Tris–HCl, pH 7.4, 1 m*M*

EDTA, and 1% Triton X-100) containing protease and phosphatase inhibitors.

- Isolate soluble fractions from cell lysates by centrifugation at 11,200 rpm for 15 min in a Microfuge. Incubate with anti-3XFLAG antibodies (1:1000, SIGMA) in binding buffer (200 m*M* NaCl, 25 m*M* Tris–HCl, pH 7.4, 1 m*M* EDTA) with constant rotation ON at 4°C.
- Add 40 μL of 50% slurry of Protein-G beads (Sigma-Aldrich) to the lysates and incubate with rotation for an additional 2 h at 4°C.
- Wash the resins five times with binding buffer and elute the samples in Laemmli buffer.
- After 7.5% acrylamide SDS-PAGE gel separation and immunoblotting, incubate with the appropriated antibodies.
- Use standard chemiluminescence methods (ECL Western Blotting Substrate, Pierce) and peroxidase-conjugated anti-rabbit or anti-mouse secondary antibodies (Calbiochem) to visualize proteins.
- Develop the membranes using a Chemidoc UVP imaging system (Ultra-Violet Products Ltd.).
- Perform densitometric quantification of protein bands images using ImageJ software (NIH).

2.4 Analysis of TFEB Subcellular Localization by Immunofluorescence

TFEB antibodies: To detect endogenous TFEB in human cells, use the anti-TFEB antibody from Cell Signaling (Cat n. 4240), following manufacturer's instructions. To detect endogenous TFEB in mouse cells, use anti-TFEB antibody from MyBiosource (MBS120432) at 1:50 dilution.

- Plate 300,000 cells on coated glass coverslips in 35-mm tissue culture dishes.
- Rinse cells with PBS once and fix them for 15 min with 4% paraformaldehyde in PBS at RT.
- Rinse the slides twice with PBS and permeabilize cells with 0.05% Triton X-100 in PBS for 5 min.
- Rinse twice with PBS and incubate with primary antibody in 5% normal donkey serum ON at 4°C.
- Rinse four times with PBS and incubate with secondary donkey antibodies (diluted 1:1000 in 5% normal donkey serum) for 45 min at RT in the dark.

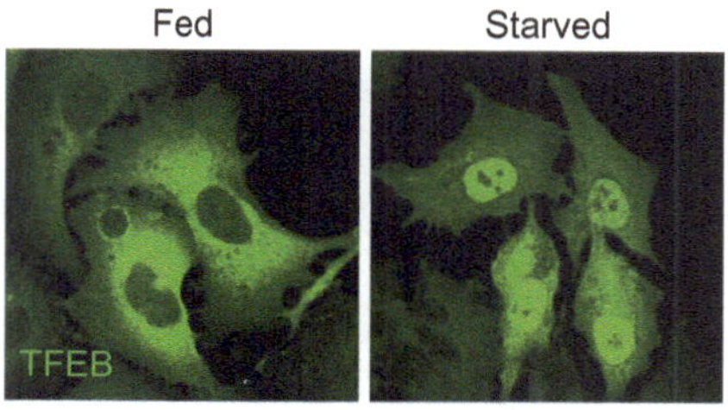

Fig. 4 Detection of TFEB subcellular localization in HeLa cells by immunofluorescence. Cells were cultured in complete media (Fed) or starved (HBSS buffer) for 1 h. Endogenous TFEB was detected using specific Cell Signaling antibodies (Cat n. 4240).

– Wash four times with PBS. Slides should be mounted onto glass coverslips, using Vectashield (Vector Laboratories) and imaged with a fluorescent microscope (Fig. 4).

2.5 Quantitative TFEB Subcellular Localization by High-Content Imaging Analysis

A setup of the immunofluorescence assay is needed to microscale the experiment to 96- or 384-well plates. This is particularly relevant in the case; this assay will be used for high-throughput screening and drug discovery (as the method below). We suggest the following of the recommendations from the Assay Guidance Manual (available http://assay.nih.gov/assay/index.php/Table_of_Contents).

Cell line: Stable HeLa-TFEB-GFP (we do not use cells older than 7 passages) obtained from an antibiotic selection of HeLa cells transfected with a vector carrying human TFEB fused to the GFP protein.

Stock plates of HeLa-TFEB-GFP cells are maintained in RPMI (10% FBS, G418 1.25 mg/mL) at 37°C, 5% CO_2.

Day 1—Seeding of assay plate:

HeLa cells are plated on 384-well plates (Perkin Elmer, CellCarrier TM 384-well, black clear bottom, TC treated, Cat n. 6007558) at a density of 4500 cells/well in 100 μL in RPMI with L-glutamine, penicillin, streptomycin, and 10% FBS, using a Multidrop-Combi (Thermo Scientific).

Day 2—Preparation of compound plate

Plate format is 384 wells. Stock compounds are 10 m*M* in DMSO. All manipulations are made using a Hamilton liquid handler.

For each plate, 3.3 μL of 28 different molecules are loaded in 96.7 μL of RPMI free (RPMI, L-glutamine, penicillin, streptomycin, NO G418,

NO serum) (final compound concentration 330 μ*M*), followed by five serial dilutions by 1/3, resulting in six final different concentrations. Negative control is RPMI free, 3% DMSO. Positive control is torin 1 (TOCRIS), 3.3 μ*M* in RPMI free. Compound plates are used as a 10 × stock solution.

Treatment of assay plate with compounds

- Cells are treated on day 2 (18 h after seeding).
- The old medium is completely removed by aspiration, using an automated plate washer (Biotek), and 90 μL RPMI free/well is added using a Multidrop-Combi (ThermoFisher).
- After this, plates are moved to an automated Hamilton STARlet liquid station, where the Compound Plate is replicated on the Assay Plate, placing 10 μL of compound in each well. Incubation with compounds is performed at 37°C for 3 h.

Fixation and nuclei counterstaining

- Incubation medium is removed using the plate washer, and cells are fixed in 50 μL of 3.7% formaldehyde in PBS added with the Multidrop.
- Following fixative removal using the plate washer, cell nuclei are stained using Hoechst 1:2000 in PBS plus 0.1% Triton X-100.
- Cells are rinsed once with PBS and left in 100 μL PBS.

Acquisition of plates and analysis of data using the OPERA system (Perkin Elmer)

- Plates are analyzed using 20 × water immersion objective. The system acquires at least 6 fields/well using two exposures (laser 405 nm for Hoechst and laser 488 nm for TFEB-GFP).
- A modification of the Acapella (Perkin Elmer) script for nuclear translocation is used to analyze the experiment. The script calculates the ratio of average nuclear GFP intensity/average cytosol GFP intensity in GFP-positive cells.

2.6 Measure of Autophagy Gene Expression Levels

TFEB is known to regulate the expression levels of several autophagy genes through direct binding of CLEAR sites in their promoter sequences. However, TFEB can also influence the expression of many autophagy genes indirectly (Settembre et al., 2013). Thus to evaluate the involvement of TFEB and more generally of a transcriptional regulation of autophagy, the expression levels of multiple autophagy genes should be analyzed.

To this aim the Autophagy RT2 Profiler PCR Array (SABiosciences, Frederick, MD) can be used, following the manufacturer's procedure.

- Extract total RNA from stimulated and nonstimulated (control) cells using TRIzol (Invitrogen) and perform a further purification using RNeasy mini kit with on-column DNAse digest (Qiagen).
- Synthesize cDNA using the RT2 First Strand Kit (SABiosciences) and perform real-time PCR using the array plates.
- Calculate fold change using SABiosciences online data analysis Web site (http://www.sabiosciences.com/pcr/arrayanalysis.php) which uses the ΔΔCt method for RT–qPCR data analysis. Average the most stable housekeeping genes included in the plate as "normalizer" genes to calculate the ΔCt value. Next, the ΔΔCt value is calculated between the "control" group and the "experimental" group. Lastly, the fold change is calculated using $2^{-\Delta\Delta Ct}$. At least three experimental replicates should be grouped to calculate the fold change.

2.7 Measurement of TFEB Binding to the Promoter of Autophagy Genes

To demonstrate that the nuclear translocation of TFEB is also associated to an enhanced TFEB activity a quantitative measure of TFEB binding to the CLEAR sites in the promoter of target genes should be performed.

- Cells stably expressing hTFEB-3XFLAG can be used to this purpose. Fifty million cells/treatment should be used. We generally use five 150-mm subconfluent dishes.
- Cross-link cells in 1% formaldehyde in PBS for 10 min (8 mL/dish).
- Quench the reaction in equal volume of 0.25 *M* glycine in PBS.
- Lyse cell on ice for 20 min in ChIP Lysis buffer (50 m*M* Tris–HCl, pH 7.5, 100 m*M* NaCl, 1% Triton X-100, 1% Tween-20). Use 1 mL lysis buffer/dish.
- Scrape and collect lysate in 2-mL tubes (Non-Stick RNase-free Microfuge Tubes, Ambion).
- Pass the lysate through G21 and G26S (four time each) and incubate 20 min on ice.
- Perform DNA digestion using MNAse Mix (2 U of MNAse from Sigma-Aldrich in 20 μL 50 m*M* $CaCl_2$) at 37°C for 5 min.
- Stop the reaction by addition of SDS and EDTA to a final concentration of 1% and 2 m*M*, respectively.

- Precipitate the unbound SDS from the cleared lysate using 50 μL of SDS-OUT (Pierce, Rockford, IL, USA).
- Dilute the supernatant 1:1 with ChIP dilution buffer (50 m*M* Tris–HCl, pH 7.5, 100 m*M* NaCl, 0.5% Triton X-100, 2 m*M* EDTA; all from Sigma-Aldrich).
- Preincubate for 30 min with 50 μL high-capacity NeutrAvidin Agarose beads (Pierce). Spin for 30 s at 4°C and remove the supernatant. Discard the beads.
- Add biotinylated FLAG antibody (2 mg ANTI-FLAG BioM2 antibody from Sigma-Aldrich) and 50 μL Neutravidin Agarose beads (previously resuspended in 1 vol. of ChIP dilution buffer supplemented with 10 mg/mL BSA). Immunoprecipitate protein–DNA complexes for 4 h at 4°C under gentle rotation.
- Spin for 30 s at 4°C and remove the supernatant (post-IP). Wash the beads with ChIP dilution buffer five times for 5 min under gentle rotation at RT.
- Elute the DNA by addition of 200 μL of 8 m*M* biotin, 1% SDS in TE buffer. Incubate 10 min under gentle rotation at RT. Spin for 1 min at RT. Collect the supernatant containing the DNA.
- Precipitate the DNA using 0.4 *M* NaCl (final concentration) at 65°C, ON. Use 1 μL of DNA for each quantitative RT-PCR.

2.8 In Vivo Autophagy Modulation via TFEB

The use of animal models, both lower organisms and mammals, has been very helpful to further elucidate TFEB function. Manipulation of TFEB levels in the liver has revealed a central role for this protein in regulating liver lipid metabolism (Settembre et al., 2013). In skeleton, TFEB regulates bone mass by controlling bone resorption by osteoclasts, the bone-resorbing cells (Ferron et al., 2013). Gain- and loss-of-function experiments in mouse skeletal muscle have shown that TFEB controls energy metabolism in this tissue (M. Mansueto, unpublished).

Studies in mice have also shown that TFEB represents an appealing therapeutic target for many human diseases that are associated with autophagic or lysosomal dysfunction and the accumulation of toxic aggregates. Indeed, induction of TFEB activity has already been successfully used as a therapeutic strategy in several disease models, such as lysosomal storage disorders, Parkinson's, Alzheimer's disease, α1-anti-trypsin deficiency, and in both diet or genetically induced obesity (reviewed in Napolitano & Ballabio, 2016).

Thus, the modulation of the activity of TFEB represents an appealing therapeutic strategy for the broad number of diseases that may potentially benefit from promoting intracellular clearance.

Below we describe the strategy we have used to overexpress TFEB in tissues such as liver, muscle, and brain of mice and the different protocols to analyze TFEB expression and activity.

2.9 TFEB Overexpression in Liver

The overexpression of TFEB in the liver of mice can be efficiently obtained using adenovirus-based strategies. We have used either adeno-associated viruses (AAVs) or helper-dependent adeno viruses (HDAds).

The HDAd-*TFEB* plasmid contained the following elements (from 5′ to 3′): a liver-restricted rat phosphoenolpyruvate carboxykinase promoter, the ApoAI intron, the human TFEB cDNA, the woodchuck hepatitis virus posttranscriptional regulatory element, the ApoE locus control region, and the human growth hormone poly(A). HDAd was produced in 116 cells with the helper virus AdNG163 as described in detail elsewhere (Pastore et al., 2013).

Hepatic transduction can be achieved by intravenous administration (tail or retro-orbital injection) of approximately 400 μL corresponding to 2×10^{12} viral particles per adult mouse (30 g weight). Age and sex-matched control mice should be infected with a transgeneless HDAd vector. Using this strategy more than 90% of the hepatocytes will express high levels of TFEB transgene.

The AAV vector serotype 8 (AAV2/8) efficiently targets the liver. To restrict transgene expression to the hepatocytes, the hepatocyte-specific thyroxine-binding globulin (TBG) promoter can be used to drive hTFEB expression in the liver of infected mice. The pAAV2.8-TBG-TFEB was produced in 293 cells as described in detail elsewhere (Doria, Ferrara, Auricchio, et al., 2013). Inject each adult mouse (30 g) with 1.25×10^{11} viral particles.

Mice can be sacrificed starting from a few days (usually 1 week) to several months after injection with either HdAD or AAV vectors.

The protocols to analyze TFEB expression and activity in liver are described below.

2.10 TFEB Overexpression in Muscle (Gastrocnemius)

The AAV serotype 1 vector can be used to efficiently transduce muscle fibers. A ubiquitous promoter (e.g., CMV or CAG) can be used to drive TFEB expression in muscle fibers.

- Inject adult mice with a total dose of 10^{11} GC of AAV2/1-CMV-hTFEB vector preparation into three sites of the right gastrocnemius (three injections of 30 μL each) using a Hamilton syringe. Inject equivalent doses of AAV2/1CMV-EGFP or equal volumes of PBS into the contralateral muscle, which will be used as nontransduced (control) sample. Mice can be sacrificed starting from a few days (usually 1 week) to several months after injection.
- Isolate gastrocnemii, free from neighboring muscles and connective tissue. The protocols to analyze TFEB expression and activity in the muscle are described below.

2.11 TFEB Overexpression in Brain

The AAV2/9 serotype showed higher transduction efficiency compared to other AAV serotypes for neuronal cells. The CMV promoter can be used to drive TFEB expression in brain.

- For newborn (P0) injection, each mouse can be injected into the lateral ventricles of both cerebral hemispheres with 4.2×10^9 total viral particles per side. Adult mice (2 months of age) can be injected to target specific brain regions according to the stereotaxic atlas of Paxinos and Franklin (2001) using the same total viral concentrations. For example, to target the cortex and hippocampus, the following coordinates can be used: AP: +1.5 mm, LAT: +1.5 mm, DV: +1.5 mm and AP: −2 mm, LAT: −1.5 mm, DV: + 1.75 mm. Mice can be analyzed starting from a few days (usually 1 week) to several months after injection. The protocols to analyze TFEB expression and activity in the brain are described below.

2.12 Isolation of Tissue Specimens From Tissues Overexpressing TFEB

- Synchronize mice before killing as follows: Prefast them for 24 h, then fed for 2 h (9 a.m. to 11 a.m.) (designated as "Fed"), and either sacrifice or refast them for 4 or 20 h (designated as "Fasted") prior to sacrifice.
- Anesthetize the mice and perform intracardiac perfusion with ice-cold PBS. Isolate tissues and weigh them.

- Cut part of the tissue into small pieces (about 0.5 cm^3) and snap freeze them in liquid nitrogen. Store them at −80°C until needed. These samples will be used to analyze TFEB mRNA and protein levels.
- Cut the remaining tissues into pieces of 1 cm^3 and fix them ON at 4°C in cold 4% PFA in PBS (50 mL). Cryoprotect the tissues in 30% sucrose in PBS solution for 24 h at 4°C, embed in Cryomatrix (Thermo Scientific), and keep them at −80°C until needed.

2.13 Analysis of TFEB mRNA Expression Levels in Transduced Tissues

- Extract total RNA from 1 piece of 0.5 cm^2 tissue using 1 mL of TRIzol (Invitrogen) according to manufacturer's protocol. Repurify RNA using RNeasy MinElute Cleanup Kit (Qiagen).
- Reverse transcription can be performed using commercially available reverse transcription reagents (e.g., TaqMan Applied Biosystems) according to manufacturer's protocols.
- Real-time qPCR can be performed using primers matching both human and mouse TFEB mRNA. The expression levels (measured as fold increase) of the human TFEB in infected tissues can be compared to the levels of the endogenous TFEB in nontransduced tissues isolated from control mice.

Primer sequences are:	h/m-TFEB For. 5′-aggagcggcagaagaaagac-3′
	h/m-TFEB Rev. 5′-caggtccttctgcatcctcc-3′

The expression levels of *Cyclophilin* and the *Ribosomal protein S16* cDNA can be used as internal reference genes, since their expression is not sensitive to starvation/refeeding conditions or to TFEB overexpression. Below are the sequences of the primers:

Primer sequences are:	m-Cycloph-For. 5′-ggcaaatgctggaccaaacacaa-3′
	m-Cycloph-Rev. 5′-gtaaaatgcccgcaagtcaaaag-3′
	m-S16-F 5′-aggagcgatttgctggtgtgg-3′
	m-S16-R 5′-gctaccagggcctttgagatg-3′

- Calculate fold change values using the ΔΔCt method.

2.14 Analysis of TFEB Protein Expression Levels in Transduced Tissues

- Lyse a 0.5 cm^3 tissue sample using a TissueLyser (Qiagen) in 0.5 mL cold RIPA buffer supplemented with 0.5% SDS, PhosSTOP, and EDTA-free protease inhibitor tablets (Roche). Samples are incubated for 30 min on ice, briefly sonicated on ice, and then the soluble fraction isolated by centrifugation at 14,000 rpm for 10 min at 4°C.
- Separate 50 μg of protein by SDS-PAGE (Invitrogen; reduced NuPAGE 4–12% bis–tris gel, MES SDS buffer) and transfer them onto a nitrocellulose membrane with an I-Blot (Invitrogen).
- Incubate with 5% nonfat milk in TBS-T buffer (TBS containing 0.05% Tween-20) for 1 h and then with the selected primary antibody (1:1000) ON at 4°C with gentle shaking. The TFEB antibody from Bethyl Laboratories (A303-672A) detects both human and mouse TFEB proteins, allowing a relative quantification of the transgene protein expression. To detect the human TFEB transgene only the rabbit anti-hTFEB polyclonal antibody from Cell Signaling (# 4240) should be used.
- Use standard chemiluminescence methods (ECL Western Blotting Substrate, Pierce) and peroxidase-conjugated anti-rabbit or anti-mouse secondary antibodies (Calbiochem) to visualize proteins.

2.15 Analysis of TFEB Protein Expression and Localization in Transduced Tissues by Immunofluorescence

The human specific TFEB antibody from Cell Signaling can be efficiently used to analyze tissue distribution and localization only of the human TFEB transgene, since no cross-reactivity with the murine TFEB can be detected using this antibody.

- Prepare 20-μm thick slices from cryopreserved tissue specimens.
- Add blocking buffer (2.5% BSA in PBS + 0.1% Triton X-100) for 2 h at RT.
- Incubate specimens for 20 h with the primary anti-TFEB antibody (Cell Signaling, Cat n. 4240) 1:100 at 4°C in a humid chamber.
- Wash three times in PBS + 0.05% TX-100, and then incubate for 3 h with secondary antibodies conjugated either with Alexafluor 488 or Alexafluor 555 (Invitrogen).
- Wash three times in PBS, one wash in H_2O and one brief wash in ethanol 70%, and dry the slides.
- Mount in Mowiol (Calbiochem) containing DAPI (0.5 μ*M*).

– Take images by confocal microscopy. Analyze the number of TFEB-positive cells (relative to total cell number = DAPI) and the ratio of average nuclear TFEB intensity/average cytosol TFEB intensity in TFEB-positive cells.

ACKNOWLEDGMENTS

We are grateful to the Fondazione Telethon; the Beyond Batten Disease Foundation; the European Research Council; the Associazione Italiana per la Ricerca sul Cancro (Italian Association for Cancer Research); and the National Institutes of Health for their generous support.

REFERENCES

Cortes, C. J. (2014). Polyglutamine-expanded androgen receptor interferes with TFEB to elicit autophagy defects in SBMA. *Nature Neuroscience*, *17*(9), 1180–1189.

Decressac, M., et al. (2013). TFEB-mediated autophagy rescues midbrain dopamine neurons from α-synuclein toxicity. *Proceedings of the National Academy of Sciences of the United States of America*, *110*(19), E1817–E1826.

Doria, M., Ferrara, A., Auricchio, A., et al. (2013). AAV2/8 vectors purified from culture medium with a simple and rapid protocol transduce murine liver, muscle, and retina efficiently. *Human Gene Therapy Methods*, *24*(6), 392–398.

Ferron, M., et al. (2013). A RANKL-PKCβ-TFEB signaling cascade is necessary for lysosomal biogenesis in osteoclasts. *Genes & Development*, *27*(8), 955–969.

Martina, J. A., Chen, Y., Gucek, M., & Puertollano, R. (2012). MTORC1 functions as a transcriptional regulator of autophagy by preventing nuclear transport of TFEB. *Autophagy*, *8*, 903–914.

Martina, J. A., Diab, H. I., Brady, O. A., & Puertollano, R. (2016). TFEB and TFE3 are novel components of the integrated stress response. *The EMBO Journal*, *35*(5), 479–495.

Martina, J. A., Diab, H. I., Li, H., & Puertollano, R. (2014). Novel roles for the MiTF/TFE family of transcription factors in organelle biogenesis, nutrient sensing, and energy homeostasis. *Cellular and Molecular Life Sciences*, *71*(13), 2483–2497.

Martina, J. A., Diab, H. I., Lishu, L., Jeong, A. L., Patange, S., Raben, N., et al. (2014). The nutrient-responsive transcription factor TFE3 promotes autophagy, lysosomal biogenesis, and clearance of cellular debris. *Science Signaling*, 7(309), ra9.

Medina, D. L., et al. (2011). Transcriptional activation of lysosomal exocytosis promotes cellular clearance. *Developmental Cell*, *21*, 421–430.

Medina, D. L., et al. (2015). Lysosomal calcium signalling regulates autophagy through calcineurin and TFEB. *Nature Cell Biology*, *17*(3), 288–299.

Moskot, M., et al. (2014). The phytoestrogen genistein modulates lysosomal metabolism and transcription factor EB (TFEB) activation. *The Journal of Biological Chemistry*, *289*(24), 17054–17069.

Napolitano, G., & Ballabio, A. (2016). TFEB at a glance. *Journal of Cell Science*, *129*(13), 2475–2481.

Palmieri, M., et al. (2011). Characterization of the CLEAR network reveals an integrated control of cellular clearance pathways. *Human Molecular Genetics*, *20*, 3852–3866.

Pastore, N., et al. (2013). Gene transfer of master autophagy regulator TFEB results in clearance of toxic protein and correction of hepatic disease in alpha-1-anti-trypsin deficiency. *EMBO Molecular Medicine*, *5*(3), 397–412.

Paxinos, G., & Franklin, K. B. J. (2001). *The mouse brain in stereotaxic coordinates*. New York: Academic Press.

Polito, V. A., et al. (2014). Selective clearance of aberrant tau proteins and rescue of neurotoxicity by transcription factor EB. *EMBO Molecular Medicine*, *6*(9), 1142–1160.
Roczniak-Ferguson, A., et al. (2012). The transcription factor TFEB links mTORC1 signaling to transcriptional control of lysosome homeostasis. *Science Signaling*, *5*, ra42.
Sardiello, M., et al. (2009). A gene network regulating lysosomal biogenesis and function. *Science*, *325*, 473–477.
Settembre, C., et al. (2011). TFEB links autophagy to lysosomal biogenesis. *Science*, *332*, 1429–1433.
Settembre, C., et al. (2012). A lysosome-to-nucleus signalling mechanism senses and regulates the lysosome via mTOR and TFEB. *The EMBO Journal*, *31*, 1095–1108.
Settembre, C., et al. (2013). TFEB controls cellular lipid metabolism through a starvation-induced autoregulatory loop. *Nature Cell Biology*, *15*(6), 647–658.
Shin, H. J., et al. (2016). AMPK-SKP2-CARM1 signalling cascade in transcriptional regulation of autophagy. *Nature*, *534*(7608), 553–557.
Song, J. X., et al. (2016). A novel curcumin analog binds to and activates TFEB in vitro and in vivo independent of MTOR inhibition. *Autophagy*, *12*(8), 1372–1389.
Spampanato, C., et al. (2013). Transcription factor EB (TFEB) is a new therapeutic target for Pompe disease. *EMBO Molecular Medicine*, *5*(5), 691–706.
Steingrímsson, E., Copeland, N. G., & Jenkins, N. A. (2004). Melanocytes and the microphthalmia transcription factor network. *Annual Review of Genetics*, *38*, 365–411.
Tsumeni, T., et al. (2012). PGC-1α rescues Huntington's disease proteotoxicity by preventing oxidative stress and promoting TFEB function. *Science Translational Medicine*, *4*(142), 142ra9.
Xiao, Q., et al. (2015). Neuronal-targeted TFEB accelerates lysosomal degradation of APP, reducing Aβ generation and amyloid plaque pathogenesis. *The Journal of Neuroscience*, *35*(35), 12137–12151.

CHAPTER SIX

Application of CRISPR/Cas9 to Autophagy Research

J. O'Prey*, J. Sakamaki*, A.D. Baudot*, M. New†, T. Van Acker†, S.A. Tooze†, J.S. Long*[,1], K.M. Ryan*[,1]

*Cancer Research UK Beatson Institute, Glasgow, United Kingdom
†The Francis Crick Institute, London, United Kingdom
[1]Corresponding authors: e-mail address: j.long@beatson.gla.ac.uk; k.ryan@beatson.gla.ac.uk

Contents

Abstract

The ability to efficiently modulate autophagy activity is paramount in the study of the field. Conventional broad-range autophagy inhibitors and genetic manipulation using RNA interference (RNAi), although widely used in autophagy research, are often limited in specificity or efficacy. In this chapter, we address the problems of conventional autophagy-modulating tools by exploring the use of three different CRISPR/Cas9 systems to abrogate autophagy in numerous human and mouse cell lines. The first system generates cell lines constitutively deleted of ATG5 or ATG7 whereas the second and third systems express a Tet-On inducible-Cas9 that enables regulated deletion of ATG5

Methods in Enzymology, Volume 588
ISSN 0076-6879
http://dx.doi.org/10.1016/bs.mie.2016.09.076

or ATG7. We observed the efficiency of autophagy inhibition using the CRISPR/Cas9 strategy to surpass that of RNAi, and successfully generated cells with complete and sustained autophagy disruption through the CRISPR/Cas9 technology.

1. INTRODUCTION

1.1 Autophagy Research and Limitations of Current Tools

Macroautophagy (hereafter referred to as autophagy) is one of the main processes involved in the degradation and turnover of cytoplasmic molecules and organelles within cells (Rosenfeldt & Ryan, 2011). This cellular process is initiated by the formation of an isolated membrane, termed a phagophore, which grows and eventually encapsulates cytoplasmic constituents or organelles to form a double-membraned vesicle called an autophagosome (Rosenfeldt & Ryan, 2011). The process proceeds with the fusing of the autophagosomes to lysosomes, wherein the autophagosome contents are finally degraded. These events are orchestrated by specific proteins encoded by a number of autophagy-related genes (*ATG*) (Mizushima, Yoshimori, & Ohsumi, 2011).

Autophagy regulates a variety of cellular processes such as cell survival, metabolism, motility, and death, and has been shown to play vital roles in both normal physiology and diseased settings (Choi, Ryter, & Levine, 2013; Ravikumar et al., 2010). The involvement of autophagy in many human conditions including aging, inflammation, cancer, and neurodegenerative disorders (Jiang & Mizushima, 2014; Rubinsztein, Marino, & Kroemer, 2011) has attracted many to the field of autophagy research. As advances in this field of research continually expand, more powerful experimental tools to modulate autophagy are required to facilitate research in this area.

One of the major limitations of the currently available tools for autophagy research is the lack of specific and efficient inhibitors of the pathway. Inhibitors that are widely used in autophagy studies, such as chloroquine, bafilomycin A1, and 3-methyladenine (3-MA), target components or aspects of the autophagic pathway, which are not exclusive to this pathway. Chloroquine and bafilomycin A1 act predominantly by impairing lysosomal function and 3-MA is an inhibitor of phosphoinositide 3-kinase (PI3K) (Wu et al., 2010; Yang et al., 2013). Therefore, these inhibitors interfere with many pathways and cellular processes involving the lysosomal compartment and PI3K, which extend beyond autophagy. Some studies have also demonstrated that these inhibitors exert their effects through regulation of additional pathways (Lin, Kuo, Wang, & Lin, 2012; Maes et al., 2014; Maycotte et al., 2012; Yuan et al., 2015).

Genetic silencing of *ATG* genes by small interfering RNA (siRNA) or short hairpin RNA (shRNA) is another common approach used to inhibit autophagy. While these strategies are more specific than the use of pharmacological inhibitors, the extent of knockdown achieved is often incomplete, resulting in the detection of residual autophagic activity. In order to achieve a complete inhibition of autophagy and to obtain a cleaner phenotype for autophagy studies, we explored the use of one of the newest genome-engineering strategies, the CRISPR/Cas9 technology, to delete two essential autophagy genes, *ATG5* and *ATG7* in a variety of cell lines.

1.2 CRISPR/Cas9 Technology in Gene Editing

The Clustered Regularly Interspaced Short Palindromic Repeats (CRISPR) Type II system, originally identified to play a role in the adaptive immunity of prokaryotes (Deveau, Garneau, & Moineau, 2010; Horvath & Barrangou, 2010), has in recent years emerged as a leading approach in genome editing. This system utilizes short RNAs, termed guide RNAs (gRNAs), to direct CRISPR-associated endonucleases (Cas9) to induce site-specific cleavage of genomic loci (Cong et al., 2013; Ran et al., 2013) (Fig. 1). Repair of the resulting double-strand break by the error-prone nonhomologous end joining (NHEJ) machinery subsequently gives rise to insertion and/or deletion (indel) mutations that cause the disruption of gene expression (Cong et al., 2013; Ran et al., 2013).

The targeting specificity of the CRISPR/Cas9 system is governed by the design of the gRNA. A gRNA consists of an invariant scaffold sequence that facilitates Cas9 binding and a customized 20-nucleotide targeting sequence (also known as the protospacer) unique to the target locus (Ran et al., 2013). A critical prerequisite of Cas9-mediated DNA cleavage is the presence of a specific Protospacer Adjacent Motif (PAM) on the target gene, without which Cas9 will not recognize the desired site of action (Fig. 1) (Sternberg, Redding, Jinek, Greene, & Doudna, 2014). DNA incision by Cas9 occurs approximately 3 bp upstream of the PAM sequence (Jinek et al., 2012). Therefore, specific Cas9 targeting can be achieved by designing a gRNA protospacer sequence incorporating the

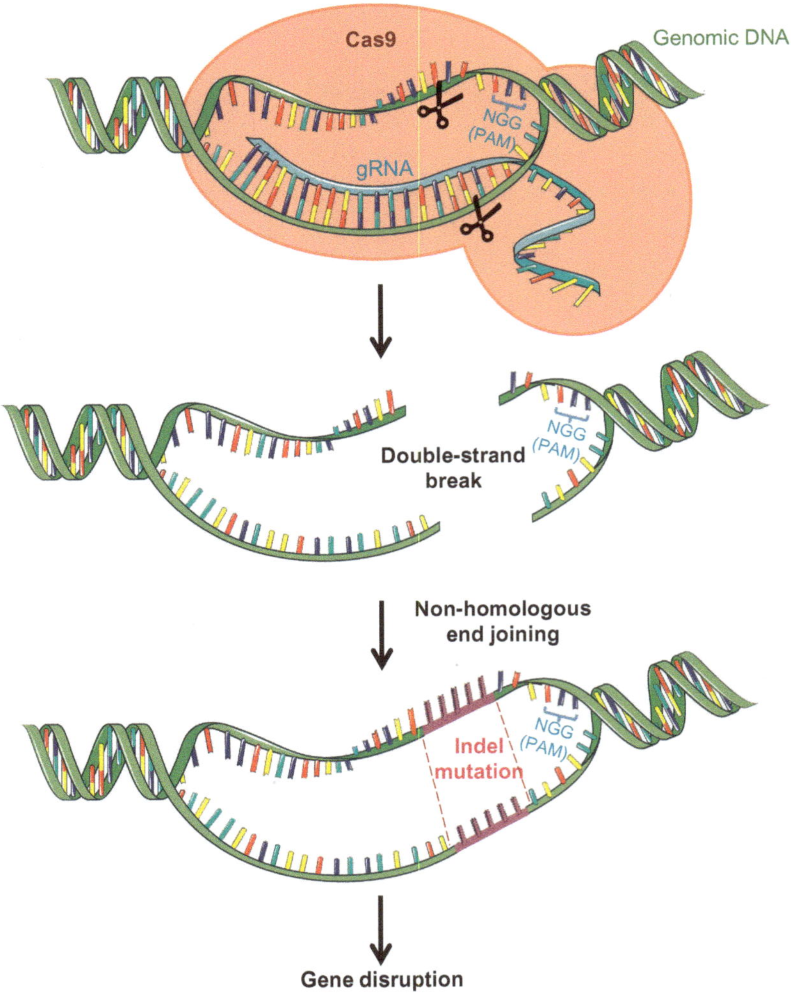

Fig. 1 Schematic of the CRISPR/Cas9 technology for genome editing. The *Streptococcus pyogenes* species of Cas9 is directed to the genomic DNA target (*green* strand) by a user-designed gRNA (*blue* strand) that is unique to the target site. The 5′-NGG-3′ PAM is indicated, and Cas9-mediated cleavage of both strands of DNA occurs 3 bp upstream of the PAM sequence. Repair of the double-strand break by the error-prone nonhomologous end joining machinery subsequently results in an indel mutation that disrupts gene expression. Image contains components from Servier Medical Art (by Servier), licensed under CC BY 3.0 (https://creativecommons.org/licenses/by/3.0/).

20-bp sequence directly upstream of a PAM site unique to the targeted region. PAM sites are found on average every 8–12 bp within the human genome, thereby providing a broad targeting spectrum for CRISPR/Cas9-mediated gene editing (Cong et al., 2013; Ran et al., 2013).

In contrast to the ability of RNA interference techniques to knockdown the expression of a gene, the CRISPR/Cas9 technology provides efficient and permanent gene deletion. The simplicity and quick design and construction of the targeting gRNAs also confer CRISPR the advantage over more labor-intensive and costly genome editing tools such as the zinc finger nucleases or transcription-activator-like effector nucleases (TALENs). CRISPR/Cas9 technology is now widely used for genetic engineering of cells and organisms, and has even been adapted for multiple gene deletions and high-throughput screenings (Cong et al., 2013; Dow et al., 2015; Ran et al., 2015; Wang, Wei, Sabatini, & Lander, 2014).

The methods detailed here demonstrate the use of CRISPR/Cas9-targeting of *ATG* genes to create various ATG-knockout cell lines devoid of autophagic activity with ease and within a short space of time. We make a comparison between the CRISPR-generated cell lines to cell lines treated with the conventional siRNA or shRNA, and observe a superior phenotype with the CRISPR/Cas9 system.

2. ESTABLISHMENT OF A CONSTITUTIVELY ACTIVE CRISPR/CAS9 SYSTEM FOR DELETION OF AUTOPHAGY IN HUMAN AND MURINE CELLS

2.1 Design of Guide RNA Sequence for Targeting ATG5 or ATG7

The first CRISPR/Cas9 system that we used to delete *ATG5* or *ATG7* was the lentiCRISPR vector (Fig. 2A; Addgene, Cambridge, USA, plasmid #52961) established by the Zhang lab (Sanjana, Shalem, & Zhang, 2014; Shalem et al., 2014). This plasmid encodes the humanized *Streptococcus pyogenes* species of Cas9 (hSpCas9) nuclease that specifically recognizes and cleaves the DNA sequence directly adjacent to the PAM sequence 5′-NGG-3′. The design and cloning of the target gRNA sequences against *ATG5* or *ATG7* was performed as recommended by the Zhang Lab GeCKO website http://www.genome-engineering.org/gecko/ and also described in this section and Section 2.2.

To identify suitable target sites for gRNA sequence design against *ATG5* or *ATG7*, the CRISPR design tool software at http://crispr.mit.edu/ was used. It was deemed important to target exonic coding sequences nearest to the start codon. Therefore, sequences of the first 250 base pairs (beginning from the ATG start codon) for *ATG5* or *ATG7* (human or mouse) were input into the software for analysis. The potential targeting sequences

A

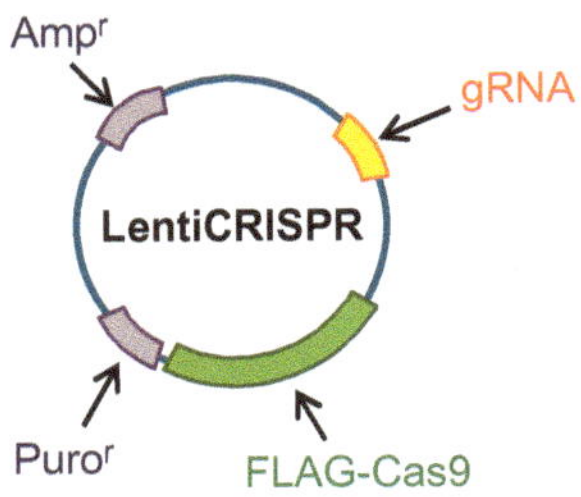

B

Guide RNA oligo design template for lentiCRISPR

5′ - CACCG**NNNNNNNNNNNNNNNNNNNN** - 3′
3′ - AAAC**NNNNNNNNNNNNNNNNNNNN** C - 5′

Human ATG5-1 forward
5′- CACCG**AACTTGTTTCACGCTATATC** -3′
Human ATG5-1 reverse
5′- AAAC**GATATAGCGTGAAACAAGTT**C -3′

Human ATG5-2 forward
5′- CACCG**AAGAGTAAGTTATTTGACGT** -3′
Human ATG5-2 reverse
5′- AAAC**ACGTCAAATAACTTACTCTT**C -3′

Human ATG7-1 forward
5′- CACCG**GAAGCTGAACGAGTATCGGC** -3′
Human ATG7-1 reverse
5′- AAAC**GCCGATACTCGTTCAGCTTC**C -3′

Human ATG7-2 forward
5′- CACCG**AACTCCAATGTTAAGCGAGC** -3′
Human ATG7-2 reverse
5′- AAAC**GCTCGCTTAACATTGGAGTT**C -3′

Control forward
5′- CACCG**GTAGCGAACGTGTCCGGCGT** -3′
Control reverse
5′- AAAC**ACGCCGGACACGTTCGCTAC**C -3′

Mouse ATG5 forward
5′- CACCG**AAGAGTCAGCTATTTGACGT** -3′
Mouse ATG5 reverse
5′- AAAC**ACGTCAAATAGCTGACTCTT**C -3′

Mouse ATG7 forward
5′- CACCG**AACTCCAACGTCAAGCGGGT** -3′
Mouse ATG7 reverse
5′- AAAC**ACCCGCTTGACGTTGGAGTT**C -3′

GFP forward
5′- CACCG**GAAGTTCGAGGGCGACACCC** -3′
GFP reverse
5′- AAAC**GGGTGTCGCCCTCGAACTTC**C -3′

Fig. 2 (A) Vector map of lentiCRISPR construct. *Puro^r*, puromycin-resistance gene; *Amp^r*, ampicillin-resistance gene; *gRNA*, guide RNA sequence. (B) gRNA oligo design template for cloning of gRNA targeting sequences into lentiCRISPR, and the list of control gRNA oligos and gRNA oligos targeting GFP, ATG5, and ATG7.

generated by the software are tabulated in Fig. S1 (http://dx.doi.org/10.1016/bs.mie.2016.09.076). The top-scoring gRNA sequences (higher scores indicate the lesser likelihood of off-target binding) were selected as targeting sequences against the respective *ATG* genes:

Human ATG5:
(1) AACTTGTTTCACGCTATATC (Score 91)—Exon 2
(2) AAGAGTAAGTTATTTGACGT (Score 77)—Exon 3
Human ATG7:
(1) GAAGCTGAACGAGTATCGGC (Score 95)—Exon 2
(2) AACTCCAATGTTAAGCGAGC (Score 91)—Exon 3
Mouse ATG5:
AAGAGTCAGCTATTTGACGT (Score 85)—Exon 3
Mouse ATG7:
AACTCCAACGTCAAGCGGGT (Score 96)—Exon 3
For controls, we used the gRNA sequence GTAGCGAACGTGTCC GGCGT, which was generated by Wang et al. (2014) or gRNA sequence targeting GFP, GAAGTTCGAGGGCGACACCC.

To clone the target sequences into the lentiCRISPR vector, oligos incorporating the target sequence and specific overhangs were ordered (as shown in Fig. 2B; target sequence in bold). Standard desalted oligos are sufficient for the cloning procedure and should be reconstituted to a working stock of 100 μ*M*.

2.2 Cloning of Guide RNA Sequence Targeting ATG5 or ATG7 Into LentiCRISPR

The cloning of the gRNA sequence into lentiCRISPR is described in this section and summarized in the flowchart in Fig. 3:

(i) Digest 5 μg of lentiCRISPR plasmid with *BsmB*I (2 h at 37°C):

5 μg	LentiCRISPR
2 μL	*BsmB*I enzyme (NEB, Herts, UK)
5 μL	10× buffer 3.1 (NEB, Herts, UK)
X μL	ddH_2O
50 μL	Total reaction volume

(ii) Run the digested plasmid on an agarose gel and gel-purify the larger band (≈11 kb) using the QIAquick Gel Extraction Kit (Qiagen, Manchester, UK, #28704). Quantify the purified product with a spectrophotometer.

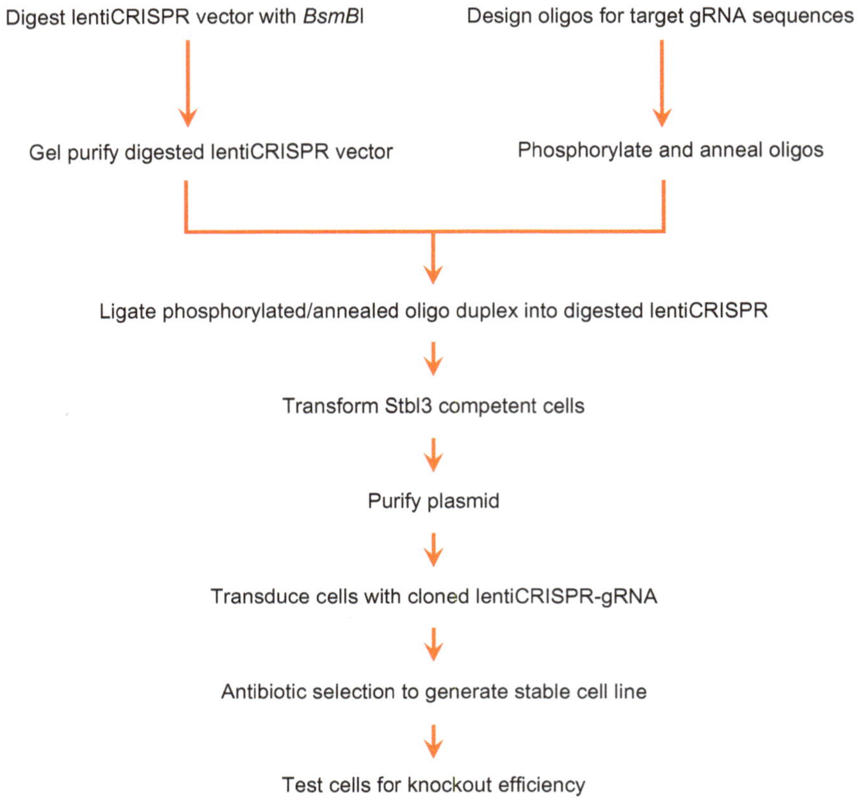

Fig. 3 Flowchart outlining the steps for cloning the target gRNA sequence into the lentiCRISPR vector, and the generation of ATG-CRISPR cells.

(iii) Set up the reaction to phosphorylate and anneal each pair of gRNA oligos:

1 μL	gRNA oligo forward (100 μ*M*)
1 μL	gRNA oligo reverse (100 μ*M*)
1 μL	T4 polynucleotide kinase 10× reaction buffer (Cambio, Cambridge, UK)
0.4 μL	ATP (25 m*M*)
0.5 μL	T4 polynucleotide kinase (Cambio, Cambridge, UK)
6.1 μL	ddH_2O
10 μL	Total reaction volume

Perform reaction in a thermocycler using the following conditions:

Step 1	37°C 30 min
Step 2	95°C 5 min and then ramp down to 25°C at 5°C/min

(iv) Dilute the annealed gRNA Oligos at a 1:200 dilution using ddH_2O.

(v) Set up a ligation reaction and incubate at room temperature for 15 min using the rapid DNA ligation kit (Roche, Burgess Hill, UK, #11 635 379 001). For comparison, set up a negative control ligation reaction with the digested lentiCRISPR vector alone and ddH_2O in place of annealed gRNA oligos.

X μL	*BsmB*I-digested lentiCRISPR plasmid from Step (ii) (50 ng)
1 μL	Diluted gRNA oligo duplex from Step (iv)
2 μL	5 × DNA dilution buffer
X μL	ddH_2O (10 μL subtotal reaction volume)
10 μL	2× T4 DNA ligation buffer
1 μL	T4 DNA ligase
21 μL	Total reaction volume

(vi) Transform Stbl3 competent cells (Thermo Fisher Scientific, Paisley, UK, #C7373-03) with 1 μL of ligation reaction. As lentiviral constructs contain long direct repeats that are prone to recombination in standard *E. coli* hosts, it is highly recommended to perform transformations using recombination-deficient strains like Stbl3.

(vii) Plate transformed Stbl3 cells onto ampicillin-containing agar plates and incubate overnight at 37°C.

(viii) Pick several colonies for each gRNA cloning and set up mini-cultures. Purify plasmid and sequence using the LKO.1 forward primer (5′-GACTATCATATGCTTACCGT-3′) to confirm successful gRNA cloning.

(ix) Amplify positive clones in a larger culture and purify plasmids for long-term storage and use.

2.3 Generation of Autophagy-Deficient Human and Murine Cells

LentiCRISPR-Control, lentiCRISPR-GFP, lentiCRISPR-ATG5 (human- or mouse-specific), or lentiCRISPR-ATG7 (human- or mouse-specific) should be transfected into the host packaging cell line, HEK 293T cells, for production of the lentivirus encoding Cas9 and the respective gRNAs. The lentivirus-containing media from HEK 293T cells is subsequently transferred onto target cell lines to generate control or *ATG*-deleted cells. The detailed procedure is as follows:

(i) Plate 1.5×10^6 HEK 293T cells onto 10 cm cell culture dishes and leave to adhere and grow overnight.

(ii) The next day, transfect HEK 293T cells using the $CaPO_4$ transfection method (or any preferred transfection method) with the following plasmids:

10 μg lentiCRISPR.

7.5 μg psPAX2 packaging plasmid.

4 μg VSV-G envelope plasmid.

(iii) Replenish media with fresh complete media (6 mL) the next day (extra care to be taken when handling and discarding old media as lentivirus production is already taking place). Plate cells targeted for *ATG*-CRISPR ($0.5–0.7 \times 10^6$ cells per 10 cm dish), including an extra plate of cells as a control for antibiotic selection.

(iv) 48 h after transfection, pass the lentivirus-containing supernatant from HEK 293T cells through a 0.45 μm PTFE filter membrane and add Polybrene (Hexadimethrine Bromide; Sigma, Gillingham, UK, #H-9268), a transduction enhancer, to a final concentration of 4 μg/mL. Subsequently, transfer the media onto target cells. Replenish the media on HEK 293T cells (6 mL) to generate more virus for a repeat infection.

(v) The next day, repeat the infection process of Step (iv) and leave cells to incubate for a further 24 h. Discard the HEK 293T cells according to local Biosafety procedures.

(vi) 24 h later, remove the virus-containing media from target cells and replenish with complete media containing antibiotic selection—puromycin (1.0–2.5 μg/mL depending on sensitivity of target cells). Replenish the puromycin-containing media every 2 days for at least 1 week to generate a stable cell line.

(vii) 2 weeks after start of infection (Step iv), cells are harvested for western blotting analysis of ATG deletion.

2.4 CRISPR Is Superior to siRNA Treatment for Autophagy Inhibition

Using the lentiCRISPR system detailed in Sections 2.1–2.3, we successfully generated KP-4 (human pancreatic ductal adenocarcinoma) cells deleted for ATG5 or ATG7 (Fig. 4A and B, lanes 3 and 4). Notably, when we separately transfected KP-4 cells with siRNAs targeting ATG5 or ATG7 (Fig. 4C), we found the extent of knockdown to be inferior to the CRISPR-mediated deletion of ATGs in the same cell line (Fig. 4A and B). This was evident by the detection of residual levels of ATG5, ATG7 and formation of the lipidated form of LC3B, LC3B-II, a critical step in the formation of

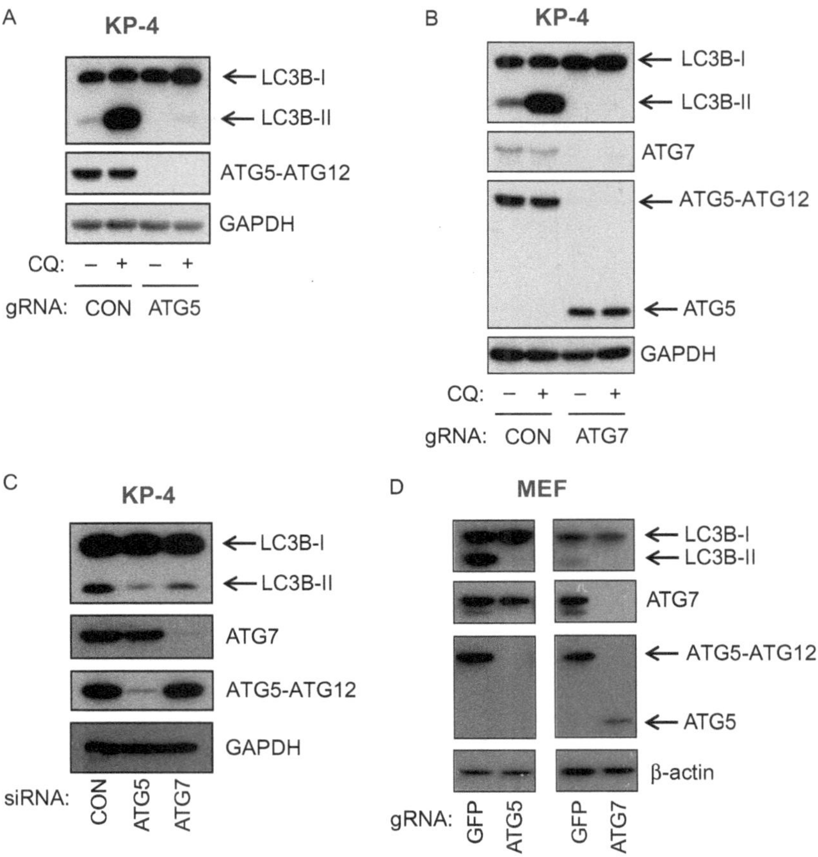

Fig. 4 (A and B) Generation of ATG5- or ATG7-deleted KP-4 cells using CRISPR/Cas9-targeting of ATG5 or ATG7. KP-4 cells stably expressing lentiCRISPR-Control (A and B), lentiCRISPR-ATG5-2 (A) or lentiCRISPR-ATG7-1 (B) were treated with or without chloroquine (50 μ*M*) for 48 h prior to harvesting. Cell extracts were subsequently analyzed by SDS-PAGE and western blotting. (C) KP-4 cells were transfected with 20 n*M* control, ATG5 or ATG7 siRNAs (control: Dharmacon, Bucks, UK, D-001810-03-20; ATG5: Dharmacon, Bucks, UK, D-004374-05; ATG7: Qiagen, Manchester, UK, SI02655373) using Lipofectamine RNAiMAX (Thermo Fisher Scientific, Paisley, UK). 72 h after transfection, cells were harvested and analyzed by SDS-PAGE and western blotting. (D) MEF cells stably expressing lentiCRISPR-GFP (as control), lentiCRISPR-ATG5 or lentiCRISPR-ATG7. Antibodies used for western blotting: LC3B (Cell signaling, NEB, Herts, UK, #2775S), ATG5 (Cell signaling, NEB, Herts, UK, #12994S), ATG7 (Santa Cruz, Germany, #sc-33211 for B; Cell signaling, NEB, Herts, UK, #8558S for C and D), GAPDH (Abcam, Cambridge, UK, #ab9484), and β-actin (Abcam, Cambridge, UK, #ab8227).

autophagosomes (Fig. 4C, lanes 2 and 3). In contrast, ATG5 or ATG7 levels were undetectable in the respective CRISPR-targeted cells (Fig. 4A and B, lanes 3 and 4). The conversion of LC3B-I to LC3B-II was also completely abolished in the CRISPR-targeted lines (Fig. 4A and B, lanes 3 and 4), even in the presence of chloroquine, which blocks any lysosome-mediated degradation of LC3B-II (Fig. 4A and B, lane 4). In addition, CRISPR-mediated deletion of ATG7 blocked the formation of the ATG5–ATG12 complex (Fig. 4B, lanes 3 and 4), in line with the essential role of ATG7 in this process (Tanida et al., 1999). Again, this was not observed upon RNAi knockdown (Fig. 4C, lane 3).

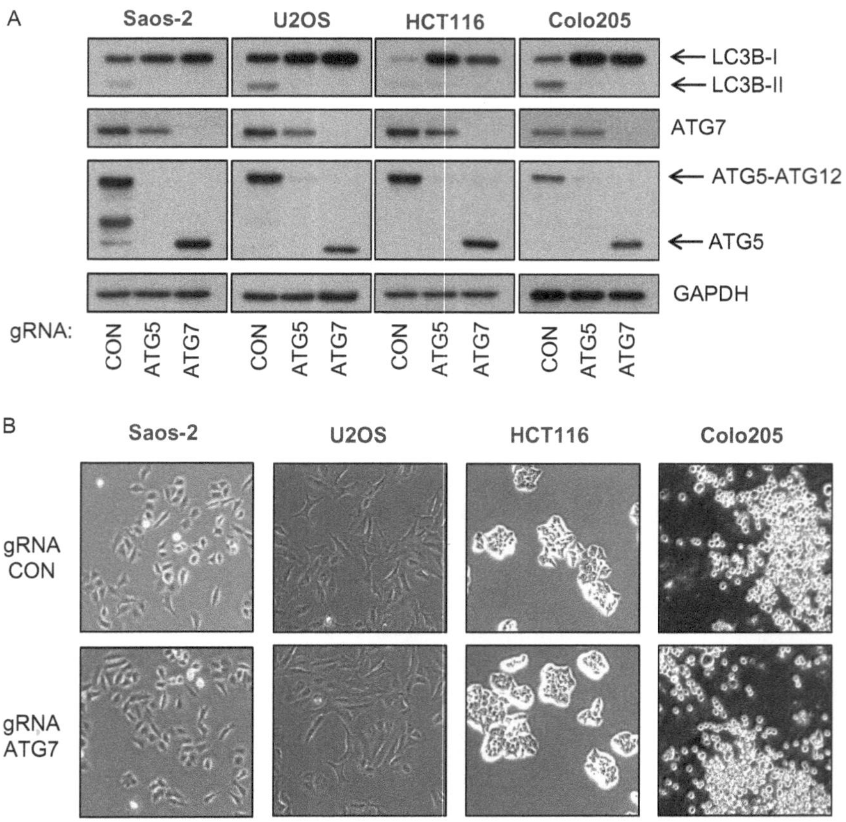

Fig. 5 (A and B) Generation of ATG5- or ATG7-deleted Saos-2, U2OS, HCT116, and Colo205 cells using CRISPR/Cas9-targeting of ATG5 or ATG7. (A) Cell extracts of Saos-2, U2OS, HCT116, and Colo205 cells expressing lentiCRISPR-Control, lentiCRISPR-ATG5-2, or lentiCRISPR-ATG7-1 were analyzed by SDS-PAGE and western blotted for LC3B (Cell signaling, NEB, Herts, UK, #2775S), ATG5 (Cell signaling, NEB, Herts, UK, #12994S), ATG7 (Cell signaling, NEB, Herts, UK, #8558S), and GAPDH (Abcam, Cambridge, UK, #ab9484). (B) Brightfield images of Saos-2, U2OS, HCT116, and Colo205 cells expressing lentiCRISPR-Control or lentiCRISPR-ATG7-1.

Importantly, we were able to delete ATG5 or ATG7 using the lentiCRISPR system in other human as well as mouse cell lines such as Saos-2 (human osteosarcoma), U2OS (human osteosarcoma), HCT116 (human colorectal carcinoma), Colo205 (human colorectal adenocarcinoma), and mouse embryonic fibroblasts (MEFs) (Figs. 4D and 5A). In general, the cells tolerated CRISPR-mediated deletion of ATG5 or ATG7 well and looked comparable to control cells (examples of Saos-2, U2OS, HCT116, and Colo205 shown in Fig. 5B), suggesting that this CRISPR system is a functionally efficient tool for abrogating autophagy in a variety of cell lines.

3. REGULATED DISRUPTION OF AUTOPHAGY USING TETRACYCLINE-INDUCIBLE CRISPR/CAS9 SYSTEMS

Conditional and inducible gene expression systems are very useful tools which allow the temporal control of target gene expression at the transcriptional level. The ability to regulate gene expression or deletion using inducible systems is advantageous not only for studying gene function and understanding complex biological pathways but also to overcome problems that arise from constitutive gene manipulations. We explored the use of two different Tet-On tetracycline-inducible-CRISPR/Cas9 systems for the regulated deletion of autophagy in cells.

3.1 Establishment of a Tet-On Inducible CRISPR/Cas9 Lentiviral System for the Regulated Disruption of Autophagy

The Tet-On inducible-CRISPR/Cas9 lentiviral system that we tested was established by Wang et al. (2014). This system consists of two separate plasmid constructs (Fig. 6A)—(1) pCW-Cas9 (Addgene, Cambridge, USA, #50661), for lentiviral expression of the inducible-hSpCas9 and (2) pLX-sgRNA (Addgene, Cambridge, USA, #50662), for lentiviral expression of targeting gRNA.

3.1.1 Design of Guide RNA Sequence for Targeting ATG5 or ATG7 in the Lentiviral Inducible-CRISPR/Cas9 System

Cloning of the target gRNA sequence into the pLX-sgRNA vector involves three polymerase chain reaction (PCR) extension steps within the *Xho*I and *Nhe*I sites of pLX-sgRNA followed by restriction digest and ligation (Fig. 6B). The target gRNA sequences used:

(i) must not contain any *Xho*I or *Nhe*I sites.

(ii) should start with a "G" to allow efficient transcription from the RNA polymerase III U6 promoter. If the sequence does not start with a "G,"

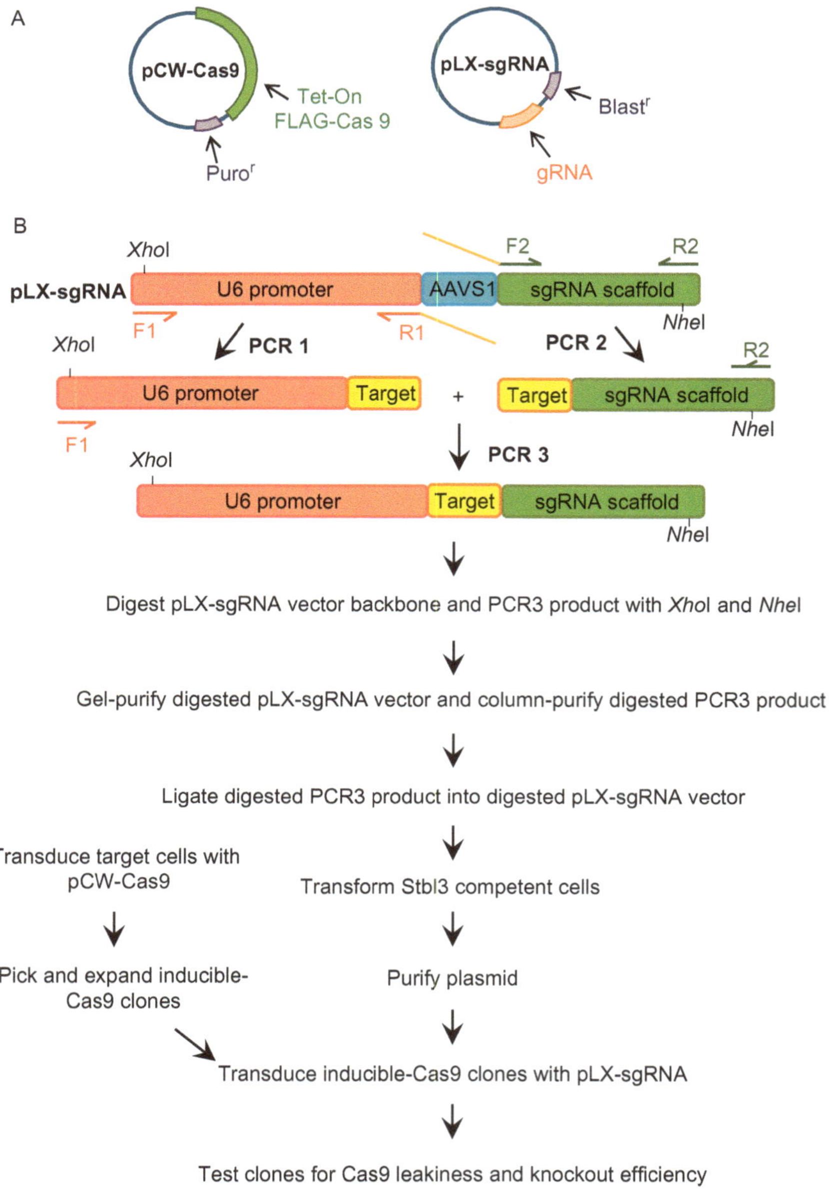

Fig. 6 (A) Vector maps of pCW-Cas9 and pLX-sgRNA constructs. *Puro^r*, puromycin-resistance gene; *Blast^r*, blasticidin resistance gene; gRNA, guide RNA sequence. (B) Flowchart outlining the steps for cloning of target gRNA sequences into the pLX-sgRNA vector and for generation of cells inducible for ATG deletion. *F1*, forward primer 1; *F2*, forward primer 2; *R1*, reverse primer 1; *R2*, reverse primer 2; *AAVS1*, gRNA targeting sequence against *AAVS1* gene; *PCR*, polymerase chain reaction.

the first base may be replaced with a "G" although this may potentially decrease targeting efficiency. Alternatively, other gRNAs that begin with a "G" can be tested.

For comparison with the constitutive lentiCRISPR system, the same gRNA sequences targeting *ATG5* or *ATG7* generated from the CRISPR design software in Section 2.1 were used to design the oligos for cloning into the pLX-sgRNA vector. The first base for gRNA sequences ATG5-1, ATG5-2, and ATG7-2 does not start with a "G" and was changed to a "G" as follows:

Human ATG5:

(1) GACTTGTTTCACGCTATATC

(2) GAGAGTAAGTTATTTGACGT

Human ATG7:

(1) GAAGCTGAACGAGTATCGGC

(2) GACTCCAATGTTAAGCGAGC

Cloning of each target gRNA sequence into the pLX-sgRNA vector requires four primers (Fig. 6B)—Primers F2 and R1 are specific for each gRNA sequence, and Primers F1 and R2 are used for all gRNA cloning in this system. The primer designs, which incorporate the target sequences for ATG5 or ATG7 and specific overhangs for pLX-sgRNA cloning are listed in Fig. 7 (target sequence in bold). Standard desalted oligos are sufficient for the cloning procedure and should be reconstituted to a working stock of 100 μ*M*.

3.1.2 Cloning of Guide RNA Sequence Targeting ATG5 or ATG7 Into pLX-sgRNA

The cloning of the gRNA sequence into pLX-sgRNA is detailed below:

(i) Perform PCR reaction 1 (with F1 and R1 primers) and PCR reaction 2 (with F2 and R2 primers):

(PCR Hotstar High Fidelity kit, Qiagen, Manchester, UK, #202602)

25 ng	pLX-sgRNA plasmid
5 μL	5 × Buffer
2.5 μL	$MgSO_4$ (25 μ*M*)
0.5 μL	F1 or F2 primer (100 μ*M*)
0.5 μL	R1 or R2 primer (100 μ*M*)
0.5 μL	Hotstar HiFi polymerase
X μL	ddH_2O
25 μL	Total reaction volume

Guide RNA oligo design template for pLX-sgRNA

F1: AAA*CTCGAG*TGTACAAAAAAGCAGGCTTTAAAG
R1: **rc(GN_{19})**GGTGTTTCGTCCTTTCC
F2: **GN_{19}**GTTTTAGAGCTAGAAATAGCAA
R2: AAA*GCTAGC*TAATGCCAACTTTGTACAAGAAAGCTG

GN_{19}= target sequence, rc(GN_{19}) = reverse complement of target sequence

Human ATG5-1 F2
GACTTGTTTCACGCTATATCGTTTTAGAGCTAGAAATAGCAA
Human ATG5-1 R1
GATATAGCGTGAAACAAGTCGGTGTTTCGTCCTTTCC

Human ATG5-2 F2
GAGAGTAAGTTATTTGACGTGTTTTAGAGCTAGAAATAGCAA
Human ATG5-2 R1
ACGTCAAATAACTTACTCTCGGTGTTTCGTCCTTTCC

Human ATG7-1 F2
GAAGCTGAACGAGTATCGGCGTTTTAGAGCTAGAAATAGCAA
Human ATG7-1 R1
GCCGATACTCGTTCAGCTTCGGTGTTTCGTCCTTTCC

Human ATG7-2 F2
GACTCCAATGTTAAGCGAGCGTTTTAGAGCTAGAAATAGCAA
Human ATG7-2 R1
GCTCGCTTAACATTGGAGTCGGTGTTTCGTCCTTTCC

Control F2
GTAGCGAACGTGTCCGGCGTGTTTTAGAGCTAGAAATAGCAA
Control R1
ACGCCGGACACGTTCGCTACGGTGTTTCGTCCTTTCC

Fig. 7 gRNA oligo design template for cloning of gRNA targeting sequences into pLX-sgRNA vector, and the list of control gRNA oligos and gRNA oligos targeting human ATG5 and ATG7.

PCR conditions:

Step 1	95°C 5 min
Step 2	94°C 30 s 58°C 1 min 72°C 30 s Repeat step (2) for 27 cycles before proceeding to Step (3)
Step 3	72°C 10 min
Step 4	4°C hold

(ii) Run both PCR products on an agarose gel and gel-purify the respective bands—PCR 1 will generate a product of 344 bp, whereas PCR 2 will generate a product of 137 bp.

(iii) Perform PCR reaction 3 using purified products from PCR reactions 1 and 2 as starting templates:

0.5 μL	Purified PCR 1 Product
0.5 μL	Purified PCR 2 Product
5 μL	5× buffer
2.5 μL	$MgSO_4$ (25 μ*M*)
0.5 μL	F1 primer (100 μ*M*)
0.5 μL	R2 primer (100 μ*M*)
0.5 μL	Hotstar HiFi polymerase
15 μL	ddH_2O
25 μL	Total reaction volume

PCR conditions:

Step 1	95°C 5 min
Step 2	94°C 30 s 58°C 1 min 72°C 40 s Repeat step (2) for 29 cycles before proceeding to Step (3)
Step 3	72°C 10 min
Step 4	4°C hold

(iv) Gel purify the product from PCR reaction 3 (product size is 462 bp). Elute in 33 μL ddH_2O.

(v) Digest pLX-sgRNA vector and purified PCR 3 product with *Nhe*I and *Xho*I (overnight at 37°C):

	pLX-sgRNA	**PCR 3 Product**
pLX-sgRNA	3 μg	—
Purified PCR 3 product	—	30 μL
Buffer 2.1 (NEB, Herts, UK)	5 μL	5 μL
*Nhe*I (NEB, Herts, UK)	2 μL	2 μL
*Xho*I (NEB, Herts, UK)	2 μL	2 μL
ddH_2O	X μL	11 μL
Total reaction volume	50 μL	50 μL

(vi) The next day, dephosphorylate the digested pLX-sgRNA vector (30 min at 37°C):

50 μL	Digested lentiCRISPR reaction mix
3 μL	Shrimp alkaline phosphatase (Promega, Southampton, UK)
53 μL	Total reaction volume

(vii) Subsequently, run the digested and dephosphorylated pLX-sgRNA reaction mix on an agarose gel and gel-purify the larger band (≈7 kb). Quantify the purified product with a spectrophotometer.

(viii) Column-purify the digested PCR 3 product using the QIAquick PCR Purification Kit (Qiagen, Manchester, UK, #28104). Quantify the purified product with a spectrophotometer.

(ix) Set up a ligation reaction and incubate at room temperature for 2 h using the Rapid DNA Ligation kit (Roche, Burgess Hill, UK, #11 635 379 001). For comparison, set up a negative control ligation reaction with the digested and dephosphorylated pLX-sgRNA vector alone and ddH_2O in place of the PCR 3 product insert.

X μL	Digested pLX-sgRNA plasmid from Step (vii) (50 ng)
X μL	Digested PCR 3 product from Step (viii) (125 ng)
2 μL	5× DNA dilution buffer
X μL	ddH_2O (10 μL subtotal reaction volume)
10 μL	2× T4 DNA ligation Buffer
1 μL	T4 DNA ligase
21 μL	Total reaction volume

(x) Transform Stbl3 competent cells (Thermo Fisher Scientific, Paisley, UK, #C7373-03) with 1 μL of ligation reaction.

(xi) Plate transformed Stbl3 cells onto ampicillin-containing agar plates and incubate overnight at 37°C.

(xii) Pick several colonies for each gRNA cloning and set up mini-cultures. Purify plasmid and sequence using the forward primer 5′-CGGGTTTATTACAGGGACAGCAG-3′ or reverse primer 5′-TACCAGTCAATCTTTCACAAATTTTGT-3′ to confirm successful gRNA cloning.

3.1.3 Generation of Cells With Inducible-Disruption of Autophagy

The generation of cells incorporating the inducible-CRISPR/Cas9 system for regulated ATG deletion is carried out in two stages (Fig. 6B)—(1) The generation of clones expressing the inducible-Cas9 followed by (2) the expression of the target gRNA in the inducible-Cas9 clones.

The inducible-Cas9 clones are established by expressing the pCW-Cas9 construct in target cells using the lentiviral infection protocol detailed in Section 2.3. Briefly, the pCW-Cas9 construct is transfected into HEK 293T packaging cells for production of lentivirus encoding the inducible-Cas9 nuclease. The lentivirus-containing media from HEK 293T cells is subsequently transferred onto the target cell line to generate cells expressing the inducible-Cas9. After puromycin selection, individual clones stably expressing the inducible-Cas9 are picked from the heterogeneous population and expanded. Subsequently, the different clones are then transduced with virus encoding the ATG-targeting gRNA, produced by HEK 293T cells transfected with the newly cloned pLX-ATGsgRNA construct from Section 3.1.2. After blasticidin selection (2–10 μg/mL, depending on the sensitivity of target cell lines) for 10 days (or until all cells in the negative control plate have died), the cells are treated with Doxycycline (Dox) to express Cas9 and induce ATG deletion. Dox is replenished every 2 days for 2 weeks, after which the cells are harvested and analyzed by SDS-PAGE and western blotting for ATG deletion.

3.1.4 Comparison of Autophagy Inhibition Between Inducible-shRNA and Inducible-CRISPR/Cas9 Targeting ATGs

To compare the knockdown efficiency between the inducible-CRISPR/Cas9 system and the inducible-shRNA system, we generated KP-4 cells expressing either the inducible-CRISPR/Cas9 lentiviral system targeting ATG5 or ATG7, or the inducible Dharmacon™TRIPZ™ lentiviral shRNA system targeting ATG5 or ATG12 (Fig. 8A and B). KP-4 cells expressing the inducible shRNA were established by expressing the pTRIPZ-ATG constructs (containing shRNA sequence targeting ATG5 or ATG12) in KP-4 cells using the lentiviral infection protocol detailed in Section 2.3.

Similar to the observations made with the siRNA-mediated knockdown of ATGs (Fig. 4C), analysis of multiple shRNAs targeting ATG5 or ATG12 revealed that we were unable to completely inhibit the conversion of LC3B-I to LC3B-II even after 19 days of Dox treatment using this approach (Fig. 8B). Conversely, Dox treatment (14 days) of cells expressing the inducible-CRISPR/Cas9 system targeting ATG5 or ATG7 resulted in

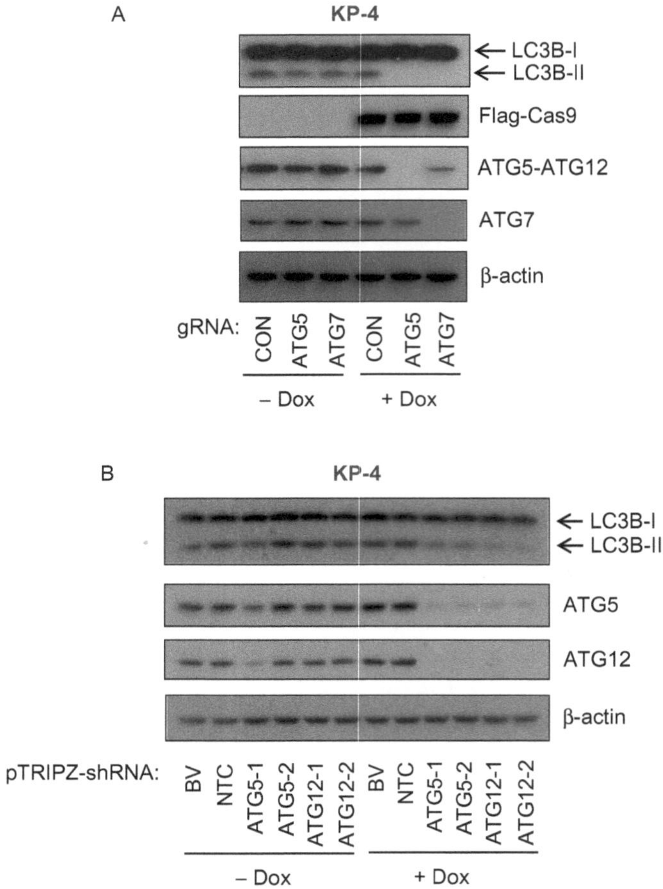

Fig. 8 (A) KP-4 cells expressing the inducible-CRISPR/Cas9 lentiviral system targeting ATG5 (ATG5-2 gRNA) or ATG7 (ATG7-2 gRNA) were treated with or without Dox (1 μg/mL) for 14 days (Dox replenished every 2 days) prior to harvesting and analysis by SDS-PAGE and western blotting. (B) KP-4 cells expressing the inducible Dharmacon™ TRIPZ™ lentiviral shRNA system targeting ATG5 [#RH4740-EG9474, Clone ID: 200702210 (ATG5-1) and 200774403 (ATG5-2)], ATG12 [#RH4740-EG9140, Clone ID: 200757071 (ATG12-1) and Clone ID: 200753541 (ATG12-2)], nontargeting control shRNA (NTC #RHS4743) or blank vector (BV #RHS4750) were treated with or without Dox (1 μg/mL) for 19 days (Dox replenished every 2 days) prior to harvesting and analysis by SDS-PAGE and western blotting. Antibodies used for western blotting: LC3B (Cell signaling, NEB, Herts, UK, #2775S), ATG5 (Cell signaling, NEB, Herts, UK, #12994S), ATG7 (Cell signaling, NEB, Herts, UK, #8558S), ATG12 (Abgent, San Diego, USA, #AP1816a), FLAG-M2 (Sigma, Gillingham, UK, #F3165), and β-actin (Abcam, Cambridge, UK, #ab8227).

the efficient deletion of the respective ATG proteins and, importantly, a complete inhibition of LC3B-II formation (Fig. 8A).

3.2 Establishment of the iCRISPR Tet-On Inducible CRISPR/Cas9 Nonviral System for the Regulated Disruption of Autophagy

As an alternative to the lentiviral-based inducible-CRISPR/Cas9 system, we also explored the use of a nonviral based system—the iCRISPR system developed by the Huangfu lab (Gonzalez et al., 2014; Zhu, Gonzalez, & Huangfu, 2014)—to target autophagy. This system was established and successfully tested for targeting of multiple genes in human pluripotent stem cells (Gonzalez et al., 2014; Zhu et al., 2014). In the iCRISPR platform, both the gRNA and hSpCas9 expression cassettes are cloned into one plasmid (Puro-iCr donor plasmid), which is then expressed together with the Neo-M2rtTA donor plasmid (Addgene, Cambridge, USA, plasmid #60843), carrying the constitutive reverse tetracycline transactivator (M2rtTA) expression module. To improve genomic integration in cells, a pair of TALENs (AAVS1-TALEN-L and AAVS1-TALEN-R, Addgene, Cambridge, USA, plasmids #59025 and 59026) are used to target the Puro-iCr donor and Neo-M2rtTA expression cassettes to the human AAVS1 locus, previously demonstrated to support robust and long-term transgene expression (Smith et al., 2008). The cloning of gRNA and Cas9 into the Puro-iCr donor, and the subsequent generation of an autophagy-targeting iCRISPR cell line are described in detail in Sections 3.2.1 and 3.2.2 and summarized in Figs. 9 and 10.

3.2.1 Design and Cloning of the Guide RNA Sequences Targeting ATG5 Into the iCRISPR System

One of the conveniences of using the iCRISPR system is that the gRNA oligos used for cloning the gRNA sequences into the lentiCRISPR system (Section 2.1) are compatible with the iCRISPR system. Hence, the same protocol used to design the gRNA oligos for the lentiCRISPR system (Section 2.1) applies to the iCRISPR system.

The cloning of the CRISPR target sequence into the iCRISPR system is carried out in two stages, according to the protocol developed by the Huangfu lab (Zhu et al., 2014)—(1) Cloning of the gRNA sequence into the piCRg entry vector containing the Cas9 expression cassette (Addgene, Cambridge, USA, plasmid #58904) and subsequently, (2) the transfer of the chimeric gRNA and Cas9 coding sequences into the Puro-iDEST plasmid

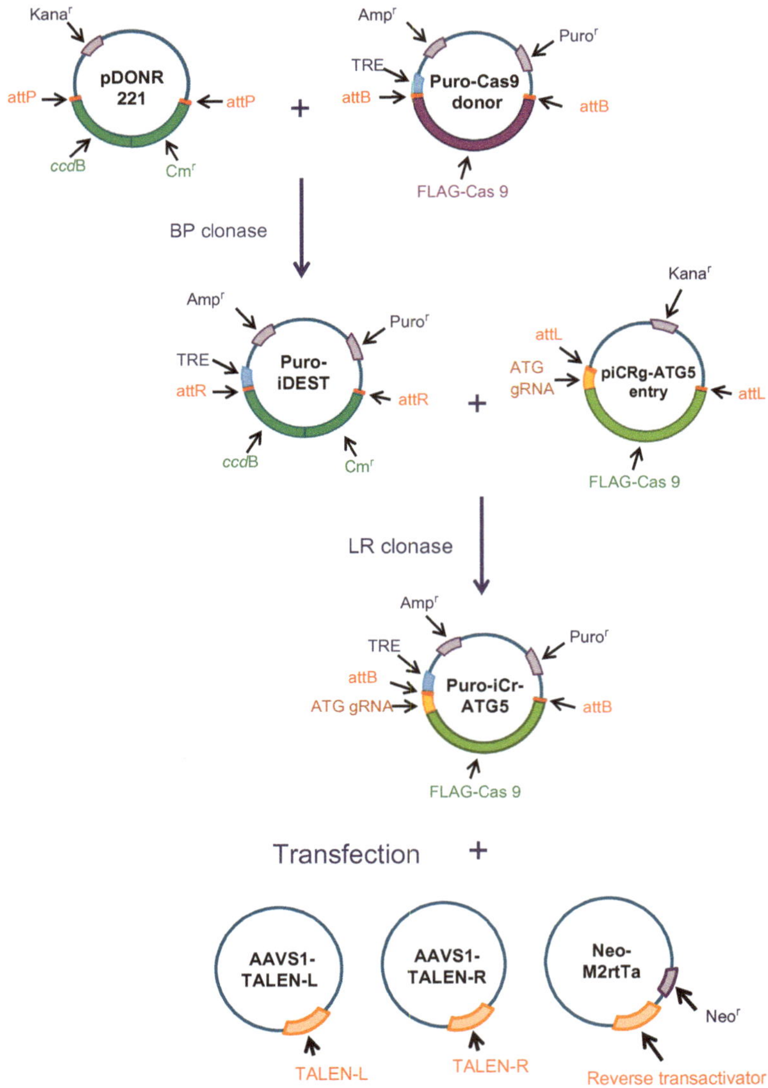

Fig. 9 (A) Generation of the Puro-iDEST destination vector by BP Clonase™ reaction between pDONR™221 and Puro-Cas9 Donor. Puro-iCr-ATG5-2 was subsequently generated by LR Clonase™ reaction between Puro-iDEST and piCRg-ATG5-2-entry plasmids. Puro-iCr-ATG5-2 is transfected into target cells along with Neo-M2rtTA, AAVS1-TALEN-L, and AAVS1-TALEN-R to generate iCRISPR-ATG5 expressing cells. *Kana^r*, kanamycin-resistance gene; *Cm^r*, chloramphenicol-resistance gene; *Puro^r*, puromycin-resistance gene; *Amp^r*, ampicillin-resistance gene; *Neo^r*, neomycin-resistance gene; *ccdB*, *ccdB* gene; *TRE*, tetracycline response element; *attP*, Gateway®attP sequence; *attB*, Gateway®attB sequence; *attR*, Gateway®attR sequence; *attL*, Gateway®attL sequence; *gRNA*, guide RNA sequence.

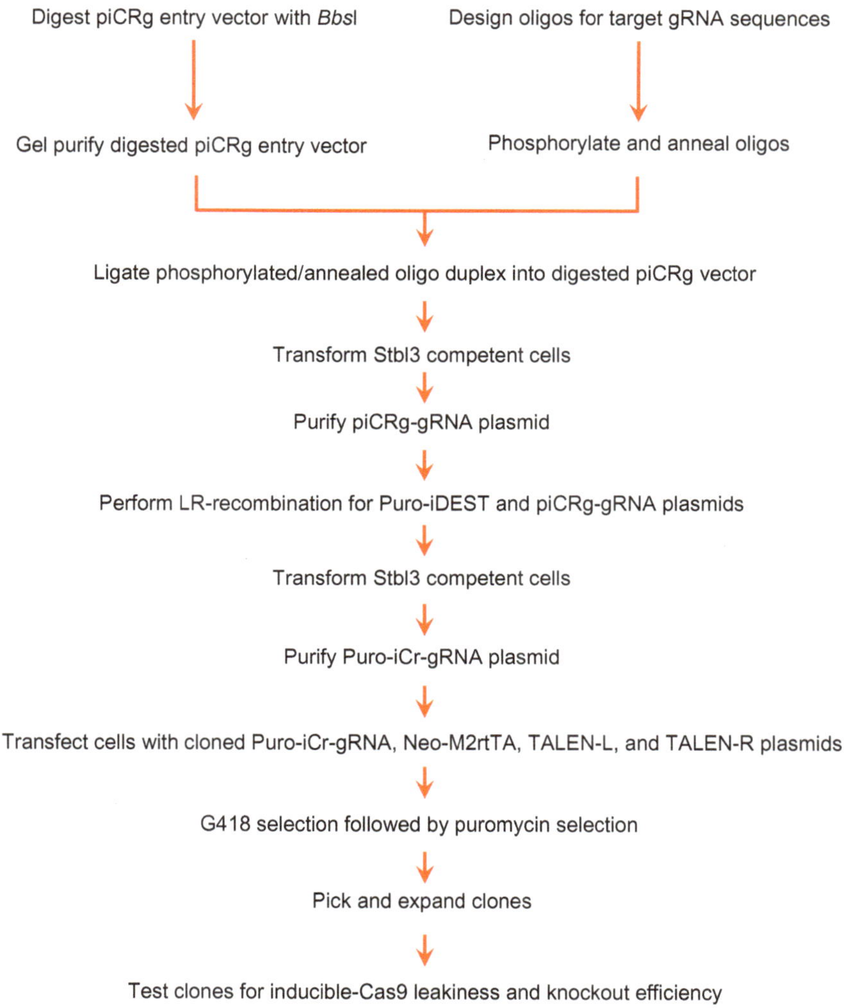

Fig. 10 Flowchart outlining the steps for cloning target gRNA sequences into the piCRg-Entry and Puro-iDEST vectors and for generation of iCRISPR-ATG5-2 cells.

through LR recombination reaction (Gateway® LR Clonase™ II Enzyme Mix, Thermo Fisher Scientific, Paisley, UK) to generate the Puro-iCr donor plasmid. These cloning steps are summarized in flowcharts in Figs. 9 and 10.

(1) Cloning of the *ATG5*-targeting gRNA sequence into the piCRg entry vector to generate piCRg-ATG5 entry plasmid.

The gRNA sequence targeting *ATG5* is cloned into the piCRg entry vector in the same way as the lentiCRISPR system (described in Section 2.2) with the exception of Step (i)—the piCRg entry vector is digested with *BbsI* instead of *BsmB*I at 37°C for 2 h. The annealing, phosphorylation, and subsequent ligation of the gRNA oligos to the *Bbs*I-digested piCRg entry vector is performed as described for the lentiCRISPR system in Section 2.2.

(2) Transfer of gRNA and Cas9 coding sequences from piCRg-ATG5 entry plasmid into Puro-iDEST through LR recombination.

The Puro-iDEST destination plasmid can be constructed as recommended by the Huangfu lab (Zhu et al., 2014) and illustrated in Fig. 9. Briefly, the Puro-iDEST plasmid is generated by removing the Cas9 expression cassette from Puro-Cas9 Donor (Addgene, Cambridge, USA, plasmid #58409) and replacing it with the Gateway destination cassette from the Gateway® pDONR™221 vector (Thermo Fisher Scientific, Paisley, UK) through a BP reaction (Gateway® BP Clonase™ II Enzyme Mix, Thermo Fisher Scientific, Paisley, UK). This creates the attR-containing Puro-iDest plasmid required for subsequent LR reaction with the attL-containing piCRg-ATG5 entry clone.

The BP reaction set up:

X μL	Puro-Cas9 Donor (150 ng)
X μL	Gateway® pDONR™221 (150 ng)
To 8 μL	TE buffer, pH 8.0
2 μL	BP Clonase™ II enzyme mix
10 μL	Total reaction volume

Incubate the reaction at 25°C for 1 h before adding 1 μL of Proteinase K to terminate the reaction. Subsequently, incubate samples at 37°C for 10 min. Transform One Shot® *ccd*B Survival™ 2T1R competent cells with 1 μL of the reaction mix and plate onto an agar plate containing both ampicillin and chloramphenicol.

To construct the Puro-iCr-ATG5-2 CRISPR expression plasmid, the chimeric gRNA and Cas9 expression sequences in the piCRg-ATG5-2-entry plasmid are transferred into Puro-iDEST through a LR recombination reaction:

X μL	piCRg-ATG5-2-entry (150 ng)
X μL	Puro-iDEST (150 ng)
To 8 μL	TE buffer, pH 8.0
2 μL	LR Clonase™ II Enzyme mix
10 μL	Total reaction volume

Incubate the reaction at 25°C for 1 h before adding 1 μL of Proteinase K to terminate the reaction. Subsequently, incubate samples at 37°C for 10 min. Transform Stbl3 competent cells with 1 μL of the reaction mix and plate onto ampicillin-containing agar plates.

3.2.2 Generation of Cells With Inducible Disruption of Autophagy Using the iCRISPR System

To generate iCRISPR-ATG5-2 expressing cells, KP-4 cells (5×10^5 cells in a 10 cm dish) are transfected with Puro-iCr-ATG5-2, Neo-M2rtTA, AAVS1-TALEN-L, and AAVS1-TALEN-R using Lipofectamine 2000 according to the manufacturer's instructions (Thermo Fisher Scientific, Paisley, UK). The ratio of plasmids used is as follows:

4 μg Puro-iCr-ATG5-2
4 μg Neo-M2rtTA
0.5 μg AAVS1-TALEN-L
0.5 μg AAVS1-TALEN-R

24 h after transfection, G418 (600 μg/mL) selection medium is added to cells. The cells are grown in G418 (replenished twice a week) until resistant colonies emerge and all the cells in the negative control plates die (approximately 12 days selection). The cells are then trypsinized and replated sparsely onto 15 cm dishes for puromycin selection (1 μg/mL). Resistant clones that emerge after 1 week of puromycin selection (replenished every 2 days) are picked, expanded and tested for ATG5-knockout efficiency.

Fig. 11 shows two different KP-4 iCRISPR-ATG5-2 clones generated through this protocol. Treatment of these clones with Dox for 14 days efficiently reduced ATG5 levels and inhibited conversion of LC3B-I to LC3B-II when compared to Dox-treated parental KP-4 cells. In an ideal setting, a system involving Dox treatment of a cell line containing Cas9 and a control gRNA would be more appropriate and we suggest researchers use such a control in their studies. This would then enable assessment of potential off-target effects from not just Dox treatment but also Cas9 expression.

4. DISCUSSION

In this chapter, we introduced several CRISPR/Cas9 strategies that we used successfully to restrict autophagy in cells. We were able to efficiently suppress autophagy either by constitutive deletion of *ATG* genes using the

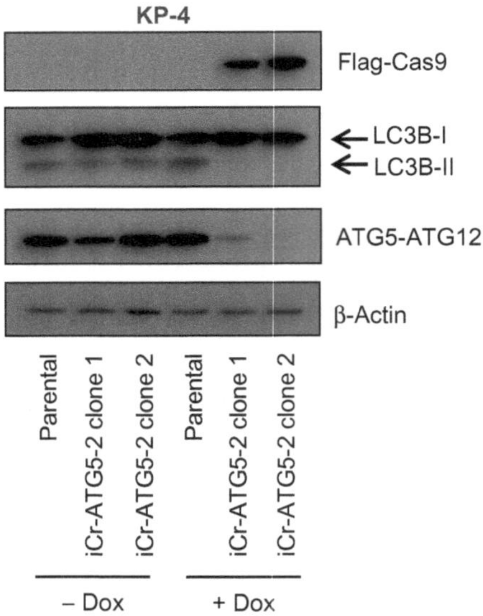

Fig. 11 Parental KP-4 cells and two different iCRISPR-ATG5-2-expressing KP-4 clones were treated with or without Dox (1 μg/mL) for 14 days (Dox replenished every 2 days) prior to harvesting and analysis by SDS-PAGE and western blotting. Antibodies used for western blotting: LC3B (Cell signaling, NEB, Herts, UK, #2775S), ATG5 (Cell signaling, NEB, Herts, UK, #12994S), FLAG-M2 (Sigma, Gillingham, UK, #F3165), and β-actin (Abcam, Cambridge, UK, #ab8227).

lentiCRISPR system, or by regulated deletion of *ATG* genes with the inducible-CRISPR/Cas9 systems.

Although the use of the CRISPR/Cas9 technology has revolutionized the efficiency and ease of gene editing, it is important to note that this technology is not faultless. One of the main concerns of the CRISPR/Cas9 system, which is also an issue common with many techniques, is the possibility of off-target activity. CRISPR-Cas9 nucleases may generate off-target mutations at sites that resemble the on-target sequence, giving rise to misleading or unintended phenotypes. In addition to the list of gRNA sequences, the CRISPR design tool at http://crispr.mit.edu/ also provides a list of potential off-target cleavage sites for each gRNA design. We recommend screening of the predicted off-target sites by sequencing to check for any undesired mutations. There are several ways to address the issue of off-target mutations, one of which has been undertaken in our study—to design and test multiple gRNA sequences targeting the same gene or pathway. Additionally, where clonal selection is required, the use of several clones is also recommended to limit the possibility that phenotypes are detected through clonal variation in the population as opposed to CRISPR-targeting of a specific gene. Modified versions of Cas9, which have been reported to exhibit higher specificity, are also currently available. These include the Cas9 D10A mutant (or Cas9 Nickase) with a mutation at the HNH nuclease domain of Cas9 (Ran et al., 2013), the eSpCas9 mutants with alanine substitutions within the HNH/RuvC groove (Slaymaker et al., 2016), and the Cas9-HF1 (Cas9-High-fidelity version 1) with four base substitutions (N497, R661, Q695, Q926) shown to reduce nonspecific DNA contact (Kleinstiver et al., 2016). Rescue experiments performed in CRISPR-targeted cells may also be considered as an additional way to substantiate any observed effects.

We have also encountered issues with leakiness of the inducible-CRISPR/Cas9 system, whereby ATG knockdown occurred even in the absence of the inducing agent, Dox. This strongly reflects the highly sensitive and irreversible nature of the CRISPR technology such that even a weak induction of Cas9 in the presence of the targeting gRNA is sufficient to cause genome editing. Leakiness may be a consequence of the integration of the inducible-CRISPR/Cas9 expression cassette in to genomic regions enriched with enhancer elements, which can be verified using commercially available integration site analysis kits. This can simply be overcome by picking and testing more clones to identify those with a tightly regulated inducible-CRISPR/Cas9. Additionally, the use of tetracycline-free serum in the culture media may alleviate the leakiness. Newer generations of Tet-On systems that exhibit tighter control of transgene expression are also available. The cloning of the Cas9 expression cassette into these systems may address this problem.

Although the CRISPR/Cas9 approach produces a more robust and sustained inhibition of autophagy when compared to conventional RNA interference methods, both technologies are equally useful for autophagy research. The CRISPR/Cas9 system generates irreversible genetic alterations, which may not be ideal for experiments requiring temporary and partial reduction of gene expression. Additionally, prolonged disruption of autophagy by CRISPR may result in a loss of cell viability or the long-term adaptation of cells to the lack of autophagy, which can complicate experimental analyses. On the contrary, some phenotypes may only present when a complete inhibition of the pathway (i.e., with CRISPR-mediated knockout) is achieved. The CRISPR/Cas9 system is also an inexpensive option for long-term autophagy studies compared to the use of siRNAs.

Furthermore, in light of recent studies showing the autophagy-independent effects of pharmacological inhibitors commonly used to modulate the autophagy pathway (Maes et al., 2014; Maycotte et al., 2012), the CRISPR/Cas9 technique will serve as a better and more specific tool for use in future autophagy studies. Ultimately, the choice of technology used depends on the nature of the experiments.

We believe that the CRISPR/Cas9 technology is a useful tool for autophagy research, one which provides a complete and specific inhibition of the pathway that is not achieved by conventional RNA interference techniques and pharmacological inhibitors, respectively. With recent advances in the use of the CRISPR/Cas9 system for in vivo gene editing (Dow et al., 2015), we can also look forward to exploring the use of CRISPR/Cas9 for autophagy targeting in vivo in the near future.

ACKNOWLEDGMENTS

This work was supported by Cancer Research UK and Astellas Pharma Inc.

Conflict of Interest. The authors declare no conflict of interest.

REFERENCES

Choi, A. M., Ryter, S. W., & Levine, B. (2013). Autophagy in human health and disease. *The New England Journal of Medicine*, *368*(7), 651–662. http://dx.doi.org/10.1056/NEJMra1205406.

Cong, L., Ran, F. A., Cox, D., Lin, S., Barretto, R., Habib, N., … Zhang, F. (2013). Multiplex genome engineering using CRISPR/Cas systems. *Science*, *339*(6121), 819–823. http://dx.doi.org/10.1126/science.1231143.

Deveau, H., Garneau, J. E., & Moineau, S. (2010). CRISPR/Cas system and its role in phage-bacteria interactions. *Annual Review of Microbiology*, *64*, 475–493. http://dx.doi.org/10.1146/annurev.micro.112408.134123.

Dow, L. E., Fisher, J., O'Rourke, K. P., Muley, A., Kastenhuber, E. R., Livshits, G., … Lowe, S. W. (2015). Inducible in vivo genome editing with CRISPR-Cas9. *Nature Biotechnology*, *33*(4), 390–394. http://dx.doi.org/10.1038/nbt.3155.

Gonzalez, F., Zhu, Z., Shi, Z. D., Lelli, K., Verma, N., Li, Q. V., & Huangfu, D. (2014). An iCRISPR platform for rapid, multiplexable, and inducible genome editing in human pluripotent stem cells. *Cell Stem Cell*, *15*(2), 215–226. http://dx.doi.org/10.1016/j.stem.2014.05.018.

Horvath, P., & Barrangou, R. (2010). CRISPR/Cas, the immune system of bacteria and archaea. *Science*, *327*(5962), 167–170. http://dx.doi.org/10.1126/science.1179555.

Jiang, P., & Mizushima, N. (2014). Autophagy and human diseases. *Cell Research*, *24*(1), 69–79. http://dx.doi.org/10.1038/cr.2013.161.

Jinek, M., Chylinski, K., Fonfara, I., Hauer, M., Doudna, J. A., & Charpentier, E. (2012). A programmable dual-RNA-guided DNA endonuclease in adaptive bacterial immunity. *Science*, *337*(6096), 816–821. http://dx.doi.org/10.1126/science.1225829.

Kleinstiver, B. P., Pattanayak, V., Prew, M. S., Tsai, S. Q., Nguyen, N. T., Zheng, Z., & Joung, J. K. (2016). High-fidelity CRISPR-Cas9 nucleases with no detectable genome-wide off-target effects. *Nature*, *529*(7587), 490–495. http://dx.doi.org/10.1038/nature16526.

Lin, Y. C., Kuo, H. C., Wang, J. S., & Lin, W. W. (2012). Regulation of inflammatory response by 3-methyladenine involves the coordinative actions on Akt and glycogen synthase kinase 3beta rather than autophagy. *Journal of Immunology, 189*(8), 4154–4164. http://dx.doi.org/10.4049/jimmunol.1102739.

Maes, H., Kuchnio, A., Peric, A., Moens, S., Nys, K., De Bock, K., … Carmeliet, P. (2014). Tumor vessel normalization by chloroquine independent of autophagy. *Cancer Cell, 26*(2), 190–206. http://dx.doi.org/10.1016/j.ccr.2014.06.025.

Maycotte, P., Aryal, S., Cummings, C. T., Thorburn, J., Morgan, M. J., & Thorburn, A. (2012). Chloroquine sensitizes breast cancer cells to chemotherapy independent of autophagy. *Autophagy, 8*(2), 200–212. http://dx.doi.org/10.4161/auto.8.2.18554.

Mizushima, N., Yoshimori, T., & Ohsumi, Y. (2011). The role of Atg proteins in autophagosome formation. *Annual Review of Cell and Developmental Biology, 27*, 107–132. http://dx.doi.org/10.1146/annurev-cellbio-092910-154005.

Ran, F. A., Cong, L., Yan, W. X., Scott, D. A., Gootenberg, J. S., Kriz, A. J., … Zhang, F. (2015). In vivo genome editing using Staphylococcus aureus Cas9. *Nature, 520*(7546), 186–191. http://dx.doi.org/10.1038/nature14299.

Ran, F. A., Hsu, P. D., Wright, J., Agarwala, V., Scott, D. A., & Zhang, F. (2013). Genome engineering using the CRISPR-Cas9 system. *Nature Protocols, 8*(11), 2281–2308. http://dx.doi.org/10.1038/nprot.2013.143.

Ravikumar, B., Sarkar, S., Davies, J. E., Futter, M., Garcia-Arencibia, M., Green-Thompson, Z. W., … Rubinsztein, D. C. (2010). Regulation of mammalian autophagy in physiology and pathophysiology. *Physiological Reviews, 90*(4), 1383–1435. http://dx.doi.org/10.1152/physrev.00030.2009.

Rosenfeldt, M. T., & Ryan, K. M. (2011). The multiple roles of autophagy in cancer. *Carcinogenesis, 32*(7), 955–963. http://dx.doi.org/10.1093/carcin/bgr031.

Rubinsztein, D. C., Marino, G., & Kroemer, G. (2011). Autophagy and aging. *Cell, 146*(5), 682–695. http://dx.doi.org/10.1016/j.cell.2011.07.030.

Sanjana, N. E., Shalem, O., & Zhang, F. (2014). Improved vectors and genome-wide libraries for CRISPR screening. *Nature Methods, 11*(8), 783–784. http://dx.doi.org/10.1038/nmeth.3047.

Shalem, O., Sanjana, N. E., Hartenian, E., Shi, X., Scott, D. A., Mikkelsen, T. S., … Zhang, F. (2014). Genome-scale CRISPR-Cas9 knockout screening in human cells. *Science, 343*(6166), 84–87. http://dx.doi.org/10.1126/science.1247005.

Slaymaker, I. M., Gao, L., Zetsche, B., Scott, D. A., Yan, W. X., & Zhang, F. (2016). Rationally engineered Cas9 nucleases with improved specificity. *Science, 351*(6268), 84–88. http://dx.doi.org/10.1126/science.aad5227.

Smith, J. R., Maguire, S., Davis, L. A., Alexander, M., Yang, F., Chandran, S., … Pedersen, R. A. (2008). Robust, persistent transgene expression in human embryonic stem cells is achieved with AAVS1-targeted integration. *Stem Cells, 26*(2), 496–504. http://dx.doi.org/10.1634/stemcells.2007-0039.

Sternberg, S. H., Redding, S., Jinek, M., Greene, E. C., & Doudna, J. A. (2014). DNA interrogation by the CRISPR RNA-guided endonuclease Cas9. *Nature, 507*(7490), 62–67. http://dx.doi.org/10.1038/nature13011.

Tanida, I., Mizushima, N., Kiyooka, M., Ohsumi, M., Ueno, T., Ohsumi, Y., & Kominami, E. (1999). Apg7p/Cvt2p: A novel protein-activating enzyme essential for autophagy. *Molecular Biology of the Cell, 10*(5), 1367–1379.

Wang, T., Wei, J. J., Sabatini, D. M., & Lander, E. S. (2014). Genetic screens in human cells using the CRISPR-Cas9 system. *Science, 343*(6166), 80–84. http://dx.doi.org/10.1126/science.1246981.

Wu, Y. T., Tan, H. L., Shui, G., Bauvy, C., Huang, Q., Wenk, M. R., … Shen, H. M. (2010). Dual role of 3-methyladenine in modulation of autophagy via different temporal patterns of inhibition on class I and III phosphoinositide 3-kinase. *The Journal of Biological Chemistry, 285*(14), 10850–10861. http://dx.doi.org/10.1074/jbc.M109.080796.

Yang, Y. P., Hu, L. F., Zheng, H. F., Mao, C. J., Hu, W. D., Xiong, K. P., … Liu, C. F. (2013). Application and interpretation of current autophagy inhibitors and activators. *Acta Pharmacologica Sinica*, *34*(5), 625–635. http://dx.doi.org/10.1038/aps.2013.5.

Yuan, N., Song, L., Zhang, S., Lin, W., Cao, Y., Xu, F., … Wang, J. (2015). Bafilomycin A1 targets both autophagy and apoptosis pathways in pediatric B-cell acute lymphoblastic leukemia. *Haematologica*, *100*(3), 345–356. http://dx.doi.org/10.3324/haematol.2014.113324.

Zhu, Z., Gonzalez, F., & Huangfu, D. (2014). The iCRISPR platform for rapid genome editing in human pluripotent stem cells. *Methods in Enzymology*, *546*, 215–250. http://dx.doi.org/10.1016/B978-0-12-801185-0.00011-8.

CHAPTER SEVEN

A Molecular Reporter for Monitoring Autophagic Flux in Nervous System In Vivo

K. Castillo[*,1], V. Valenzuela[†,‡,§], M. Oñate[†,‡,§], C. Hetz[†,‡,§,¶,||,1]
[*]Centro Interdisciplinario de Neurociencia de Valparaíso, Facultad de Ciencias, Universidad de Valparaíso, Valparaíso, Chile
[†]Biomedical Neuroscience Institute, Faculty of Medicine, University of Chile, Santiago, Chile
[‡]Center for Geroscience, Brain Health and Metabolism, Santiago, Chile
[§]Program of Cellular and Molecular Biology, Institute of Biomedical Sciences, Center for Molecular Studies of the Cell, University of Chile, Santiago, Chile
[¶]Buck Institute for Research on Aging, Novato, CA, United States
[||]Harvard School of Public Health, Boston, MA, United States
[1]Corresponding authors: e-mail addresses: karen.castillo@cinv.cl; chetz@med.uchile.cl; chetz@hsph.harvard.edu

Contents

Abstract

The relevance of autophagy in neuronal health has been extensively reported in a plethora of conditions affecting the nervous system, such as neurodegenerative diseases, cancer, diabetes, and tissue injury, where altered autophagic activity may contribute to the pathological process. Autophagy is a dynamic pathway involving the formation of a membrane surrounding and enclosing cargoes that are delivered to lysosomal compartments for degradation. Cargoes can include large protein aggregates,

Methods in Enzymology, Volume 588
ISSN 0076-6879
http://dx.doi.org/10.1016/bs.mie.2016.09.077

organelles, or even pathogens. Traditionally, autophagy assessment relies on the measurement of LC3-II protein levels or the visualization of LC3-positive puncta. However, these approaches represent a static measurement of autophagy markers, making difficult the dissection of the actual changes in the autophagy process (activation, inhibition, or no effects), due to the dynamic regulation of LC3 viral levels. To circumvent this limitation, we previously developed an adeno-associated vector (AAV) to deliver a molecular autophagy sensor to the neuronal compartment in vivo. Here, we describe the detailed design and methods to use an engineered AAV harboring the monomeric tandem mCherry-GFP-LC3 to determine autophagic fluxes in the nervous system. Key methodological details to succeed in the use of this reporter are provided.

1. INTRODUCTION

Macroautophagy (here referred to as autophagy) is a catabolic pathway involved in the degradation of cellular components such as damaged or superfluous organelles, abnormal protein aggregates, and other cytosolic components (Sica et al., 2015). Autophagy begins by the formation of an initiation membrane called phagophore which grows surrounding cargoes and fuses, forming double-membrane vesicles that engulf cargoes, called autophagosomes. Further fusion with lysosomes leads to autophagolysosome formation and cargo degradation (Green & Levine, 2014). Two ubiquitin-like conjugation pathways regulate autophagy (see detailed review in Feng, Yao, & Klionsky, 2015). One involves the covalent binding of Atg12 to Atg5, at the phagophore level, and its dissociation after autophagosome consolidation. The other mechanism implicates the conjugation of microtubule-associated protein 1 light chain 3 (LC3) to phosphatidylethanolamine to form LC3-II. Lipidated LC3 binds to the expanding phagophore and remains associated with autophagosomes even after fusion with lysosomes (Feng et al., 2015). Under resting conditions LC3 protein remains cytosolic, but upon autophagy induction it becomes lipidated and anchor to autophagosomal membranes, an association that remains virtual during the whole autophagic process up to cargo degradation in lysosomes. Monitoring LC3 turnover, LC3 subcellular distribution, or its flux through the autophagy pathway are the gold standards for measuring autophagy activity (Klionsky et al., 2016).

Accumulating evidence indicates that an altered balance between mechanisms that assemble the autophagic machinery and those involved in the degradation of cargoes generates defects in cellular homeostasis (Menzies, Fleming, & Rubinsztein, 2015; Schneider & Cuervo, 2014). Thus, maintaining a balanced integrity of the autophagy pathway is critical to sustain cell health. Dysregulation of autophagy results in accumulation of misfolded proteins, superfluous or damaged organelles, a phenomenon that could result in cellular death and disease (Sridhar, Botbol, Macian, & Cuervo, 2012). In fact, genetic inactivation of autophagy in the nervous system causes spontaneous neurodegeneration (Hara et al., 2006; Komatsu et al., 2006). Accordingly, an emerging pathological feature of neurodegenerative diseases is the alteration of the autophagy capacity, offering new targets for disease intervention (Vidal, Matus, Bargsted, & Hetz, 2014). This is particularly relevant for neurodegenerative diseases involving protein misfolding and abnormal aggregation like ALS, Alzheimer's disease, Huntington's disease, and Parkinson's disease. In addition, autophagy impairment has been proposed as a molecular hallmark of the aging process, the major risk factor to develop neurodegenerative diseases (Kaushik & Cuervo, 2015). Besides, autophagosome accumulation has been reported in other pathological conditions affecting the nervous system, including brain trauma, spinal cord injury, brain cancer, and stroke (see examples in Adhami et al., 2006; Kaza, Kohli, & Roth, 2012; Liu, Chen, Dietrich, & Hu, 2008; Sadasivan, Dunn, Hayes, & Wang, 2008).

In order to define optimal targets for disease intervention, it is essential to identify which specific steps of the autophagy pathway are altered in selected diseases. In general, analysis of autophagy levels in the nervous system has relied on the static measurement of LC3-II protein levels or its distribution using histological analysis (Matus, Valenzuela, & Hetz, 2014). Although these are very useful tools to determine possible alterations in autophagy levels, they do not allow discerning if an accumulation or decrease of LC3 content is due to alterations in its biosynthesis or clearance. To overcome this limitation, a dynamic autophagy flux sensor was developed consisting on tandem construct of mCherry (or RFP), coupled to GFP-LC3 (Kimura, Noda, & Yoshimori, 2007; Pankiv et al., 2007). This method is based on the inactivation of GFP fluorescence in acidic compartments, such as lysosomes (see Fig. 1A). Then, the incorporation of LC3 to autophagosomal membranes is monitored by quantifying the ratio between mCherry/RFP and GFP-positive puncta (Kimura et al., 2007; Pankiv et al., 2007). As illustrated in Fig. 1A, increased content of red puncta is indicative of augmented autophagic flux (delivery of cargo into lysosomes), whereas

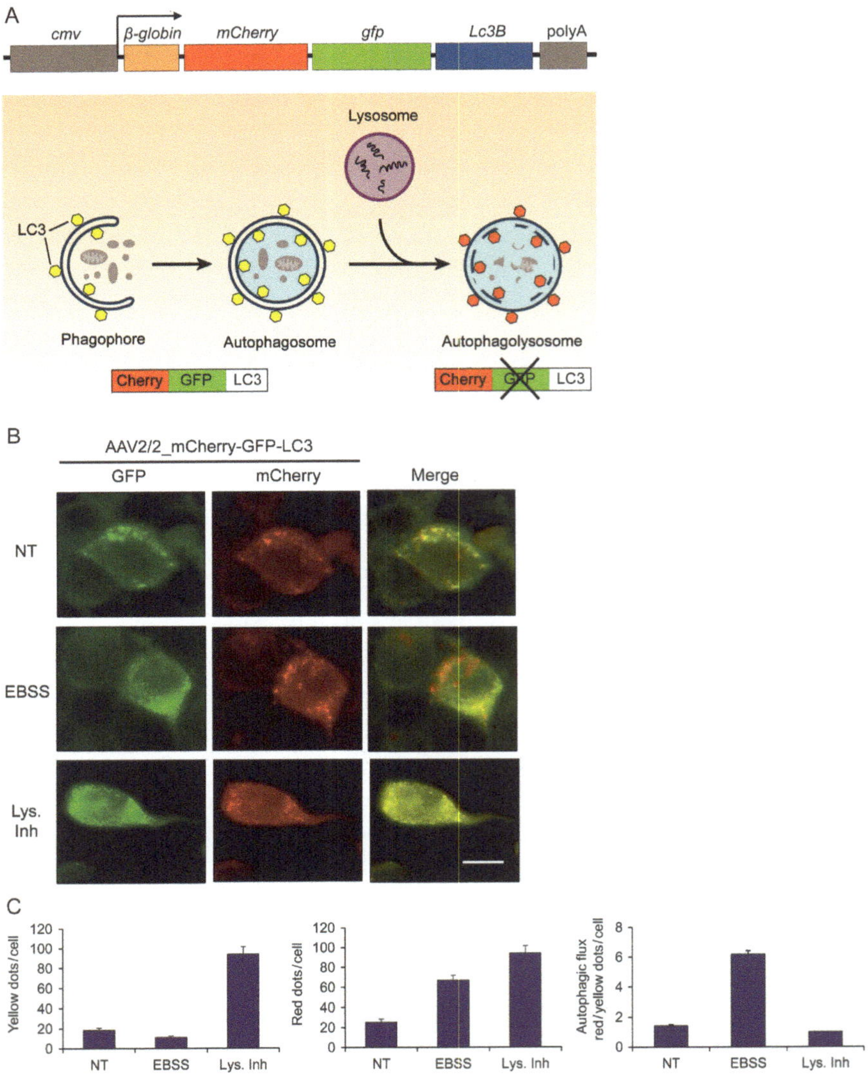

Fig. 1 Description and analysis of autophagy flux using AAV_mCherry-GFP-LC3. (A) Diagram of mCherry-GFP-LC3 construct used to engineer AAV2/2 vector with the autophagic reporter (*upper*). Upon activation of autophagy, LC3 binds to phagophore to form the autophagosome, where the fluorescence of mCherry-GFP-LC3 generates *yellow* puncta. Then, autophagosomes fuse with lysosomes where the acidic environment quenches the *green* GFP fluorescence, and *red dots* are visualized under fluorescence microscope as illustrated in compartment labeled in *red* (*lower*). (B) Representative images of NSC34 cells transduced with AAV2/2_mCherry-GFP-LC3 for 72 h, followed by exposure to EBSS media to induce autophagy by severe nutrient starvation. As control, cells were treated with a cocktail of lysosome inhibitors (bafilomycin, pepstatin, and E64D; Lys. Inh.) during 4 h to monitor basal autophagy flux in these cells. *NT*, nontreated. (C) Quantification of representative pictures showing the functionality of the sensor. Manual counting of *red* and *yellow dots* was performed. Autophagic flux is estimated as the ratio between *red* and *yellow dots*.

yellow puncta (a merge of green and red channels) are indicative of autophagosome formation and accumulation.

Several methods have been developed to monitor autophagy flux in vivo, highlighting the use of lysosomal inhibitors to monitor LC3-II accumulation in different tissues like liver and muscle (Esteban-Martinez & Boya, 2015; Haspel et al., 2011; Iwai-Kanai et al., 2008), or the use of transgenic flies to express LC3-GFP or LC3-mCherry (Nagy, Varga, Kovacs, Takats, & Juhasz, 2015; Nezis et al., 2010). However, due to poor blood–brain barrier permeability, pharmacological approaches are not suitable to measure LC3 fluxes in the nervous system. We recently developed a strategy to deliver a mCherry-GFP-LC3 fusion construct to the central and peripheral nervous system of mice by the injection of AAVs (Castillo et al., 2013). Using this method, we were able to measure autophagy flux after pharmacological induction of autophagy with rapamicyn or trehalose. In addition, we succeeded in determining autophagy flux in neuropathological settings such as sciatic nerve damage and spinal cord injury. Furthermore, we applied this tool to measure autophagic vesicle trafficking in axons ex vivo. Here, we provide a detailed protocol for the use of AAV_mCherry-GFP-LC3 for determining autophagy flux in the nervous system using mouse models, including intracerebroventricular (ICV) delivery of AAVs particles, histological details, and the image processing procedure needed to accurely quantify autophagy flux. We also discuss potential applications of this strategy to monitor autophagy in central and peripheral nervous system.

2. ASSESSING AUTOPHAGY FLUX IN VITRO

2.1 Design of AAV2_mCherry-GFP-LC3 Vectors

The whole mCherry-GFP-LC3 expression cassette was excised from a plasmid kindly provided by Terje Johansen's laboratory, and inserted into a shuttle proviral plasmid, containing both the AAV2/2 inverted terminal repeats and the 1.6-kb cytomegalovirus enhancer promoter (Fig. 1A). Serotype 2 was chosen because of a preferential tropism for neurons under our delivery method. In addition, different serotypes have been widely used for intraventricular injections of neonatal mice to obtain widespread transduction (along CNS) or specific transduction (cell-type directed) (Dirren et al., 2014; Gholizadeh, Tharmalingam, Macaldaz, & Hampson, 2013). Large-scale preparations of AAV can be obtained by triple transfection of 293T cells using a rep/cap plasmid and pHelper, followed by purification by column affinity chromatography as previously described (Anderson, Macdonald, Corbett, Whiteway, & Prentice, 2000). Viral titers can be quantified by real-time TaqMan PCR assay using primers that are specific for the bovine growth hormone polyadenylation (*bgh*-PolyA) sequence (Navarro, Serrano-Heras, Castano, & Solera, 2015). AAV preparations and quantifications can be obtained as a service in multiple core facilities and companies at a low cost.

2.2 Cell Culture and Transduction Verification

The first step is to corroborate the infectivity of the AAV preparation and confirm the expression of the autophagic reporter. In our previous experiments, we have chosen the motoneuron-like NSC34 cells, a hybrid cell line generated by fusion of neuroblastoma with mouse motoneuron-enriched primary spinal cord cells (Cashman et al., 1992). This cell line shows low basal levels of autophagy, representing a suitable tool for autophagy assessment after stimulation. However, although most of the laboratories test the activity of AAVs using 293T cells, we think that this tool can be applicable at a broad number of cell types. Indeed, we tested the reporter successfully in stable MEF cell lines, N2A, and in HEK cells (not shown). To use the reporter in cell lines, a plasmid containing the tandem mCherry-GFP-LC3 protein can also be used by regular transfection protocols. However, the AAV-mediated delivery of the autophagy flux sensor allows higher transduction levels comparing plasmid transfection. Therefore due to high transduction capacity of AAV-mediated gene delivery, we suggest this methodology to be used in several cell types and tissues, for other studies, such as cancer cell lines or tumors by local injections (Li et al., 2006). Additional to in vivo approaches, this sensor can be applied in primary culture and neuronal or glial cell lines (Lawlor, Bland, Mouravlev, Young, & During, 2009; Malik, Maronski, Dichter, & Watson, 2012), hopefully with a noticeable soma size, to ease puncta counting. It can be also applied in neuronal and/or glial primary culture, always considering promoter used to obtain cell-type-specific infection.

1. Cells are maintained in DMEM supplemented with 5% fetal bovine serum. 300,000 cells/mL are seeded in 6-well plates for Western blot analysis and 100,000 cell/mL for imaging analysis. After 24 h, cells are washed with filtered PBS, and media replaced. Cells are transduced with

a dilution 1:500 of AAV2/2 particles from 1×10^{13} DNAse-resistant particles (DRPs)/mL stock preparations. This dilution works in our experimental conditions, although we recommend to test the adequate concentration of viral particles to detect transduction for specific experimental demands.

2. After 72 h of transduction, cells are exposed to starvation, with nutrient-free EBSS media for 1, 4, or 6 h to induce autophagy (autophagic activation by nutrient deprivation).
3. As control, NSC34 cells can be treated with a lysosome inhibitor cocktail containing 200 n*M* bafilomycin A_1, 10 μg/mL protease inhibitors, 10 μg/mL pepstatin, and 10 μg/mL E64D. This treatment generates a blockage of fusion with lysosomes, leading to autophagic vesicle formation and LC3 accumulation.
4. At the end of the treatment, cells are placed on ice to slow down cellular activity and washed with cold sterile PBS. Total cell extracts are prepared in cold RIPA buffer (50 m*M* Tris, 150 m*M* NaCl, 0.1% SDS, 0.5% sodium deoxycholate, 1% Triton X-100), supplemented with a protease inhibitors cocktail, and then centrifuged at 12,500 rpm. Cell pellets are sonicated with a stem sonicator in a cold (~4°C) chamber. 20–50 μg of total proteins are loaded into 15% SDS-PAGE gels and LC3 electrophoretic pattern analyzed by Western blot, using standard methods (Hetz et al., 2009).
5. For imaging analysis, cells are washed with cold PBS, fixed with 4% PFA for 10 min, washed three times with sterile PBS, and then mounted with fluoromount for visualization by fluorescence microscopy. The presence of autophagic structures is augmented in response to starvation, and cells are populated mainly by red puncta, when dots are observed in merged channels (Fig. 1B, middle panel). This indicates that mCherry-GFP-LC3 successfully reached the lysosome, as GFP signal is absent. In contrast, cells treated with lysosome inhibitors exhibit an accumulation of yellow dots when merged images are visualized, suggesting that the flux is interrupted and cargoes are not delivered into the lysosomal compartment (Fig. 1B, lower panel). The ratio between the number of red and yellow dots per cell represents an estimation of the autophagic flux as shown in Fig. 1C.

Analysis of images is performed by counting puncta both manually or using an image software analysis like ImageJ (see below). Once the method is validated in vitro, experimental procedures are carried out in mice.

3. IN VIVO MEASUREMENTS OF AUTOPHAGIC FLUX

In this section we describe the procedures to deliver the mCherry-GFP-LC3 autophagy reporter by ICV injection of the AAV2/2_mCherry-GFP-LC3 construct followed by tissue processing and microscopy analysis. Detailed videos on the injection protocol can be observed in Glascock et al. (2011).

3.1 Materials

1. AAV2/2_mCherry-GFP-LC3 preparation aliquot on ice
2. 1 mL filtered PBS on ice
3. 1% FastGreen, used as dye to track the success of injection
4. Insulin syringe
5. Masking tape
6. P20 and P200 micropipette tips
7. Surgery board covered with aluminum paper. Generate a small pillow of aluminum paper on the board for the pup's head (Fig. 2A)
8. Parafilm
9. Cold light lamp
10. Fine marker pencil
11. Low adhesion microcentrifugue tubes to handle AAVs

Use all biosafety considerations to work with recombinant viruses such as suit, double gloves, chlorinated container for disposal of material in contact with AAVs, and biosafety cabin (see rules in https://web.stanford.edu/dept/EHS/prod/researchlab/bio/docs/Working_with_Viral_Vectors.pdf).

3.2 AAVs Preparation

1. Prepare a suitable volume of 0.08% FastGreen in sterile PBS (recommended final volume of 30–100 μL).
2. Add 2.5 μL of FastGreen dilution to 10 μL of AAV prepared from 1×10^{13} stock.
3. Injection loading volume is 2.5 μL per newborn pup.

3.3 Injection Procedure

1. Prepare a cold light lamp and the surgery board over the bench. Before the injection, place an ice pack over the surgery board to keep it cold.

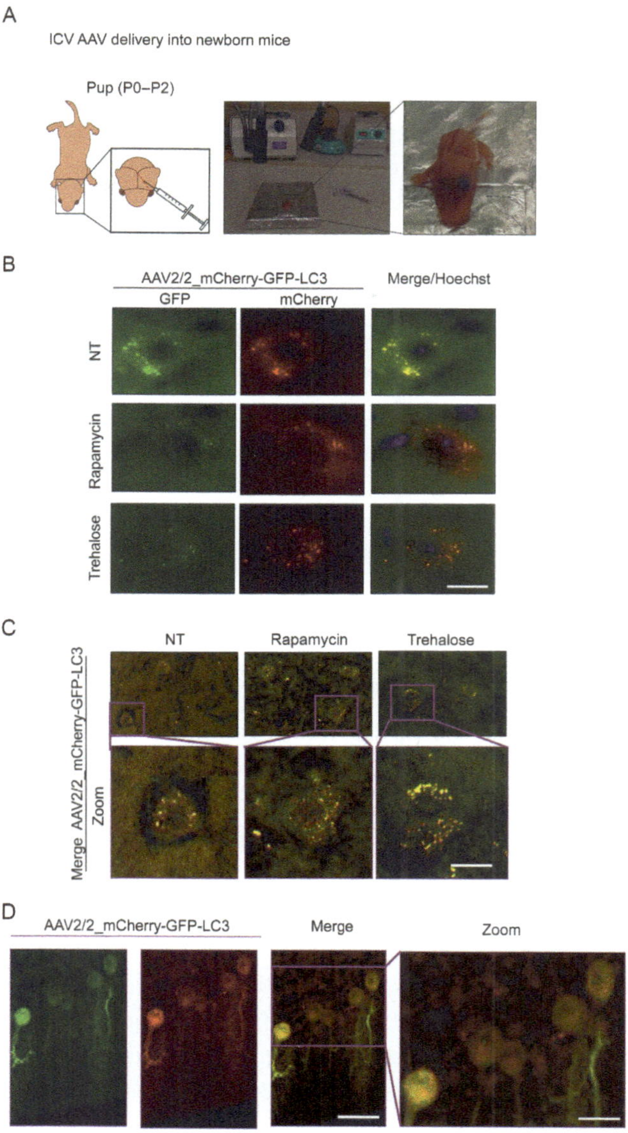

Fig. 2 Examples of tissues transduced with the AAV2/2_mCherry-GFP-LC3 after autophagy stimulation in vivo. (A) Scheme and pictures of the injection procedure employed to transduce neonatal mice with AAV-based autophagic reporter, by ICV injection. After anesthesia, animals are put on a cold aluminum-covered platform with a little pillow to accommodate the head of the animal. The place of injection is marked with a point. See text for details. (B) Representative spinal cord motoneurons images acquired in spinning disk microscope, for nontreated (NT), rapamycin-treated, and trehalose-treated animals. *Yellow* and *red dots* in merge channel are easily distinguishable. (C) Spinal cord motoneurons images acquired in a super-zoom microscope. A focal plane with a group of motoneurons is showed for animals treated with rapamycin and trehalose and non-treated animals (*upper*). In the *lower panel* a zoom of the indicated cell is showed. As in the previous case *yellow* and *red dots* are evident. (D) Representative images of Purkinje's cell of cerebellum from an animal treated with rapamycin, taken in a spinning disk microscope. A zoom of the merge image is provided to highlight autophagic dots in detail.

2. Cut a small piece of parafilm and place it besides the surgery table and draw a circle on it by using the marker (Fig. 2A). This is the place where the drop with 2.5 μL of the final vector preparation will be added and loaded into the syringe, and where the syringe needle will stand before the injection.
3. Obtain neonatal mice between P0 and P2. The manipulation of pups must be carried out carefully to avoid stress of the mother (it may kill the pups after the injections are done). A desirable option is to impregnate gloves with the smell of the bedding material (wood shaving in our facility) of the same cage before picking up the pup. To take out the pup from the litter, hit the cage in the opposite side of the nest to force the mother to leave it, and remove immediately only one pup at a time. Put the pup over a precooled aluminum paper over ice and wait until it completely stops moving as an anesthesia protocol (3–5 min approximately).
4. Add 2.5 μL of final AAVs preparation into the drawn circle of the parafilm paper and load the syringe slowly avoiding the generation of bubbles.
5. Place the pup over the surgery table (dorsal up) and hold it with two pieces of transparent masking tape, one over the back and the other over its nose. The face of the pup must look to the front of you (Fig. 2A).
6. Draw a dot on the head of the pup over lambda with a fine marker pencil, and another dot in the middle distance between lambda and the eyeball from the same side of the animal. This point corresponds to the site of ventricular AAV injection (see Fig. 2A).
7. Display the loaded syringe over the point of injection with the bezel pointing to the center. Turn the syringe approximately 10 degrees to the right (out) and the 10 degrees to you. Insert the syringe into the head between 2 and 3 mm. Inject the content fast and secure. Remove the syringe maintaining the angle and keeping a slight positive pressure. The diffusion of the dye through the ventricles is easily visualized as a defined dark region (because of dye) in the injection region. If the injection does not reach the ventricle, the staining will appear subcutaneously as a defined dark dot. In that case this animal should be discarted.
8. Immediately place the pup in a heat source (i.e., a regular heater) until it starts to open the mouth (after the first limbs movements are observed).

Then place the pup back into the cage in the opposite place of the nest, half buried in shaving wood. The mother will start looking after the pup. It is a good sign when the mother picks up the pup and put it back in the nest.

9. Eliminate all waste that had contact with the virus in a glass chlorinated container, followed by UV radiation of all materials used.

It is important to note that only pups in which the ventricle was filled up with the mix of viral preparation must be included in the future analysis. P0–P2 pups can receive one injection of 2 μL AAV2/2_ mCherry-GFP-LC3 into the cerebral lateral ventricle of one of the two hemispheres (regularly their left hemispheres). However, if it is necessary, a second injection can be performed on a consecutive day. The AAV solution is injected at full strength with a total concentration of 1×10^{13} DRPs/mL. This method of AAV-mediated transgene delivery should allow massive transduction of the nervous system after 2 weeks of injection, as reported before (Ayers et al., 2015; Dirren et al., 2014; Glascock et al., 2011; Gong et al., 2015; McLean et al., 2014; Passini & Wolfe, 2001).

Here, we present a protocol for free-hand delivery of AAV transgene. Besides, the AAV delivery can be supported by a stereotaxic apparatus. Both cases, including a nontoxic color dye in the injection volume, help to visualize the intraventricular injection site immediately after injection to check delivery accuracy. As the AAV2 vectors exhibit broad tropism, it is expectable to observe prolonged and widespread similar patterns of transduction and nervous system expression by these two approaches of AAV delivery, into newborn animals (for additional details of stereotaxic delivery, see Kim, Grunke, & Jankowsky, 2016).

4. PHARMACOLOGICAL INDUCTION OF AUTOPHAGY

To control the ability to monitor autophagy fluxes in the nervous system, we recommend performing positive controls with a pharmacological approach. To this aim, rapamycin or trehalose could be administrated through intraperitoneal (IP) injections (see examples in Castillo et al., 2013; Davies, Sarkar, & Rubinsztein, 2006; Erlich, Alexandrovich, Shohami, & Pinkas-Kramarski, 2007; Liu et al., 2008; Ravikumar et al., 2004; Rodriguez-Navarro et al., 2010; Sarkar, Ravikumar, Floto, &

Rubinsztein, 2009; Sekiguchi, Kanno, Ozawa, Yamaya, & Itoi, 2012; Zhang et al., 2012). Many other drugs are also available to stimulate autophagy in the brain (Rubinsztein, Codogno, & Levine, 2012; Vidal et al., 2014). To note, a LC3 mutant (G120A) deficient in lipidation can be used as a negative control for the method.

4.1 Drug Treatments

One-month-old mice are treated with IP injections of 2 g/kg of trehalose or 2 mg/kg of rapamycin, three times per week during 1 or 2 weeks. Rapamycin solution should be prepared at stock solution of 25 mg/mL in DMSO and stored at −20°C. Before injections, an aliquot of rapamycin is diluted immediately with 0.5 mL aqueous solution containing 5% polyethylene glycol 400 (PEG400) and 5% Tween 80. Volume is completed with filtered PBS. Trehalose can be prepared at 500 mg/mL in filtered PBS, aliquoted, and stored at −20°C until use. In addition, trehalose could be supplemented in the drinking water at 3% w/v ad libitum, and should be daily replaced for a new fresh bottle.

5. TISSUE PROCESSING AND HISTOLOGY

1. In order to process and collect different tissues for analysis after autophagy stimulation, AAV_mCherry-GFP-LC3-transduced mice are euthanized and perfused transcardially with 4% PFA prepared in PBS. Target tissues such as spinal cord, sciatic nerve, and brain are removed and postfixed for 3 h in 4% PFA prepared in filtered PBS adjusted to pH 7.4.

   Note that adjusting pH in buffers used to process tissues is important to maintain the stability of pH and avoid putative GFP fluorescence reemission after quenching in lysosomes, which could lead to an underestimation of experimental data.
2. After fixation, tissues are rinsed with filtered PBS pH 7.4 to remove PFA.
3. Then, tissues are embedded during successive days into an increasing sucrose gradient by changing the solution every day in the followed order: 5%, 10%, and 30% sucrose, prepared in sterile PBS, pH 7.4. Sucrose–PBS solutions should be filtered before use.

4. After sucrose treatment, tissues must be protected before freezing, with cryoembedding solution (optimal cutting temperature compound, e.g., Tissue-Tek), and then fast freeze in liquid nitrogen.
5. Tissues derived from brain and spinal cord are sliced by using cryostat generating sections of 10–25 μm width and are mounted on Superfrost glass slides. Nuclei staining is achieved by directly adding 1:5000 Hoechst nuclei staining (dilution in Vectashield mounting medium) and then mounted for microscopy analysis. Sample tissue sections are visualized under 40 × magnification, at least, under fluorescent microscopes to ensure dots visualization, as exemplified in Fig. 2B–D.
6. To obtain single sciatic nerve fibers, tissue is removed and fixed for 1 h in filtered 4% PFA prepared in PBS pH 7.4, and then dissected to obtain single fibers. The dissociation of the sciatic nerve is performed mechanically with a tungsten needle. Then, fibers are mounted with Hoechst nuclei staining 1:5000 directly in Vectashield, and tissue images are acquired using a focus stacking (*z*-stack) capture microscope, with at least 40 × magnification objectives. A super-zoom microscope using 5 × and 43 × objectives or spinning disk using 40 × and 60 × objectives can be also used as we previous reported (Castillo, Valenzuela, et al., 2013).
7. *Z*-stack sections of 1 μm could be used to obtain better signals and perform a *z*-projection of images in green and red channels at high resolution (1024 × 1024 pixels). Three images for focal plane are taken and averaged. Increasing the number of pictures per focal plane, to obtain averaged fluorescent images, is needed to decrease background tissue noise and thus favor the clear presence of red and green dots. The repetitive laser scanning over the tissues could produce quenching of the fluorescent signal. In our hands three images are enough to obtain clean pictures, avoiding a significant fluorescence quenching.

Of note, the diffuse fluorescence of GFP and mCherry could mask the appearance of puncta if an epifluorescence microscope is used. Slides must be kept out of light and stored at 4°C. The fluorescent signal is relatively stable; however, for quantification of autophagy by fluorescence analysis, it is important to collect images during the first month after tissue management.

6. QUANTIFICATION OF FLUORESCENT PUNCTA

The autophagy reporter used here serves to visualize autophagosomes and autophagolysosomes in cells of the nervous systems, representing an opportunity for monitoring autophagy activity in nervous system tissues. For quantitative estimation of autophagy, fixed tissues must be imaged under appropriate filters for fluorescence emission in red and green channels. For dot visualization with good resolution, we recommend acquiring images at a minimum of 40 × magnification, as shown in Fig. 2. Regular confocal or spinning disk images of cells and tissues are used to determine the fluorescence intensity of GFP-LC3 and mCherry-LC3 puncta and bulk distribution, using an image analysis software such as ImageJ (http://rsbweb.nih.gov/ij/).

1. Make a *z*-projection of tissue sections images taken from brain or spinal cord. This procedure generates a single file for green channel (GFP), and another for red (mCherry) channel. If the stack is made every 1 μm in a tissue section of 20 μm, the resulting file is a composition of 20 images per channel.
2. Images should be pseudocolored (for red and green) if they are acquired in 16-bit (gray scale), or directly combined if they are acquired in color channels. A merge file image is generated by combining pictures from GFP and mCherry channels, for the subsequent analysis of the three image files.

   Note that puncta can be counted manually because dots are clearly detected with the methods (Figs. 1 and 2), but when many images and treatments are analyzed, an automatization of the process is desirable. Then, we provide an alternate quantification method using ImageJ.
3. First of all, all images of each experimental group need to be converted into 16-bit (binarized) files, for both GFP and mCherry focal field (raw images), as shown in Fig. 3A.
4. On binary images, background signal is excluded by determining a threshold over which fluorescent intensity is considered positive in relation to negative controls (such as noninjected animals) in the maximum slope of intensity for each channel.
5. Then, obtain the integrated density of signals in the image files in which background was subtracted, for each channel using ImageJ plugins including: (i) raw image, (ii) particle analysis, and (iii) area coverage.

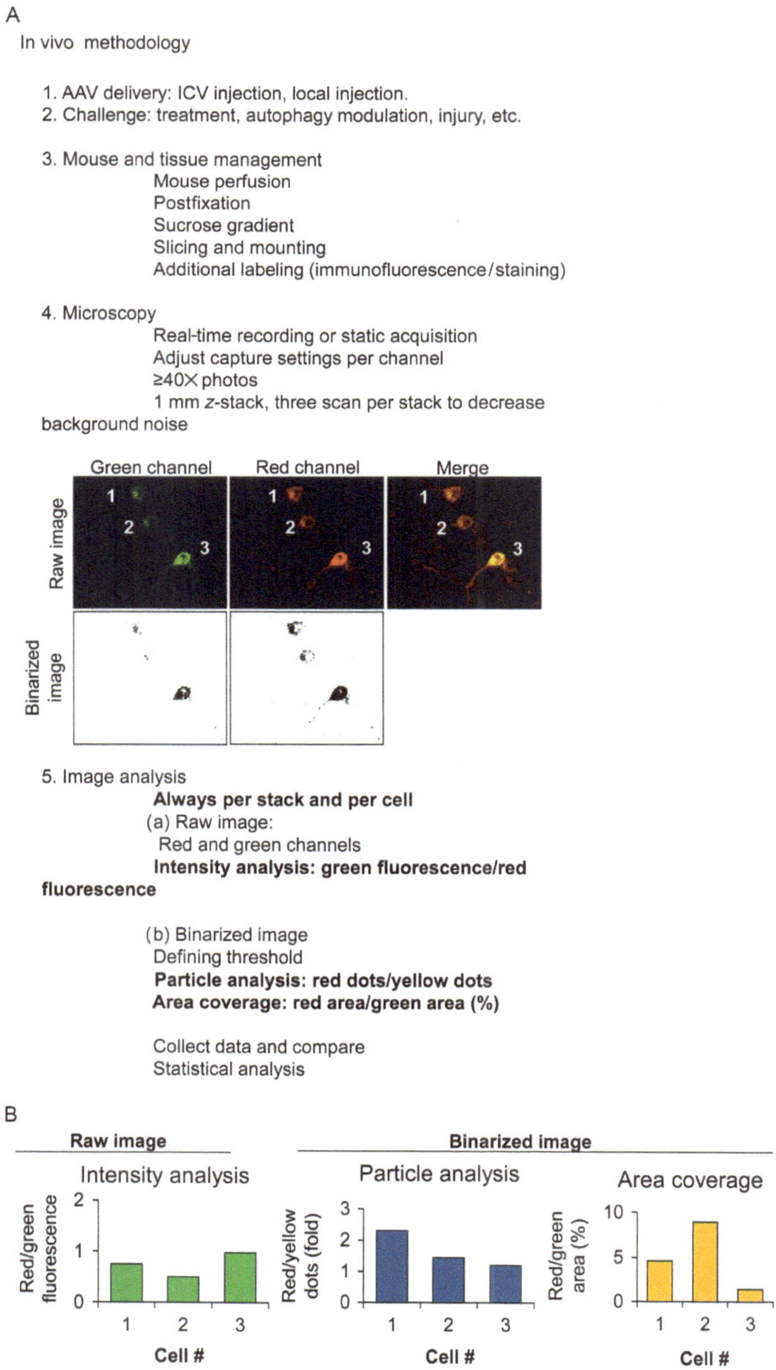

Fig. 3 See legend on opposite page.

6. Repeat the process for images of tissues derived from different animals and treatments.
7. Autophagic flux determination is obtained by the ratio between integrated density data corresponding to processed images for red vs green channels (Fig. 3B).
8. Data obtained in the three different modes (raw, particle, and area coverage) are plotted, compared, averaged, and analyzed for statististics to determine the differences in LC3 flux among experimental groups.

7. EX VIVO ANALYSIS OF LC3 VESICLE TRAFFICKING

In order to apply this method to determine the dynamic of autophagic vesicles in axons, explants of dorsal root ganglia (DRG) can be transduced with AAV_mCherry-GFP-LC3 particles.

1. Isolate DRGs of embryonic E16 Sprague Dawley rats in ice-cold Petri dishes containing L15 medium as described (Villegas et al., 2014).
2. Plate each DRG explant on a well of a 6-well plate and culture them for 4 days with a 1:100 dilution of AAV_mCherry-GFP-LC3 reporter (from stock preparations) in Neurobasal medium (2% B27, 0.3% L-glutamine, 1% streptomycin/penicillin, 4 μ*M* aphidicolin, 7.5 μg/mL of fluoro-2-deoxyuridine, and 50 ng/mL of NGF).
3. To stimulate autophagy, after 96 h treat DRG explants with 400 n*M* rapamycin and visualize axons in a spinning disk microscope in a time-lapse mode, acquiring images every 30 s.
4. Autophagy vesicle trafficking could be estimated by measuring the position of an autophagic vesicle over time by applying the kymograph analysis of ImageJ software. For each frame of each channel a ROI (region of

Fig. 3 Protocol to image analysis of mCherry-GFP-LC3 fluorescence. (A) Summarized protocol of monitoring autophagy flux by AAV2_mCherry-GFP-LC3 method. A representative image of a group of three Purkinje's cells is displayed to show the semiautomatic method of quantification of fluorescence intensity in *green*, *red*, and *yellow* channels. Florescence images are converted to 16-bit images. (B) Quantification of autophagy flux by three different methods, from *left* to *right*: (i) intensity analysis obtained from a raw image, (ii) particle analysis, and (iii) area coverage or the ratio of area coverage (*red/green*), for cells 1, 2, and 3 from images showed in (A). This procedure should be repeated for several images, for each experimental condition, to proceed with statistical analysis.

interest) is specified on a defined axonal segment. If a structure is moving along the ROI, it will appear as a contrast edge with different angles depending on velocity in the time–space (kymograph) plot.

Image analysis procedures should be performed for green, red, and merge channels in all images (stacks). With this approach it is possible to obtain the velocities of autophagic vesicles moving along the axons (kymograph description and plugins used to calculate particle velocities can be downloaded from http://www.embl.de/eamnet/html/body_kymograph.html).

8. CONCLUDING REMARKS

Autophagy is emerging as a relevant target for drug discovery because of its involvement in physiology and pathophysiology. Despite the fact that this research area has evolved very fast, the dynamic measurement of autophagic activity has been difficult to assess. Pharmacological and genetic manipulations that enhance autophagy activity in cell culture models are usually correlated with the effects in vivo to infer the role of autophagy in the nervous system. The method described here to monitor autophagy activity in the central and peripheral nervous system is simple and allows measuring autophagy flux by combining three different strategies: (i) the use of high titers of AAVs, (ii) the analysis of dynamic fluorescent reporter mCherry-GFP-LC3, and (iii) the delivery of the particles to CNS and PNS using ICV injection in newborn animal. mCherry-GFP-LC3-positive puncta and autophagy flux are easily detected with epifluorescence, confocal, and spinning disk microscopy, upon activation of the pathway with injury or pharmacological inducers. As possible applications of this method to disease conditions, we previously measured LC3 flux after spinal cord injury or peripheral nerve crush using the sciatic nerve (Castillo, Valenzuela, et al., 2013). Our methodological principles can be applied to any mouse model of diseases; it can be used to test the activity of small molecules in vivo, or to define the effect of certain genetic manipulations on autophagy levels in different neuronal populations. The method described here is simple and can be applied by any laboratory since AAVs production is commercially available, and protocols for ICV injections are simple. An important point to consider is the presence of high levels of green background in tissue samples from the nervous system that complicates the detection of green dots. This issue may represent a caveat for analysis when poor transduction is obtained.

Averaging multiple images scanned per tissue sections could help to decrease the background noise.

AAVs have been extensively used to deliver transgenes to the CNS in preclinical models and even in several clinical trials (Wang & Gao, 2014a, 2014b). In animal models, ICV injections can be also performed to rats as demonstrated (Shenouda, Johns, Kintsurashvili, Gavras, & Gavras, 2006), and even other mammals additionally to mice. AAVs also allow delivering this reporter into adult mice by performing local injections in the brain or spinal cord by stereotaxis. Adult mouse delivery of AAVs to the CNS allows mainly focal transduction; however, this methodology can be adapted according to the tissue-specific desirable distribution of transduction, for example, to inject autophagy flux sensor in striatum or substantia nigra in Parkinson's disease models or hippocampus to evaluate autophagy in memory loss in Alzheimer's disease models.

In addition, depending of the serotype used, different tissues and cell types of the body can be transduced (Bourdenx, Dutheil, Bezard, & Dehay, 2014; Gao, Vandenberghe, & Wilson, 2005; McCown, 2011; Snyder et al., 2011). It is important to consider the well-known delay of transgene expression when AAVs are used (Aschauer, Kreuz, & Rumpel, 2013). In our hands we have observed transgenic mRNA expression 5 days postdelivery in intraspinal cord injections and 7 days using ICV method. The development of newly engineered AAVs allows the generation of higher transduction efficiency by the ICV method and also modifies the delivery method. For example, the development of modified AAV9 allows a spread transduction in the CNS by intravenous delivery (Choudhury et al., 2016; Dayton, Wang, & Klein, 2012). Moreover, the development of variants of AAVs serotypes has allowed to extend the administration such as intramuscular or intravenous routes to deliver transgenes into the CNS (Bourdenx et al., 2014).

Autophagy activity in the nervous system has been indirectly inferred through a variety of techniques, including detection of autophagosomes by electron microscopy, the measure of static levels of LC3-II by Western blot, by the assessment of the distribution of autophagy-related proteins to vesicular structures, or the decrease in the levels of autophagy substrates such as p62. Because monitoring these markers can lead to misleading interpretations, there is a growing need for simple methods to measure autophagy flux in neurons in vivo. Our strategy allows monitoring different rates of autophagy activation in subpopulations of cells and the identification of particular cell types undergoing autophagy activation or inhibition. Moreover, our method enables to measure autophagosome dynamics in vivo in the context of the 3D structure of the brain in real time by using multiphoton microscopy and study single-particle features by superresolution microscopy. In summary, the use of ICV injection of the molecular reporter AAV2/2_mCherry-GFP-LC3 represents a valuable tool in the advancement toward defining the contribution of autophagy to diverse physiological and pathological conditions affecting the nervous system.

ACKNOWLEDGMENTS

This work is funded by FONDAP 15150012, the Frick Foundation 20014-15, ALS Therapy Alliance 2014-F-059, Muscular Dystrophy Association 382453, CONICYT-USA 2013-0003, Michael J Fox Foundation for Parkinson's Research—Target Validation Grant No. 9277, COPEC-UC Foundation 2013.R.40, Ecos-Conicyt C13S02, FONDECYT No. 1140549, Office of Naval Research Global (ONR-G) N62909-16-1-2003, ALSRP Therapeutic Idea Award AL150111, and Millennium Institute No. P09-015-F (C.H.). Doctoral fellow was supported by CONICYT Nos. 21130843 (M.O.) and 21120411 (V. V.), and Millennium Institute CINV P09-022-F (K.C.).

REFERENCES

Adhami, F., Liao, G., Morozov, Y. M., Schloemer, A., Schmithorst, V. J., Lorenz, J. N., … Kuan, C. Y. (2006). Cerebral ischemia-hypoxia induces intravascular coagulation and autophagy. *The American Journal of Pathology*, *169*(2), 566–583. http://dx.doi.org/10.2353/ajpath.2006.051066.

Anderson, R., Macdonald, I., Corbett, T., Whiteway, A., & Prentice, H. G. (2000). A method for the preparation of highly purified adeno-associated virus using affinity column chromatography, protease digestion and solvent extraction. *Journal of Virological Methods*, *85*(1–2), 23–34.

Aschauer, D. F., Kreuz, S., & Rumpel, S. (2013). Analysis of transduction efficiency, tropism and axonal transport of AAV serotypes 1, 2, 5, 6, 8 and 9 in the mouse brain. *PLoS One*, *8*(9), e76310. http://dx.doi.org/10.1371/journal.pone.0076310.

Ayers, J. I., Fromholt, S., Sinyavskaya, O., Siemienski, Z., Rosario, A. M., Li, A., … Levites, Y. (2015). Widespread and efficient transduction of spinal cord and brain following neonatal AAV injection and potential disease modifying effect in ALS mice. *Molecular Therapy*, *23*(1), 53–62. http://dx.doi.org/10.1038/mt.2014.180.

Bourdenx, M., Dutheil, N., Bezard, E., & Dehay, B. (2014). Systemic gene delivery to the central nervous system using adeno-associated virus. *Frontiers in Molecular Neuroscience*, 7, 50. http://dx.doi.org/10.3389/fnmol.2014.00050.

Cashman, N. R., Durham, H. D., Blusztajn, J. K., Oda, K., Tabira, T., Shaw, I. T., … Antel, J. P. (1992). Neuroblastoma x spinal cord (NSC) hybrid cell lines resemble developing motor neurons. *Developmental Dynamics*, *194*(3), 209–221. http://dx.doi.org/10.1002/aja.1001940306.

Castillo, K., Nassif, M., Valenzuela, V., Rojas, F., Matus, S., Mercado, G., … Hetz, C. (2013). Trehalose delays the progression of amyotrophic lateral sclerosis by enhancing autophagy in motoneurons. *Autophagy*, *9*(9), 1308–1320. http://dx.doi.org/10.4161/auto.25188.

Castillo, K., Valenzuela, V., Matus, S., Nassif, M., Onate, M., Fuentealba, Y., … Hetz, C. (2013). Measurement of autophagy flux in the nervous system in vivo. *Cell Death & Disease*, 4, e917. http://dx.doi.org/10.1038/cddis.2013.421.

Choudhury, S. R., Harris, A. F., Cabral, D. J., Keeler, A. M., Sapp, E., Ferreira, J. S., … Sena-Esteves, M. (2016). Widespread central nervous system gene transfer and silencing after systemic delivery of novel AAV-AS vector. *Molecular Therapy*, *24*(4), 726–735. http://dx.doi.org/10.1038/mt.2015.231.

Davies, J. E., Sarkar, S., & Rubinsztein, D. C. (2006). Trehalose reduces aggregate formation and delays pathology in a transgenic mouse model of oculopharyngeal muscular dystrophy. *Human Molecular Genetics*, *15*(1), 23–31.

Dayton, R. D., Wang, D. B., & Klein, R. L. (2012). The advent of AAV9 expands applications for brain and spinal cord gene delivery. *Expert Opinion on Biological Therapy*, *12*(6), 757–766. http://dx.doi.org/10.1517/14712598.2012.681463.

Dirren, E., Towne, C. L., Setola, V., Redmond, D. E., Jr., Schneider, B. L., & Aebischer, P. (2014). Intracerebroventricular injection of adeno-associated virus 6 and 9 vectors for cell type-specific transgene expression in the spinal cord. *Human Gene Therapy*, *25*(2), 109–120. http://dx.doi.org/10.1089/hum.2013.021.

Erlich, S., Alexandrovich, A., Shohami, E., & Pinkas-Kramarski, R. (2007). Rapamycin is a neuroprotective treatment for traumatic brain injury. *Neurobiology of Disease*, *26*(1), 86–93.

Esteban-Martinez, L., & Boya, P. (2015). Autophagic flux determination in vivo and ex vivo. *Methods*, *75*, 79–86. http://dx.doi.org/10.1016/j.ymeth.2015.01.008.

Feng, Y., Yao, Z., & Klionsky, D. J. (2015). How to control self-digestion: Transcriptional, post-transcriptional, and post-translational regulation of autophagy. *Trends in Cell Biology*, *25*(6), 354–363. http://dx.doi.org/10.1016/j.tcb.2015.02.002.

Gao, G., Vandenberghe, L. H., & Wilson, J. M. (2005). New recombinant serotypes of AAV vectors. *Current Gene Therapy*, *5*(3), 285–297.

Gholizadeh, S., Tharmalingam, S., Macaldaz, M. E., & Hampson, D. R. (2013). Transduction of the central nervous system after intracerebroventricular injection of adeno-associated viral vectors in neonatal and juvenile mice. *Human Gene Therapy Methods*, *24*(4), 205–213. http://dx.doi.org/10.1089/hgtb.2013.076.

Glascock, J. J., Osman, E. Y., Coady, T. H., Rose, F. F., Shababi, M., & Lorson, C. L. (2011). Delivery of therapeutic agents through intracerebroventricular (ICV) and intravenous (IV) injection in mice. *Journal of Visualized Experiments: JoVE*, *56*, 2968. https://www.ncbi.nlm.nih.gov/pmc/articles/PMC3227174/.

Gong, Y., Mu, D., Prabhakar, S., Moser, A., Musolino, P., Ren, J., … Eichler, F. S. (2015). Adenoassociated virus serotype 9-mediated gene therapy for x-linked adrenoleukodystrophy. *Molecular Therapy*, *23*(5), 824–834. http://dx.doi.org/10.1038/mt.2015.6.

Green, D. R., & Levine, B. (2014). To be or not to be? How selective autophagy and cell death govern cell fate. *Cell*, *157*(1), 65–75. http://dx.doi.org/10.1016/j.cell.2014.02.049.

Hara, T., Nakamura, K., Matsui, M., Yamamoto, A., Nakahara, Y., Suzuki-Migishima, R., … Mizushima, N. (2006). Suppression of basal autophagy in neural cells causes neurodegenerative disease in mice. *Nature*, *441*(7095), 885–889. http://dx.doi.org/10.1038/nature04724.

Haspel, J., Shaik, R. S., Ifedigbo, E., Nakahira, K., Dolinay, T., Englert, J. A., & Choi, A. M. (2011). Characterization of macroautophagic flux in vivo using a leupeptin-based assay. *Autophagy*, 7(6), 629–642.

Hetz, C., Thielen, P., Matus, S., Nassif, M., Court, F., Kiffin, R., … Glimcher, L. H. (2009). XBP-1 deficiency in the nervous system protects against amyotrophic lateral sclerosis by increasing autophagy. *Genes & Development*, *23*(19), 2294–2306. http://dx.doi.org/10.1101/gad.1830709.

Iwai-Kanai, E., Yuan, H., Huang, C., Sayen, M. R., Perry-Garza, C. N., Kim, L., & Gottlieb, R. A. (2008). A method to measure cardiac autophagic flux in vivo. *Autophagy*, *4*(3), 322–329.

Kaushik, S., & Cuervo, A. M. (2015). Proteostasis and aging. *Nature Medicine*, *21*(12), 1406–1415. http://dx.doi.org/10.1038/nm.4001.

Kaza, N., Kohli, L., & Roth, K. A. (2012). Autophagy in brain tumors: A new target for therapeutic intervention. *Brain Pathology*, *22*(1), 89–98. http://dx.doi.org/10.1111/j.1750-3639.2011.00544.x.

Kim, J. Y., Grunke, S. D., & Jankowsky, J. L. (2016). Widespread neuronal transduction of the rodent CNS via neonatal viral injection. *Methods in Molecular Biology*, *1382*, 239–250. http://dx.doi.org/10.1007/978-1-4939-3271-9_17.

Kimura, S., Noda, T., & Yoshimori, T. (2007). Dissection of the autophagosome maturation process by a novel reporter protein, tandem fluorescent-tagged LC3. *Autophagy*, *3*(5), 452–460.

Klionsky, D. J., Abdelmohsen, K., Abe, A., Abedin, M. J., Abeliovich, H., Acevedo Arozena, A., … Zughaier, S. M. (2016). Guidelines for the use and interpretation of assays for monitoring autophagy (3rd edition). *Autophagy*, *12*(1), 1–222. http://dx.doi.org/10.1080/15548627.2015.1100356.

Komatsu, M., Waguri, S., Chiba, T., Murata, S., Iwata, J., Tanida, I., … Tanaka, K. (2006). Loss of autophagy in the central nervous system causes neurodegeneration in mice. *Nature*, *441*(7095), 880–884. http://dx.doi.org/10.1038/nature04723.

Lawlor, P. A., Bland, R. J., Mouravlev, A., Young, D., & During, M. J. (2009). Efficient gene delivery and selective transduction of glial cells in the mammalian brain by AAV serotypes isolated from nonhuman primates. *Molecular Therapy*, *17*(10), 1692–1702. http://dx.doi.org/10.1038/mt.2009.170.

Li, J., Zhou, J., Chen, G., Wang, H., Wang, S., Xing, H., … Ma, D. (2006). Inhibition of ovarian cancer metastasis by adeno-associated virus-mediated gene transfer of nm23H1 in an orthotopic implantation model. *Cancer Gene Therapy*, *13*(3), 266–272. http://dx.doi.org/10.1038/sj.cgt.7700899.

Liu, C. L., Chen, S., Dietrich, D., & Hu, B. R. (2008). Changes in autophagy after traumatic brain injury. *Journal of Cerebral Blood Flow and Metabolism*, *28*(4), 674–683. http://dx.doi.org/10.1038/sj.jcbfm.9600587.

Malik, S. Z., Maronski, M. A., Dichter, M. A., & Watson, D. J. (2012). The use of specific AAV serotypes to stably transduce primary CNS neuron cultures. *Methods in Molecular Biology*, *846*, 305–319. http://dx.doi.org/10.1007/978-1-61779-536-7_26.

Matus, S., Valenzuela, V., & Hetz, C. (2014). A new method to measure autophagy flux in the nervous system. *Autophagy*, *10*(4), 710–714. http://dx.doi.org/10.4161/auto.28434.

McCown, T. J. (2011). Adeno-associated virus (AAV) vectors in the CNS. *Current Gene Therapy*, *11*(3), 181–188.

McLean, J. R., Smith, G. A., Rocha, E. M., Hayes, M. A., Beagan, J. A., Hallett, P. J., & Isacson, O. (2014). Widespread neuron-specific transgene expression in brain and spinal cord following synapsin promoter-driven AAV9 neonatal intracerebroventricular injection. *Neuroscience Letters*, *576*, 73–78. http://dx.doi.org/10.1016/j.neulet.2014.05.044.

Menzies, F. M., Fleming, A., & Rubinsztein, D. C. (2015). Compromised autophagy and neurodegenerative diseases. *Nature Reviews. Neuroscience*, *16*(6), 345–357. http://dx.doi.org/10.1038/nrn3961.

Nagy, P., Varga, A., Kovacs, A. L., Takats, S., & Juhasz, G. (2015). How and why to study autophagy in Drosophila: It's more than just a garbage chute. *Methods*, *75*, 151–161. http://dx.doi.org/10.1016/j.ymeth.2014.11.016.

Navarro, E., Serrano-Heras, G., Castano, M. J., & Solera, J. (2015). Real-time PCR detection chemistry. *Clinica Chimica Acta*, *439*, 231–250. http://dx.doi.org/10.1016/j.cca.2014.10.017.

Nezis, I. P., Shravage, B. V., Sagona, A. P., Lamark, T., Bjorkoy, G., Johansen, T., … Stenmark, H. (2010). Autophagic degradation of dBruce controls DNA fragmentation in nurse cells during late Drosophila melanogaster oogenesis. *The Journal of Cell Biology*, *190*(4), 523–531. http://dx.doi.org/10.1083/jcb.201002035.

Pankiv, S., Clausen, T. H., Lamark, T., Brech, A., Bruun, J. A., Outzen, H., … Johansen, T. (2007). p62/SQSTM1 binds directly to Atg8/LC3 to facilitate degradation of ubiquitinated protein aggregates by autophagy. *The Journal of Biological Chemistry*, *282*(33), 24131–24145. http://dx.doi.org/10.1074/jbc.M702824200.

Passini, M. A., & Wolfe, J. H. (2001). Widespread gene delivery and structure-specific patterns of expression in the brain after intraventricular injections of neonatal mice with an adeno-associated virus vector. *Journal of Virology*, *75*(24), 12382–12392. http://dx.doi.org/10.1128/JVI.75.24.12382-12392.2001.

Ravikumar, B., Vacher, C., Berger, Z., Davies, J. E., Luo, S., Oroz, L. G., … Rubinsztein, D. C. (2004). Inhibition of mTOR induces autophagy and reduces toxicity of polyglutamine expansions in fly and mouse models of Huntington disease. *Nature Genetics*, *36*(6), 585–595. http://dx.doi.org/10.1038/ng1362.

Rodriguez-Navarro, J. A., Rodriguez, L., Casarejos, M. J., Solano, R. M., Gomez, A., Perucho, J., … Mena, M. A. (2010). Trehalose ameliorates dopaminergic and tau pathology in parkin deleted/tau overexpressing mice through autophagy activation. *Neurobiology of Disease*, *39*(3), 423–438.

Rubinsztein, D. C., Codogno, P., & Levine, B. (2012). Autophagy modulation as a potential therapeutic target for diverse diseases. *Nature Reviews. Drug Discovery*, *11*(9), 709–730. http://dx.doi.org/10.1038/nrd3802.

Sadasivan, S., Dunn, W. A., Jr., Hayes, R. L., & Wang, K. K. (2008). Changes in autophagy proteins in a rat model of controlled cortical impact induced brain injury. *Biochemical and Biophysical Research Communications*, *373*(4), 478–481. http://dx.doi.org/10.1016/j.bbrc.2008.05.031.

Sarkar, S., Ravikumar, B., Floto, R. A., & Rubinsztein, D. C. (2009). Rapamycin and mTOR-independent autophagy inducers ameliorate toxicity of polyglutamine-expanded huntingtin and related proteinopathies. *Cell Death and Differentiation*, *16*(1), 46–56. http://dx.doi.org/10.1038/cdd.2008.110.

Schneider, J. L., & Cuervo, A. M. (2014). Autophagy and human disease: Emerging themes. *Current Opinion in Genetics & Development*, *26*, 16–23. http://dx.doi.org/10.1016/j.gde.2014.04.003.

Sekiguchi, A., Kanno, H., Ozawa, H., Yamaya, S., & Itoi, E. (2012). Rapamycin promotes autophagy and reduces neural tissue damage and locomotor impairment after spinal cord injury in mice. *Journal of Neurotrauma*, *29*(5), 946–956.

Shenouda, S. M., Johns, C., Kintsurashvili, E., Gavras, I., & Gavras, H. (2006). Long-term inhibition of the central alpha(2B)-adrenergic receptor gene via recombinant AAV-delivered antisense in hypertensive rats. *American Journal of Hypertension*, *19*(11), 1135–1143. http://dx.doi.org/10.1016/j.amjhyper.2006.04.001.

Sica, V., Galluzzi, L., Bravo-San Pedro, J. M., Izzo, V., Maiuri, M. C., & Kroemer, G. (2015). Organelle-specific initiation of autophagy. *Molecular Cell*, *59*(4), 522–539. http://dx.doi.org/10.1016/j.molcel.2015.07.021.

Snyder, B. R., Gray, S. J., Quach, E. T., Huang, J. W., Leung, C. H., Samulski, R. J., … Federici, T. (2011). Comparison of adeno-associated viral vector serotypes for spinal cord and motor neuron gene delivery. *Human Gene Therapy*, *22*(9), 1129–1135. http://dx.doi.org/10.1089/hum.2011.008.

Sridhar, S., Botbol, Y., Macian, F., & Cuervo, A. M. (2012). Autophagy and disease: Always two sides to a problem. *The Journal of Pathology*, *226*(2), 255–273. http://dx.doi.org/10.1002/path.3025.

Vidal, R. L., Matus, S., Bargsted, L., & Hetz, C. (2014). Targeting autophagy in neurodegenerative diseases. *Trends in Pharmacological Sciences*, *35*(11), 583–591. http://dx.doi.org/10.1016/j.tips.2014.09.002.

Villegas, R., Martinez, N. W., Lillo, J., Pihan, P., Hernandez, D., Twiss, J. L., & Court, F. A. (2014). Calcium release from intra-axonal endoplasmic reticulum leads to axon degeneration through mitochondrial dysfunction. *The Journal of Neuroscience: The Official Journal of the Society for Neuroscience*, *34*(21), 7179–7189. http://dx.doi.org/10.1523/JNEUROSCI.4784-13.2014.

Wang, D., & Gao, G. (2014a). State-of-the-art human gene therapy: Part I. Gene delivery technologies. *Discovery Medicine*, *18*(97), 67–77.

Wang, D., & Gao, G. (2014b). State-of-the-art human gene therapy: Part II. Gene therapy strategies and clinical applications. *Discovery Medicine*, *18*(98), 151–161.

Zhang, J., Zhang, Y., Li, J., Xing, S., Li, C., Li, Y., … Zeng, J. (2012). Autophagosomes accumulation is associated with beta-amyloid deposits and secondary damage in the thalamus after focal cortical infarction in hypertensive rats. *Journal of Neurochemistry*, *120*(4), 564–573. http://dx.doi.org/10.1111/j.1471-4159.2011.07496.x.

CHAPTER EIGHT

Magnetic Resonance Spectroscopy to Study Glycolytic Metabolism During Autophagy

Y.-L. Chung*[,1], M.O. Leach*, T.R. Eykyn*[,†]
*Cancer Research UK Cancer Imaging Centre, The Institute of Cancer Research, London, United Kingdom
[†]The Rayne Institute, St Thomas' Hospital, King's College London, London, United Kingdom
[1]Corresponding author: e-mail address: ylichung@icr.ac.uk

Contents

Abstract

Cancer cells undergoing starvation- and treatment-induced autophagy were found to exhibit reduced intracellular lactate, reduced rates of steady-state lactate excretion and reduced real-time pyruvate–lactate exchange rates, indicating that glycolytic metabolism was altered in autophagic cells. In this chapter, we describe the technical details of the use of ^{1}H-magnetic resonance spectroscopy (MRS) to measure endogenous cellular

Methods in Enzymology, Volume 588
ISSN 0076-6879
http://dx.doi.org/10.1016/bs.mie.2016.09.078

concentrations of lactate and glucose in autophagic cells and tissues, how to measure the rate of steady-state lactate excretion and glucose uptake by ^{1}H-MRS in autophagic cells, and details of the real-time measurement of [1-^{13}C] pyruvate to lactate exchange in autophagic cells by ^{13}C-MRS-DNP (dynamic nuclear polarization).

1. INTRODUCTION

Glucose is metabolized by cells via glycolysis to produce pyruvate. The efficiency of this process is dependent on the rate of glucose uptake which is controlled by the glucose transporters (GLUTs) and the activity of the various enzymes along the glycolytic pathway. Pyruvate is then converted to acetyl-CoA by pyruvate dehydrogenase (PDH) or converted to oxaloacetate by pyruvate carboxylase (PC) and enters the TCA cycle to produce substrates for oxidative phosphorylation in the mitochondria. Alternative metabolic fates for pyruvate include transamination to form alanine or reduction in the cytosol to form lactate. The rate of reduction of pyruvate to lactate is dependent on the expression level and activity of lactate dehydrogenase (LDH) and cofactors such as oxidized (NAD^{+}) and reduced nicotinamide adenine dinucleotide (NADH) (Fig. 1), as well as

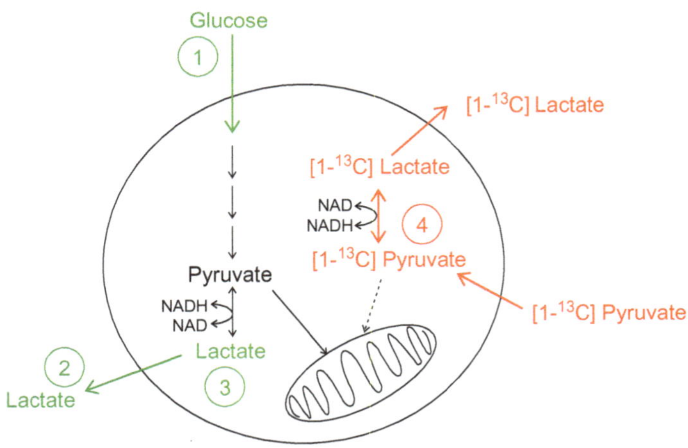

Fig. 1 Schematic pathway diagram showing the uptake of glucose mediated via the GLUTs and subsequent conversion via glycolysis to produce pyruvate and lactate. Pyruvate and lactate are both subject to monocarboxylate transport into and out of the cell. Measurements in *green* correspond to steady-state ^{1}H MRS analysis of (1) glucose uptake, (2) lactate excretion, and (3) intracellular lactate. Measurements in *red* correspond to real-time ^{13}C MRS analysis of (4) pyruvate–lactate exchange kinetics.

the monocarboxylate (MCT)-mediated transport of substrates into and out of the cell. If the availability of oxygen is restricted, then preferential conversion of pyruvate to lactate and concomitant transport of lactate out of the cell regenerates NAD^+ in the cytosol, which allows glycolysis to continue even in the absence of oxygen, such as in ischemia or acute exercise.

A shift from oxidative phosphorylation to glycolysis and the conversion of pyruvate to lactate for energy production is favored by cancer cells, even in the presence of freely available oxygen. In cancer, there is increased transcriptional regulation of a number of glycolytic and mitochondrial enzymes, including lactate dehydrogenase-A (LDH-A), pyruvate dehydrogenase kinase, among many others. This reprogramming of energy metabolism is called the "Warburg effect" and has become a widely accepted hallmark of cancer (Hanahan & Weinberg, 2000, 2011; Warburg, 1956).

Glycolytic metabolism was found to be altered in starvation (amino acid and serum deprived)- and treatment (PI103 and dichloroacetate)-induced autophagic cancer cells (Lin, Andrejeva, et al., 2014; Lin, Hill, et al., 2014). In these experiments, induction of autophagy was confirmed in colorectal HT29 and HCT116 Bax-ko cancer cells by overexpression of LC3II in Western blots and the presence of double-membrane autophagic vesicles by electron microscopy, while the absence of apoptosis was confirmed by the lack of cleaved poly(ADP-ribose) polymerase (PARP) or change in caspase 3 expression. Reductions in the rate of lactate excretion and intracellular lactate were observed, as well as reductions in pyruvate–lactate exchange kinetics measured in real time. It was also shown that by replacing the amino acid- and serum-deprived media with full media, or by stopping the treatment, there was a reversal of this phenomenon, which reported on cellular recovery from autophagy (Lin, Andrejeva, et al., 2014; Lin, Hill, et al., 2014). In these studies, magnetic resonance spectroscopy (MRS) methods were used to examine the rates of lactate production or exchange in steady state and in real time, respectively. ^{1}H-MRS analysis was used to measure cellular lactate levels in cell extracts and lactate secretion and glucose uptake in cell culture media. Dynamic nuclear polarization (DNP) and ^{13}C-MRS of ^{13}C-labeled pyruvate were used to measure the rate of pyruvate exchange with lactate in vitro and in vivo in real time. The mechanisms behind these changes were found to be associated with decreased LDH activity in starvation-induced autophagy and an increased NAD^+/NADH ratio in treatment-induced autophagy, suggesting a possible shift in glycolytic metabolism during autophagy (Lin, Andrejeva, et al., 2014; Lin, Hill, et al., 2014).

This chapter describes the technical details of how to measure endogenous cellular lactate and glucose in autophagic cells and tissues by ^{1}H-MRS, how to measure the rate of lactate secretion and glucose uptake in autophagic cells by ^{1}H-MRS, and the measurement of the real-time exchange of ^{13}C-labeled pyruvate to lactate in cells by ^{13}C-MRS-DNP and the derivation of simple kinetic parameters from these measurements. The methods described here are generally applicable in cells and tissues to investigate cellular processes, cellular metabolism alterations following genetic perturbations or in disease state, and monitoring treatment response (Chung, Basetti, & Griffiths, 2015; Chung & Griffiths, 2011).

1.1 Magnetic Resonance Spectroscopy

MRS is limited by low signal strength at thermal equilibrium due to low spin polarization. Nevertheless, MRS and MRS imaging offer chemically specific analysis of metabolite concentrations in body fluids, cell or tissue extracts, intact tissues, or in vivo. In combination with metabolomics and statistical methods such as principal component analysis, MRS has become a widely used tool to measure a wide range of metabolites in cell extracts, whole cells, or tissue biopsy samples (Beckonert et al., 2007; Chung et al., 2015; Chung & Griffiths, 2011). Furthermore, the ability to perform metabolic imaging in vivo using MRS (de Graaf, 2007) and the translation of these techniques clinically in humans allow the development of molecular biomarkers of response to therapeutics that act directly on metabolic enzymes and transporters or through the inhibition of cell signaling pathways that transcriptionally control and regulate metabolic enzymes.

MRS exploits an intrinsic property of the nucleus known as *spin*. A wide range of biologically important isotopes possess a nonzero spin, including ^{1}H, ^{13}C, ^{31}P, ^{23}Na, and ^{15}N, among others. When placed in an external magnetic field of the spectrometer, the spin states (angular momentum) of these nuclei become quantized. When excited by a radiofrequency pulse, a characteristic spectrum is generated depending on the gyromagnetic ratio of the nucleus studied and the electronic environment of the molecule. Thus, unique spectral peaks with distinct resonance frequencies known as *chemical shift* correspond to unique molecular environments. In addition, MRS is quantitative at thermal equilibrium in the sense that the peak integral is proportional to molecular concentration and can be measured directly via the addition of a reference compound of known concentration. Despite limited sensitivity, ^{1}H MRS is able to measure metabolite concentrations within the micromole range. In addition, incorporation of stable isotopes such as ^{13}C can report on the steady-state activities and fluxes through different metabolic pathways. ^{13}C isotopomer incorporation can be measured by MRS or mass spectrometry and is dependent on the respective labeling positions in starting substrates and their relative rates of incorporation through different pathways into reaction products and intermediates (Buescher et al., 2015; DeBerardinis et al., 2007).

1.2 Dynamic Nuclear Polarization

The sensitivity limitations of MRS can be overcome by using hyperpolarization techniques to transiently increase the spin polarization and thereby increase sensitivity. Techniques such as DNP enable significant MRS signal enhancements by a factor greater than 10,000 for low gyromagnetic ratio nuclei such as ^{13}C and ^{15}N in a range of endogenous biological metabolites (Ardenkjaer-Larsen et al., 2003). The ability to distinguish parent and downstream metabolites by virtue of a difference in chemical shift allows the measurement of the interconversion of metabolites in real time in cellular systems, whole organ preparations, as well as in vivo, and thereby reports on the activity of endogenous enzymes and membrane transporters that catalyze their kinetic interconversion (Day et al., 2007). For a more extensive discussion of DNP, including some of the challenges and limitations of the technique, a number of excellent reviews in this area have recently been published (Chaumeil, Najac, & Ronen, 2015; Comment & Merritt, 2014; Kurhanewicz et al., 2011).

To date, the most widely employed metabolic substrate for hyperpolarized ^{13}C-MR imaging has been [1-^{13}C] pyruvate (Golman, in't Zandt, Lerche, Pehrson, & Ardenkjaer-Larsen, 2006). This is largely due to the very favorable spin lattice relaxation time (T_1) of the C1 and C2 carbonyl carbons, which is reported to be in the range of 40–70 s depending on the magnetic field strength (Keshari & Wilson, 2014), as well as the fast rates of MCT-mediated entry of pyruvate into the cell. Pyruvate is located at the end point of glycolysis being subject to a number of metabolic fates, including LDH-mediated exchange with lactate, alanine transaminase-mediated exchange with alanine, one-way decarboxylation mediated by PDH to form CO_2 or one-way carboxylation mediated by PC to form oxaloacetate. The ability to probe the metabolic fates of hyperpolarized pyruvate, as well as its associated kinetic rates and metabolic fluxes, is therefore of great current interest for imaging metabolic processes in vivo and is complementary to steady-state metabolomics-type techniques. The apparent exchange rate constant of hyperpolarized [1-^{13}C] pyruvate to lactate (k_{PL}) has previously been shown to provide a potential metabolic biomarker for diagnosis and for assessing treatment response (Golman, in't Zandt, & Thaning, 2006; Golman & Petersson, 2006; Park et al., 2010; Ward et al., 2010). This apparent rate has been shown to decrease following drug-induced cell death, attributed to apoptosis with the activation of PARP and depletion of the cofactors NAD(H) (Day et al., 2007).

Apparent rate constants have also been shown to be dependent on NAD^+/NADH ratios (Christensen, Karlsson, Winther, Jensen, & Lerche, 2014), the expression and activity of LDH (Ward et al., 2010), as well as on the activity of the MCT transporter family (Harris, Eliyahu, Frydman, & Degani, 2009), which mediate pyruvate and lactate transport into and out of the cell. The rates of exchange-mediated pyruvate–lactate conversion have further been shown to correlate with FDG uptake following etoposide treatment in an EL4 tumor model with concomitant decreases in NADH levels and GLUT3 expression (Witney et al., 2009). Kinetic assays using hyperpolarized pyruvate are therefore sensitive to a range of physicochemical properties of the cell that report on glycolytic fluxes, as well as being complementary to more conventional imaging techniques for probing glycolysis, such as FDG PET imaging.

2. MEASUREMENT OF INTRACELLULAR LACTATE AND GLUCOSE LEVELS IN AUTOPHAGIC CELLS OR TUMOR EXTRACTS

Intracellular lactate and glucose metabolites can be extracted from cultured cells or tissues using a dual-phase extraction method detailed by Tyagi, Azrad, Degani, and Salomon (1996).

2.1 Dual-Phase Extraction of Cultured Cells

(1) Seed and treat the cells according to your desired experimental schedule. At least 10 million cells are ideally needed, but using fewer cells is also possible by acquiring the NMR spectra for a longer period.

(2) Set up two extra parallel flasks of cells per condition for cell counting.

(3) Remove the culture medium from the cell culture flask at a given point in time using a pipette and store 1 mL of the medium in a –20°C freezer. Media samples are later analyzed by ^{1}H-MRS to obtain information on the level of lactate secretion and glucose uptake (see Section 3).

(4) Trypsinize the parallel flasks of cells and count the number of cells per flask using a standard cell counting protocol.

(5) For the cell extraction experiment, wash the cells three times with 10 mL of ice-cold saline and remove each wash carefully with a pipette.

(6) Put 3 mL of ice-cold methanol into each flask (for T75 flasks) to cover the cells and keep the flasks on ice for 5–10 min. The cells are then scraped using a cell scraper, and the cell/methanol suspensions are placed into a clean centrifuge tube.

(7) Add 3 mL of ice-cold chloroform into the tube and vortex the tube vigorously for 30 s.

(8) Pipette 3 mL of ice-cold deionized water into the tube and vortex the tube vigorously again for 30 s.

(9) Centrifuge the samples for about 5000 × *g* at 4°C for 20 min for phase separation. The final chloroform:methanol:water ratio should be 1:1:1 (v/v/v). The upper methanol–water phase contains the water-soluble cellular metabolites, the middle phase contains the protein pellet and the bottom chloroform phase contains the cellular lipids.

(10) Keep and store the cell pellets (the middle phase) at –80°C (for protein concentration determination if required).

(11) Pipette the upper methanol–water phase into a clean centrifuge tube and add about 5 mg of Chelex 100 to remove divalent ions.

(12) Centrifuge the sample for about 5000 × *g* at 4°C for 5 min to separate the beads from the solution, and then transfer the clear supernatant to a clean centrifuge tube.

(13) Add 10 μL of universal pH indicator solution.

(14) Store the supernatant in a –80°C freezer until freeze-drying occurs.

2.2 Dual-Phase Extraction of Tissues

(1) Weigh the freeze-clamped or snap-frozen tissue sample and record the wet tissue weight.

(2) Grind the tissue sample into fine powder in liquid nitrogen using a Pestle and Mortar and place the ground tissue sample in a clean centrifuge tube.

(3) Put 3 mL of ice-cold methanol into a clean centrifuge tube with the ground tissue sample and vortex the tube vigorously for 30 s.

(4) Continue with the extraction procedures as described in Section 2.1 from step 5 onward.

2.3 Preparation of Extracted Cell or Tissue Samples for ^{1}H-MRS Analysis

(1) Freeze-dry the methanol–water phase of the cell or tumor extract samples (directly from the –80°C freezer) in a freeze-dryer until the samples are in powder form.

(2) Reconstitute the freeze-dried sample in 650 μL of D_2O and 50 μL of 0.75% sodium 3-trimethylsilyl-2,2,3,3-tetradeuteropropionate (TSP) in D_2O and centrifuge the sample for about 5000 × *g* at 4°C for 5 min.

(3) Put 600 μL of the sample into a 5-mm NMR tube and adjust the sample to pH7 using 0.1 *M* of potassium hydroxide and 0.6% perchloric acid.

2.4 ^{1}H-MRS of Extracted Cell and Tissue Samples

(1) ^{1}H-MR spectra of the extracted samples are acquired at 25°C using a pulse-acquired MR sequence with water suppression (1D NOESY presat sequence) in a broad-band-inverse (BBI) NMR probe on a 500- or 600-MHz NMR system. For cell extract and tissue extract samples, the typical NMR acquisition parameters are 7500 Hz spectral width, 32,768 time domain points, 2.7 s repetition time, and 256 scans.

(2) A ^{1}H-MR spectrum from a cell extract sample is shown in Fig. 2A.

(3) Commercially available software packages, such as Bruker Topsin-3.2 (Coventry, United Kingdom) and MestRe-C-4.9.9.6 (Santiago de Compostela, Spain), are used to analyze the MR spectra.

(4) Spectra are first processed by using exponential multiplication with a line-broadening factor (lb = 0.3 Hz), followed by Fourier-transform, zero- and first-order phase correction, baseline correction, and spectral peak integration of lactate and glucose.

(5) Calculate cellular lactate or glucose concentrations from a ^{1}H-MR spectrum of cell or tumor extract:

$$[M]_{\text{sample}} = \frac{N_{\text{TSP}} \cdot I_{\text{sample}} \cdot [\text{TSP}]}{N_{\text{Met}} \cdot W_{\text{sample}} \cdot I_{\text{TSP}}},$$

where $[M]_{\text{sample}}$ is the cellular concentration of lactate or glucose; N_{TSP} is the number of protons giving rise to the signal integral of TSP, $N_{\text{TSP}} = 9$; N_{Met} is the number of protons giving rise to the signal integral of lactate or glucose ($N_{\text{MET}} = 3$ for lactate and $N_{\text{MET}} = 1$ for

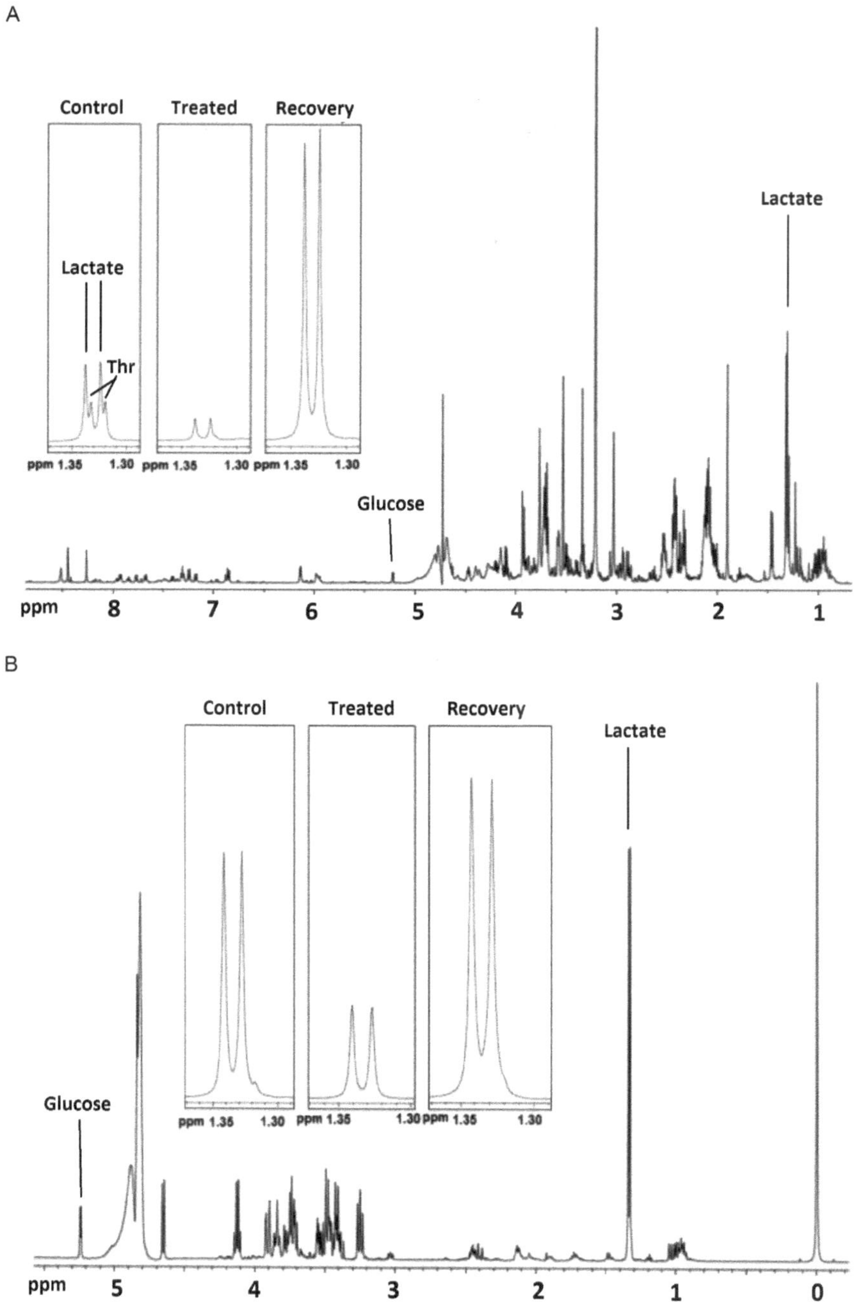

Fig. 2 See legend on next page.

glucose); I_{TSP} and I_{sample} are the signal integrals of TSP and lactate or glucose, respectively; [TSP] is the concentration of TSP; and W_{sample} is the sample cell number, protein concentration, or tissue wet weight.

(6) The mean, standard deviation and standard error of metabolite concentrations for control and treated groups (minimum $n = 3$ per group) are calculated. A 2-tails student T-test on control and treatment groups to assess statistical significance is performed. In this, $^{*}p < 0.05$ is taken to be significant.

3. MEASUREMENTS OF THE RATE OF LACTATE SECRETION AND GLUCOSE UPTAKE IN CULTURED CELLS

3.1 Preparation of Media Samples for ^{1}H-MRS Analysis

(1) Put 500 μL of the culture medium (see Section 2.1, step 2) into a 5-mm NMR tube.

(2) Add 50 μL of D_2O and 50 μL of 0.75% TSP in D_2O in the sample.

3.2 ^{1}H-MRS of Media Samples

(1) ^{1}H-MR spectra of the media samples are acquired at 25°C using a pulse-acquired MR sequence with water suppression (a 1D NOESY presat sequence) in a BBI NMR probe on a 500- or 600-MHz NMR system. For cell culture media samples, the typical NMR acquisition parameters are 7500 Hz spectral width, 32,768 time domain points, 2.7 s repetition time, and 64 scans. NMR spectra from fresh media samples are also acquired under the same conditions.

(2) A ^{1}H-MR spectrum from a culture media sample is shown in Fig. 2B.

(3) Commercially available software packages, such as Bruker Topsin-3.2 or MestRe-C-4.9.9.6, are used to analyze the MR spectra.

(4) Spectra are first processed by using exponential multiplication with a line-broadening factor (lb = 0.3 Hz), and then followed by Fourier-transform,

Fig. 2 An example of ^{1}H-MRS spectra of (A) a cell extract sample and (B) a culture media sample from HCT116 Bax-ko cells. Expanded lactate signals are shown for control, 24-h starved, and 48-h recovered HCT116 Bax-ko cells. *Thr*, threonine. *This figure is adapted from Lin, G., Andrejeva, G., Te Fong, A. C. W., Hill, D. K., Orton, M. R., Parkes, H. G., et al. (2014). Reduced Warburg effect in cancer cells undergoing autophagy: Steady-state H-1-MRS and real-time hyperpolarized C-13-MRS studies.* Plos One, 9.

zero- and first-order phase correction, baseline correction, and spectral peak integration of lactate and glucose.

(5) The rates of lactate being secreted into the culture media or glucose being taken up from the media are measured from a ^{1}H-MR spectrum of the culture medium that has been incubated with the cells for a given time with respect to the fresh culture medium:

$$K_{\text{Uptake/Secretion}} = \frac{(I_{\text{SM}} - I_{\text{CM}}) \cdot N_{\text{TSP}} \cdot V_{\text{Tot}} \cdot [\text{TSP}]}{N_{\text{Met}} \cdot W_{\text{sample}} \cdot I_{\text{TSP}} \cdot V_{\text{tube}} \cdot T},$$

where $K_{\text{Uptake/Secretion}}$ is the rate of glucose uptake from or lactate secreted into the culture media by the cells. A positive value indicates the rate of glucose being taken up from the media by the cells, and a negative value shows the rate of lactate being secreted into the media by the cells. I_{SM} is the signal integral of lactate or glucose in the starting media. I_{CM} is the signal integral of lactate or glucose in the media incubated with the cells. I_{TSP} is the signal integral of TSP. N_{TSP} is the number of protons giving rise to the signal integral of TSP, $N_{\text{TSP}} = 9$. N_{Met} is the number of protons giving rise to the signal integral of lactate or glucose; $N_{\text{MET}} = 3$ for lactate and $N_{\text{MET}} = 1$ for glucose. [TSP] is the concentration of TSP. W_{sample} is the sample cell number or protein concentration. V_{Tot} is the total volume of media incubated with the cells. V_{tube} is the volume of media placed into the NMR tube (500 μL, as described in Section 3.1). Finally, T is the length of time since the last media change.

(6) The mean, standard deviation, and standard error of the rates of excretion or uptake for control and treated groups (minimum $n = 3$ per group) are calculated; and 2-tails student T-test on control and treatment groups to assess statistical significance are performed. Here, $^{*}p < 0.05$ is taken to be significant.

As an addendum, we further note that the protocols outlined in Sections 2.1–3.2 offer a general methodology for extracting metabolites from cells or tissues, as well as their measurement in culture media by MRS. Therefore, the protocols may readily be extended to steady-state ^{13}C-labeling experiments, which require similar extraction protocols to those described here. We do not discuss this methodology further in the current context, but it also would be of interest in the study of autophagic cells, and would be complementary to the real-time hyperpolarized ^{13}C-MR labeling experiments, as discussed later.

4. HYPERPOLARIZATION METHODS FOR DISSOLUTION DNP USING PYRUVIC ACID

DNP exploits the high spin polarization of unpaired electrons in the form of stable-free radicals at a very low temperature. When placed in a strong magnetic field and cooled in liquid helium to temperatures <4.2 K, unpaired electrons are polarized to near unity owing to their large gyromagnetic ratio. By irradiating the sample close to the electron paramagnetic resonance frequency ($\omega_e \pm \omega_C$) (where ω_e and ω_C are the Larmor frequencies of the electron and carbon, respectively, at a field strength of 3.35 T), polarization is transferred from unpaired electrons to surrounding nuclei. This generates a large population difference or hyperpolarization of the associated nuclear spins, as well as a significant enhancement of the NMR signal. Ardenkjaer-Larsen et al. (2003) showed that hyperpolarization generated by DNP at low temperature in the solid state could be conserved if the sample was rapidly melted and dissolved using an aqueous physiological buffer (Ardenkjaer-Larsen et al., 2003; Golman, in't Zandt, Lerche, et al., 2006; Golman, in't Zandt, & Thaning, 2006; Wolber et al., 2004). The concentration of the hyperpolarized molecule in solution is typically in the millimolar range following dissolution, depending on the amount of substrate polarized and the volume of buffer employed for the dissolution. The enhanced signal enables the acquisition of ^{13}C spectra with a single acquisition and therefore the derivation of kinetic rate constants for the interconversion of metabolites in real time.

Samples are polarized in a commercially available Hypersense system (Oxford Instruments, Abington, United Kingdom). This system consists of a superconducting magnet operating at 3.35 T. In the bore of the magnet, there is a variable temperature insert (VTI) containing a microwave cavity fed via a microwave guide tube from a 94-GHz source. The sample chamber within the VTI is filled with liquid helium supplied from the main cryostat of the magnet and fed along a capillary tube with feedback control from a needle valve operated from a stepper motor outside the magnet. The VTI is connected to a rotary vacuum pump to reduce the pressure within it and achieve an operating temperature of 1.4 K.

4.1 Sample Preparation for DNP

[1-^{13}C] pyruvic acid (99% isotopically enriched) is prepared as a stock solution containing 15 m*M* of trityl-free radical OX63 [tris

[8-carboxyl-2,2,6,6-benzo(1,2-*d*:4,5-*d*)-bis(1,3)dithiole-4-yl] methyl sodium salt] and 1 m*M* of gadolinium MRI contrast agent in the form of Dotarem (Gadoteric acid: gadolinium complex of 1, 4, 7, 10 tetraazacyclododedane-*N*,*N*′,*N*″,*N*‴ tetraacetic acid).

(1) A 1-g bottle of pyruvic acid is used to prepare a stock (Sigma-Aldrich, Poole, United Kingdom, 99% [1-^{13}C], MW = 89 g/mol). Measure the total volume using a Gilson pipette (about 850 μL/g) and put it into an Eppendorf.

(2) Weigh 18.4 mg of trityl radical (OX063, MW = 1426.8 g/mol) into a separate Eppendorf using an analytical balance and dissolve by adding the pyruvic acid and vortexing for 2 min.

(3) Make a stock of Dotarem (gadoterate meglumine; Guerbet, Villepinte, France, 0.5 *M* concentration) and dilute 1:5 with deionized water to make a 100 m*M* solution.

(4) Add 8.5 μL of 100 m*M* Dotarem stock to the pyruvic acid stock and vortex for 2 min to achieve a final concentration of 1 m*M* [Gd].

(5) Wrap the stock polarization solution in foil and keep in a −20°C freezer in the dark to prevent photolysis of the radical.

(6) Thaw the stock when required for an experiment and vortex to ensure a homogeneous solution.

(7) For in vitro cell experiments, pipette and weigh 14 μL (18 mg) of the polarization solution into a sample cup corresponding to 200 μmol of ^{13}C labeled pyruvic acid.

4.2 Operation of the DNP Polarizer

(1) Initialize the Hypersense system by pressing the Cooldown tab. The pump is turned on to evacuate the VTI, and once a vacuum is achieved for 10–20 s, the needle valve gradually opens from 0% (closed) to 99% (fully open) to start filling the VTI with liquid helium.

(2) Within about 5 min, the system achieves operating conditions with the VTI filled to 65% helium and a temperature of 1.4 K. The system is ready to insert a sample.

(3) Attach the sample cup containing polarization solution to the end of the insertion stick.

(4) Press the Insert Sample button. An overpressure of helium gas is blown into the VTI to prevent contamination from condensation of air. The sample holder is pneumatically raised out of the helium bath, and the access valve opens to allow insertion of the sample.

(5) Insert the sample stick through the access valve to the bottom of the sample holder and pull the sample release button from the stick to leave the sample cup in situ. Remove the sample stick and press Finished.

(6) Once inserted, the access valve is closed to reequilibrate to the preset helium level (65%) and temperature (1.4 K).

(7) Run the Hypersense polarimeter and press Solid State Build-up to enable build-up monitoring of the polarization using the integrated ^{13}C NMR coil for detection.

(8) Select Polarize Sample and enter a measurement interval (typically 300 s) and check Finished. The microwave source is turned on with the selected frequency of $\omega_e - \omega_C = 94.1$ GHz. Polarization build-up is recorded every 5 min. For [1-^{13}C] pyruvic acid polarization typically takes about 1 h (build-up time constant in the order of 15 min).

4.3 Dissolution

(1) For in vitro cell experiments employing pyruvic acid, make up a stock aqueous phosphate buffer containing 1 m*M* of EDTA and 50 m*M* of sodium lactate (Sigma Aldrich, MW = 112G). The in vitro cell experiment includes unlabeled lactate in the buffer (adapted from that presented by Day et al. (2007) to ensure that the LDH dynamics are at or close to equilibrium so that both forward and reverse reactions are taking place and therefore that both NAD and NADH are concomitantly interconverted.

(2) Inject 4 mL of aqueous buffer into the solvent heater vessel at the top of the dissolution stick.

(3) For in vitro cell experiments, pipette 20 μL of 10 *M* NaOH into the bottom of the collection vessel. Attach the collection vessel to the exit tube of the polarizer.

(4) Position the dissolution stick at the entrance port to the VTI. Press the Run Dissolution button. The buffer is heated to a high temperature (about 200°C) while gradually elevating the pressure to 10 bar.

(5) When the solvent pressure reaches 10 bar, the dissolution stick is pneumatically driven into the VTI (while applying an overpressure of helium gas to prevent contamination) and engages with the sample cup, and there is instantaneous mixing of the frozen, hyperpolarized sample and hot buffer. The solution is ejected under pressure into the collection vessel, followed by a chase gas to ensure the complete retrieval of hyperpolarized solution.

(6) The hyperpolarized ^{13}C signal decays with its characteristic solution state T_1; therefore, it must be used immediately. During transfer of the solution between the magnetic field of the polarizer and the field of the MR spectrometer, the sample is subjected to significant magnetic field gradients, and zero field crossing points should be avoided or the sample maintained in a constant field during transit.

4.4 In Vitro Cell Assays

In the following protocol, we describe an in vitro cancer cell suspension assay for assessing the rates of ^{13}C-pyruvate to lactate exchange. Such approaches have been widely used in the research literature. However, some drawbacks should be noted, including the possibility of cell settling during the experiment, which could lead to possible changes in field homogeneity over the course of an experiment. There is also a time delay between harvesting the cells and performing the assay, which should be kept to a minimum to ensure good cellular viability while maintaining temperature constant at 37°C. In addition, in the assay described here, cells are mixed with hyperpolarized pyruvate outside the bore of the magnet, and the temporal data are therefore missing the initial 5–10 s of the kinetic time points, which could potentially contain important information for the fitting. Some of these issues have been addressed with the introduction of bioreactors that allow in situ measurement of hyperpolarized ^{13}C kinetics from cells grown on beads or encapsulated cells (Harris et al., 2009; Keshari et al., 2010; Ward et al., 2010).

(1) The NMR spectrometer is set up with a broad-band-observe (BBO) NMR probe and is tuned and matched on a nonhyperpolarized ^{13}C-pyruvate solution in the same dissolution buffer with 10% D_2O for the lock signal. After shimming on the deuterium lock signal, a dummy thermal ^{13}C-MR spectrum is acquired with the same parameters as the hyperpolarized experiment (later) and a receiver gain of 1024 to ensure functioning of the spectrometer. The test sample is then removed, and the lock and sweep are turned off prior to the hyperpolarized experiment.

(2) Harvest the cultured cells by trypsinization. Wash the cell pellet with culture media (without serum or pyruvate added) twice and spin down the cell pellet after each wash. Perform a cell count measurement.

(3) Resuspend the cells in 500 μL of culture media (serum and pyruvate free) in an Eppendorff tube and maintain at 37°C. Use fresh cells within 10 min of harvesting to ensure good viability.

(4) Using a 1 mL of Gilson, pipette 500 μL of cell suspension into a clean, 5-mm NMR tube. Maintain the cell temperature at 37°C using a heater block.

(5) Create a new data set and start the NMR acquisition prior to inserting the sample. ^{13}C spectra are acquired every 2 s in a pseudo-two-dimensional experiment using a single scan and a 10-degree flip angle and 128 spectra acquired in 4 min 16 s total acquisition time. ^{13}C spectra are acquired with WALTZ16 ^{1}H decoupling, a sweep width of 150 ppm (18,797 Hz), carrier frequency positioned at 150 ppm, a time domain of 16,384 data points in the FID, yielding an acquisition time of 0.436 s, a prescan delay of 1.564 s to yield a repetition time of 2 s, and the receiver gain set to 32.

(6) Following dissolution, immediately inject 100 μL of hyperpolarized solution into the NMR tube. Mix vigorously by inverting the tube and shaking to yield a homogenous suspension of cells with a final concentration of 8 m*M* hyperpolarized [1-^{13}C] pyruvate and 8 m*M* unpolarized ^{12}C lactate.

(7) Insert the sample into the NMR spectrometer.

(8) At the end of the experiment, spectra are Fourier-transformed with an exponential line-broadening factor (lb = 3 Hz) and phase and baseline corrected. Integrate the peak areas of lactate and pyruvate over the time course of the experiment using Bruker Topsin-3.2 software.

(9) An example of fitting the temporal data of pyruvate and lactate from a HCT116 Bax-ko cell suspension is shown in Fig. 3A.

(10) Using Topspin software, the sum projection of the time series of spectra are calculated to yield a single 1D ^{13}C-MR spectrum, and the spectrum are saved in a separate protocol.

(11) An example of sum spectra showing the sum over the entire dynamic time series from a control HCT116 Bax-ko cell experiment, following 6 h of starvation with HBSS media, 24 h of starvation with HBSS media, and 24 h of starvation, followed by 48 h cell recovery in full media, is illustrated in Fig. 3B.

4.5 Kinetic Modeling

Kinetic modeling is carried out by fitting the time series of peak areas of pyruvate and lactate with the modified Bloch equations, as described in Hill et al. (2013). A number of alternative methods have been proposed in the literature for the kinetic analysis of hyperpolarized data, either in vitro (such as the

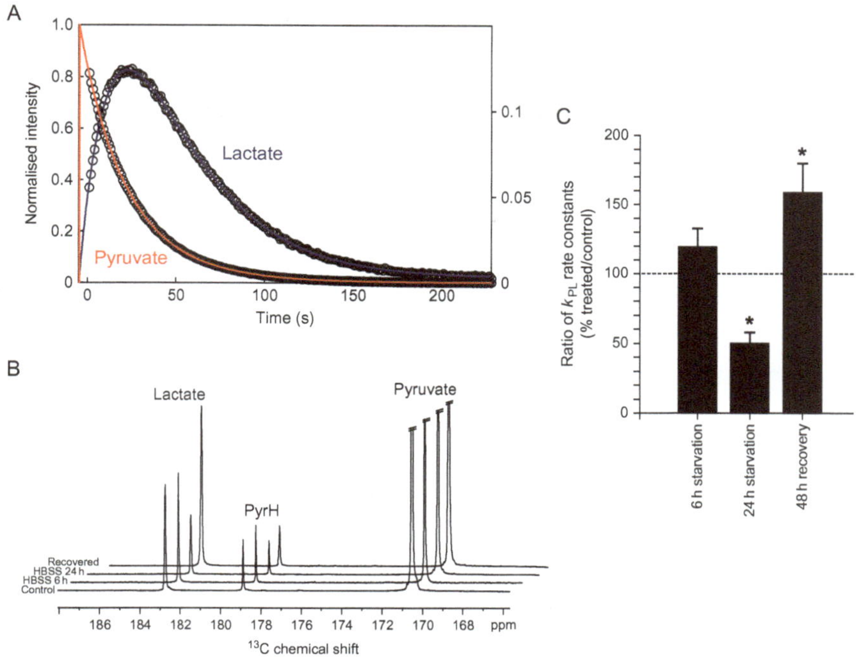

Fig. 3 Hyperpolarized ^{13}C-data from an HCT116 Bax-ko cell assay. (A) Example fitting of the temporal data of pyruvate (*red*—left-hand scale) and lactate (*blue*—right-hand scale), from which apparent exchange rate constants can be derived. (B) Example of sum spectra showing the sum over the entire dynamic time series from a control cell experiment, following 6 h of starvation with HBSS media, 24 h of starvation with HBSS media, and 24 h of starvation, followed by 48 h cell recovery in full media. The spectrum displays peaks from pyruvate, lactate, and pyruvate hydrate (PyrH). (C) The mean ratios (treated/24 h vehicle control) of the apparent forward reaction rate of pyruvate to lactate exchange k_{PL} (derived from fitting of the experimental data and normalized to cell number) (± s.e.m.) for 6 h of HBSS starvation, 24 h of HBSS starvation, and 24 h starvation, followed by 48 h recovery in HCT116 Bax-ko cells. Minimum $n=3$ in each group. Statistically significant changes are indicated ($^{*}p \leq 0.05$). *This figure is adapted from Lin, G., Andrejeva, G., Te Fong, A. C. W., Hill, D. K., Orton, M. R., Parkes, H. G., et al. (2014). Reduced Warburg effect in cancer cells undergoing autophagy: Steady-state H-1-MRS and real-time hyperpolarized C-13-MRS studies.* Plos One, 9.

experiments presented here), *ex vivo* in perfused organs, or in vivo. A recent paper has compared different quantitative and semiquantitative methods for assessing ^{13}C-pyruvate metabolism in vivo using a SpinLab system and imaging performed on a clinical scanner (Daniels et al., 2016). These include fitting directly to the differential equations describing a given reaction model (which may assume one-way or two-way kinetics), fitting to the integrated solution to a given differential model, or semiquantitative indices such as the time-to-peak, ratio of the lactate to pyruvate peak heights, or ratios of the areas under the curve (AUCs) of the lactate and pyruvate peaks. The methods available for characterizing hyperpolarized kinetics are usually considered to be semiquantitative, particularly for in vivo experiments, since it is difficult to measure absolute local concentrations of hyperpolarized substrate delivered nor the endogenous pool sizes of metabolites or cofactors involved in a given enzyme reaction. For in vitro experiments, we are able to relate the kinetics to the absolute concentration of substrate delivered, which can be informative compared with steady-state measurements of endogenous pools of intracellular and excreted metabolites as described earlier.

(1) The pyruvate and lactate temporal data are fitted simultaneously using a program written in MATLAB (Mathworks, Natick, Massachusetts). Apparent rate constants of the reaction are derived by kinetic modeling to a first-order, two-site exchange model (Day et al., 2007; Spielman et al., 2009; Zierhut et al., 2010). The program yields estimates of the forward (k_{PL}) and reverse (k_{LP}) rate constants, as well as estimates of the relaxation times of pyruvate (T_{1P}) and lactate (T_{1L}). Divide the fitted rates by the number of cells in the sample to yield an effective rate constant per million cells. The rate of pyruvate to lactate exchange in controls and autophagic cells can then be compared (see the example in Fig. 3C).

(2) The lactate and pyruvate peaks are integrated from the summed ^{13}C-MR spectra derived in Section 4.4, step 7. The ratio of the AUC corresponding to the ratio of the integrals of lactate/pyruvate is calculated. The AUC is divided by the number of cells in the sample to yield an effective AUC per million cells.

(3) The mean, standard deviation, and standard error of k_{PL} derived from the fitting and AUC for different treatment groups (minimum $n=3$ per group) are calculated. The 2-tails student T-test applied to vehicle and treatment groups is performed to assess statistical significance. Here, $^{*}p<0.05$ is taken to be significant.

5. DISCUSSION

We have described in this chapter the use of in vitro steady-state ^{1}H-MRS methods and real-time hyperpolarized ^{13}C-MRS for probing glycolytic metabolism in cells undergoing autophagy. The methods report on the steady-state rates of glucose uptake and lactate excretion, endogenous concentration of lactate, and the rate constants for the exchange of pyruvate to lactate. In turn, these measurements are sensitive to a range of cellular parameters, including GLUT, downstream rates of glycolysis, LDH expression and activity, the ratio of the cofactors NAD^+ and NADH, and MCT-mediated influx and efflux of pyruvate and lactate. The link between autophagy and metabolism is poorly understood, but it is likely that autophagy induces multiple cellular metabolic adaptations in cancer cells, one manifestation of which may be a reduced reliance on glycolysis, as reported by Lin, Hill, et al. (2014) and Lin, Andrejeva, et al. (2014). The experiments that we have proposed here give information on glycolytic changes that may reflect not only the autophagic cell processes per se, but also any direct effects that the treatments may have on metabolism. Thus, observations using these techniques should be validated with standard biochemical assays. The methods that we describe here in the context of autophagy have a general applicability in other cell types and will be sensitive to a range of cellular processes and treatment regimes.

ACKNOWLEDGMENTS

This work is supported by funding received from the CR-UK Cancer Imaging Centre, in association with the MRC and Department of Health (England) Grant C1060/A10334, C1060/A16464, and NHS funding to the NIHR Biomedical Research Centre.

REFERENCES

Ardenkjaer-Larsen, J. H., Fridlund, B., Gram, A., Hansson, G., Hansson, L., Lerche, M. H., et al. (2003). Increase in signal-to-noise ratio of >10,000 times in liquid-state NMR. *Proceedings of the National Academy of Sciences of the United States of America, 100*, 10158–10163.

Beckonert, O., Keun, H. C., Ebbels, T. M. D., Bundy, J. G., Holmes, E., Lindon, J. C., et al. (2007). Metabolic profiling, metabolomic and metabonomic procedures for NMR spectroscopy of urine, plasma, serum and tissue extracts. *Nature Protocols, 2*, 2692–2703.

Buescher, J. M., Antoniewicz, M. R., Boros, L. G., Burgess, S. C., Brunengraber, H., Clish, C. B., et al. (2015). A roadmap for interpreting C-13 metabolite labeling patterns from cells. *Current Opinion in Biotechnology, 34*, 189–201.

Chaumeil, M. M., Najac, C., & Ronen, S. M. (2015). Studies of metabolism using (13)C MRS of hyperpolarized probes. *Methods in Enzymology, 561*, 1–71.

Christensen, C. E., Karlsson, M., Winther, J. R., Jensen, P. R., & Lerche, M. H. (2014). Non-invasive in-cell determination of free cytosolic [NAD+]/[NADH] ratios using hyperpolarized glucose show large variations in metabolic phenotypes. *The Journal of Biological Chemistry, 289*, 2344–2352.

Chung, Y.-L., Basetti, M., & Griffiths, J. R. (2015). Metabolism and metabolomics by MRS. *eMagRes, 4*, 689–698.

Chung, Y. L., & Griffiths, J. R. (2011). Metabolomic studies on cancer and on anticancer drugs by NMR ex vivo. In R. K. Harris & R. E. Wasylishen (Eds.), *Encyclopedia of magnetic resonance*. Chichester: John Wiley. http://dx.doi.org/10.1002/9780470034590.emrstm1093.

Comment, A., & Merritt, M. E. (2014). Hyperpolarized magnetic resonance as a sensitive detector of metabolic function. *Biochemistry*, *53*, 7333–7357.

Daniels, C. J., McLean, M. A., Schulte, R. F., Robb, F. J., Gill, A. B., McGlashan, N., et al. (2016). A comparison of quantitative methods for clinical imaging with hyperpolarized C-13-pyruvate. *NMR in Biomedicine*, *29*, 387–399.

Day, S. E., Kettunen, M. I., Gallagher, F. A., Hu, D. E., Lerche, M., Wolber, J., et al. (2007). Detecting tumor response to treatment using hyperpolarized C-13 magnetic resonance imaging and spectroscopy. *Nature Medicine*, *13*, 1382–1387.

de Graaf, R. A. (2007). *In vivo NMR spectroscopy* (2nd ed.). Chichester: John Wiley & Sons, Ltd.

DeBerardinis, R. J., Mancuso, A., Daikhin, E., Nissim, I., Yudkoff, M., Wehrli, S., et al. (2007). Beyond aerobic glycolysis: Transformed cells can engage in glutamine metabolism that exceeds the requirement for protein and nucleotide synthesis. *Proceedings of the National Academy of Sciences of the United States of America*, *104*, 19345–19350.

Golman, K., in't Zandt, R., Lerche, M., Pehrson, R., & Ardenkjaer-Larsen, J. H. (2006a). Metabolic imaging by hyperpolarized C-13 magnetic resonance imaging for in vivo tumor diagnosis. *Cancer Research*, *66*, 10855–10860.

Golman, K., in't Zandt, R., & Thaning, M. (2006b). Real-time metabolic imaging. *Proceedings of the National Academy of Sciences of the United States of America*, *103*, 11270–11275.

Golman, K., & Petersson, J. S. (2006). Metabolic imaging and other applications of hyperpolarized C-13. *Academic Radiology*, *13*, 932–942.

Hanahan, D., & Weinberg, R. A. (2000). The hallmarks of cancer. *Cell*, *100*, 57–70.

Hanahan, D., & Weinberg, R. A. (2011). Hallmarks of cancer: The next generation. *Cell*, *144*, 646–674.

Harris, T., Eliyahu, G., Frydman, L., & Degani, H. (2009). Kinetics of hyperpolarized C-13 (1)-pyruvate transport and metabolism in living human breast cancer cells. *Proceedings of the National Academy of Sciences of the United States of America*, *106*, 18131–18136.

Hill, D. K., Orton, M. R., Mariotti, E., Boult, J. K. R., Panek, R., Jafar, M., et al. (2013). Model free approach to kinetic analysis of real-time hyperpolarized C-13 magnetic resonance spectroscopy data. *PloS One*, *8*, e71996. http://dx.doi.org/10.1371/journal.pone.0071996. eCollection 2013.

Keshari, K. R., Kurhanewicz, J., Jeffries, R. E., Wilson, D. M., Dewar, B. J., Van Criekinge, M., et al. (2010). Hyperpolarized C-13 spectroscopy and an NMR-compatible bioreactor system for the investigation of real-time cellular metabolism. *Magnetic Resonance in Medicine*, *63*, 322–329.

Keshari, K. R., & Wilson, D. M. (2014). Chemistry and biochemistry of C-13 hyperpolarized magnetic resonance using dynamic nuclear polarization. *Chemical Society Reviews*, *43*, 1627–1659.

Kurhanewicz, J., Vigneron, D. B., Brindle, K., Chekmenev, E. Y., Comment, A., Cunningham, C. H., et al. (2011). Analysis of cancer metabolism by imaging hyperpolarized nuclei: Prospects for translation to clinical research. *Neoplasia*, *13*, 81–97.

Lin, G., Andrejeva, G., Te Fong, A. C. W., Hill, D. K., Orton, M. R., Parkes, H. G., et al. (2014). Reduced Warburg effect in cancer cells undergoing autophagy: Steady-state H-1-MRS and real-time hyperpolarized C-13-MRS studies. *PloS One*, *9*, e92645. http://dx.doi.org/10.1371/journal.pone.0092645. eCollection 2014.

Lin, G., Hill, D. K., Andrejeva, G., Boult, J. K. R., Troy, H., Fong, A. C. L. F. W. T., et al. (2014). Dichloroacetate induces autophagy in colorectal cancer cells and tumours. *British Journal of Cancer*, *111*, 375–385.

Park, I., Larson, P. E. Z., Zierhut, M. L., Hu, S., Bok, R., Ozawa, T., et al. (2010). Hyperpolarized C-13 magnetic resonance metabolic imaging: Application to brain tumors. *Neuro-Oncology*, *12*, 133–144.

Spielman, D. M., Mayer, D., Yen, Y. F., Tropp, J., Hurd, R. E., & Pfefferbaum, A. (2009). In vivo measurement of ethanol metabolism in the rat liver using magnetic resonance spectroscopy of hyperpolarized [1-C-13]pyruvate. *Magnetic Resonance in Medicine, 62*, 307–313.

Tyagi, R. K., Azrad, A., Degani, H., & Salomon, Y. (1996). Simultaneous extraction of cellular lipids and water-soluble metabolites: Evaluation by NMR spectroscopy. *Magnetic Resonance in Medicine, 35*, 194–200.

Warburg, O. (1956). On the origin of cancer cells. *Science, 123*, 309–314.

Ward, C. S., Venkatesh, H. S., Chaumeil, M. M., Brandes, A. H., VanCriekinge, M., Dafni, H., et al. (2010). Noninvasive detection of target modulation following phosphatidylinositol 3-kinase inhibition using hyperpolarized (13)C magnetic resonance spectroscopy. *Cancer Research, 70*, 1296–1305.

Witney, T. H., Kettunen, M. I., Day, S. E., Hu, D. E., Neves, A. A., Gallagher, F. A., et al. (2009). A comparison between radiolabeled fluorodeoxyglucose uptake and hyperpolarized C-13-labeled pyruvate utilization as methods for detecting tumor response to treatment. *Neoplasia, 11*, 574–582.

Wolber, J., Ellner, F., Fridlund, B., Gram, A., Johannesson, H., Hansson, G., et al. (2004). Generating highly polarized nuclear spins in solution using dynamic nuclear polarization. *Nuclear Instruments & Methods in Physics Research Section A, 526*, 173–181.

Zierhut, M. L., Yen, Y. F., Chen, A. P., Bok, R., Albers, M. J., Zhang, V., et al. (2010). Kinetic modeling of hyperpolarized 13C1-pyruvate metabolism in normal rats and TRAMP mice. *Journal of Magnetic Resonance, 202*, 85–92.

CHAPTER NINE

Assessment of Glycolytic Flux and Mitochondrial Respiration in the Course of Autophagic Responses

V. Sica[*,†,‡,§,¶,||], **J.M. Bravo-San Pedro**[*,†,‡,§,¶], **F. Pietrocola**[*,†,‡,§,¶], **V. Izzo**[*,†,‡,§,¶], **M.C. Maiuri**[*,†,‡,§,¶], **G. Kroemer**[†,‡,§,¶,#,**,††,1], **L. Galluzzi**[*,†,‡,§,¶,‡‡,1]

[*]Gustave Roussy Cancer Campus, Villejuif, France
[†]INSERM, U1138, Paris, France
[‡]Equipe 11 labellisée par la Ligue Nationale contre le Cancer, Centre de Recherche des Cordeliers, Paris, France
[§]Université Paris Descartes/Paris V, Sorbonne Paris Cité, Paris, France
[¶]Université Pierre et Marie Curie/Paris VI, Paris, France
[||]Faculté de Medicine, Université Paris Saclay/Paris XI, Le Kremlin-Bicêtre, France
[#]Pôle de Biologie, Hôpital Européen Georges Pompidou, AP-HP, Paris, France
[**]Metabolomics and Cell Biology Platforms, Gustave Roussy Cancer Campus, Villejuif, France
[††]Karolinska University Hospital, Stockholm, Sweden
[‡‡]Weill Cornell Medical College, New York, NY, United States
[1]Corresponding authors: e-mail address: kroemer@orange.fr; deadoc@vodafone.it

Contents

Abstract

Autophagy is an evolutionarily conserved process that mediates prominent homeostatic functions, both at the cellular and organismal level. Indeed, baseline autophagy not only ensures the disposal of cytoplasmic entities that may become cytotoxic upon accumulation, but also contributes to the maintenance of metabolic fitness in physiological conditions. Likewise, autophagy plays a fundamental role in the cellular and organismal adaptation to homeostatic perturbations of metabolic, physical, or chemical

Methods in Enzymology, Volume 588
ISSN 0076-6879
http://dx.doi.org/10.1016/bs.mie.2016.09.079

nature. Thus, the molecular machinery for autophagy is functionally regulated by a broad panel of sensors that detect indicators of metabolic homeostasis. Moreover, increases in autophagic flux have a direct impact on core metabolic circuitries including (but not limited to) glycolysis and mitochondrial respiration. Here, we detail a simple methodological approach to monitor these two processes in cultured cancer cells that mount a proficient autophagic response to stress.

1. INTRODUCTION

Autophagy is an evolutionarily ancient process that culminates with the lysosomal degradation of superfluous, old, or damaged (and hence potentially dangerous) cytoplasmic structures (Codogno, Mehrpour, & Proikas-Cezanne, 2012). One specific form of autophagy, i.e., macroautophagy, involves the sequestration of autophagic substrates by double-membraned organelles that are known as "autophagosomes." Upon closure, autophagosomes fuse with lysosomes (to form so-called autolysosomes), followed by lysosomal acidification, activation of lysosomal hydrolases, and digestion of autophagic substrates (Codogno et al., 2012; Lamb, Yoshimori, & Tooze, 2013). For the sake of simplicity, we will refer to macroautophagy as to autophagy from here onward.

In physiological conditions, autophagy operates at baseline levels to mediate prominent homeostatic functions, at both the cellular and organismal levels (Ma, Galluzzi, Zitvogel, & Kroemer, 2013; Menzies, Fleming, & Rubinsztein, 2015). Thus, basal autophagy prevents the accumulation of potentially cytotoxic products of ordinary cellular activities, including (but not limited to) damaged mitochondria and redox-active protein aggregates (Green, Galluzzi, & Kroemer, 2011). Moreover, autophagy contributes to the preservation of physiological metabolic fitness (Kaur & Debnath, 2015). In line with this notion, the whole-body deletion of autophagy related 5 (*Atg5*), which encodes a central component of the molecular machinery for autophagy, is linked to perinatal lethality with 100% penetrance (Kuma et al., 2004). Along similar lines, the tissue-specific or heterozygous deletion of other genes that are required for proficient autophagic responses, like *Atg7*, beclin 1 (*Becn1*), or *Becn2*, imposes robust metabolic alterations and accelerates the pathogenesis of various diseases (He et al., 2013; Jung et al., 2008).

Autophagy is also a key player in adaptation to stress (Kroemer, Marino, & Levine, 2010; Rubinsztein, Codogno, & Levine, 2012). In particular, proficient autophagic responses generally support cellular viability in the presence of homeostatic perturbations, including metabolic/nutritional, physical, and chemical cues. Accordingly, pharmacological agents as well as genetic manipulations that compromise autophagic proficiency normally accelerate the demise of cells responding to stress (Galluzzi et al., 2015). Such adaptive autophagic responses are regulated by sensors that continuously monitor indicators of cellular fitness at specific subcellular compartments (Galluzzi, Pietrocola, Levine, & Kroemer, 2014; Sica et al., 2015). The majority of autophagic sensors operate through signal transduction cascades that hinge on the activation state of mechanistic target of rapamycin (MTOR) complex 1 (MTORC1), a key suppressor of autophagy (Zoncu, Efeyan, & Sabatini, 2011), and several of them actually respond to metabolic parameters, including ATP/AMP ratio, NAD^+/NADH ratio, and many others (Galluzzi et al., 2014). Thus, multiple perturbations of metabolic homeostasis are capable of increasing or decreasing autophagic flux (Galluzzi, Kepp, Vander Heiden, & Kroemer, 2013).

Furthermore, autophagic activity has a direct impact on various core metabolic circuitries, including (but presumably not limited to) glycolysis, mitochondrial respiration, lipolysis, glycogen breakdown, and fatty acid oxidation (Kaur & Debnath, 2015; Kim & Lee, 2014). Thus, lung tumors driven by mutant KRAS or BRAF cannot progress to adenocarcinoma but rather form relatively benign oncocytomas in the absence of *Atg7*, and this is associated with reduced fatty acid oxidation, lipid accumulation, and compromised mitochondrial function (Guo et al., 2013; Strohecker et al., 2013). Similarly, mouse embryonic fibroblasts lacking *Atg5* accumulate triglycerides owing to the defective autophagosomal transfer of lipids from lipid droplets to lysosomes (Singh et al., 2009). These examples reinforce the notion that autophagy is intimately connected with intermediate metabolism, both as it responds to metabolic fluctuations and as it impacts on various metabolic circuitries.

Here, we provide a detailed description of a simple assay for quantitatively monitoring glycolysis and mitochondrial respiration in cultured cancer cells that mount an adaptive autophagic response to stress. In this setting, flux analysis is employed to measure extracellular acidification rate (ECAR) as an indicator of glycolysis-dependent lactate secretion, and oxygen consumption rate (OCR) as an indicator of mitochondrial respiration. With some modifications, this method can be easily adapted to assess glycolysis and mitochondrial respiration in a wide panel of cells responding to autophagy-promoting stimuli in vitro.

2. CELL CULTURE AND TREATMENTS

1. Upon thawing, human colorectal carcinoma HCT 116 cells display optimal adherent growth in McCoy's 5A medium containing 3.0 g/L D-glucose and 1.50 m*M* L-glutamine, supplemented with 1 m*M* sodium pyruvate, 100 m*M* 2-[4-(2-hydroxyethyl)piperazin-1-yl]ethanesulfonic acid (HEPES) buffer, and 10% fetal calf serum (see Note 1). Cells are routinely maintained in standard culture conditions (37°C, 5% CO_2) in conventional 75-cm^2 cell culture flasks (see Note 2).
2. Before full confluence is attained, HCT 116 cells are gently washed with prewarmed PBS, detached with 0.25% trypsin/EDTA, diluted, and reseeded (see Notes 3–8). This allows for the propagation of maintenance cultures (see Notes 9 and 10) and provides cells for experimental determinations.
3. For the latter, 18×10^3 HCT 116 cells are seeded in XF96 Polystyrene Cell Culture Microplates (Seahorse Bioscience, Billerica, MA) in 80 μL complete culture medium, and allowed to adapt and recover normal growth behavior for at least 24 h (see Notes 11 and 12). Background correction wells (A1, A12, H1, H12) should NOT contain cells, but medium only.
4. Exhausted culture medium is discarded and replaced with 100 μL fresh medium (see Note 13), 100 μL fresh medium supplemented with 10 μ*M* rapamycin (see Note 14), 100 μL Hanks' balanced salt solution (HBSS) (see Note 14), and/or other experimental conditions that are expected to promote autophagy.
5. After 6–12 h (see Note 15), cells are processed for the assessment of glycolytic flux (see Sections 5 and 8) or mitochondrial respiration (see Sections 7 and 8).

3. COMMON PROCEDURES

On the day prior to the assay:

1. Fill each well of the Utility Plate with 200 μL of XF Calibrant.
2. Place the Sensor Cartridge upside down onto the Utility Plate, making sure that sensors are fully immersed in XF Calibrant (see Note 16).
3. Incubate overnight at 37°C in a humidified incubator (see Note 17).
4. Turn the XF^e Extracellular Flux Analyzer on and allow it to warm for at least 5 h prior to the initiation of the assay (see Note 18).

4. ASSESSMENT OF GLYCOLYTIC FLUX: PRINCIPLES

Glycolysis, which is a cytoplasmic process, begins with the phosphorylation of glucose into glucose-6-phosphate (G6P), an ATP-dependent and ubiquitous reaction that is catalyzed by hexokinase (or glucokinase, in some tissues). G6P undergoes several other enzymatic transformations that eventually produce two molecules of pyruvate, two molecules of NADH, and two molecules of ATP (net production). In aerobic conditions, pyruvate is promptly taken up by mitochondria and converted into acetyl-coenzyme A (acetyl-CoA) by pyruvate dehydrogenase, hence fueling the Krebs's cycle (Pietrocola, Galluzzi, Bravo-San Pedro, Madeo, & Kroemer, 2015). Conversely, in anaerobic conditions pyruvate is reduced into lactate by cytosolic lactate dehydrogenase, an NADH-dependent reaction that preserves the cytosolic NAD^+ pool (which is required for glycolysis to persist). Lactate is rapidly expulsed into the extracellular microenvironment via various members of the monocarboxylate transporter protein family, which operate driven by the transmembrane H^+ gradient. Since lactate is a weak acid, the net effect of glycolysis is the acidification of the extracellular microenvironment (Kaur & Debnath, 2015).

Some other cellular processes, however, tend to acidify the milieu, which is a potential source of bias when extracellular pH is monitored as an indicator of glycolytic proficiency. The Seahorse XF Glycolysis Stress Test (Seahorse Bioscience, Billerica, MA) provides a simple and convenient method to measure glycolytic proficiency based on the timed assessment of ECAR coupled to the sequential administration of these molecules: glucose, oligomycin, and 2-deoxyglucose (2-DG). The ECAR is initially measured in the complete absence of glucose, when it reflects nonglycolytic extracellular acidification. The addition of glucose rapidly initiates glycolysis and is employed to determine the normal glycolytic capacity. Since cells are in aerobic conditions; however, a significant fraction of glycolytic pyruvate is diverted to the Krebs' cycle. To quantify maximal glycolytic capacity, oligomycin (an inhibitor of respiratory complex V, see later) (Galluzzi et al., 2013) is administered. Finally, the hexokinase inhibitor 2-DG (Galluzzi et al., 2013) is employed to completely abolish glycolytic flux. Thus, the ECAR values measured throughout the assay directly represent or can be employed to calculate: (1) nonglycolytic extracellular acidification (ECAR at baseline or ECAR after the addition of 2-DG); (2) glycolytic capacity (ECAR after the addition of glucose − ECAR at baseline); (3) maximal glycolytic capacity (ECAR after the addition of oligomycin − ECAR after the addition of 2-DG); and (4) glycolytic reserve (ECAR after the addition of oligomycin − ECAR after the addition of glucose). Here below, we detail a simple protocol to interrogate glycolytic flux in cultured cells by means of the Seahorse XF Glycolysis Stress Test.

5. ASSESSMENT OF GLYCOLYTIC FLUX: PROTOCOL

On the day of the assay:

1. Reconstitute powdered Dulbecco's modified Eagle's medium (DMEM; #D5030; Sigma-Aldrich, St. Louis, MO) in 990 mL of tissue culture-grade water and supplement with 143 m*M* NaCl, 3 mg/L phenol red, and 2 m*M* L-glutamine to generate Assay Medium (see Note 19).
2. Warm the Assay Medium to 37°C, adjust the pH to 7.35 ± 0.05, sterilize by passing it through a 0.2 μm Ø filter, and maintain it at 37°C until use.
3. Discard exhausted culture medium and rinse cells two times with 300 μL Assay Medium per well (see Note 20).
4. Add 175 μL Assay Medium per well and place cells at 37°C in a humidified incubator for 45–60 min (see Note 17).
5. Place the B/C Loading Guide on top of the Sensor Cartridge and orient it so that the letter "B" is located in the upper left corner, opposed to the triangular notch (see Notes 21 and 22).
6. Load ports B of the Sensor Cartridge with 80 m*M* glucose (25 μL per port per well), freshly prepared in Assay Medium and warmed to 37°C (see Notes 23–26).
7. Reorient the B/C Loading Guide so that the letter "C" is located in the upper left corner, opposed to the triangular notch (see Note 22).
8. Load ports C of the Sensor Cartridge with 9 μ*M* oligomycin (25 μL per port per well), freshly prepared in Assay Medium and warmed to 37°C (see Notes 23–26).
9. Place the A/D Loading Guide on top of the Sensor Cartridge and orient it so that the letter "D" is located in the upper left corner, opposed to the triangular notch (see Notes 21 and 22).
10. Load ports D of the Sensor Cartridge with 100 m*M* 2-DG (25 μL per port per well), freshly prepared in Assay Medium and warmed to 37°C (see Notes 23–26).

11. Discard the Loading Guide, check all ports for equal loading (see Notes 27–30), and proceed to calibration and run (see Section 8).

6. ASSESSMENT OF MITOCHONDRIAL RESPIRATION: PRINCIPLES

The mitochondrial respiratory chain (also known as electron transport chain, ETC) is a highly conserved multicomponent system of the inner mitochondrial membrane (IMM). The ETC is composed of five distinct supramolecular complexes and two electron shuttles that operate in concert to (1) generate an electrochemical gradient across the IMM (complexes I–IV; coenzyme Q, cytochrome *C*) and (2) dissipate such gradient in a controlled manner to catalyze ATP synthesis (complex V). The two main substrates of the ETC are NADH and succinate, both of which are generated in the mitochondrial matrix by the Krebs' cycle (Pietrocola et al., 2015). NADH is oxidized by complex I into NAD, while succinate is oxidized by complex II into fumarate. Both these reactions generate electrons that are shuttled to complex III by coenzyme Q. However, only the oxidation of NADH (but not that of succinate) results in the extrusion of protons from the mitochondrial matrix. The electrons derived from coenzyme Q are passed on to cytochrome *C* via complex III, along with the translocation of additional protons into the mitochondrial intermembrane space (IMS). Cytochrome *C* delivers electrons to complex IV, which reduces molecular O_2 into H_2O while transferring other protons into the IMS. Finally, complex V (also known as F_1F_O-ATPase) harnesses the electrochemical gradient generated by complexes I–IV to condense ADP and inorganic phosphate into ATP (Galluzzi, Kepp, Trojel-Hansen, & Kroemer, 2012).

In physiological conditions, the activity of complexes I–IV and complex V is finely coupled, implying that mitochondrial O_2 consumption reflects the rate of ATP synthesis by the F_1F_O-ATPase (Galluzzi et al., 2012). However, several other cellular enzymes employ O_2 as a substrate, calling for the development of a strategy that takes into account this potential source of bias. The Seahorse XF Cell Mito Stress Test (Seahorse Bioscience, Billerica, MA) provides a simple and convenient approach to interrogate mitochondrial respiration in cultured cells, based on the timed assessment of OCR coupled to the sequential administration of these specific ETC inhibitors: oligomycin, carbonyl cyanide-*p*-trifluoromethoxyphenylhydrazone (CCCP), and rotenone plus antimycin A. Oligomycin (see earlier) (Galluzzi et al., 2013) is employed to identify the percentage of basal OCR (which reflects both mitochondrial and extramitochondrial O_2 consumption) that depends on mitochondrial ATP synthesis. CCCP disrupts the integrity of the IMM and hence disrupts the electrochemical gradient created by respiratory complexes I–IV (Fulda, Galluzzi, & Kroemer, 2010). This results in maximal O_2 consumption by complex IV and is employed to identify the maximal respiratory capacity of the cells. Finally, rotenone and antimycin A inhibit complex I and complex II, respectively, which abolishes mitochondrial respiration (Galluzzi et al., 2013). In the presence of oligomycin, CCCP, rotenone, and antimycin A, the OCR only reflects extramitochondrial O_2 consumption. Thus, the OCR values measured throughout the assay directly represent or can be employed to calculate: (1) basal mitochondrial respiration (global OCR at baseline − extramitochondrial respiration); (2) ATP synthesis-dependent mitochondrial respiration (OCR at baseline − OCR after the addition of oligomycin); (3) proton leak across the IMM (OCR after the addition of oligomycin − OCR at the end of the assay); (4) maximal mitochondrial respiration (OCR after the addition of CCCP − OCR at the end of the assay); (5) spare mitochondrial respiratory capacity (OCR after the addition of CCCP − OCR at baseline); and (6) extramitochondrial respiration (OCR after the addition of rotenone plus antimycin A). Here below, we detail a simple protocol to interrogate mitochondrial respiration in cultured cells by means of the Seahorse XF Cell Mito Stress Test.

7. ASSESSMENT OF MITOCHONDRIAL RESPIRATION: PROTOCOL

On the day of the assay:

1. Supplement XF Base Medium with 10 m*M* D-glucose, 1 m*M* sodium pyruvate, and 1.5 m*M* L-glutamine to generate Assay Medium, and warm it to 37°C (see Notes 19 and 31).
2. Adjust the pH of the Assay Medium to 7.35 ± 0.05, sterilize by passing it through a 0.2 μm Ø filter, and maintain it at 37°C until use.
3. Substitute culture medium in XF96 Polystyrene Cell Culture Microplates with 175 μL of Assay Medium and place cells at 37°C in a humidified incubator for 45–60 min (see Note 17).
4. Place the B/C Loading Guide on top of the Sensor Cartridge and orient it so that the letter "B" is located in the upper left corner, opposed to the triangular notch (see Notes 21 and 22).

5. Load ports B of the Sensor Cartridge with 8 μ*M* oligomycin (25 μL per port per well), freshly prepared in Assay Medium and warmed to 37°C (see Notes 23–26).
6. Reorient the B/C Loading Guide so that the letter "C" is located in the upper left corner, opposed to the triangular notch (see Note 22).
7. Load ports C of the Sensor Cartridge with 9 μ*M* CCCP (25 μL per port per well), freshly prepared in Assay Medium and warmed to 37°C (see Notes 23–26).
8. Place the A/D Loading Guide on top of the Sensor Cartridge and orient it so that the letter "D" is located in the upper left corner, opposed to the triangular notch (see Notes 21 and 22).
9. Load ports D of the Sensor Cartridge with 12.5 μ*M* rotenone plus 5 μ*M* antimycin A (25 μL per port per well), freshly prepared in Assay Medium and warmed to 37°C (see Notes 23–26).
10. Discard the Loading Guide, check all ports for equal loading (see Notes 27–30), and proceed to calibration and run (see Section 8).

8. CALIBRATION AND RUN

1. Open the Design File of choice with the proprietary Seahorse software Wave (see Note 32), select the "Review and Run" tab, and then click "Start Run."
2. When prompted, transfer the Sensor Cartridge and Utility Plate to the XF^e Extracellular Flux Analyzer, click "I'm ready," and wait for the completion of calibration (see Note 33).
3. Remove the Utility Plate, transfer the XF96 Polystyrene Cell Culture Microplate to the XF^e Extracellular Flux Analyzer, and click "I'm ready" to initiate the assay (see Note 34).
4. The Seahorse XF Stress Test Report Generator automatically calculates the Seahorse XF Cell Glycolysis Stress Test parameters and Seahorse XF Cell Mito Stress Test parameters, respectively (see Sections 4 and 6, and Notes 35–37), from data in Microsoft Excel format (Figs. 1 and 2).

9. CONCLUDING REMARKS

With minor variations, the method described in this chapter offers a straightforward approach to monitor glycolytic and respiratory proficiency in cells of human or nonhuman origin that mount autophagic responses to

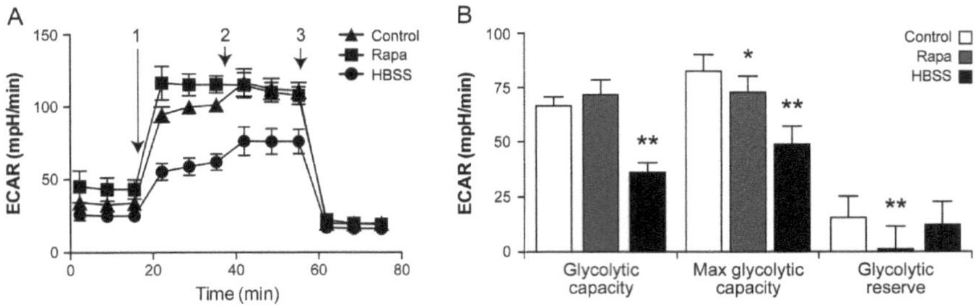

Fig. 1 Glycolytic flux in HCT 116 cells responding to rapamycin and starvation. Human colorectal carcinoma HCT 116 cells were seeded in Seahorse XF96 Polystyrene Cell Culture Microplates, allowed to adapt for 24 h and then placed in fresh culture medium (control), fresh culture medium supplemented with 10 μ*M* rapamycin (Rapa) or Hanks' balanced salt solution (HBSS) for 6 h, as described in Section 2. Thereafter, cells were processed for the assessment of glycolytic flux, as detailed in Sections 5 and 8. Extracellular acidification rate (ECAR) is reported (mean ± SD, $n = 10$ parallel assessments). *Arrows* indicate glucose (1), oligomycin (2), and 2-deoxyglucose (3) injections. *$p < 0.05$, **$p < 0.01$ as compared to cells maintained in control conditions (paired Student's *t* test).

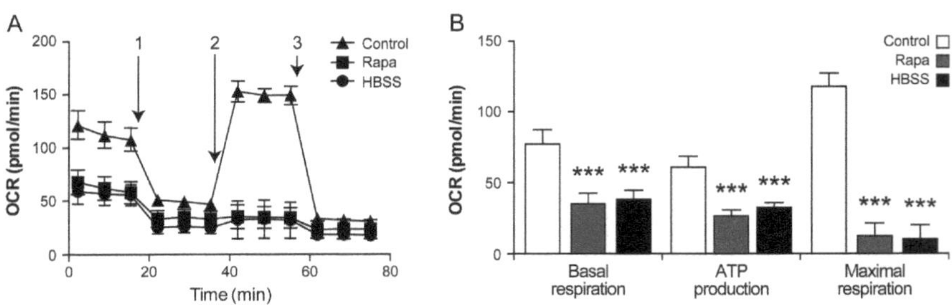

Fig. 2 Mitochondrial respiration in HCT 116 cells responding to rapamycin and starvation. Human colorectal carcinoma HCT 116 cells were seeded in Seahorse XF96 Polystyrene Cell Culture Microplates, allowed to adapt for 24 h, and then placed in fresh culture medium (control), fresh culture medium supplemented with 10 μ*M* rapamycin (Rapa) or Hanks' balanced salt solution (HBSS) for 6 h, as detailed in Section 2. Thereafter, cells were processed for the assessment of mitochondrial respiration, as described in Sections 7 and 8. Oxygen consumption rate (OCR) is reported (mean ± SD, $n = 10$ parallel assessments). *Arrows* indicate oligomycin (1), CCCP (2), and rotenone plus antimycin A (3) injections. ***$p < 0.001$ as compared to cells maintained in control conditions (paired Student's *t* test).

stress in vitro. This protocol relies on extracellular flux analysis, which is particularly convenient to monitor bioenergetic metabolism in living cells, as well as respiratory functions in nonisolated mitochondria within permeabilized cells (Ferrick, Neilson, & Beeson, 2008; Nadanaciva et al., 2012; Salabei, Gibb, & Hill, 2014). The relatively limited distribution of XF[e]

Extracellular Flux Analyzers may represent the main limitation of this technique. However, the number of laboratories equipped with XF^e Extracellular Flux Analyzers is steadily increasing, and no less than 25 Core Research Facilities acquainted with this technology are currently available worldwide (http://www.seahorsebio.com/learning/core-facilities.php).

Thoroughly studying glycolytic flux and mitochondrial respiration in cells that undergo autophagy in response to stress may not only shed new light on the intimate crosstalk between autophagy and metabolism but also pave the way to the development of novel therapeutic agents targeting metabolic liabilities that specifically arise in the course of autophagic responses.

10. NOTES

1. Based on the recommendations of the American Type Culture Collection (ATCC, Manassas, VA). Please note that optimal culture conditions (i.e., culture medium, supplements) for other cell lines may differ quite considerably.
2. To adapt to specific laboratory requirements (e.g., limited space within incubators, elevated number of experimental determinations), cell cultures can be properly maintained in smaller or larger supports (e.g., 25- or 175-cm^2 flasks). Indicatively, 25-, 75-, and 175-cm^2 flasks displaying 75% confluence contain approximately $2.0–2.5 \times 10^6$, $6.5–7.0 \times 10^6$, and $14.5–15.0 \times 10^6$ HCT 116 cells, respectively.
3. Washing cells with PBS removes serum leftovers, which are known to inactivate trypsin and hence limit detachment. To avoid a considerable loss of cells, however, washing should be gentle. Some cell types including HCT 116 are indeed exquisitely prone to detachment.
4. TrypLE™ Express (from Gibco®-Thermo Fisher Scientific, Carlsbad, CA) can be used as an alternative to trypsin/EDTA. This recombinant trypsin-like proteolytic enzyme exhibits increased stability (at both usage and storage temperatures) and does not require inactivation.
5. The time required for trypsin/EDTA to completely detach a cellular monolayer varies considerably with cell type and degree of confluence. Most malignant cell lines of human and murine origin are efficiently detached upon 1–3 min incubation (at 37°C).
6. Detachment should be inspected on light microscopy upon gentle agitation of flasks. Clumps that may form at this passage can be easily dissolved by repeatedly passing the cell suspension though a 5- or 10-mL pipette.

7. Although over-trypsinization normally does not affect viability, it should be avoided as it can cause cellular damage while altering several membrane-related processes (e.g., growth factor signaling).
8. The ATCC recommends to passage HCT 116 cells at a 1:3–1:8 subcultivation ratio. Based on personal experience, diluting HCT 116 cells up to a 1:15 subcultivation ratio does not cause noticeable morphological or behavioral shifts.
9. Both over- and under-confluence should be avoided, as the former may impose a metabolic and oxidative stress on the cell culture, while the latter may favor genetic drifts in the population (see also Note 10).
10. To ensure the stability of maintenance cultures (see also Note 9), and hence the reproducibility of results over time, cells should be constantly kept under exponential growth conditions and subcultured for a relatively low, predetermined number of passages. This good laboratory practice requires a large stock of cryovials, which should be generated prior to experimental determinations with early-passage, clean cells.
11. The optimal number of cells for seeding on the day prior to the assay may vary to a considerable extent with cell type. Preliminary experiments should be performed to identify optimal seeding conditions. As a note, the seeding surface of XF96 Polystyrene Cell Culture Microplates is slightly smaller (0.106 cm^2) than that of standard 96-well plate. Typical ECAR values of reference (min–max) on the day of the assay are 20–90 mpH/min.
12. Adaptation varies to considerable degrees with cell type. HCT 116 cells are particularly sensitive in this respect and require at least 24 h to resume exponential growth. Conversely, other cells including human cervical carcinoma HeLa cells and human nonsmall cell lung carcinoma A549 cells display normal morphology and growth rate after 8–12 h.
13. This provides appropriate negative control conditions. Since the autophagic machinery is very sensitive to fluctuations in nutrient availability, it is crucial to replace exhausted medium with fresh one in this setting.
14. Both rapamycin and HBSS provide appropriate positive control conditions. Rapamycin potently activates autophagy upon direct inhibition of MTORC1. As a nutrient-free milieu, HBSS promotes autophagy via multiple signal transduction cascades that ultimately suppress MTORC1 signaling (Galluzzi et al., 2014).
15. The time required for autophagic responses to fully develop depends on several parameters, including cell type, stimulus, and microenvironmental conditions (e.g., oxygen tension, pH, growth factor availability,

etc). Indicatively, HCT 116 cells exhibit robust activation of autophagy as soon as 4–6 h following exposure to rapamycin or nutrient-free conditions.

16. Since sensors may not work properly if suboptimally hydrated, complete immersion should be carefully verified.
17. A controlled CO_2 atmosphere is not required, but humidification must be ensured to avoid evaporation.
18. This allows all components of the XF^e Extracellular Flux Analyzer to reach 37°C.
19. Indicatively, 30 mL are required to perform the Seahorse XF Glycolysis Stress Test or the Seahorse XF Cell Mito Stress Test on a single XF96 Polystyrene Cell Culture Microplate.
20. This procedure produces a >1000-fold dilution of the culture medium, which is the minimum recommended to minimize potential bias from the medium buffer capacity.
21. Ports on the Sensor Cartridge should be loaded while the Sensor Cartridge is maintained on the Utility Plate.
22. Caution should be taken to avoid the dislodgement of the Loading Guide during pipetting.
23. Indicatively, 2 mL of working solutions are sufficient to load the Sensor Cartridge and perform the Seahorse XF Glycolysis Stress Test or the Seahorse XF Cell Mito Stress Test on a single XF96 Polystyrene Cell Culture Microplate.
24. The Sensor Cartridge should be loaded with maximal care. In particular, caution should be employed to prevent the formation of bubbles, and shaking/tapping should be avoided to reduce the risk of spillage and cross-port contamination.
25. For how the XF^e Extracellular Flux Analyzer operates in this configuration, ports B must be loaded with 8 × working solutions, ports C with 9 × working solutions, and ports D with 10 × working solutions.
26. Please note that the optimal concentrations of reagents dispensed during the Seahorse XF Glycolysis Stress Test and Seahorse XF Cell Mito Stress Test may vary to considerable degree with cell type. Preliminary experiments are therefore required to determine the optimal concentrations of oligomycin, CCCP, rotenone, and antimycin A.
27. An additional port (port A) per well is available, which can be conveniently employed to perform an experimental intervention before the initiation of the assay. In such an alternative configuration, ports A must be loaded with 8 × working solutions of the reagent of choice, while

ports B, C, and D must be loaded with 9×, 10×, and 11× working solutions of assay reagents, respectively.

28. It is highly recommendable to carefully annotate unevenly loaded ports, if any, to facilitate data analysis and trouble-shooting.

29. Ports must be loaded on all wells, and each series of ports must contain the same solution.

30. To limit the possibility of spillage and cross-port contamination, it is recommendable not to transport the loaded Sensor Cartridge from one zone of the lab to another.

31. To ensure the accurate, functional measurement of metabolic phenotypes, the manufacturer recommends the use of nonbuffered medium.

32. Design Files should be prepared before the initiation of the assay. They indicate to the XF^e Extracellular Flux Analyzer: (1) plate layout; (2) injection sequence and timing, (3) type of measurement; and (4) number of cycles.

33. Calibration runtime is approximately 20 min.

34. Assay runtime is approximately 90 min.

35. The Seahorse XF Stress Test Report Generator can be installed with the Wave package or downloaded later at http://www.seahorsebio.com/support/software/stress-test-generator.php.

36. The Seahorse XF Stress Test Report Generator allows for the calculation of standard (see also Sections 4 and 6) as well as customizable parameters.

37. If experimental interventions are expected to differentially affect the number of living cells present in each well at the moment of the assay, results should be normalized. One convenient way to do so is to recover XF96 Polystyrene Cell Culture Microplates at the end of the assay, discard the Assay Medium, wash once with PBS, fix with 4% paraformaldehyde in PBS supplemented with 1 μg/mL Hoechst 33342, and quantify cells/field in each well by (automated) fluorescence microscopy.

ACKNOWLEDGMENTS

We are indebted with Oliver Kepp (Gustave Roussy Cancer Campus; Villejuif, France) for help with manuscript preparation. G.K. is supported by the Ligue contre le Cancer (équipe labellisée); Agence National de la Recherche (ANR)—Projets blancs; ANR under the frame of E-Rare-2, the ERA-Net for Research on Rare Diseases; Association pour la recherche sur le cancer (ARC); Cancéropôle Ile-de-France; Institut National du Cancer (INCa); Fondation Bettencourt-Schueller; Fondation de France; Fondation pour la Recherche

Médicale (FRM); the European Commission (ArtForce); the European Research Council (ERC); the LabEx Immuno-Oncology; the SIRIC Stratified Oncology Cell DNA Repair and Tumor Immune Elimination (SOCRATE); the SIRIC Cancer Research and Personalized Medicine (CARPEM); the Swiss Bridge Foundation, ISREC; and the Paris Alliance of Cancer Research Institutes (PACRI).

REFERENCES

Codogno, P., Mehrpour, M., & Proikas-Cezanne, T. (2012). Canonical and non-canonical autophagy: Variations on a common theme of self-eating? *Nature Reviews. Molecular Cell Biology*, *13*, 7–12.

Ferrick, D. A., Neilson, A., & Beeson, C. (2008). Advances in measuring cellular bioenergetics using extracellular flux. *Drug Discovery Today*, *13*, 268–274.

Fulda, S., Galluzzi, L., & Kroemer, G. (2010). Targeting mitochondria for cancer therapy. *Nature Reviews. Drug Discovery*, *9*, 447–464.

Galluzzi, L., Bravo-San Pedro, J. M., Vitale, I., Aaronson, S. A., Abrams, J. M., Adam, D., et al. (2015). Essential versus accessory aspects of cell death: Recommendations of the NCCD 2015. *Cell Death and Differentiation*, *22*, 58–73.

Galluzzi, L., Kepp, O., Trojel-Hansen, C., & Kroemer, G. (2012). Mitochondrial control of cellular life, stress, and death. *Circulation Research*, *111*, 1198–1207.

Galluzzi, L., Kepp, O., Vander Heiden, M. G., & Kroemer, G. (2013). Metabolic targets for cancer therapy. *Nature Reviews. Drug Discovery*, *12*, 829–846.

Galluzzi, L., Pietrocola, F., Levine, B., & Kroemer, G. (2014). Metabolic control of autophagy. *Cell*, *159*, 1263–1276.

Green, D. R., Galluzzi, L., & Kroemer, G. (2011). Mitochondria and the autophagy-inflammation-cell death axis in organismal aging. *Science*, *333*, 1109–1112.

Guo, J. Y., Karsli-Uzunbas, G., Mathew, R., Aisner, S. C., Kamphorst, J. J., Strohecker, A. M., et al. (2013). Autophagy suppresses progression of K-ras-induced lung tumors to oncocytomas and maintains lipid homeostasis. *Genes & Development*, *27*, 1447–1461.

He, C., Wei, Y., Sun, K., Li, B., Dong, X., Zou, Z., et al. (2013). Beclin 2 functions in autophagy, degradation of G protein-coupled receptors, and metabolism. *Cell*, *154*, 1085–1099.

Jung, H. S., Chung, K. W., Won Kim, J., Kim, J., Komatsu, M., Tanaka, K., et al. (2008). Loss of autophagy diminishes pancreatic beta cell mass and function with resultant hyperglycemia. *Cell Metabolism*, *8*, 318–324.

Kaur, J., & Debnath, J. (2015). Autophagy at the crossroads of catabolism and anabolism. *Nature Reviews. Molecular Cell Biology*, *16*, 461–472.

Kim, K. H., & Lee, M. S. (2014). Autophagy—A key player in cellular and body metabolism. *Nature Reviews. Endocrinology*, *10*, 322–337.

Kroemer, G., Marino, G., & Levine, B. (2010). Autophagy and the integrated stress response. *Molecular Cell*, *40*, 280–293.

Kuma, A., Hatano, M., Matsui, M., Yamamoto, A., Nakaya, H., Yoshimori, T., et al. (2004). The role of autophagy during the early neonatal starvation period. *Nature*, *432*, 1032–1036.

Lamb, C. A., Yoshimori, T., & Tooze, S. A. (2013). The autophagosome: Origins unknown, biogenesis complex. *Nature Reviews. Molecular Cell Biology*, *14*, 759–774.

Ma, Y., Galluzzi, L., Zitvogel, L., & Kroemer, G. (2013). Autophagy and cellular immune responses. *Immunity*, *39*, 211–227.

Menzies, F. M., Fleming, A., & Rubinsztein, D. C. (2015). Compromised autophagy and neurodegenerative diseases. *Nature Reviews. Neuroscience*, *16*, 345–357.

Nadanaciva, S., Rana, P., Beeson, G. C., Chen, D., Ferrick, D. A., Beeson, C. C., et al. (2012). Assessment of drug-induced mitochondrial dysfunction via altered cellular respiration and acidification measured in a 96-well platform. *Journal of Bioenergetics and Biomembranes*, *44*, 421–437.

Pietrocola, F., Galluzzi, L., Bravo-San Pedro, J. M., Madeo, F., & Kroemer, G. (2015). Acetyl coenzyme A: A central metabolite and second messenger. *Cell Metabolism*, *21*, 805–821.

Rubinsztein, D. C., Codogno, P., & Levine, B. (2012). Autophagy modulation as a potential therapeutic target for diverse diseases. *Nature Reviews. Drug Discovery*, *11*, 709–730.

Salabei, J. K., Gibb, A. A., & Hill, B. G. (2014). Comprehensive measurement of respiratory activity in permeabilized cells using extracellular flux analysis. *Nature Protocols*, *9*, 421–438.

Sica, V., Galluzzi, L., Bravo-San Pedro, J. M., Izzo, V., Maiuri, M. C., & Kroemer, G. (2015). Organelle-specific initiation of autophagy. *Molecular Cell*, *59*, 522–539.

Singh, R., Kaushik, S., Wang, Y., Xiang, Y., Novak, I., Komatsu, M., et al. (2009). Autophagy regulates lipid metabolism. *Nature*, *458*, 1131–1135.

Strohecker, A. M., Guo, J. Y., Karsli-Uzunbas, G., Price, S. M., Chen, G. J., Mathew, R., et al. (2013). Autophagy sustains mitochondrial glutamine metabolism and growth of BrafV600E-driven lung tumors. *Cancer Discovery*, *3*, 1272–1285.

Zoncu, R., Efeyan, A., & Sabatini, D. M. (2011). mTOR: From growth signal integration to cancer, diabetes and ageing. *Nature Reviews. Molecular Cell Biology*, *12*, 21–35.

CHAPTER TEN

Methods to Assess Mitochondrial Morphology in Mammalian Cells Mounting Autophagic or Mitophagic Responses

S. Marchi, M. Bonora, S. Patergnani, C. Giorgi, P. Pinton[1]
Section of Pathology, Oncology and Experimental Biology, Laboratory for Technologies of Advanced Therapies (LTTA), University of Ferrara, Ferrara, Italy
[1]Corresponding author: e-mail address: paolo.pinton@unife.it

Contents

Abstract

It is widely acknowledged that mitochondria are highly active structures that rapidly respond to cellular and environmental perturbations by changing their shape, number, and distribution. Mitochondrial remodeling is a key component of diverse biological

Methods in Enzymology, Volume 588
ISSN 0076-6879
http://dx.doi.org/10.1016/bs.mie.2016.09.080

processes, ranging from cell cycle progression to autophagy. In this chapter, we describe different methodologies for the morphological study of the mitochondrial network. Instructions are given for the preparation of samples for fluorescent microscopy, based on genetically encoded strategies or the employment of synthetic fluorescent dyes. We also propose detailed protocols to analyze mitochondrial morphometric parameters from both three-dimensional and bidimensional datasets. Finally, we describe a protocol for the visualization and quantification of mitochondrial structures through electron microscopy.

1. INTRODUCTION

Mitochondria are highly dynamic organelles that often change shape and intracellular distribution during their life course. Both number and morphology of mitochondria are regulated by specifically controlled rates of organelle fusion and fission (Chan, 2012; Youle & van der Bliek, 2012). Therefore, the molecular machinery involved in the control of mitochondrial dynamics include both proteins responsible of mitochondrial fusion, such as the outer mitochondrial membrane proteins mitofusin 1 (Mfn1) and mitofusin 2 (Mfn2) (Santel & Fuller, 2001) and the inner membrane protein optic atrophy 1 (Opa1) (Meeusen et al., 2006), and pro-fission factors, such as dynamin-related protein 1 (Drp1) (Bleazard et al., 1999), Fis1 (Yoon, Krueger, Oswald, & McNiven, 2003), and the mitochondrial fission factor (Mff) (Otera et al., 2010). A high number of papers have described dramatic remodeling of the mitochondrial network during a wide range of pathophysiological conditions, which include differentiation (Kasahara & Scorrano, 2014), cell cycle progression (Mitra, Wunder, Roysam, Lin, & Lippincott-Schwartz, 2009), mitochondrial respiratory chain function (Benard & Rossignol, 2008), Ca^{2+} transfer (Szabadkai et al., 2004), apoptosis (Karbowski & Youle, 2003; Morciano, Pedriali, et al., 2016), and autophagy (Galluzzi et al., 2015; Okamoto & Kondo-Okamoto, 2012). Upon autophagy induction by nutrient depletion, mitochondria seem to elongate in order to supply ATP in times of starvation, especially at early phases of the autophagic process (Gomes, Di Benedetto, & Scorrano, 2011; Rambold, Kostelecky, Elia, & Lippincott-Schwartz, 2011). Conversely, it is widely accepted as mitochondrial fragmentation events occurred before selective clearance of mitochondria by mitophagy, which allowed the engulfment of damaged and potentially harmful mitochondria into autophagosomes (Rimessi et al., 2013). Thus, changes in mitochondrial morphology represented a key aspect in the regulation of both autophagy and mitophagy. Here, we describe different assays for monitoring and quantifying the state of the mitochondrial network before and after autophagy/mitophagy induction. Importantly, most of these methods can be exploited for morphometric analyses of the mitochondrial compartment that might occur upon stimuli of various nature or different pathological contexts.

2. HIGH-RESOLUTION IMAGING OF MITOCHONDRIA IN LIVE CELLS BASED ON FLUORESCENT PROTEIN VARIANTS

The proposed protocol allows the measurement of mitochondrial morphology and dynamics with recombinant variants of the green fluorescent protein (GFP), obtained from jellyfish *Aequorea victoria* in the early 1990s (Heim & Tsien, 1996). Optimal analysis of mitochondrial network is primary obtained by using a mitochondrially targeted GFP (mtGFP) (Rizzuto et al., 1998) or a monomeric red fluorescent protein mCherry fused with the mitochondrial target sequence from subunit VIII of human cytochrome *c* oxidase (mCherry-Mito-7; Addgene plasmid #55102) (Olenych, Claxton, Ottenberg, & Davidson, 2007).

2.1 Equipment Setup

Mitochondrial structures could be imaged and recorded by using a spinning disk confocal microscope (or equivalent competitors) equipped with the appropriate filter set. The system should be set in order to reach a configuration as close as possible to the Nyquist rate (https://svi.nl/Nyquistcalculator). Typically, correct sampling can be achieved by using 60 × lens or higher with numerical aperture (N.A.) > 1.3, and high-resolution CCD or CMOS cameras (with pixel size <8 μm). In this protocol, we will illustrate experiments performed on a Nikon LiveScan Swept Field Confocal Microscope (SFC) Eclipse Ti with a 60 × magnification and equipped with NIS-Elements microscope imaging software (Nikon Instruments, Melville, NY, USA) and Andor iXon EM-CCD camera (Belfast, Northern Ireland) (Marchi et al., 2015; Patergnani et al., 2013).

2.2 Reagents Setup

- Transfection reagents.
- Dulbecco's Modified Eagle Medium (DMEM) supplemented of FBS 10% or Modified Krebs-ringer solution (KRB): 125 m*M* NaCl,

5 mM KCl, 1 mM KH_2PO_4, 1 mM $MgSO_4$, 1 mM $CaCl_2$, 5.5 mM glucose, and 20 mM HEPES, pH 7.4.
- Glass coverslips 24 mm in diameter.
- Attofluor chamber for microscopy with stage holder for mounting.

2.3 Sample Preparation and Transfection

The cells are seeded on glass coverslips (24 mm in diameter) and allowed to grow until 50% confluence. After seeding the cells, wait at least 24 h. The cells are then transfected using appropriate transfection methods with mitochondrially targeting fluorescent proteins.

Conventionally cancerous and immortalized cell lines may be easily transfect with Ca^{2+}-phosphate or lipoamines transfection methods. Primary cultures, which are notably "hard-to-transfect cells," request electroporation or adenoviral vectors. Cultures obtained from human beings may be reluctant to the transfection. In order to perform mitochondrial network analysis in these cell samples, please refer to Section 4.

2.4 Measurements

- Mount the glass coverslips into the Attofluor chamber and add 500–1000 μL of DMEM or KRB.
- Place the chamber on the 37°C thermo-controlled stage of the microscope.
- Prepare the appropriate camera setting. For high-resolution imaging, we recommend to use the standard binning mode 1 × 1 (meaning that each logical pixel is equal to one physical pixel) and to collect a *z*-stack of images (commonly 51) at 0.2 μm each, in order to capture the whole mitochondrial network of the cell. *Important note:* in case of low speed microscope system or highly flat cells, the operator might avoid capturing *z*-stacks of the images, thereby limiting the acquisitions to a single plan. In such case, see Section 4.2 for quantification of the results.
- Acquire images in multiple points mode by using motorized stage. If microscope is not equipped, record single-point pictures. We recommend to store all acquisitions with the same settings, for both camera and microscope. In order to obtain statistical significance, a number of at least 40–50 images per condition should be collected.
- Stop the analysis system, remove the chamber from the microscope stage, and clean well the objective from the immersion medium.

- Clean well the Attofluor chamber with 70% ethanol and then with Milli-Q water, before perform a new experiment.

3. HIGH-RESOLUTION IMAGING OF MITOCHONDRIAIN LIVE CELLS WITH MITOCHONDRIAL-SPECIFIC FLUORESCENT DYES

The very limited photobleaching of GFP provides the primary advantage of the fluorescent protein over the dyes. Nevertheless, the employment of such plasmids is not always possible. Several primary cultures are reluctant to the transfection with the conventional methods (including electroporation and viral vectors). The main cell populations "hard-to-transfect cells" are those obtained from human beings, such as fibroblasts. Considering this aspect, alternative approaches to stain the mitochondrial network of these cells are needed. We propose a detailed protocol aimed to describe the use of the most common mitochondrial fluorescent chemical probes to analyze mitochondrial dynamics and network.

3.1 Equipment and Reagents Setup

- Confocal microscope (please refer to Section 2.1).
- Mitochondrial fluorescent dyes: MitoTracker Green, MitoTracker Orange, MitoTracker Red, and MitoTracker Deep Red versions.

3.2 Reagents Setup

- DMEM supplemented of FBS 10% (DMEM) or Modified KRB: 125 m*M* NaCl, 5 m*M* KCl, 1 m*M* KH_2PO_4, 1 m*M* $MgSO_4$, 1 m*M* $CaCl_2$, 5.5 m*M* glucose, and 20 m*M* HEPES, pH 7.4.
- Glass coverslips 24 mm in diameter.
- Attofluor chamber for microscopy with standard 35-mm diameter stage holder for mounting.
- MitoTracker dyes. We recommend preparing a stock solution in a range of 100–500 μ*M*. Working solution may be utilized up to 1–2 months and conserved at −20°C.

3.3 Sample Preparation and Measurements

- Seed the cells on glass coverslips (24 mm in diameter) and allow them to reach 50–70% confluence. After seeding cells, wait at least 24 h.
- Load the cells with an appropriate MitoTracker dye in agreement with the microscope filter setting available. We recommend using the probes

at final concentration in a range of 1–100 n*M*. The appropriate final concentration depends on cell type and should be achieved by the operator. To obtain an optimal dye loading, we suggest to incubate cells for 30–60 min at 37°C, 5% CO_2.

- Wash cells with KRB or DMEM for at least three times.
- Mount the glass coverslips into the Attofluor chamber and add 500–1000 μL of DMEM or KRB.
- Place the chamber on the 37°C thermo-controlled stage of the microscope.
- Prepare the appropriate camera setting. For high-resolution imaging, we recommend to use the standard binning mode 1 × 1 (meaning that each logical pixel is equal to one physical pixel) and collect a *z*-stack of images (commonly 51) at 0.2 μm each, in order to capture the whole mitochondrial network of the cell. *Important note:* in case of low speed microscope system or highly flat cells, the operator might avoid capturing *z*-stacks of the images, thereby limiting the acquisitions to a single plan. In such case, see Section 4.2 for quantification of the results.
- Acquire images in multiple points mode by using motorized stage. If microscope is not equipped, record single-point pictures. We recommend storing all acquisitions with the same settings, for both camera and microscope. In order to obtain statistical significance, a number of at least 40–50 images per condition should be collected.
- Stop the analysis system, remove the chamber from the microscope stage, and clean well the objective from the immersion medium.
- Clean well the Attofluor chamber with 70% ethanol and then with Milli-Q water, before starting a new experiment.

Notes:

- Mitochondrial structures may be also imaged and recorded after fixation procedures. Several fixatives are available. The most common are aldehydes (such as formaldehyde or glutaraldehyde) and organic solvents, such as methanol, ethanol, and acetone. Among them, we recommended to fix cells with paraformaldehyde (PFA). The optimal PFA concentration depends from the cellular type. We suggest to use 4% PFA for immortalized cells and 2% for primary cells.
- It is very important to consider that not all chemical probes staining the mitochondrial population are retained after fixation procedures. For fixed cells, we recommend using MitoTracker Orange or MitoTracker Deep Red versions. Such fluorescent dyes are well retained after aldehyde fixation.

- All the mitochondrial fluorescent dyes reported here are very sensitive to variations in mitochondrial membrane potential (MMP). Thus, when using compounds that could induce autophagy/mitophagy through perturbation of the MMP, such as Antimycin A or FCCP (carbonyl cyanide-4-(trifluoromethoxy)phenylhydrazone), we suggest to mark the mitochondrial compartment through antibodies raised against highly expressed mitochondrial proteins, rather than using fluorescent chemical compounds. We propose four antibodies that work very well in immunofluorescence technique and could allow a reliable quantification of the mitochondrial morphology:
  1. Anti-TOM20 [outer mitochondrial membrane protein; Santa Cruz Biotechnology (Santa Cruz, CA, USA) (sc-11415); dilution: 1:100–1:1000].
  2. Anti-HSP60 [matrix protein; Santa Cruz Biotechnology (sc-13115); dilution: 1:100–1:1000].
  3. Anti-TIM23 [mitochondrial inner membrane protein; BD Transduction Laboratories (Franklin Lakes, NJ, USA) (code: 611222); dilution: 1:100].
  4. Anti-ATP5B [matrix protein; Millipore (Darmstadt, Germany) (MAB3494); dilution: 1:100–1:1000].

4. ANALYSIS OF THE RESULTS

4.1 Analysis of Mitochondrial Morphology on a Three-Dimensional Dataset (Fig. 1)

- Open Fiji software (Schindelin et al., 2012) and load the files output from microscope software.
- Denoise images by applying a Gaussian filter with dimension size of 1 pixel (Menu Process → Filters → Gaussian...).
- Remove background setting the rolling ball to a size of approximately 1 μm (Menu Process → Subtract Background...).
- Generate segmented image (Menu Analyze → 3D Objects Counter). An automated threshold algorithm will find the best threshold value, nonetheless this could require minor adjustment to include all the mitochondria. Set the filters to a minimum of 8 pixels to a maximum of 1,000,000 pixels. Check the option "Statistics." A result table will appear.
- Collect the volume measurement column in a spreadsheet file and calculate average volume (average mitochondrial size), the sum of all the volume (mitochondrial content), and number of objects (mitochondrial counts).

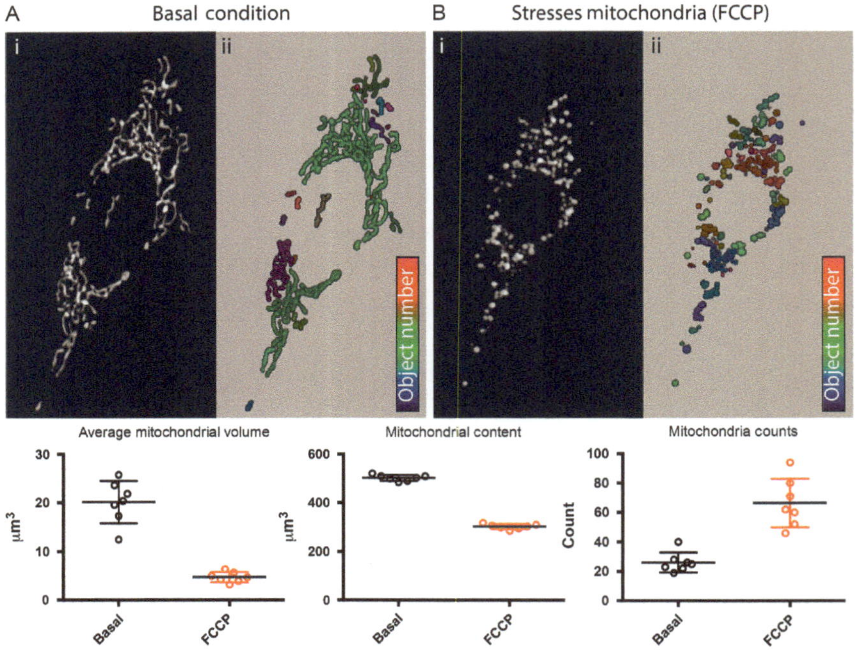

Fig. 1 Representative volume render (Ai and Bi) and segmented image (Aii and Bii) of mitochondrial network derived from mtGFP expressing CHO cells during basal conditions (A) or after FCCP exposure (B).

- Repeat the procedure for all the collected samples.
- Generate three data table, for independent collection of the three measurements.
- Choose the appropriate statistical method to compare volume measurement and object number within groups. A list of the different outputs that could arise from the analysis and of the mitochondrial processes that are associated to the morphological rearrangements are showed in Table 1.

Notes: The present procedure refers to the open source software for image analysis Fiji. Comparable results could be obtained with other open source (e.g., Icy) or commercial available software (e.g., Bitplane Imaris, Perkin Elmer Volocity).

4.2 Analysis of Mitochondrial Morphology on a Bidimensional Dataset (Fig. 2)

- Open Fiji software (Schindelin et al., 2012) and load the files output from microscope software.

Table 1 Representative Interpretation of Most Common Phenotypes That Could Be Associated to Different Processes That Regulate Mitochondrial Morphology

Process	Observed Phenotype	Refs.
Mitochondrial fission	Increase in mitochondrial counts and reduction in average mitochondrial size	Bonora et al. (2013)
Mitochondrial fusion	Reduction in mitochondrial counts and increase in average mitochondrial size	Morciano, Giorgi, et al. (2016)
Mitophagy	Reduction in mitochondrial content and reduction in average mitochondrial size	Sumpter et al. (2016)
Mitochondrial biogenesis	Increase in mitochondrial content and increase in average mitochondrial size (not always)	Bianchi et al. (2006)

- Denoise images by applying a Gaussian filter with dimension size of 1 pixel (Menu Process → Filters → Gaussian...).
- Remove Background setting the rolling ball to a size of approximately 1 μm (Menu Process → Subtract Background...).
- Threshold image to isolate mitochondria from background (Menu Image → Adjust → Threshold...). Choose the most appropriate thresholding algorithms (usually "Otsu" or "Triangle" are the most performing).
- Set measurement type (Menu Analyze → Set measurements...). Check the options "Area," "Perimeter," "Feret's diameter," "Limit to threshold").
- Measure mitochondria (Menu Analyze → Analyze particles). Set a minimum filter size to 2, check the "Pixel units," and "Display results" options.
- Collect the measurement columns in a spreadsheet file and calculate the aspect ratio and form factor according to equations displayed in Fig. 2Bi
- Repeat the procedure for all the collected samples.
- Generated three data table, for independent collection of the three measurements.
- Choose the appropriate statistical method to compare form indexes.

Notes:

The thresholding method should not be modified between samples.

The present procedure refers to the open source software for image analysis Fiji. Comparable results could be obtained with other open source (e.g., Icy) or commercial available software (e.g., Molecular Devices Metamorph, Perkin Elmer Volocity, Nikon NIS-Elements, Zeiss Zen, Leica LAS X).

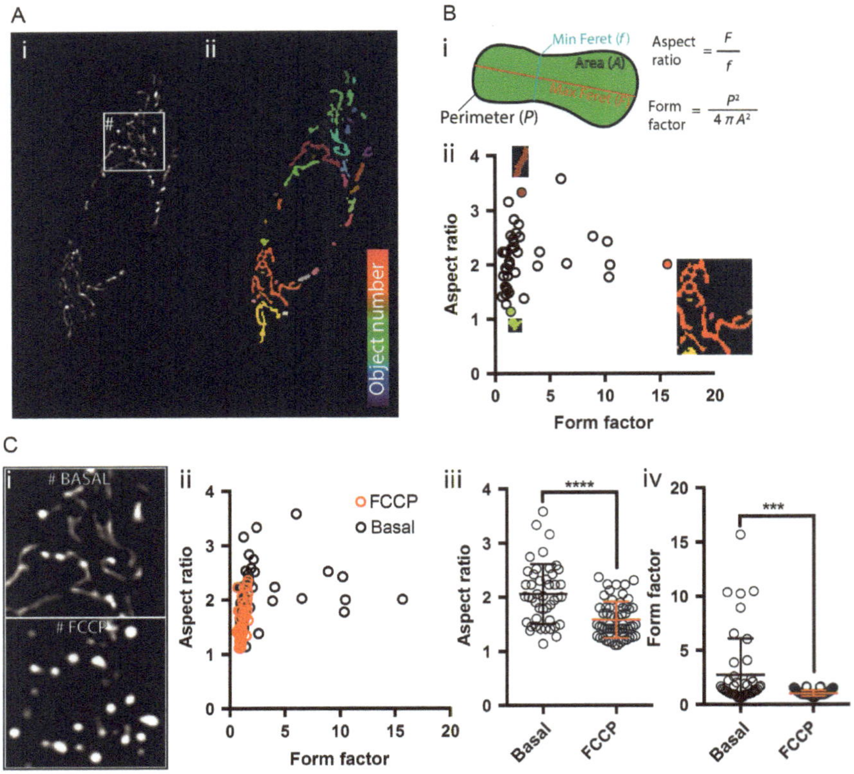

Fig. 2 Representative confocal section (Ai) and segmented image (Aii) of mitochondrial network in mtGFP expressing CHO cells during basal conditions. Diagrammatic representation of morphological meaning of aspect ratio and form factor indexes (Bi) and relative distribution in scatter plot (Bii). Zoom image of inset from image (Ai) before and after administration of FCCP (Ci). Representative scatter plot of form indexes distribution before and after FCCP administration (Cii) also statistical analysis of both indexes are displayed in (Ciii) and (Civ) (***$p < 0.005$ and ****$p < 0.001$ for Student's *T*-test).

5. ASSESSMENT OF MITOCHONDRIAL MORPHOLOGY USING ELECTRON MICROSCOPY

The proposed protocol is referred to cells grown in culture dishes. The same protocol could be exploited also to cells in suspension or tissue sections. Notably, a detailed protocol for determination of mitochondrial ultrastructures in tissue samples by electron microscopy (EM) has been described by Sasaki (2010).

For the visualization of mitochondrial structure, we suggest the employment of a transmission electron microscope (TEM), with a magnification of at least 20,000 ×.

5.1 Reagents Setup

- Fixative: 2% glutaraldehyde and 0.1 *M* sodium cacodylate, pH 7.4.
- 1% Osmium tetroxide (OsO_4), freshly prepared.
- 0.1 *M* Sodium cacodylate, pH 7.4, kept at room temperature (RT) for up to 6 months.
- Ethanol at different percentages, all stored at 4°C.

5.2 Sample Preparation

- Seed cells on 3- or 6-cm culture dishes and allow cells to grow until reach approximately 70% confluence. You should obtain a visible pellet, reaching an approximate volume of 10–30 μL.
- Fix cells with 2% (w/v) glutaraldehyde buffered with 0.1 *M* sodium cacodylate, pH 7.4, for 2 h at RT.
- Carefully detach the fixed cells using a plastic cell scraper.
- Collect the cells into Eppendorf tubes, and centrifuge (5 min at 500 × *g*) to obtain the pellet.
- Wash three times with 0.1 *M* sodium cacodylate, pH 7.4.
- Postfix the pellets in 1% osmium tetroxide at RT for 1 h.
- Wash three times in distilled water.
- Stain with 0.5% uranyl acetate overnight at 4°C.
- Dehydrate the pellets in graded steps of ethanol (50%, 70%, 90%, and 100%) two times with 100% acetone (each of 15–30 min at 4°C).
- Embed the pellets into Epon™ (Sigma-Aldrich).
- Cut sections (60 nm thick) using an ultramicrotome.
- Electron microscopy examination.

5.3 Analysis of EM Images

The first critical step in quantitative microscopy is sampling, which must absolutely be unbiased. Moreover, every region of the sample should have an equal chance to be included in the counting, based on the principle of uniform random sampling (Yla-Anttila, Vihinen, Jokitalo, & Eskelinen, 2009). Due to the peculiar architecture of mitochondria, a double-membrane system with a wide cristae compartment, is quite easy to manually annotate mitochondrial regions on EM images. However, automated or semiautomatic methods that allow both shape information and regional statistics to segment mitochondria in EM images have been recently described (Higaki et al., 2015; Seyedhosseini, Ellisman, & Tasdizen, 2013).

The analysis of mitochondrial compartments on EM samples could be carried out manually or through the support of ImageJ (or Fiji) software.

A manual quantification of mitochondrial morphology is based on their appearance, especially referring to the shape of mitochondrial cristae. Four degrees of mitochondrial conformations have been proposed to classify the state and healthy of the organelles (Cogliati, Enriquez, & Scorrano, 2016; Rimessi et al., 2015; Scorrano et al., 2002). This classification includes

- Class I mitochondria: normal mitochondria with a well-defined outer membrane and a numerous narrow pleomorphic cristae (appearing in TEM as small electron-transparent areas) in a contiguous electron-dense matrix space.
- Class II mitochondria: remodeled mitochondria with widen cristae junctions with a serpentine electron-transparent intracristal compartment interrupted by electron-dense matrix spaces.
- Class III mitochondria: evident morphological defects, with partial rupture of the outer membrane.
- Class IV mitochondria: High morphological derangements, consisting in definitively swollen and ruptured mitochondria with undistinguishable cristae structure.

Morphological parameters of mitochondria in EM images could be also analyzed using the ImageJ software, as described by Picard et al. to characterize and quantify the differences in morphological descriptors of subsarcolemmal and intermyofibrillar mitochondria (Picard, White, & Turnbull, 2013). This analysis includes the assessment of the major mitochondrial shape factors (please refer to Section 4.2 and Fig. 2), such as Feret's diameter (representing the longest distance, expressed in μm, between any two points within a given mitochondrion), aspect ratio (measure of length), form factor (reflecting the complexity and branching aspect of mitochondria), and perimeter.

6. SUMMARY

In the last years, numerous findings sustained the close relationship existing between mitochondria and autophagy (Okamoto & Kondo-Okamoto, 2012; Sica et al., 2015). Alterations in the molecular mechanisms that regulate autophagy could result in mitochondrial remodeling and, at the same time, disturbance in mitochondrial homeostasis could contribute to malfunction of autophagy. The mutual regulation between the two processes finds its higher expression when we focus on the morphological changes that occur at mitochondrial level. Indeed, MEFs depleted for Drp1, which displayed a mitochondrial hyperfused state, strongly suppressed Parkin-mediated degradation of depolarized mitochondria (Tanaka et al.,

2010). Accordingly, liver-specific *Atg7*-knockout and systematically mosaic *Atg5*-deficient mice showed mitochondrial swelling and decreased respiration (Inami et al., 2011; Takamura et al., 2011). Again, loss of AMPK or ULK1 in liver cells resulted in accumulation of aberrant mitochondria (Egan et al., 2011). Therefore, a meticulous analysis of the mitochondrial network is required to understand the pivotal role of mitochondria in the regulation of the autophagic process or if pathological conditions that are characterized by altered autophagy could be worsened by deregulation of the mitochondrial compartment.

ACKNOWLEDGMENTS

P.P. is grateful to Camilla degli Scrovegni for continuous support. This research was supported by: the Italian Ministry of Education, University and Research (COFIN n. 20129JLHSY_002, FIRB n. RBAP11FXBC_002, and Futuro in Ricerca n. RBFR10EGVP_001), and Telethon (GGP15219/B); the Italian Cystic Fibrosis Research Foundation (#19/2014) to P.P.; the Italian Association for Cancer Research (AIRC: IG-14442 and MFAG-13521 to P.P. and C.G.); and the Italian Ministry of Health to C.G. and P.P. S.P. was supported by a research fellowship FISM—Fondazione Italiana Sclerosi Multipla—cod. 2014/B3.

REFERENCES

Benard, G., & Rossignol, R. (2008). Ultrastructure of the mitochondrion and its bearing on function and bioenergetics. *Antioxidants & Redox Signaling*, *10*(8), 1313–1342. http://dx.doi.org/10.1089/ars.2007.2000.

Bianchi, K., Vandecasteele, G., Carli, C., Romagnoli, A., Szabadkai, G., & Rizzuto, R. (2006). Regulation of Ca^{2+} signalling and Ca^{2+}-mediated cell death by the transcriptional coactivator PGC-1alpha. *Cell Death and Differentiation*, *13*(4), 586–596. http://dx.doi.org/10.1038/sj.cdd.4401784.

Bleazard, W., McCaffery, J. M., King, E. J., Bale, S., Mozdy, A., Tieu, Q., … Shaw, J. M. (1999). The dynamin-related GTPase Dnm1 regulates mitochondrial fission in yeast. *Nature Cell Biology*, *1*(5), 298–304. http://dx.doi.org/10.1038/13014.

Bonora, M., Bononi, A., De Marchi, E., Giorgi, C., Lebiedzinska, M., Marchi, S., … Pinton, P. (2013). Role of the c subunit of the FO ATP synthase in mitochondrial permeability transition. *Cell Cycle*, *12*(4), 674–683. http://dx.doi.org/10.4161/cc.23599.

Chan, D. C. (2012). Fusion and fission: Interlinked processes critical for mitochondrial health. *Annual Review of Genetics*, *46*, 265–287. http://dx.doi.org/10.1146/annurev-genet-110410-132529.

Cogliati, S., Enriquez, J. A., & Scorrano, L. (2016). Mitochondrial cristae: Where beauty meets functionality. *Trends in Biochemical Sciences*, *41*(3), 261–273. http://dx.doi.org/10.1016/j.tibs.2016.01.001.

Egan, D. F., Shackelford, D. B., Mihaylova, M. M., Gelino, S., Kohnz, R. A., Mair, W., … Shaw, R. J. (2011). Phosphorylation of ULK1 (hATG1) by AMP-activated protein kinase connects energy sensing to mitophagy. *Science*, *331*(6016), 456–461. http://dx.doi.org/10.1126/science.1196371.

Galluzzi, L., Pietrocola, F., Bravo-San Pedro, J. M., Amaravadi, R. K., Baehrecke, E. H., Cecconi, F., … Kroemer, G. (2015). Autophagy in malignant transformation and cancer

progression. *The EMBO Journal, 34*(7), 856–880. http://dx.doi.org/10.15252/embj.201490784.
Gomes, L. C., Di Benedetto, G., & Scorrano, L. (2011). During autophagy mitochondria elongate, are spared from degradation and sustain cell viability. *Nature Cell Biology, 13*(5), 589–598. http://dx.doi.org/10.1038/ncb2220.
Heim, R., & Tsien, R. Y. (1996). Engineering green fluorescent protein for improved brightness, longer wavelengths and fluorescence resonance energy transfer. *Current Biology, 6*(2), 178–182.
Higaki, T., Kutsuna, N., Akita, K., Sato, M., Sawaki, F., Kobayashi, M., … Hasezawa, S. (2015). Semi-automatic organelle detection on transmission electron microscopic images. *Scientific Reports, 5*, 7794. http://dx.doi.org/10.1038/srep07794.
Inami, Y., Waguri, S., Sakamoto, A., Kouno, T., Nakada, K., Hino, O., … Komatsu, M. (2011). Persistent activation of Nrf2 through p62 in hepatocellular carcinoma cells. *The Journal of Cell Biology, 193*(2), 275–284. http://dx.doi.org/10.1083/jcb.201102031.
Karbowski, M., & Youle, R. J. (2003). Dynamics of mitochondrial morphology in healthy cells and during apoptosis. *Cell Death and Differentiation, 10*(8), 870–880. http://dx.doi.org/10.1038/sj.cdd.4401260.
Kasahara, A., & Scorrano, L. (2014). Mitochondria: From cell death executioners to regulators of cell differentiation. *Trends in Cell Biology, 24*(12), 761–770. http://dx.doi.org/10.1016/j.tcb.2014.08.005.
Marchi, S., Corricelli, M., Trapani, E., Bravi, L., Pittaro, A., Delle Monache, S., … Pinton, P. (2015). Defective autophagy is a key feature of cerebral cavernous malformations. *EMBO Molecular Medicine*, 7(11), 1403–1417. http://dx.doi.org/10.15252/emmm.201505316.
Meeusen, S., DeVay, R., Block, J., Cassidy-Stone, A., Wayson, S., McCaffery, J. M., & Nunnari, J. (2006). Mitochondrial inner-membrane fusion and crista maintenance requires the dynamin-related GTPase Mgm1. *Cell, 127*(2), 383–395. http://dx.doi.org/10.1016/j.cell.2006.09.021.
Mitra, K., Wunder, C., Roysam, B., Lin, G., & Lippincott-Schwartz, J. (2009). A hyperfused mitochondrial state achieved at G1-S regulates cyclin E buildup and entry into S phase. *Proceedings of the National Academy of Sciences of the United States of America, 106*(29), 11960–11965. http://dx.doi.org/10.1073/pnas.0904875106.
Morciano, G., Giorgi, C., Balestra, D., Marchi, S., Perrone, D., Pinotti, M., & Pinton, P. (2016). Mcl-1 involvement in mitochondrial dynamics is associated with apoptotic cell death. *Molecular Biology of the Cell, 27*(1), 20–34. http://dx.doi.org/10.1091/mbc.E15-01-0028.
Morciano, G., Pedriali, G., Sbano, L., Iannitti, T., Giorgi, C., & Pinton, P. (2016). Intersection of mitochondrial fission and fusion machinery with apoptotic pathways: Role of Mcl-1. *Biology of the Cell, 108*, 279–293. http://dx.doi.org/10.1111/boc.201600019.
Okamoto, K., & Kondo-Okamoto, N. (2012). Mitochondria and autophagy: Critical interplay between the two homeostats. *Biochimica et Biophysica Acta, 1820*(5), 595–600. http://dx.doi.org/10.1016/j.bbagen.2011.08.001.
Olenych, S. G., Claxton, N. S., Ottenberg, G. K., & Davidson, M. W. (2007). The fluorescent protein color palette. *Current Protocols in Cell Biology, 5*, 1–34. http://dx.doi.org/10.1002/0471143030.cb2105s36. Chapter 21, Unit 21.
Otera, H., Wang, C., Cleland, M. M., Setoguchi, K., Yokota, S., Youle, R. J., & Mihara, K. (2010). Mff is an essential factor for mitochondrial recruitment of Drp1 during mitochondrial fission in mammalian cells. *The Journal of Cell Biology, 191*(6), 1141–1158. http://dx.doi.org/10.1083/jcb.201007152.
Patergnani, S., Marchi, S., Rimessi, A., Bonora, M., Giorgi, C., Mehta, K. D., & Pinton, P. (2013). PRKCB/protein kinase C, beta and the mitochondrial axis as

key regulators of autophagy. *Autophagy*, *9*(9), 1367–1385. http://dx.doi.org/10.4161/auto.25239.

Picard, M., White, K., & Turnbull, D. M. (2013). Mitochondrial morphology, topology, and membrane interactions in skeletal muscle: A quantitative three-dimensional electron microscopy study. *Journal of Applied Physiology (Bethesda, Md.: 1985)*, *114*(2), 161–171. http://dx.doi.org/10.1152/japplphysiol.01096.2012.

Rambold, A. S., Kostelecky, B., Elia, N., & Lippincott-Schwartz, J. (2011). Tubular network formation protects mitochondria from autophagosomal degradation during nutrient starvation. *Proceedings of the National Academy of Sciences of the United States of America*, *108*(25), 10190–10195. http://dx.doi.org/10.1073/pnas.1107402108.

Rimessi, A., Bezzerri, V., Patergnani, S., Marchi, S., Cabrini, G., & Pinton, P. (2015). Mitochondrial Ca^{2+}-dependent NLRP3 activation exacerbates the *Pseudomonas aeruginosa*-driven inflammatory response in cystic fibrosis. *Nature Communications*, *6*, 6201. http://dx.doi.org/10.1038/ncomms7201.

Rimessi, A., Bonora, M., Marchi, S., Patergnani, S., Marobbio, C. M., Lasorsa, F. M., & Pinton, P. (2013). Perturbed mitochondrial Ca^{2+} signals as causes or consequences of mitophagy induction. *Autophagy*, *9*(11), 1677–1686. http://dx.doi.org/10.4161/auto.24795.

Rizzuto, R., Pinton, P., Carrington, W., Fay, F. S., Fogarty, K. E., Lifshitz, L. M., … Pozzan, T. (1998). Close contacts with the endoplasmic reticulum as determinants of mitochondrial Ca^{2+} responses. *Science*, *280*(5370), 1763–1766.

Santel, A., & Fuller, M. T. (2001). Control of mitochondrial morphology by a human mitofusin. *Journal of Cell Science*, *114*(Pt. 5), 867–874.

Sasaki, S. (2010). Determination of altered mitochondria ultrastructure by electron microscopy. *Methods in Molecular Biology*, *648*, 279–290. http://dx.doi.org/10.1007/978-1-60761-756-3_19.

Schindelin, J., Arganda-Carreras, I., Frise, E., Kaynig, V., Longair, M., Pietzsch, T., … Cardona, A. (2012). Fiji: An open-source platform for biological-image analysis. *Nature Methods*, *9*(7), 676–682. http://dx.doi.org/10.1038/nmeth.2019.

Scorrano, L., Ashiya, M., Buttle, K., Weiler, S., Oakes, S. A., Mannella, C. A., & Korsmeyer, S. J. (2002). A distinct pathway remodels mitochondrial cristae and mobilizes cytochrome c during apoptosis. *Developmental Cell*, *2*(1), 55–67.

Seyedhosseini, M., Ellisman, M. H., & Tasdizen, T. (2013). Segmentation of mitochondria in electron microscopy images using algebraic curves. *Proceedings/IEEE International Symposium on Biomedical Imaging: From Nano to Macro*, *2013*, 860–863. http://dx.doi.org/10.1109/ISBI.2013.6556611.

Sica, V., Galluzzi, L., Bravo-San Pedro, J. M., Izzo, V., Maiuri, M. C., & Kroemer, G. (2015). Organelle-specific initiation of autophagy. *Molecular Cell*, *59*(4), 522–539. http://dx.doi.org/10.1016/j.molcel.2015.07.021.

Sumpter, R., Jr., Sirasanagandla, S., Fernandez, A. F., Wei, Y., Dong, X., Franco, L., … Levine, B. (2016). Fanconi anemia proteins function in mitophagy and immunity. *Cell*, *165*(4), 867–881. http://dx.doi.org/10.1016/j.cell.2016.04.006.

Szabadkai, G., Simoni, A. M., Chami, M., Wieckowski, M. R., Youle, R. J., & Rizzuto, R. (2004). Drp-1-dependent division of the mitochondrial network blocks intraorganellar Ca^{2+} waves and protects against Ca^{2+}-mediated apoptosis. *Molecular Cell*, *16*(1), 59–68. http://dx.doi.org/10.1016/j.molcel.2004.09.026.

Takamura, A., Komatsu, M., Hara, T., Sakamoto, A., Kishi, C., Waguri, S., … Mizushima, N. (2011). Autophagy-deficient mice develop multiple liver tumors. *Genes & Development*, *25*(8), 795–800. http://dx.doi.org/10.1101/gad.2016211.

Tanaka, A., Cleland, M. M., Xu, S., Narendra, D. P., Suen, D. F., Karbowski, M., & Youle, R. J. (2010). Proteasome and p97 mediate mitophagy and degradation of mitofusins induced by Parkin. *The Journal of Cell Biology*, *191*(7), 1367–1380. http://dx.doi.org/10.1083/jcb.201007013.

Yla-Anttila, P., Vihinen, H., Jokitalo, E., & Eskelinen, E. L. (2009). Monitoring autophagy by electron microscopy in Mammalian cells. *Methods in Enzymology*, *452*, 143–164. http://dx.doi.org/10.1016/S0076-6879(08)03610-0.

Yoon, Y., Krueger, E. W., Oswald, B. J., & McNiven, M. A. (2003). The mitochondrial protein hFis1 regulates mitochondrial fission in mammalian cells through an interaction with the dynamin-like protein DLP1. *Molecular and Cellular Biology*, *23*(15), 5409–5420.

Youle, R. J., & van der Bliek, A. M. (2012). Mitochondrial fission, fusion, and stress. *Science*, *337*(6098), 1062–1065. http://dx.doi.org/10.1126/science.1219855.

CHAPTER ELEVEN

Monitoring Mitophagy in Mammalian Cells

L. Chen*, K. Ma*, J. Han*, Q. Chen*,†,1, Y. Zhu*,1
*State Key Laboratory of Medicinal Chemical Biology, Tianjin Key Laboratory of Protein Science, College of Life Sciences, Nankai University, Tianjin, China
†State Key Laboratory of Membrane Biology, Institute of Zoology, Chinese Academy of Sciences, Beijing, China
[1]Corresponding authors: e-mail address: chenq@ioz.ac.cn; zhuys@nankai.edu.cn

Contents

Abstract

In eukaryotes, physiological cell functions rely on the preservation of the size and activity of the mitochondrial network. Mitophagy provides a key contribution in this setting by ensuring the removal of permeabilized or supernumerary mitochondria. Throughout the past decade, mitophagy has attracted considerable attention from both fundamental researchers and translational investigators, both of whom have called for the development of techniques that allow the precise quantification of mitophagy (as opposed to general autophagy). In this chapter, we present morphological, biochemical, and fluorescence-based approaches to measure autophagy in mammalian cells, and discuss recent progress in mouse models for the assessment of mitophagy in vivo.

Methods in Enzymology, Volume 588
ISSN 0076-6879
http://dx.doi.org/10.1016/bs.mie.2016.10.038

1. INTRODUCTION

Mitochondria are vital organelles that can control cell life and death. It is thus essential for a cell to maintain a sufficient quantity of high-quality mitochondria for normal cellular functions (Chan, 2006; Karbowski & Youle, 2003; Schon & Przedborski, 2011; Suen, Norris, & Youle, 2008; Youle & Narendra, 2011). Even under basal conditions, mitochondria undergo continuous cycles of fusion and fission and of biogenesis and degradation that serve to produce daughter mitochondria and remove unwanted or defective mitochondria. It is now clear that mitochondria can be selectively removed through mitochondrial autophagy, or mitophagy, although mitochondria are also degraded during nonselective bulk cytoplasmic degradation through general autophagy (Feng et al., 2013). Both receptor- and nonreceptor-mediated mitophagy have been described and extensively reviewed (Ding & Yin, 2012; Feng et al., 2013; Liu et al., 2014; Matsuda & Tanaka, 2010; Scarffe et al., 2014; Stolz, Ernst, & Dikic, 2014; Winklhofer, 2014; Zhu, Dagda, & Chu, 2011). Given the significant progress and the explosion of information in this exciting field, it is important for the community to measure mitophagy properly. Here, we will discuss the current techniques that have been used to monitor mitophagy in mammalian cells. As mitophagy is a specific and selective form of autophagy, it is important to measure the involvement of both general autophagy and the mitochondrial-specific type. In addition, as with general autophagy, both morphological and biochemical parameters are required for monitoring the occurrence of mitophagy.

2. TRANSMISSION ELECTRON MICROSCOPY

The morphological hallmark of general autophagy is enclosed the misfolded proteins, protein aggregates, damaged organelles, lipid droplets, and even nuclear components by a double-membrane structure called an *autophagosome*. The autophagosomes will fuse with lysosomes to form mature autolysosomes to degradation, which are single-membrane structures. To detect both early autophagosomes and late autolysosomes, transmission electron microscopy (TEM) remains the best and most effective tool. Similarly, the ultrastructural assay of mitophagy using TEM is a direct and powerful method for confirming mitochondrial autophagy or clearance. The early stage of mitophagy (autophagosomes with engulfed mitochondria) can be easily detected through the identification of unique

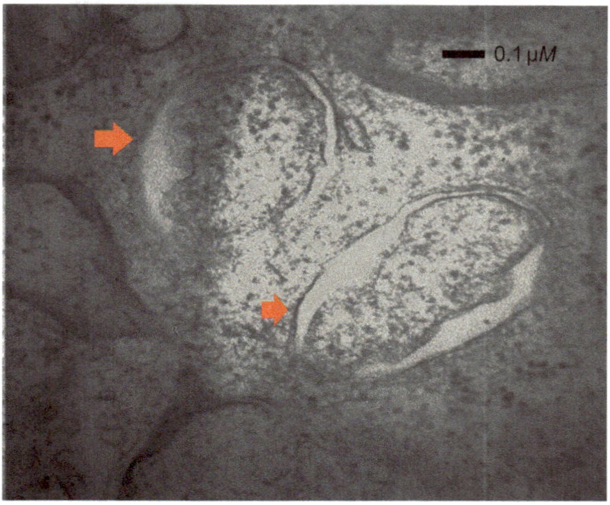

Fig. 1 Analysis of mitophagy by electron microscopy. HeLa cells were treated with FCCP for 2 h, and the samples were analyzed. *Arrows* indicate autophagosomes that contain mitochondria.

mitochondrial morphology structures such as the cristae or double-membrane structure. The late stage of mitophagy may be reflected by single-membrane autolysosomes with residual mitochondria, which can be identified on the basis that they have similar electron density to mitochondria (Klionsky et al., 2012) (see Fig. 1). To check mitophagy by mitochondria morphology, it is better to perform immunogold-labeled EM for specific mitochondrial protein markers, such as TOMM20, TIMM23, or VDAC1 (Ding et al., 2012; Liu et al., 2012), to confirm the nature of the mitochondria in the late stage of mitophagy. It should be noted that TEM is sometimes problematic due to misinterpretations that mostly derive from methodological artifacts. It is thus important to take proper care of the sample to be analyzed right from the beginning of the preparation and the fixation procedures. The fixation of *in vitro* samples is relatively straightforward (see the protocol given later in this chapter). However, the fixation of tissue samples requires care in order to avoid problems of nonrepresentative or uninformative tissue sections. Fresh samples should be handled in a timely manner to avoid changes in cellular structures that may occur during the manipulations; the samples should be fixed immediately with freshly prepared fixatives. For *in vitro* cultured cells, they can be grown in plastic chips placed in a culture medium such as DMEM 1640, and the fixation and sectioning can be carried out without trypsin digestion to keep the cells intact during TEM. For fresh tissue samples, perfusion fixation should be used as much as possible. It is important to be aware that autophagosomal maturation and autophagosome-lysosome fusion are relatively rapid processes, and it may be very difficult to recognize mitochondrial morphology within autophagosomes. Prior to prepare TEM samples, application of the autophagic inhibitor bafilomycin A1 (BFA), an inhibitor of vacuolar-type H(+)-ATPase and autophagosome–lysosome fusion, may prevent autophagosome maturation, thus trapping newly formed autophagosomes. This may facilitate the identification of cargo such as mitochondria, as well as providing comparative information on the rates of autophagy progression.

Details

1. Cells were cultured in a medium containing 10% FBS and 0.1 mg/mL penicillin–streptomycin at 37°C under 5% CO_2. When the cell density was 60–80%, the cells were placed under conditions that induce mitophagy. For example, hypoxia is a typical mitophagy inducer, and hypoxic conditions can be achieved by flushing the cells with a preanalyzed gas mixture of 1% O_2, 5% CO_2, and 94% N_2 in a hypoxia chamber (Billups-Rothenberg Inc., San Diego, California). Carbonyl cyanide *p*-trifluoromethoxyphenylhydrazone (FCCP), carbonyl cyanide *m*-chlorophenylhydrazone (CCCP), and other mitochondrial toxins can also be employed to induce mitophagy. BFA (10–20 n*M*) was added for 2–4 h prior to harvesting cells for easy detection of autophagosomal structures.
2. The cells were rinsed three times with phosphate-buffered saline (PBS) and fixed for at least 60 min with freshly prepared 2.5% glutaraldehyde at room temperature or overnight at 4°C.
3. After fixation, the cell monolayers were washed three times in PBS and then fixed in aqueous 2% OsO_4 for 1 h.
4. The cells were then dehydrated with sequential washes in 50%, 70%, 90%, 95%, and 100% ethanol and then embedded in PolyBed 812 epoxy resin.
5. Ultrathin (60-nm) sections were collected on copper grids and stained with 2% uranyl acetate in 50% methanol for 10 min, followed by incubation in 1% lead citrate for 7 min.
6. The sections were photographed using TEM at 80 kV.

3. WESTERN BLOTTING ANALYSIS OF MITOPHAGY

Mitophagy can be monitored biochemically through the degradation of mitochondrial proteins, together with the appearance of autophagy markers such as LC3-II (microtubule-associated protein 1 light chain 3) and degradation of SQSTM1/p62. This can be achieved by analyzing the biochemical hallmarks such as mitochondrial inner-, inter-, and outer-membrane proteins or mitochondrial matrix proteins. The LC3 protein is an ubiquitinlike protein that can be conjugated to phosphatidylethanolamine (PE). The total levels of LC3 are maintained in a balanced state, and under stress conditions, there may be increased conversion of LC3-I (19 kDa) to LC3-II (17 kDa) by PE-conjugated. The decrease in LC3-II relative to LC3-I in the case of the degradation of LC3-II via lysosomal turnover is particularly rapid (see Fig. 2). Thus, Western blotting can be easily used to monitor changes in LC3 amounts during mitophagy, and when LC3 flux is inhibited by BFA (Chen et al., 2014). Western blotting analysis of the levels of mitochondrial proteins from different mitochondrial subcompartments is valuable for validating the data from EM studies. However, more recent studies suggest that most outer mitochondrial membrane proteins are degraded by the ubiquitin proteasome system or mitochondria-derived vesicle pathway, whereas the matrix proteins are degraded through proteasome pathway and autophagy (McLelland et al., 2014; Yoshii et al., 2011). Therefore, a simultaneous analysis of outer and inner mitochondrial membrane proteins and matrix proteins, such as TOMM20, TIMM23, CypD, and HSP60, would be more appropriate for monitoring mitophagy. In our opinion, VDAC1 may not be a suitable candidate protein because

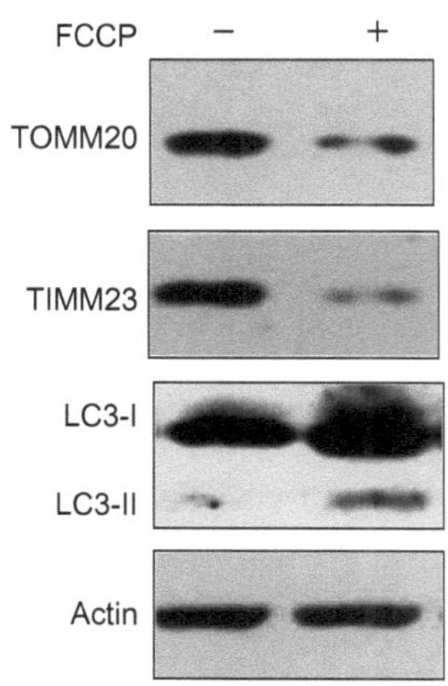

Fig. 2 FCCP induces mitophagy. HeLa cells were treated with 10 μ*M* of FCCP for 6 h. The mitochondrial inner membrane protein TOMM20, the outer-membrane protein TIMM23, and LC3 were detected via Western blotting.

of its response to many different types of stress. For general autophagy, an autophagy inhibitor, such as 3-MA or BFA, will be supplement with the cells to block the mitochondrial protein degradation process to detect mitophagy. The levels of Atg proteins may not change significantly during mitophagy. However, knockdown of *Atg5* or *Atg7* expression is normally able to prevent mitochondrial protein degradation, whereas knockdown of Beclin 1 may fail to prevent mitophagy in some conditions.

The degradation of SQSTM1/p62 has been widely used as a biochemical marker for general autophagy. SQSTM1/p62 was reported to act as a cargo receptor for ubiquitinated protein aggregates that are intended for selective autophagy (Geisler et al., 2010; Kirkin et al., 2009a, 2009b; Kraft, Peter, & Hofmann, 2010; Lamark et al., 2009). It was found that p62 translocates to mitochondria via interaction with ubiquitin, which is significantly increased in response to mitochondrial stresses. SQSTM1/p62 was also found to bind other critical signaling intermediates such as TRAF6 (TNF receptor-associated factor 6), which is essential for the regulation of NF-κB during osteoclastogenesis and bone homeostasis, as well as in cancer (Mathew et al., 2009; Moscat & Diaz-Meco, 2009; Sanz et al., 2000; Yu et al., 2009). Whether ubiquitin/p62/LC3 complexing is sufficient for Parkin-mediated mitophagy remains a subject of debate and requires further investigation. We recently showed that p62 is a substrate of Parkin, which mediates ubiquitination and proteasomal degradation (Song et al., 2016), and TRIM21 can directly interact with and ubiquitylate SQSTM1/p62 via K63-linkage to regulate its sequestration and cellular redox homeostasis (Pan et al., 2016). SQSTM1/p62 may be a useful marker for mitophagy, but care should be taken to interpret the data properly because it is such a multifunctional protein and is subjected to proteasomal degradation.

Details

(1) Cells were treated with mitophagy-inducing conditions such as hypoxia or FCCP (during the treatment, the culture medium must be replaced with fresh medium or medium supplemented with mitophagy inducers because in the old medium, nutrients are utilized, which causes energy deprivation and general autophagy). The cells were then treated with BFA (5–10 n*M*) for 2–4 h before the sample collection. Cells were scraped and immediately washed once in PBS once, then kept on ice.

(2) Cells were lysed in lysis buffer for 40 min on ice (20 m*M* Tris, pH 7.4, 137 m*M* NaCl, 2 m*M* EDTA, 10% glycerol, 1% NP-40, cocktail inhibitors). After centrifugation at 12,000 × *g* for 15 min, the supernatant was boiled in sample buffer (4% SDS, 20 m*M* Tris, pH 6.8, 20% glycerol, 0.1% bromophenol blue, 200 m*M* β-mercaptoethanol) for 5 min.

(3) Equivalent quantities of protein (10–20 μg) were subjected to sodium dodecyl sulfate–polyacrylamide gel electrophoresis (SDS-PAGE) and then transferred to nitrocellulose membranes.

(4) The membranes were probed with the indicated primary antibodies (TOMM20, BD Biosciences 612278; TIMM23, BD Biosciences 611222; LC3, Sigma L7543 or MBL PM036/M186; p62, MBL PM045; Actin, Sigma A5441), followed by the appropriate HRP-conjugated secondary antibodies (KPL, Gaithersburg, Maryland).

(5) Immunoreactive bands were visualized with a chemiluminescence kit (Millipore, Billerica, Massachusetts).

4. DETECTED MITOPHAGY BY ANALYSIS MITOCHONDRIAL MASS

The simplest method of monitoring mitophagy is to analyze the mitochondrial mass. This method can be employed to study the efficiency of the degradation of dysfunctional, aged, or impaired mitochondria. Mitochondrial mass can be assayed either in a population of cells using a fluorescence-activated cell sorting (FACS) technique with acridine orange 10-nonyl bromide (NAO, Invitrogen A1372) or in single cells by immunocytochemistry using MitoTracker Green FM (Invitrogen M7514) and mitochondrial marker protein antibodies (as described previously). EM can also be used to verify the loss of all the mitochondria, and polymerase chain reaction or fluorescence microscopy can be used to quantify mitochondrial DNA by analysis of mitochondrial genes such as 16S rRNA, COX1 (Complex IV subunit), CYTB (Complex III subunit), or nuclear DNA by analysis of nuclear genes such as hexokinase, H19, or beta-1 globin. The mitochondrial-to-nuclear DNA ratio in each sample can then be calculated by dividing the value for mitochondrial DNA by that for nuclear DNA (Lagouge et al., 2006).

5. FLUORESCENCE PROBES

5.1 Fluorescence Methods for Analyzing Mitophagy

Fluorescence microscopy can be widely used to monitor mitophagy by observing colocalization of the autophagy marker GFP-LC3 with mitochondrial proteins in cells (Chen et al., 2014; Liu et al., 2012). The GFP-LC3 plasmid can be cotransfected into cells with mitochondrial localized DsRed (Mito-DsRed), or mitochondria can be stained with MitoTracker or immunostained with antibodies against TOMM20, TIMM23, VADC, or cytochrome *c*. Observed colocalization of LC3 and mitochondria is revealed as yellow puncta in merged images. Sometimes colocalized LC3 and mitochondria do not appear yellow (the overlap color) due to differences in the intensity of staining. A high-resolution confocal microscope can clearly show that the mitochondria (red) are enclosed in GFP-LC3-labeled autophagosomes (see Fig. 3). The average number or the surface of GFP-LC3 puncta are as colocalized with mitochondria per cell can be quantified by counting no fewer than 100 cells (Chen et al., 2014; Liu et al., 2012) in three separate experiments. LC3, ULK1, and p62, are known to translocate onto mitochondria to induce mitophagy (Li et al., 2015). And ULK1 plays a more specific role in mitophagy. An *Ulk1*-knockout animal model demonstrated defects in the autophagic clearance of mitochondria during erythroid maturation (Kundu et al., 2008). ULK1 is recruited to fragmented mitochondria in response to hypoxia or FCCP (Itakura et al., 2012). The translocated ULK1 interacts

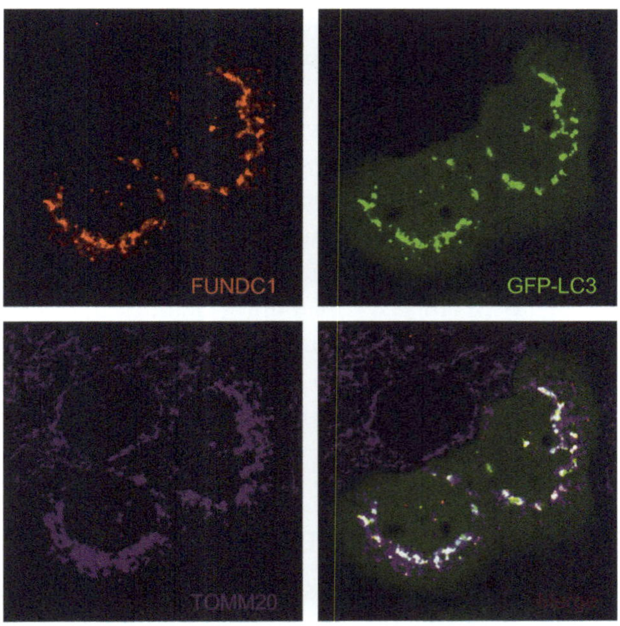

Fig. 3 Analysis of mitophagy by fluorescence microscopy. FUNDC1-Myc and GFP-LC3 were ectopically expressed in HeLa cells for 24 h; the cells were then fixed and immunostained with Myc (*red*) and TOMM20 (*blue*).

with its substrate FUNDC1 to phosphorylate FUNDC1 (Wu et al., 2014). And found that ULK1 can be ubiquitated by E3-ligase MUL1 to induced mitophagy in response to reactive oxygen species (Li et al., 2015).

Details

(1) Cells were grown on glass coverslips. Following treatment, the cells were washed three times with PBS and fixed with freshly prepared 4% formaldehyde at 37°C for 30 min.

(2) Antigen accessibility was increased by treatment with 0.2% Triton X-100 for 10 min on ice. The cells were then incubated with primary antibodies for 2 h, and, after washing five times with PBS, they were stained with a secondary antibody for an additional 1 h (donkey anti-mouse Alex Flour 555, Invitrogen A31570; donkey anti-rabbit Alex Flour 555, Invitrogen A3157; donkey anti-goat-Cy5, Invitrogen A21082; donkey anti-mouse 488, Invitrogen A21202).

(3) Cell images were captured with a confocal microscope or high-resolution confocal microscope. The area and number of LC3 aggregates were measured in approximately 200 cells using Image J software.

5.2 Mitochondrial-Targeting Probes Assay Mitophagy

Checking the level of mitochondrial proteins by Western blotting is a widely used method to measure mitophagy. TEM is more direct and definitive, but that method suffers from a lack of sensitivity and specificity, so many scientists have set up direct immunofluorescence methods.

In the general immunofluorescence assay, colocalization of GFP-LC3 (the marker of autophagosomes and autolysosomes) with dysfunctional mitochondria shows that mitophagy has been initiated. However, mitochondrial engulfment takes less than 10 min, and the GFP-LC3 fluorescence is lost soon after acidification in the low-pH environment once lysosomal fusion occurs. Such a dynamic process is hard to monitor efficiently. In the past, other more reliable assays have been employed to detect mitophagy, like mt-Keima (mitochondria-targeted Keima), MitoTimer, MitoTracker, and GFP-Cherry-Mito. These assays are often based on markers with fluorescent emission, which is stable in a lower-pH environment or changes in different pH surroundings. The markers are used in tandem with a mitochondrial-targeting signal (inner- or outer-membrane sorting signal).

5.2.1 Mt-Keima Assay

Keima is a naturally occurring fluorescent protein that is resistant to lysosomal proteases and shows a characteristic shift in its excitation wavelength depending upon the pH of the environment (Katayama et al., 2011; Rosado et al., 2008). Recently, Katayama et al. developed a sensitive and quantitative technique using mt-Keima, a ratiometric, pH-sensitive fluorescent protein that is targeted into the mitochondrial matrix. A low ratio of mt-Keima-derived fluorescence (543 nm/458 nm) indicates a neutral environment, whereas a high ratio indicates an acidic pH. Thus, mt-Keima enables the differential imaging of mitochondria in the cytoplasm and mitochondria in acidic lysosomes. Because mt-Keima is resistant to lysosomal proteases, it enables measurement of the cumulative lysosomal delivery of mitochondria over time (Katayama et al., 2011).

The group of Kanki developed HeLa cells that express tetracycline (Tet)-suppressible mitochondrial matrix-localized Keima (mito-Keima), and measured mitophagy in response to stresses such as starvation, hypoxia, and CCCP. Starvation and hypoxia, but not CCCP, induced a certain number of acidic puncta (Hirota et al., 2015). Other groups overexpressed mt-Keima, mitochondrial FRB-Fis1, and PINK1Δ110-YFP-2×FKBP in HeLa cells, and found that 7.98% of the cells initiated mitophagy after 24 h of treatment with rapalog when analyzed by sensitive FACS (Hirota et al., 2015).

In order to evaluate mitochondrial autophagy in the mouse heart in vivo, Shirakabe and colleagues generated adeno-associated virus (AAV) serotype 9 harboring either Mito-Keima or Lamp1-YFP. The mice were administered intravenously with AAV9-Mito-Keima and AAV9-Lamp1-YFP and fasted for 48 h, and then were killed and prepared the heart slices in cold PBS were subjected to confocal microscopic analyses, and the acidic mt-Keima dots were reported to be colocalized with Lamp1-YFP dots (Shirakabe et al., 2016). Time-lapse imaging of mt-Keima and GFP-Parkin was used to simultaneously observe Parkin recruitment to mitochondria and their subsequent autophagic degradation after membrane depolarization. This method was recently employed to determine whether USP30 opposes Parkin-mediated mitophagy (Bingol et al., 2014).

5.2.2 MitoTimer for Mitochondrial Turnover

Mitochondrial biogenesis and mitophagy are processes that determine mitochondrial turnover, along with fusion and fission. Recently, Hernandez and colleagues developed a new probe called MitoTimer for detecting mitochondrial turnover. MitoTimer is a mutant of the red fluorescent protein DsRed, in which the fluorescence shifts irreversibly from green to red as the protein is oxidized. They targeted the probe to the mitochondrial matrix with a mitochondrial-targeting sequence. MitoTimer can continuously monitor mitochondrial dynamics within cells and is of particular value where mitophagy is closely matched with biogenesis. Coupled with organelle flow cytometry, MitoTimer offers new opportunities to investigate mitochondrial subpopulations using biochemical or proteomic methods (Ferree et al., 2013; Hernandez et al., 2013; Laker et al., 2014). In contrast to mt-Keima, which reflects the final phase of mitophagy (delivery of mitochondria to the lysosome), MitoTimer provides information on biogenesis and, by using time-restricted expression, can also provide information on mitophagy.

5.2.3 MitoTracker Assay

MitoTracker is able to label live mitochondria in different mammalian cells, and it also can be used to demonstrate colocalization of mitochondria with autophagosomes or lysosomes. MitoTracker (Deep Red) has been used to measure mitophagy through flow cytometry, and MitoTracker Green has also been used to indicate mitochondrial mass. However, it has been shown that these two dyes are both affected by mitochondrial potential or ROS. Because of these limitations, quantitative assessment of mitophagy using MitoTracker should be always interpreted along with other mitophagy assays.

5.2.4 GFP-Cherry-Mito Assay

When checking the mitophagy level using immunofluorescence, we often overexpress GFP-LC3, which is recruited to mitochondria if mitophagy is initiated. However, the green fluorescent protein (GFP) signal is quenched easily in a lower-pH environment; in addition, GFP-LC3 cannot specifically label mitophagy when macroautophagy is occurring at the same time. It is more reliable to check the level of mitophagy using a specific signal attached to the mitochondria. The recombinant GFP-Cherry-Mito protein uses the transmembrane domain of Bcl-xL to attach a tandem GFP and Cherry signal onto the outer mitochondrial membrane. It sensitively monitors mitophagy when the mitophagy adaptor recruits lysosomes, which results in fluorescence quenching of the GFP signal, but not the Cherry signal (see Fig. 4).

Details

(1) Cells were grown on glass coverslips for 24 h, and then transfected with the appropriate GFP-Cherry-Mito plasmid (0.5 μg or less per cell of a six-cell-plate).

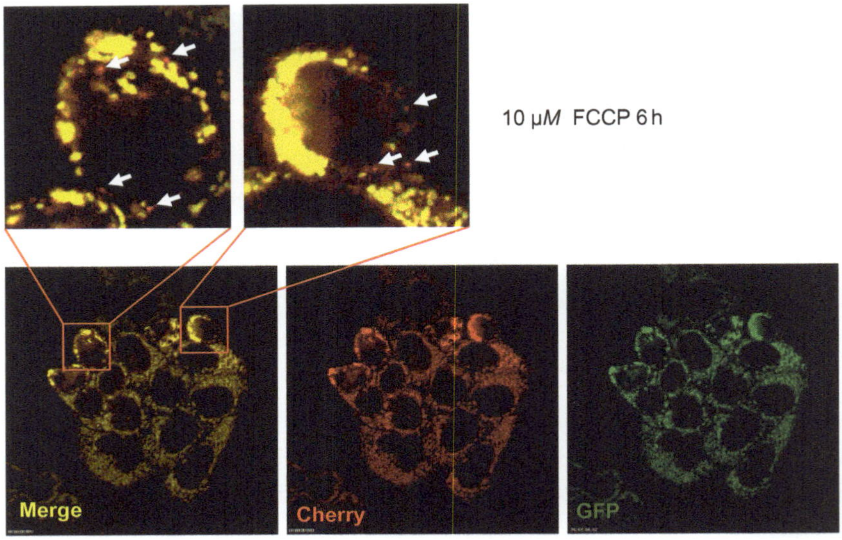

Cherry-GFP-Mito

Fig. 4 HeLa cells are transfected with Mito-GFP-Cherry-Mito for 24 h (or stable expressed cells) and then were treated with FCCP for 6–12 h, next fixed and checked for mitophagy with a confocal microscope. The *red signal* is from Cherry, the *green signal* is from GFP, and the *blue signal* is from FUNDC1 detected with anti-Myc antibodies. In the zoomed image, the *white arrow* indicates mitochondria that are engulfed by lysosomes.

(2) The next day, cells were treated with mitophagy inducers such as FCCP and hypoxia.

(3) Following treatment, the cells were washed three times with PBS, and then fixed with freshly prepared 4% formaldehyde at 37°C for 30 min.

(4) Formaldehyde was removed by washing three times with PBS, and then the cells were finally transferred to coverslips. Cell images were captured with a confocal microscope or high-resolution confocal microscope and statistical analysis was performed on approximately 100–200 positive cells.

5.2.5 Om45-GFP Assay

The Om45-GFP methods was first reported by the Klionsky group to monitor mitophagy in yeast (Kanki, Kang, & Klionsky, 2009). Om45 is a mitochondrial outer-membrane protein, and a GFP-tagged Om45 (GFP at the C-terminus of Om45) is correctly localized on mitochondria. Following the uptake of mitochondria into the vacuole, Om45-GFP is degraded, releasing the intact form of GFP, which can be detected by Western blotting. Thus, mitophagy can be monitored and quantified according to the appearance and intensity of free GFP on the immunoblot. A similar GFP-tagged mitochondrial protein-processing assay was also developed in mammalian cells. Yoshii et al. (2011) developed a flow cytometry-based assay using cell lines that stably express GFP proteins fused to either a mitochondrial matrix protein (subunit 9 of F0-ATPase, Su9-GFP) or an outer mitochondrial membrane protein (GFP-Omp25) (Yoshii et al., 2011). They found that the Su9-GFP and GFP-Omp25 signals were largely unaffected in Parkin-untransfected mouse embryo fibroblasts (MEFs) but were significantly decreased in Parkin-transfected wild-type MEFs in a time-dependent manner following CCCP treatment, and this decrease was inhibited in the absence of FIP200 or Atg5 (Yoshii et al., 2011). These studies identified that the degradation of matrix proteins can be used as a quantitative methods to monitor mitophagy in mammalian cells. In addition, Tandem mCherry–GFP tags were targeted to the outer mitochondrial membrane by signal of the protein FIS1 (residues 101–152) and were used to monitor the delivery of mitochondria to lysosomes by the color change during mitophagy (Allen et al., 2013). Under normal conditions, mitochondria displayed both red and green fluorescence. Upon mitophagy, the mitochondria were delivered to lysosomes, in which the low-pH quenches the signal from GFP, but not from mCherry (Allen et al., 2013).

6. MITOPHAGY MOUSE MODELS

6.1 Mt-Keima Transgenic Mouse

Finkel's group has generated a reporter mouse containing the fluorescent reporter mt-Keima, which allows for the in vivo assessment of mitophagy in various tissues under normal conditions. Using the mt-Keima mouse, we can assess how mitophagy is altered following changes in diet, oxygen availability, genetic perturbations, and normal aging. The results show that mitophagy varies between and within tissues, and hepatic mitophagy is inhibited under chronic high-fat feeding but increased under hypoxia (Sun et al., 2015). This model provides the first robust assessment of in vivo mitophagy under a wide range of experimental conditions (Sun et al., 2015).

6.1.1 Mouse Tissue Collection

Fresh tissues were used for both confocal and stimulated emission depletion microscopy studies. For the brain, tissues were rapidly dissected out, rinsed with cold PBS, and then further processed into coronal sections (0.5–2 mm).

These sections were placed into cold PBS and immediately analyzed at selected time points. For other tissues, following a rinse with cold PBS, tissues were cut into small pieces (approximately 100 mm^3), then placed onto a 35-mm coverglass #1.5 bottom microwell dish and analyzed immediately (Sun et al., 2015). Fluorescence of mt-Keima was imaged in two channels via two sequential excitations (458 nm, green; 561 nm, red) and using a 570- to 695-nm emission range. Laser power was set at the lowest output that would allow clear visualization of the mt-Keima signal and was individualized for each experimental condition. Imaging settings were maintained with the same parameters for comparison between different experimental conditions (Sun et al., 2015).

6.1.2 Western Blotting Assay

Fresh tissues (or tissue saved in −80°C or liquid nitrogen) were used to assay mitophagy by Western blotting.

The details of the process are as follows: Thirty micrograms of tissue samples were homogenized in 1 mL of RIPA tissue lysis buffer [50 m*M* Tris (pH 7.4), 150 m*M* NaCl, 1% NP-40, 0.5% sodium deoxycholate, 0.1% SDS, with PMSF freshly added] with a Kimble Dounce homogenizer, then disrupted five times in a sonicator on ice under 30% strength for 5 s each time. After centrifugation for 10 min at 12,000 rpm, the supernatant was transferred to a new tube. 100 μL of 6 × loading buffer (20 m*M* Tris–HCl, pH 6.8, 4% SDS, 0.1% bromophenol blue, 20% glycerol, and 200 m*M* β-mercaptoethanol) was mixed with 500 μL supernatant and incubated for 20 min in boiling water. 20-μg samples were subjected to SDS-PAGE and Western blotting for LC3, p62, ubiquitin, or mitophagy proteins such as TOMM20 and TIMM23.

6.2 GFP-LC3 Transgene Model

Mizushima et al. have generated GFP-LC3 transgenic mice and examined the occurrence of autophagy in various tissues. These GFP-LC3-expressing mice (CAG-GFP-LC3-mice) have been useful tools for detecting autophagosomes systemically (Mizushima et al., 2004). Next, Kim et al. reported the characterization of the dynamics of mitochondrial turnover by mitophagy in hepatocytes isolated from GFP-LC3 transgenic mice during nutrient deprivation (Kim & Lemasters, 2011). However, it is difficult to differentiate increased formation of autophagosomes from decreased clearance of autophagosomes (Lwai-Kanai et al., 2008). Terada et al. generated double-transgenic mice by crossing GFP-LC3-transgenic mice with αMyHC-mCherry-LC3-transgenic mice (under control of the αMyHC promoter). The differential resistance of GFP and mCherry to acidic conditions overcame this limitation. The authors suggest that both GFP/mCherry-LC3-mice and GFP-LC3-mice treated with chloroquine are useful to show that starvation increases autophagic activity. However, there are some advantages in using GFP/mCherry-LC3-mice. Only autophagosomes can be detected in GFP-LC3-mice, while both autophagosomes and autolysosomes can be detected and distinguished simultaneously using GFP/mCherry-LC3-mice. This approach can overcome the limitation of monitoring autophagic activity in the heart. Because dysfunction during the fusion of autophagosomes with lysosomes might contribute to heart disease, this new method can be used to evaluate the effect of impaired fusion under various stresses, such as ischemia/reperfusion and pressure-overload (Terada et al., 2010).

7. MAM ASSAY

The MAM (ER–mitochondria contact site) domain is important in many cellular functions, including efficient transfer of Ca^{2+} from the endoplasmic reticulum (ER) to mitochondria, lipid metabolism, and controlling mitochondrial biogenesis and intracellular trafficking. Importantly, MAM membrane sources are essential for the initiation of mitophagy (Lamb, Yoshimori, & Tooze, 2013; Rowland & Voeltz, 2012). Also, new data have shown that both Drp1 and FUNDC1 are essential for hypoxia-induced mitophagy (Wu et al., 2016).

Details

(1) 30 confluent plates (15 cm in diameter) of HeLa cells were collected and resuspended in ice-cold IB cells-1 buffer [225 m*M* of mannitol, 75 m*M* of sucrose, 0.1 m*M* of EGTA, and 30 m*M* of Tris–HCl, pH 7.4].

(2) Cells were gently homogenized with a Dounce homogenizer about 200 times.

(3) The homogenate was centrifuged at 600 × *g* for 5 min at 4°C. The supernatants were again centrifuged at 600 × *g* for 5 min at 4°C (post-nuclear supernatant—PNS). The supernatant was transferred to a new 50-mL tube and centrifuged at 10,000 × *g* for 10 min at 4°C to obtain the crude mitochondrial pellet.

(4) The supernatant was further centrifuged at 100,000 × *g* for 1 h and obtained the supernatant or isolation of ER (pellet) and cytosolic (supernatant) fractions.

(5) The crude mitochondrial pellet was resuspended gently in 2 mL of ice-cold mineral resource buffer (MRB) (250 m*M* of mannitol, 5 m*M* of HEPES with pH 7.4, and 0.5 m*M* of EGTA) and added on top of 8 mL of Percoll medium in an ultracentrifuge tube.

(6) MRB solution was then layered gently on top of the mitochondrial suspension to fill the centrifuge tube.

(7) Centrifugation was carried out at 95,000 × *g* for 60 min at 4°C to isolate the MAMs and pure mitochondria.

(8) The MAMs and mitochondria were lysed in NP40 buffer for detection of MAM marker proteins such as Mfn2, March5, VDAC1, Calreticulin, MFF, FUNDC1, STX17, FACL4, and IP3Rs.

8. MITOPHAGY INDUCERS AND INHIBITORS

An increasing number of inducers are used to trigger mitophagy. These include ROS, mitochondrial toxins (FCCP/CCCP, rotenone, antimycin A, valinomycin, oligomycin, 1-methyl-4-phenyl-1,2,3,6-tetrahydropyridine (MPTP), salinomycin, and 6-hydroxydopamine), hypoxia, ceramides, and selenite. It appears that the perturbation of mitochondrial adenosine triphosphate (ATP) production and ROS generation can lead to defects in mitochondria and the activation of mitophagy.

It has been demonstrated that increased intracellular ROS can activate both general autophagy and mitophagy. The specific induction of mitochondrial ROS was also found to induce mitophagy. The in situ induction of ROS was performed using a genetically encoded mitochondrial-targeted photosensitizer, KillerRed, which is targeted to mitochondria (mtKR) to examine the effect of a short burst of ROS in the mitochondria (Wang et al., 2012). KillerRed is a homolog of GFP that produces 1000-fold more ROS than GFP upon photoactivation with 561 nm of light (Roy et al., 2010). KillerRed can be readily expressed in cultured cells and can be targeted to mitochondria by attaching two mitochondrial-targeting sequences to the N-terminus of KillerRed (mtKR) (Bulina et al., 2006). Using live-cell imaging, it was demonstrated that the photoactivation of mtKR produces ROS to directly cause mitochondrial damage, which leads to PARK2-dependent mitophagy (Wang et al., 2012). The overexpression of the mitochondrial antioxidant protein superoxide dismutase 2 suppresses the mitophagy induced by ROS. Studies have demonstrated that ROS can mediate the selective mitophagy induced by 6-OHDA, rotenone, and staurosporine in neuron cells, which can induce the externalization of cardiolipin for initiating mitophagy (Chu et al., 2013).

Many chemicals, such as sodium selenite, diquat, paraquat, and retigeric acid B, have been reported to trigger mitophagy in response to induction of oxidative stress in mitochondria (Kralova et al., 2012; Li et al., 2015; Liu et al., 2013). Sodium selenite, which can induce mitophagy in human glioma cells but not astrocytes (Kim et al., 2007), induced mitophagy by promoting an ROS-dependent activation of the E3 ubiquitin ligase MUL1 (Li et al., 2015). Finally, sodium selenite can activate both Ub-dependent and receptor-driven mitophagy.

Mitochondria, the central organelles that utilize oxygen for energy production, are highly sensitive to cellular oxygen levels. It is not surprising that hypoxia induces mitophagy mediated by the FUNDC1 or BNIP3L/NIX pathway. Specifically, we found that FUNDC1, a new mitophagy receptor localized at the outer mitochondrial membrane protein, has an atypical LIR (LC3 interaction region) motif that specifically interacts with LC3 for subsequent mitophagy. Under normal conditions, FUNDC1 is highly phosphorylated by Src kinase and CK2 at distinct sites. In response to hypoxia, Src kinase and CK2 are inactivated and a phosphatase, such as PGAM5, is activated to dephosphorylate FUNDC1, which then has significantly higher affinity for LC3 binding and mitophagy induction. NIX/BNIP3L and BNIP3 are also responsive to hypoxia and are phosphorylated under stress conditions. We have thus proposed that monitoring the phosphorylation status of mitophagy receptors such as FUNDC1, NIX/BNIP3L, and BNIP3 by specific antiphosphoric antibodies will be useful for quantitatively assaying mitophagy levels. Both wild-type FUNDC1 and its S13A mutant can induce mitophagy in the absence of Parkin (Chen et al., 2014; Liu et al., 2012).

Many mitochondrial complex inhibitors have been employed to induce mitophagy in cells. The complex I inhibitor rotenone (Chu et al., 2013), the complex III inhibitor antimycin A (McLelland et al., 2014), the potassium ionophore valinomycin, and the ATP synthase inhibitor oligomycin A can induce mitophagy in a PINK1/Parkin-dependent fashion (Bhatia-Kissova & Camougrand, 2013; Yamano et al., 2014). CCCP/FCCP is a lipid-soluble weak acid and a potent mitochondrial uncoupling agent that increases the proton permeability across the mitochondrial inner membrane, thus dissipating the transmembrane potential and depolarizing the mitochondria. CCCP/FCCP can induce the lysosomal removal of depolarized mitochondria, which is considered to be mediated by mitophagy. CCCP/FCCP

(10–20 μ*M*) was used as an inducer in Parkin/PINK1-, NIX/BNIP3L-, and FUNDC1-mediated mitophagy when the cells were treated for 0.5–24 h (Chen et al., 2014; Ding et al., 2010; Liu et al., 2012; Narendra et al., 2008). NIX/BNIP3L was found to be involved in the FCCP-induced autophagic degradation of mitochondria (Sandoval et al., 2008). MEF cells with NIX deficiency had a lower ROS level. ROS and the mitophagy receptor NIX are important in the induction and initiation of mitophagy by enhancing the translocation of Parkin to the damaged mitochondria (Ding et al., 2010). Increased numbers of depolarized mitochondria and mitophagy were observed in fibroblasts derived from patients who were deficient in coenzyme (CoQ), a small lipophilic molecule critical for the transport of electrons from complexes I and II to complex III in the mitochondrial respiratory chain. Interestingly, CoQ supplementation or CsA treatment attenuated mitophagy in these cells, suggesting that these mitochondrial defects are specifically induced by CoQ deficiency and that mitochondrial permeability transition can regulate this type of mitochondrial damage as well as mitophagy (Cotan et al., 2011; Rodriguez-Hernandez et al., 2009).

Ceramide is a bioactive sphingolipid that mediates cell death (Ogretmen & Hannun, 2004; Salazar et al., 2009). The role of ceramide and other sphingolipids in general autophagy has been widely reported. Recent studies have demonstrated that C_{18}-ceramide induces lethal autophagy by selectively targeting mitochondria through a direct interaction between ceramide and LC3B-II on mitochondrial membranes (Jiang & Ogretmen, 2013; Sentelle et al., 2012). Accordingly, CerS1/C_{18}-ceramide was shown to be necessary and sufficient for the induction of lethal mitophagy and tumor suppression in vivo (Sentelle et al., 2012).

9. FUTURE PERSPECTIVES

Mitophagy is a highly complex and dynamic cellular process that regulates both mitochondrial quality and quantity. Both morphological and biochemical hallmarks have been described, and the use of a combination of approaches to measure mitophagy is strongly recommended. No single parameter is sufficient to document mitophagy. Despite the significant recent advances in our knowledge of selective mitophagy, reliable quantitative assays to monitor mitophagy need to be developed. In addition, monitoring mitophagy in vivo remains a challenge. Given that many of the mitophagy players (such as Parkin, p62, and ULK1, and mitophagy receptors, including FUNDC1, NIX/BNIP3L, BNIP3, and Atg32) are phosphorylated in response to many distinct mitophagy stimuli, we recently suggested that the phosphorylation status of these mitophagy players could be used in a quantitative approach and can be applied in vivo (Chen et al., 2014). This requires developing specific antibodies to detect the specific phosphorylation sites. It is also important to identify the kinases and phosphatases that are responsible for regulating mitophagy.

As described in this chapter, ceramide and many compounds, including mitochondrial toxins, are used to induce mitophagy. These compounds are useful for studying mitophagy in cultured cells. It should be noted that these compounds may have off-target effects in addition to mitophagy. More specific mitochondrial-targeting compounds may be useful in the study of mitophagy. Compounds that target to mitophagy receptors or regulators are important for the study of mitophagy and for drug development. When possible, physiologically relevant stimuli, such as hypoxia, are recommended. In addition, an expanding number of players have been reported to be involved in mitophagy. Future studies should address whether there are common or converging pathways to mediate mitophagy in response to distinct mitochondrial stresses. Better qualitative and quantitative methods for assaying mitophagy in physiologically relevant settings will increase our understanding of molecular mitophagy and of possible routes for drug development.

ACKNOWLEDGMENTS

This research was supported by the Natural Science Foundation of China (31471300, 31671441, 31271529, 81130045).

REFERENCES

Allen, G. F., et al. (2013). Loss of iron triggers PINK1/Parkin-independent mitophagy. *EMBO Reports*, *14*(12), 1127–1135.

Bhatia-Kissova, I., & Camougrand, N. (2013). Mitophagy is not induced by mitochondrial damage but plays a role in the regulation of cellular autophagic activity. *Autophagy*, *9*(11), 1897–1899.

Bingol, B., et al. (2014). The mitochondrial deubiquitinase USP30 opposes parkin-mediated mitophagy. *Nature*, *510*, 370–375.

Bulina, M. E., et al. (2006). A genetically encoded photosensitizer. *Nature Biotechnology*, *24*(1), 95–99.

Chan, D. C. (2006). Dissecting mitochondrial fusion. *Developmental Cell*, *11*(5), 592–594.

Chen, G., et al. (2014). A regulatory signaling loop comprising the PGAM5 phosphatase and CK2 controls receptor-mediated mitophagy. *Molecular Cell*, *54*(3), 362–377.

Chu, C. T., et al. (2013). Cardiolipin externalization to the outer mitochondrial membrane acts as an elimination signal for mitophagy in neuronal cells. *Nature Cell Biology*, *15*(10), 1197–1205.

Cotan, D., et al. (2011). Secondary coenzyme Q10 deficiency triggers mitochondria degradation by mitophagy in MELAS fibroblasts. *The FASEB Journal, 25*(8), 2669–2687.

Ding, W. X., & Yin, X. M. (2012). Mitophagy: Mechanisms, pathophysiological roles, and analysis. *Biological Chemistry, 393*(7), 547–564.

Ding, W. X., et al. (2010). Nix is critical to two distinct phases of mitophagy, reactive oxygen species-mediated autophagy induction and Parkin-ubiquitin-p62-mediated mitochondrial priming. *The Journal of Biological Chemistry, 285*(36), 27879–27890.

Ding, W. X., et al. (2012). Parkin and mitofusins reciprocally regulate mitophagy and mitochondrial spheroid formation. *The Journal of Biological Chemistry, 287*(50), 42379–42388.

Feng, D., et al. (2013). Molecular signaling toward mitophagy and its physiological significance. *Experimental Cell Research, 319*(12), 1697–1705.

Ferree, A. W., et al. (2013). MitoTimer probe reveals the impact of autophagy, fusion, and motility on subcellular distribution of young and old mitochondrial protein and on relative mitochondrial protein age. *Autophagy, 9*(11), 1887–1896.

Geisler, S., et al. (2010). PINK1/Parkin-mediated mitophagy is dependent on VDAC1 and p62/SQSTM1. *Nature Cell Biology, 12*(2), 119–131.

Hernandez, G., et al. (2013). MitoTimer: A novel tool for monitoring mitochondrial turnover. *Autophagy, 9*(11), 1852–1861.

Hirota, Y., et al. (2015). Mitophagy is primarily due to alternative autophagy and requires the MAPK1 and MAPK14 signaling pathways. *Autophagy, 11*(2), 332–343.

Itakura, E., et al. (2012). Structures containing Atg9A and the ULK1 complex independently target depolarized mitochondria at initial stages of Parkin-mediated mitophagy. *Journal of Cell Science, 125*(Pt 6), 1488–1499.

Jiang, W., & Ogretmen, B. (2013). Ceramide stress in survival versus lethal autophagy paradox: Ceramide targets autophagosomes to mitochondria and induces lethal mitophagy. *Autophagy, 9*(2), 258–259.

Kanki, T., Kang, D., & Klionsky, D. J. (2009). Monitoring mitophagy in yeast: The Om45-GFP processing assay. *Autophagy, 5*(8), 1186–1189.

Karbowski, M., & Youle, R. J. (2003). Dynamics of mitochondrial morphology in healthy cells and during apoptosis. *Cell Death and Differentiation, 10*(8), 870–880.

Katayama, H., et al. (2011). A sensitive and quantitative technique for detecting autophagic events based on lysosomal delivery. *Chemistry & Biology, 18*(8), 1042–1052.

Kim, I., & Lemasters, J. J. (2011). Mitochondrial degradation by autophagy (mitophagy) in GFP-LC3 transgenic hepatocytes during nutrient deprivation. *American Journal of Physiology. Cell Physiology, 300*, C308–C317.

Kim, E. H., et al. (2007). Sodium selenite induces superoxide-mediated mitochondrial damage and subsequent autophagic cell death in malignant glioma cells. *Cancer Research, 67*(13), 6314–6324.

Kirkin, V., et al. (2009a). A role for ubiquitin in selective autophagy. *Molecular Cell, 34*(3), 259–269.

Kirkin, V., et al. (2009b). NBR1 cooperates with p62 in selective autophagy of ubiquitinated targets. *Autophagy, 5*(5), 732–733.

Klionsky, D. J., et al. (2012). Guidelines for the use and interpretation of assays for monitoring autophagy. *Autophagy, 8*(4), 445–544.

Kraft, C., Peter, M., & Hofmann, K. (2010). Selective autophagy: Ubiquitin-mediated recognition and beyond. *Nature Cell Biology, 12*(9), 836–841.

Kralova, V., et al. (2012). Selenite-induced apoptosis and autophagy in colon cancer cells. *Toxicology In Vitro, 26*(2), 258–268.

Kundu, M., et al. (2008). Ulk1 plays a critical role in the autophagic clearance of mitochondria and ribosomes during reticulocyte maturation. *Blood, 112*(4), 1493–1502.

Lagouge, M., et al. (2006). Resveratrol improves mitochondrial function and protects against metabolic disease by activating SIRT1 and PGC-1alpha. *Cell, 127*(6), 1109–1122.

Laker, R. C., et al. (2014). A novel MitoTimer reporter gene for mitochondrial content, structure, stress, and damage in vivo. *The Journal of Biological Chemistry*, *289*(17), 12005–12015.
Lamark, T., et al. (2009). NBR1 and p62 as cargo receptors for selective autophagy of ubiquitinated targets. *Cell Cycle*, *8*(13), 1986–1990.
Lamb, C. A., Yoshimori, T., & Tooze, S. A. (2013). The autophagosome: Origins unknown, biogenesis complex. *Nature Reviews. Molecular Cell Biology*, *14*(12), 759–774.
Li, J., et al. (2015). Mitochondrial outer-membrane E3 ligase MUL1 ubiquitinates ULK1 and regulates selenite-induced mitophagy. *Autophagy*, *11*(8), 1216–1229.
Liu, L., et al. (2012). Mitochondrial outer-membrane protein FUNDC1 mediates hypoxia-induced mitophagy in mammalian cells. *Nature Cell Biology*, *14*(2), 177–185.
Liu, Y. Q., et al. (2013). Retigeric acid B-induced mitophagy by oxidative stress attenuates cell death against prostate cancer cells in vitro. *Acta Pharmacologica Sinica*, *34*(9), 1183–1191.
Liu, L., et al. (2014). Receptor-mediated mitophagy in yeast and mammalian systems. *Cell Research*, *24*(7), 787–795.
Lwai-Kanai, E., et al. (2008). A method to measure cardiac autophagic flux in vivo. *Autophagy*, *4*(3), 322–329.
Mathew, R., et al. (2009). Autophagy suppresses tumorigenesis through elimination of p62. *Cell*, *137*(6), 1062–1075.
Matsuda, N., & Tanaka, K. (2010). Uncovering the roles of PINK1 and parkin in mitophagy. *Autophagy*, *6*(7), 952–954.
McLelland, G. L., et al. (2014). Parkin and PINK1 function in a vesicular trafficking pathway regulating mitochondrial quality control. *The EMBO Journal*, *33*(4), 282–295.
Mizushima, N., et al. (2004). In vivo analysis of autophagy in response to nutrient starvation using transgenic mice expressing a fluorescent autophagosome marker. *Molecular Biology of the Cell*, *15*(3), 1101–1111.
Moscat, J., & Diaz-Meco, M. T. (2009). To aggregate or not to aggregate? A new role for p62. *EMBO Reports*, *10*(8), 804.
Narendra, D., et al. (2008). Parkin is recruited selectively to impaired mitochondria and promotes their autophagy. *The Journal of Cell Biology*, *183*(5), 795–803.
Ogretmen, B., & Hannun, Y. A. (2004). Biologically active sphingolipids in cancer pathogenesis and treatment. *Nature Reviews. Cancer*, *4*(8), 604–616.
Pan, J. A., et al. (2016). TRIM21 ubiquitylates SQSTM1/p62 and suppresses protein sequestration to regulate redox homeostasis (vol 61, pg 720, 2016). *Molecular Cell*, *62*(1), 149–151.
Rodriguez-Hernandez, A., et al. (2009). Coenzyme Q deficiency triggers mitochondria degradation by mitophagy. *Autophagy*, *5*(1), 19–32.
Rosado, C. J., et al. (2008). Rosella—A fluorescent pH-biosensor for reporting vacuolar turnover of cytosol and organelles in yeast. *Autophagy*, *4*(2), 205–213.
Rowland, A. A., & Voeltz, G. K. (2012). Endoplasmic reticulum-mitochondria contacts: Function of the junction. *Nature Reviews. Molecular Cell Biology*, *13*(10), 607–615.
Roy, A., et al. (2010). Diffusion pathways of oxygen species in the phototoxic fluorescent protein KillerRed. *Photochemical & Photobiological Sciences*, *9*(10), 1342–1350.
Salazar, M., et al. (2009). Cannabinoid action induces autophagy-mediated cell death through stimulation of ER stress in human glioma cells. *The Journal of Clinical Investigation*, *119*(5), 1359–1372.
Sandoval, H., et al. (2008). Essential role for Nix in autophagic maturation of erythroid cells. *Nature*, *454*(7201), 232–235.
Sanz, L., et al. (2000). The atypical PKC-interacting protein p62 channels NF-kappaB activation by the IL-1-TRAF6 pathway. *The EMBO Journal*, *19*(7), 1576–1586.

Scarffe, L. A., et al. (2014). Parkin and PINK1: Much more than mitophagy. *Trends in Neurosciences*, *37*(6), 315–324.
Schon, E. A., & Przedborski, S. (2011). Mitochondria: The next (neurode)generation. *Neuron*, *70*(6), 1033–1053.
Sentelle, R. D., et al. (2012). Ceramide targets autophagosomes to mitochondria and induces lethal mitophagy. *Nature Chemical Biology*, *8*(10), 831–838.
Shirakabe, A., et al. (2016). Evaluating mitochondrial autophagy in the mouse heart. *Journal of Molecular and Cellular Cardiology*, *92*, 134–139.
Song, P., et al. (2016). Parkin promotes proteasomal degradation of p62: Implication of selective vulnerability of neuronal cells in the pathogenesis of Parkinson's disease. *Protein & Cell*, *7*(2), 114–129.
Stolz, A., Ernst, A., & Dikic, I. (2014). Cargo recognition and trafficking in selective autophagy. *Nature Cell Biology*, *16*(6), 495–501.
Suen, D. F., Norris, K. L., & Youle, R. J. (2008). Mitochondrial dynamics and apoptosis. *Genes & Development*, *22*(12), 1577–1590.
Sun, N., et al. (2015). Measuring in vivo mitophagy. *Molecular Cell*, *60*(4), 685–696.
Terada, M., et al. (2010). Double transgenic mice crossed GFP-LC3 transgenic mice with alpha MyHC-mCherry-LC3 transgenic mice are a new and useful tool to examine the role of autophagy in the heart. *Circulation Journal*, *74*(1), 203–206.
Wang, Y., et al. (2012). ROS-induced mitochondrial depolarization initiates PARK2/PARKIN-dependent mitochondrial degradation by autophagy. *Autophagy*, *8*(10), 1462–1476.
Winklhofer, K. F. (2014). Parkin and mitochondrial quality control: Toward assembling the puzzle. *Trends in Cell Biology*, *24*(6), 332–341.
Wu, W., et al. (2014). ULK1 translocates to mitochondria and phosphorylates FUNDC1 to regulate mitophagy. *EMBO Reports*, *15*(5), 566–575.
Wu, W., et al. (2016). FUNDC1 regulates mitochondrial dynamics at the ER-mitochondrial contact site under hypoxic conditions. *The EMBO Journal*, *35*(13), 1368–1384.
Yamano, K., et al. (2014). Mitochondrial Rab GAPs govern autophagosome biogenesis during mitophagy. *eLife*, *3*, e01612.
Yoshii, S. R., et al. (2011). Parkin mediates proteasome-dependent protein degradation and rupture of the outer mitochondrial membrane. *The Journal of Biological Chemistry*, *286*(22), 19630–19640.
Youle, R. J., & Narendra, D. P. (2011). Mechanisms of mitophagy. *Nature Reviews. Molecular Cell Biology*, *12*(1), 9–14.
Yu, H. B., et al. (2009). Sequestosome-1/p62 is the key intracellular target of innate defense regulator peptide. *The Journal of Biological Chemistry*, *284*(52), 36007–36011.
Zhu, J., Dagda, R. K., & Chu, C. T. (2011). Monitoring mitophagy in neuronal cell cultures. *Methods in Molecular Biology*, *793*, 325–339.

CHAPTER TWELVE

Cytofluorometric Assessment of Mitophagic Flux in Mammalian Cells and Tissues

L. Esteban-Martinez[1], B. Villarejo-Zori[1], P. Boya[2]
Centro de Investigaciones Biológicas, CSIC, Madrid, Spain
[2]Corresponding author: e-mail address: pboya@cib.csic.es

Contents

Abstract

Mitophagy is the selective autophagic degradation of mitochondria inside lysosomes and is the only mechanism capable of eliminating whole mitochondria that are either damaged or no longer required. Mitochondrial dysfunction is a common hallmark in many pathological conditions and has been implicated in cellular damage in aged tissues. The limited availability of mitophagy research methods underscores the need for more robust, quantitative, and objective tools in order to better understand this process. Here we describe a flow cytometry-based method using MitoTracker Deep Red for mitophagy assessment that can be applied to cells and tissues. Moreover, we demonstrate that when used in conjunction with lysosomal inhibitors, our method provides a novel means of assessing mitophagic flux, which is the most reliable indicator of whether mitochondria are truly delivered to lysosomes for degradation.

1. INTRODUCTION

Mitophagy is a form of selective autophagy whereby mitochondria are specifically targeted for degradation inside lysosomes following their

[1] Equal contribution.

Methods in Enzymology, Volume 588
ISSN 0076-6879
http://dx.doi.org/10.1016/bs.mie.2016.09.081

engulfment within autophagosomes. In recent years, this process has attracted significant research attention due to an improved understanding of the molecular mechanisms that regulate mitophagy and the fact that defective mitochondria build up in humans is a common hallmark of both aging and many pathological processes from cancer to degenerative diseases.

Progress in any field of research is dependent on the availability of reliable investigative methods and tools that can be applied to a variety of biological and experimental scenarios. While the tools and methods used to study autophagy have been refined over time (Klionsky et al., 2016), the range of assays available for mitophagy assessment remains limited. Mitophagy can be "observed" by ultrastructural analysis, as first described during the process of kidney development (Clark, 1957). However, in the case of electron microscopy techniques, a high level of expertise is required to distinguish autophagosomes and identify the lumen content of mitochondria in varying states of degradation. Moreover, ultrastructural analyses are usually nonquantitative.

Mitophagy is most commonly evaluated using fluorescence microscopy. This approach combines immunofluorescence for mitochondrial, autophagosomal, and lysosomal antibodies with determination of the colocalization of mitochondria-specific dyes and the autophagosomal marker GFP-LC3 in living or fixed cells (Dolman, Chambers, Mandavilli, Batchelor, & Janes, 2013). However, given the very short mean life span of an autophagosome (minutes), the number of autophagosomes at a specific time point does not provide useful information about autophagy per se, and unless lysosomal degradation is blocked, mitochondrial turnover occurs very rapidly and colocalization events are rare (Twig et al., 2008). Moreover, these approaches require sophisticated image analysis techniques to detect a sufficient number of mitochondria/autophagosome colocalization events. Finally, given that defects in lysosomal degradation also increase colocalization, the aforementioned methods do not provide a measure of effective mitophagy, where mitochondria are degraded inside lysosomes.

Thus, flux through the system must be analyzed in order to properly asses mitophagy in order to avoid confusing increases in mitophagosome number caused by autophagy induction with those caused by alterations in lysosomal function. Therefore, autophagic and mitophagic flux are monitored by comparing autophagosome numbers in the absence and presence of lysosomal inhibitors (Klionsky et al., 2016). However, quantifying images and identifying colocalization events entail tedious and time-consuming analyses of large numbers of cells to obtain statistically relevant sample sizes.

The development of a simple, quantitative method to assess mitophagic flux would constitute a major breakthrough in the field.

We recently developed a simple, rapid, and quantitative method to determine mitophagy in cells and tissues by flow cytometry using mitochondrion-selective probes (Mauro-Lizcano et al., 2015). When combined with lysosomal inhibitors, this approach allows for the determination of mitophagic flux. This method can be applied to a variety of experimental systems and thus represents an important advance in the analysis of mitophagic flux in physiological and pathological conditions (Domenech et al., 2015; Mauro-Lizcano et al., 2015; Swiader et al., 2016).

Here we present two protocols for the assessment of mitophagy in a neuroblastoma cell line and in retinal tissue dissociated into single cells.

2. MITOPHAGY EVALUATION IN CELL LINES

Mitochondrial depolarization is the most common strategy used to induce mitophagy in vitro. This can be achieved using the mitochondrial uncoupler CCCP (carbonyl cyanide *m*-chlorophenyl hydrazone), which reduces the mitochondrial membrane potential, or by blocking mitochondrial respiration with a combination of oligomycin and antimycin A (AO) (Lazarou et al., 2015; Narendra, Tanaka, Suen, & Youle, 2008). We treat cells of the SHSH-5Y human neuroblastoma cell line with these two mitophagy inducers for short (3 h) and long (18 h) periods and assess mitochondrial levels using the mitochondrion-selective probe MitoTracker Deep Red (MTDR). To confirm that a decrease in labeling is due to autophagic degradation, we block lysosomal degradation by treating cells with the classical lysosomal inhibitor hydroxychloroquine (HCQ) (Boya et al., 2005) and prevent mitophagy using cyclosporine A (CsA), which blocks the mitochondrial permeability transition pore and has been previously used to prevent mitophagy by us and other research groups (Carreira, Lee, Ghochani, Gustafsson, & Gottlieb, 2010; Domenech et al., 2015; Elmore, Qian, Grissom, & Lemasters, 2001; Kang & Hwang, 2009; Mauro-Lizcano et al., 2015; Rodriguez-Enriquez, Kim, Currin, & Lemasters, 2006).

The detailed protocol is described later:

1. Seed cells (8×10^5) in a 24-well plate with 500 mL DMEM, high glucose (Biowest, L0104-500), supplemented with 220 m*M* glutamine (Sigma, G8540), 17 m*M* penicillin–streptomycin (Gibco, 15140122), and 10% fetal calf serum (complete DMEM).

2. Prepare the 25 m*M* stock CCCP solution (Sigma, C2759-100) in DMSO (Sigma, D8418). Prepare the cocktail of AO in DMSO, CsA at 5 m*M* in DMSO, and HCQ at 30 mg/mL in water.
3. Incubate cells for 3 and 18 h at 37°C and 5% CO_2 with 10 μ*M* CCCP, 10 μ*M* antimycin A, and 1 μ*M* oligomycin. CsA is used at 5 μ*M* and HCQ at 30 μg/mL.
4. Aspirate the supernatant and transfer to flow cytometry tubes.
5. Wash cells with 500 μL of PBS.
6. Add 200 μL of trypsin (Gibco, 25300054) and incubate for 5 min or until the cells begin to detach from the well.
7. Add 500 μL of complete DMEM and homogenize the cells by gentle pipetting.
8. Add the cell suspension to a flow cytometry tube or leave it in the flow cytometry plate for plaque analysis.
9. Centrifuge at 770 × *g* for 5 min to eliminate the trypsin solution.
10. Resuspend the cells in 500 μL of 10 n*M* MTDR solution (Invitrogen, M22426) in DMEM in flow cytometry tubes and incubate for 15 min at 37°C and 5% CO_2.
11. Gently homogenize the cells and acquire at least 3000 cells in the FL4 channel. Analyze the cells with an FC500 flow cytometer (Beckman Coulter, L'Hospitalet de Llobregat, Barcelona).
12. Plot mean fluorescence of the viable cell population in the FL4 channel and normalize against that of untreated cells.
13. Data shown in the resulting figures represent the average of at least three experiments performed in triplicate and are presented as the mean ± standard error of the mean. Differences between treatments are analyzed using the Student's *t*-test or, in the case of nonnormally distributed data, the nonparametric Mann–Whitney *U*-test (SPSS 17.0). Statistical significance is set at $P < 0.05$.

Fig. 1A shows two representative histograms of cells treated with AO and CCCP for 18 h. Fig. 1B shows the reduction in fluorescence as a function of time in cells treated with CCCP and AO. Fig. 1C and D shows MTDR labeling in cells treated for 3 and 18 h with the lysosomal inhibitor HCQ and the mitophagy inhibitor CsA, respectively.

3. MITOPHAGY EVALUATION IN TISSUE

The retina is a commonly used model for the study of the central nervous system. It can be cultured ex vivo, allowing the preservation of cell–cell

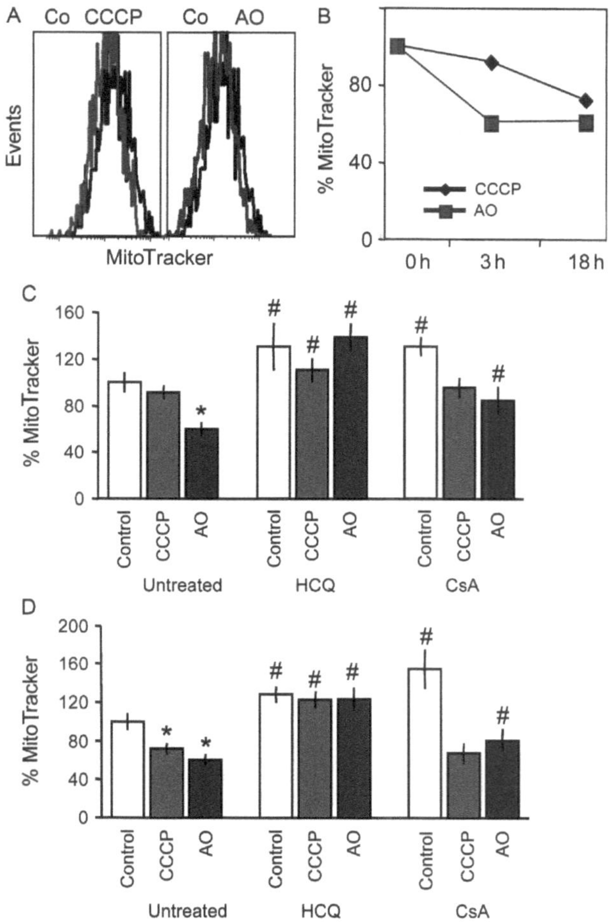

Fig. 1 Mitophagy assessment in SHSY-5Y cells. (A) Representative histogram of SH-SY5Y cells treated for 18 h with AO or CCCP, stained with MTDR, and analyzed by flow cytometry. (B) Decrease in MTDR fluorescence in cells treated as described in (A), plotted as a function of time. (C) Cells treated with CCCP and AO for 3 h (C) and 18 h (D) in the absence or presence of HCQ and CsA, incubated with MTDR, and analyzed by flow cytometry. $^{*}P < 0.05$ with respect to corresponding control group. $^{\#}P < 0.05$ with respect to corresponding group incubated in the absence of HCQ or CsA (*untreated*).

and cell–matrix interactions. The following is the procedure we use to assess mitophagy in embryonic retinas isolated from E13.5 embryos.

1. Dissect the retinas at the appropriate embryonic stage.
2. Dissociate the retina by incubating in 0.5 mg/mL trypsin (Worthington, LS0033703) for 5 min at 37°C.

3. Culture in DMEM:HAM/F12 (Gibco, 42400) supplemented with 10 m*M* sodium bicarbonate (Sigma, S7561), 1.1 mg/mL sodium pyruvate (Sigma, P2256), 0.1 mg/mL apotransferrin (Sigma, T2252), 0.0052 mg/mL sodium selenite (Sigma, S5261), 0.32 mg/mL putrescine (Sigma, P7580), progesterone (Sigma, P8783), and gentamycin (Gibco, 15710), adjusted to pH 7.4. Supplement the medium with 10 n*M* human recombinant insulin (Sigma, I2643) just before culture.
4. Incubate for 6 h with 10 μ*M* antimycin A and 1 μ*M* oligomycin, EBSS (Sigma, E2888), 30 mg/mL HCQ (prepared as above), and 5 μ*M* CsA (prepared as above).
5. Dissociate by gentle pipetting to produce a single-cell suspension.
6. Add 10% fetal calf serum to inactivate trypsin activity.
7. Centrifuge at 770 × *g* and wash the cells with 300 mL of HBSS (Gibco, 14170-088) with 10 n*M* MTDR (Invitrogen, M22426).
8. Gently homogenize the cells and acquire 10,000 cells in the FL4 channel.
9. Plot mean fluorescence intensity of the viable cell population in the FL4 channel and normalize against that of untreated cells.

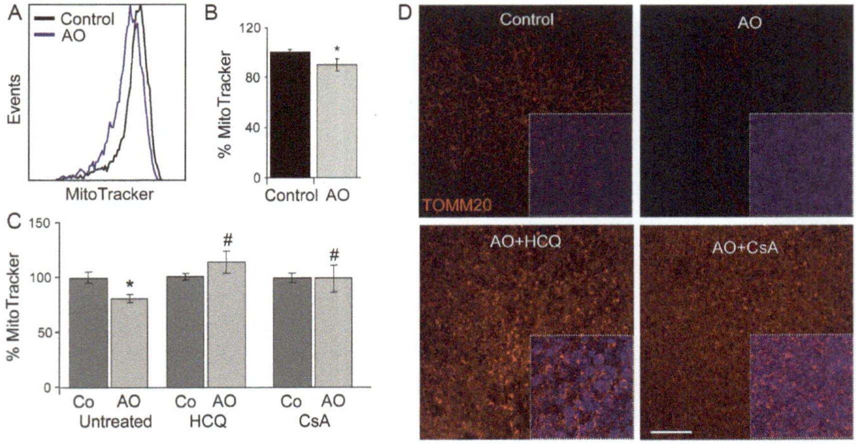

Fig. 2 Mitophagy assessment in E13.5 embryonic retinas treated with AO. (A) Representative histogram of E13.5 retinas treated for 6 h with AO, dissociated, and stained with MTDR for flow cytometry analysis. (B) Retinas cultured for 6 h with the indicated treatments, dissociated, stained with MTDR, and assessed by flow cytometry. (C) The percentage mean fluorescence intensity vs control cells is shown. $^{*}P<0.05$ with respect to corresponding control group. $^{\#}P<0.05$ with respect to corresponding group incubated in the absence of HCQ or CsA (*untreated*). (D) TOMM20 staining in retinal flatmounts of E13.5 retinas cultured for 6 h and treated as indicated. *Insets* show the magnification of TOMM20 and nuclei stained with DAPI.

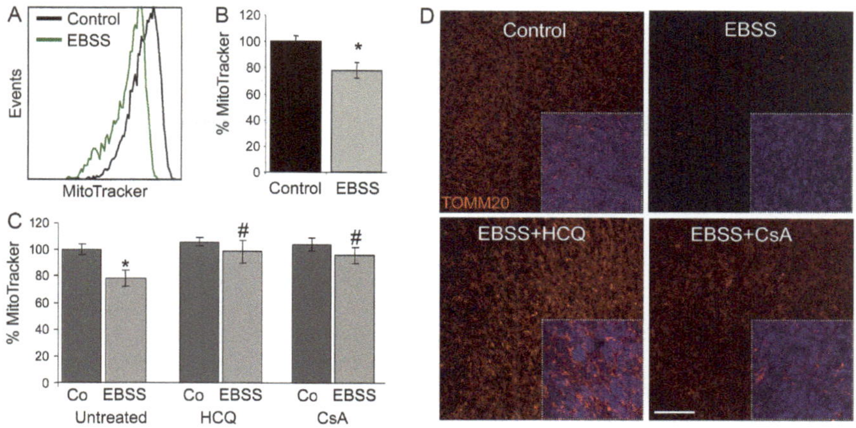

Fig. 3 Mitophagy assessment in E13.5 embryonic retinas treated with EBSS. (A) Representative histogram of E13.5 retinas cultured for 6 h in EBSS, dissociated, and stained with MTDR for flow cytometry analysis. (B) Retinas cultured for 6 h with the indicated treatments, dissociated, stained with MTDR, and assessed by flow cytometry. (C) The percentage mean fluorescence intensity vs control cells is shown. $^{*}P<0.05$ with respect to corresponding control group. $^{\#}P<0.05$ with respect to corresponding group incubated in the absence of HCQ or CsA (*untreated*). (D) TOMM20 staining in retinal flatmounts of E13.5 retinas cultured for 6 h and treated as indicated. *Insets* show the magnification of TOMM20 and nuclei stained with DAPI.

A representative histogram of E13.5 retinas cultured for 6 h with AO is shown in Fig. 2A. Fig. 2B and C shows the quantification of the percentage mean florescence intensity in retinas cultured for 6 h with the indicated treatments. Confocal images of retinal flatmounts of retinas treated as described earlier (Santa Cruz Biotechnology, sc-11415) are shown in Fig. 2D; mitochondria are labeled with anti-TOMM20 antibody. As shown in Fig. 3, amino acid starvation also induces mitophagy in embryonic retinas, while blockade of lysosomal activity with HCQ and of mitophagy with CsA attenuates the decrease in mitochondrial staining as determined by flow cytometry with MTDR staining and by confocal microscopy analysis in retinal flatmounts stained for TOMM20.

4. ASSESSMENT OF MITOPHAGIC FLUX

We define mitophagic flux as the ratio of MTDR fluorescence in the presence of lysosomal inhibitors (HCQ) to that in the absence of inhibitors, normalized to the corresponding value in control cells. As shown in Fig. 4, CCCP and AO both display mitophagic flux, although with distinct kinetic

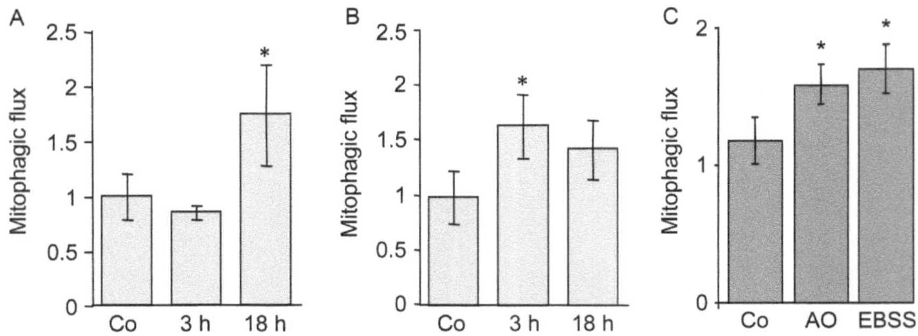

Fig. 4 Determination of mitophagic flux. Mitophagic flux is calculated by the following formula: % vs control (MFI in the presence of inhibitor/MFI in the absence of inhibitor). Mitophagic flux in SHSY-5Y cells treated with CCCP (A) or AO (B) in the presence of HCQ. (C) Mitophagic flux in E13.5 retinas treated with AO or EBSS in the presence of HCQ for 6 h.

profiles (Fig. 4A and B). Amino acid starvation and a cocktail of oligomycin + antimycin A also result in productive mitophagic flux in embryonic retinas (Fig. 4C).

5. SUMMARY

We have developed and validated a highly reproducible method of mitophagy assessment using MTDR. This method enables the determination of mitophagic flux when used in combination with inhibitors of autophagy and mitophagy, thus allowing direct comparison of different inducers and inhibitors in the quantitative evaluation of mitophagy. The use of those inhibitors such as HCQ is very consistent under different experimental conditions. Moreover, this method is effective in primary culture of cells and tissues and constitutes a new tool for the evaluation of mitochondrial degradation by autophagy in a variety of experimental settings.

ACKNOWLEDGMENTS

This work was supported by grants from MINECO (Spain), SAF2012-36079, and BFU2015-65623-R. B.V-.Z. is a recipient of a predoctoral grant from the Fundación Tatiana de Guzmán el Bueno.

REFERENCES

Boya, P., Gonzalez-Polo, R. A., Casares, N., Perfettini, J., Dessen, P., Larochette, N., et al. (2005). Inhibition of macroautophagy triggers apoptosis. *Molecular and Cellular Biology*, *25*, 1025–1040.

Carreira, R. S., Lee, Y., Ghochani, M., Gustafsson, A. B., & Gottlieb, R. A. (2010). Cyclophilin D is required for mitochondrial removal by autophagy in cardiac cells. *Autophagy*, *6*, 462–472.

Clark, S. L., Jr. (1957). Cellular differentiation in the kidneys of newborn mice studied with the electron microscope. *The Journal of Biophysical and Biochemical Cytology*, *3*, 349–362.

Dolman, N. J., Chambers, K. M., Mandavilli, B., Batchelor, R. H., & Janes, M. S. (2013). Tools and techniques to measure mitophagy using fluorescence microscopy. *Autophagy*, *9*, 1653–1662.

Domenech, E., Maestre, C., Esteban-Martinez, L., Partida, D., Pascual, R., Fernandez-Miranda, G., et al. (2015). AMPK and PFKFB3 mediate glycolysis and survival in response to mitophagy during mitotic arrest. *Nature Cell Biology*, *17*, 1304–1316.

Elmore, S. P., Qian, T., Grissom, S. F., & Lemasters, J. J. (2001). The mitochondrial permeability transition initiates autophagy in rat hepatocytes. *The FASEB Journal*, *15*, 2286–2287.

Kang, H. T., & Hwang, E. S. (2009). Nicotinamide enhances mitochondria quality through autophagy activation in human cells. *Aging Cell*, *8*, 426–438.

Klionsky, D. J., Abdelmohsen, K., Abe, A., Abedin, M. J., Abeliovich, H., Acevedo Arozena, A., et al. (2016). Guidelines for the use and interpretation of assays for monitoring autophagy (3rd edition). *Autophagy*, *12*, 1–222.

Lazarou, M., Sliter, D. A., Kane, L. A., Sarraf, S. A., Wang, C., Burman, J. L., et al. (2015). The ubiquitin kinase PINK1 recruits autophagy receptors to induce mitophagy. *Nature*, *524*, 309–314.

Mauro-Lizcano, M., Esteban-Martinez, L., Seco, E., Serrano-Puebla, A., Garcia-Ledo, L., Figueiredo-Pereira, C., et al. (2015). New method to assess mitophagy flux by flow cytometry. *Autophagy*, *11*, 833–843.

Narendra, D., Tanaka, A., Suen, D. F., & Youle, R. J. (2008). Parkin is recruited selectively to impaired mitochondria and promotes their autophagy. *The Journal of Cell Biology*, *183*, 795–803.

Rodriguez-Enriquez, S., Kim, I., Currin, R. T., & Lemasters, J. J. (2006). Tracker dyes to probe mitochondrial autophagy (mitophagy) in rat hepatocytes. *Autophagy*, *2*, 39–46.

Swiader, A., Nahapetyan, H., Faccini, J., D'Angelo, R., Mucher, E., Elbaz, M., et al. (2016). Mitophagy acts as a safeguard mechanism against human vascular smooth muscle cell apoptosis induced by atherogenic lipids. *Oncotarget*, 7, 28821–28835.

Twig, G., Elorza, A., Molina, A. J., Mohamed, H., Wikstrom, J. D., Walzer, G., et al. (2008). Fission and selective fusion govern mitochondrial segregation and elimination by autophagy. *The EMBO Journal*, *27*, 433–446.

CHAPTER THIRTEEN

Automated Analysis of Fluorescence Colocalization: Application to Mitophagy

A. Sauvat[*,†,‡,§], H. Zhou[*,†,‡,§,¶], M. Leduc[*,†,‡,§], L.C. Gomes-da-Silva[*,†,‡,§,||], L. Bezu[*,†,‡,§,¶], K. Müller[*,†,‡,§], S. Forveille[*,†,‡,§], P. Liu[*,†,‡,§,¶], L. Zhao[*,†,‡,§,¶], G. Kroemer[*,†,‡,§,¶,||,#,**,1], O. Kepp[*,†,‡,§,1]

[*]Metabolomics and Cell Biology Platforms, Gustave Roussy Comprehensive Cancer Institute, Villejuif, France
[†]Equipe 11 labellisée Ligue contre le Cancer, Centre de Recherche des Cordeliers, Paris, France
[‡]Université Paris Descartes, Sorbonne Paris Cité, Paris, France
[§]Université Pierre et Marie Curie, Paris, France
[¶]University of Paris Sud XI, Kremlin Bicêtre, France
[||]University of Coimbra, Coimbra, Portugal
[#]Pôle de Biologie, Hôpital Européen Georges Pompidou, AP-HP, Paris, France
[**]Karolinska Institute, Karolinska University Hospital, Stockholm, Sweden
[1]Corresponding authors: e-mail address: kroemer@orange.fr; captain.olsen@gmail.com

Contents

Abstract

Mitophagy is a peculiar form of organelle-specific autophagy that targets mitochondria. This process ensures cellular homeostasis, as it fosters the disposal of aged and damaged mitochondria that otherwise would be prone to produce reactive oxygen species and hence endanger genomic stability. Similarly, autophagic clearance of depolarized mitochondria plays a fundamental role in organismal homeostasis as exemplified by the link between Parkinson disease and impaired function of the mitophagy-mediating proteins PINK1 and Parkin. Here, we detail an image-based approach for the quantification of mitochondrial Parkin translocation, which is mechanistically important for the initiation of mitophagy.

Methods in Enzymology, Volume 588
ISSN 0076-6879
http://dx.doi.org/10.1016/bs.mie.2016.09.082

1. INTRODUCTION

Autophagy constitutes a phylogenetically conserved adaptive stress response that allows cells to safeguard viability in response to homeostatic perturbations such as malnutrition, shear stress, or chemical cues. This adaptive stress response allows for liberating energy resources on demand and hence safeguards homeostasis via targeted degradation of superfluous, old, or damaged (and hence potentially dangerous) cellular structures (Galluzzi et al., 2015; Galluzzi, Pietrocola, Levine, & Kroemer, 2014; Madeo, Eisenberg, & Kroemer, 2009; Marino, Madeo, & Kroemer, 2011; Sica et al., 2015). The autophagic process involves the formation of double-membraned autophagosomes that engulf cellular material and serve as containment for its lysosomal degradation before metabolic recycling of the autophagic products (Feng, He, Yao, & Klionsky, 2014; Klionsky et al., 2011; Youle & Narendra, 2011).

Mitophagy is a peculiar form of autophagy that culminates in the selective autophagic clearance of damaged mitochondria in response to various mitochondrial stressors leading to the depolarization and/or malfunctioning of the organelle. Similar to macroautophagy, this process involves the sequestration of substrates by double-membraned structures, their closure, and final fusion with lysosomes. The resulting lysosomal acidification and the activation of lysosomal hydrolases drives the cleavage and final recycling of the substrates (Codogno, Mehrpour, & Proikas-Cezanne, 2012; Lamb, Yoshimori, & Tooze, 2013). The recessive Parkinson's disease-linked genes coding for PTEN-induced kinase 1 (PINK1) and PARK2 (better known as Parkin) are functionally important for the clearance of malfunctioning mitochondria. Thus, mutations affecting either PINK1 or Parkin have been found in hereditary forms of Parkinson's disease that are accompanied by the accumulation of dysfunctional mitochondria. PINK1 is continuously imported into healthy mitochondria via the TOMM complex and then degraded. In response to mitochondrial dysfunction and depolarization, PINK1 becomes stabilized and hence more abundant in the outer membrane of mitochondria. PINK1 then recruits the cytoplasmic E3 ubiquitin ligase Parkin to the damaged organelle (Amadoro et al., 2014; Narendra, Tanaka, Suen, & Youle, 2008; Narendra & Youle, 2011) Parkin-dependent ubiquitination of various outer mitochondrial membrane proteins drives the selective targeting of mitochondria by LC3-containining isolation membranes. The engulfment of targeted mitochondria is facilitated by mitochondrial fission

(Buhlman et al., 2014; Twig et al., 2008) that can be reversed by promoting fusion in some models (Deng, Dodson, Huang, & Guo, 2008; Poole et al., 2008). Mitophagosomes fuse with lysosomes to form mitophagolysosomes that specifically localize in perinuclear regions (Eid, Ito, Maemura, & Otsuki, 2013). Lysosomal degradation finally liberates catabolic substrates to counteract cellular stress and reestablish homeostasis.

Here, we detail an exemplary workflow for the microscopic quantitation of mitophagy in Parkin-expressing biosensor cells that mount an adaptive response to mitochondrial depolarization. We applied an R-based algorithm to measure the onset of mitophagy by monitoring the translocation of Parkin to mitochondria. The colocalization of the biosensor with chemically stained mitochondrial structures was accurately quantified based on the global correlation of pixels. This assay may be readily used for the high-throughput assessment of mitophagy in vitro.

2. CELL CULTURE AND TREATMENTS

1. Human osteosarcoma U-2 OS cells stably expressing mCherry-Parkin (Narendra et al., 2008; Zhou et al., 2016), display optimal adherent growth in McCoys medium supplemented with nonessential amino acids, and 10% fetal calf serum (FCS) (see Note 1). Cells are routinely cultured in standard cell culture conditions (37°C, 5% CO_2) in conventional 75 or 175 cm^2 cell culture vessels (see Note 2).
2. Before the monolayer becomes confluent, cells are detached with 0.25% trypsin/EDTA (see Notes 3–9) and seeded in µclear 96-well microplates (Greiner bio-one, Kremsmünster, Austria) (see Notes 10 and 11) in 100 µL of supplemented culture medium. Cells are then allowed to adapt for 24 h.
3. A staining mix is prepared by diluting MitoTracker Green FM (MTG; Molecular Probes, Eugene, OR, USA) in complete medium to a final concentration of 150 n*M*.
4. Cell supernatants are discarded and the cells are incubated with the MitoTracker dye for 20 min at 37°C. Following the cells are rinsed twice with prewarmed complete medium.
5. Cell supernatants are discarded and replaced with 100 µL media supplemented or not with 10 µ*M* or 100 µ*M* carbonyl cyanide m-chlorophenylhydrazone (CCCP; Sigma-Aldrich, St. Louis, MO,

USA) or 5 μ*M* rapamycin (RAPA, R&D Biosystems, Minneapolis, MN, USA).

6. Cells are transferred into an ImageXpress Micro XL automated microscope (Molecular Devices, Sunnyvale, CA, USA) equipped with environmental control (that maintains a humidified atmosphere containing 5% CO_2 at a temperature of 37°C) (see Note 12). All preselected wells are imaged with a 20 min interval for a total of 12 h.

3. ACQUISITION PROCEDURE

1. Perform the imaging using a 20 × objective (see Note 13) and acquire a minimum of four view fields per well (see Note 14).
2. MTG is monitored using a GFP filter set (see Note 15). Adjust the exposure time to achieve an optimal signal-to-noise ratio while avoiding saturated signals.
3. The mCherry-Parkin fusion protein is imaged using a TexasRed filter set (see Note 16). Adjust the acquisition parameters as detailed earlier.
4. Adjust the focal plane for both fluorescences to the same value (see Note 17).
5. The interval for live cell acquisition is set to 20 min (see Note 18) and the acquisition is launched (representative images in Fig. 1A).

4. ASSESSMENT OF COLOCALIZATION: PRINCIPLES

The colocalization of two fluorophores, employed to label different molecules, can be assessed based on two different methods: (i) cooccurrence and (ii) correlation. The first approach consists in quantifying the degree of spatial cooccurrence by enumerating overlapping pixels on the images obtained from one specimen with different optical filters sets, regardless of variations in pixel intensity, whereas the latter assesses the degree of correlation by quantifying the degree of similarity in intensity variations between the fluorescence channels. Results from both methods have to be interpreted with care as cooccurrence is extremely sensitive to image segmentation and can simply reflect close physicochemical properties (e.g., hydrophobic molecules that would attach to membranes), while correlation is very sensitive to background and noise, specifically when the signal-to-noise ratio is low or decreases with time (Adler & Parmryd, 2010).

There are certain limitations to most currently employed methods, i.e., colocalization studies often are performed on a single cell or, in the best case

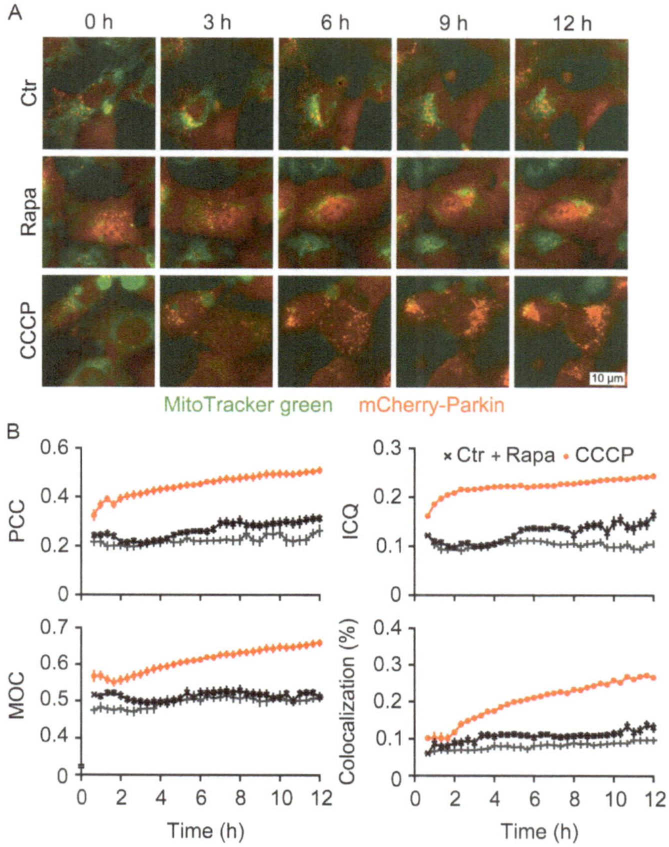

Fig. 1 Quantitative assessment of Parkin translocation in live cell imaging. Mitochondria of human osteosarcoma U-2 OS cells stably expressing Parkin-mCherry were counterstained with MitoTracker green. Thereafter, cells were treated with the mTOR inhibitor rapamycin or the protonophore CCCP at the indicated concentrations. The translocation of Parkin was monitored by live cell microscopy by means of an automated widefield imager at an interval of 20 min for 12 h. Please note that there is an offset in time due to technicalities. Following acquisition, images were analyzed using a custom-built algorithm and different coefficients were calculated. (A) A series of representative images are depicted for each treatment at 0, 3, 6, 9, and 12 h after treatment (size bar equals 10 μm). In panel (B), time-dependent changes in the Pearson correlation coefficient (PCC), Manders overlay coefficient (MOC), intensity correlation quotient (ICQ), and the percent of colocalization in the UNION compartment for mCherry-Parkin and MitoTracker green are reported (mean ± SD, $n = 16$).

scenario, multiple cells in a single image. In both cases, results can hardly be validated independently due to the statistical insignificant number of samples. Furthermore, methods utilized thus far are largely incompatible with the analysis of image-based high-throughput screening- (HTS) and time-lapse videomicroscopic approaches, due to the extensive number of images generated in both approaches. In addition, the signal-to-noise ratio of many widefield imaging systems' is not compatible with most of the colocalization software available today.

In order to adapt colocalization studies to widefield image-based HTS, we decided to develop an algorithm (coded in R) that allows for the efficient assessment of both correlation and cooccurrence in a large set of images acquired on an automated widefield imager. Using the algorithm that was designed for multithreading, thousands of images obtained from videomicroscopic acquisition were analyzed within only few hours (see Note 19).

5. THE R ENVIRONMENT

1. Create the necessary programming environment by installing the free open-source R programming language on your computer (see Notes 20–22).
2. Install the free open-source integrated development interface RStudio (see Notes 23 and 24).
3. Several libraries are necessary to execute the colocalization algorithm. The following libraries can be downloaded and installed within RStudio as packages available from cran.r-project.org : tiff; natural sort; parallel; foreach; doParallel; iterators; reshape. The library EBImage can be downloaded and installed from bioconductor.org by executing the code provided on the website.
4. The source code of the colocalization algorithm is available for download at https://github.com/kroemerlab/HTSColoc. The code has to be executed in the R-environment described earlier, according to the instructions given in the readme.md file retrievable on the website.

6. ASSESSMENT OF COLOCALIZATION: PROTOCOL

1. The database, that contains the image location, together with an unique identifier for the database entry, which in our case was the plate we imaged from on a widefield HTS microscope, are entered as arguments into the algorithm (see Notes 25 and 26). If the analysis computer has no database connection, or the acquired images are not managed by any

database structure, the path to the folder, which contains the images to be analyzed can be furnished instead.

2. Storage location and identifier are loaded into the R environment.
3. For each image a log operator is applied to the original pixel matrix (ORI), and the pixel intensities are then normalized and ranged between 0 and 1 (LOG).
4. Extreme intensity values from ORI are removed so that the distribution of pixel intensities in ORI is adjusted to 0.1–99.9% meaning that only 99.8% of pixels intensities compose the resulting image (NORM).
5. Following a Gaussian filter with a standard deviation of 50 is applied to NORM (GAUSS).
6. GAUSS is subtracted from ORI, while negative pixels are set to 0, and the image is normalized between 0 and 1. (CLEAN).
7. A Top Hat filter, using a structural element defined as a 29-pixels diameter disk, is applied to LOG. Negative pixels are set to 0, and the image is normalized between 0 and 1 (TOP).
8. TOP is binarized using a threshold determined with Otsu's Method. Background pixels are set to 0 and object pixels are set to 1. Single-pixels (connexity 8) are removed from the obtained image (SEG).
9. The Boolean operation OR is applied to SEGs derived from the two channels between which colocalization has to be quantified (UNION).
10. The equivalent operation is realized by using the Boolean operator AND (INTER).
11. Numeric elements from NORM (for the two compared images) are taken where UNION equals 1, and Pearson correlation coefficient (PCC), intensity correlation quotient (ICQ), and Manders overlap coefficient (MOC) are calculated (Fig. 1B).
12. The same operation is realized with INTER, and the overlapping coefficients are calculated (Fig. 1B) (see Notes 27 and 28).

7. CONCLUDING REMARKS

The method presented earlier provides an automated and reliable approach for quantifying mitophagic events by means of monitoring the translocation of Parkin to the mitochondrial membrane in cultured cells responding to chemical or biological cues. The PINK1-dependent aggregation of Parkin at the mitochondrial membrane is one of the initial, functionally important steps of mitophagy. At the next step, the organelle is included by LC3-containining isolation membranes that form a reaction compartment for the lysosome dependent dissociation of the cargo (Narendra & Youle, 2012). This process is sensitive to the blockage of autophagic degradation by lysosomal fusion inhibitors such as bafilomycin A1 (Yamamoto et al., 1998). Thus in order to underpin the importance of mitophagy and to validate potential hit compounds from HTS campaigns, the decrease in mitochondrial mass that mandatorily follows an increase in mitophagy should be assessed (Narendra et al., 2008). The decrease in mitochondrial mass reflects a direct consequence of lysosomal activity and can therefore be interfered with by lysosomal fusion inhibitors like bafilomycin A1 (Klionsky et al., 2016). The sum of this additional validation steps hence allows for the final interpretations of results obtained from primary screening campaigns. Irrespective of this second line experiments, the described method is suitable to identify chemicals or biologicals with mitophagy-modulatory potential within large libraries and/or with a high temporal resolution.

8. NOTES

1. Optimal cell culture conditions (i.e., culture medium, supplements, and detachment method) for other cell lines might differ from the ones described here. Please refer to your provider.
2. According to the number of experimental conditions and adapting to the given space for maintaining cells, the dimensions of the vessels for cell culture can differ from the ones described here. The ATCC (American tissue type culture collection) recommends T-75 flasks yet to our experience others vessel work equally well.
3. The highest efficacy of trypsin is achieved by minimizing serum content. Rinsing the cells with either PBS or trypsin solution ameliorates detachment in response to trypsin/EDTA.
4. The enzymatic reaction of trypsin works optimal at 37°C and should yield results within 1–3 min. It should be taken care to avoid extended exposure to trypsin in order to minimize cellular stress as it might affect several membrane-related processes.
5. Detachment should be visualized via light microscopy and might vary considerably with cell type and confluency.
6. According to the recommendations of the ATCC U-2 OS cells should be passaged at a ratio of 1:3 to 1:6 with a medium renewal 2–3 times per week to achieve optimal growth (see Note 7).

7. Cells should be kept under exponential growth conditions throughout their culture. Underconfluence should be avoided to circumvent clonal outgrowth (see Note 8). Similarly, overconfluence could lead to oxidative or metabolic stress and has to be avoided.
8. The passage number from an initial stock should not exceed a relatively low, predefined number of passages. New subcultures should be launched from another aliquot of the initial stock to avoid genetic drifts in the cellular population.
9. Maintenance cultures of U-2 OS cell stably expressing mCherry-Parkin were kept under minimal selection pressure by means of geneticin in order to maintain homogenous expression in the cellular population.
10. Five thousand U-2 OS cell stably expressing Parkin fused to mCherry were seeded in μclear 96-well microplates (Greiner bio-one, Kremsmünster, Austria). Geneticin was withdrawn from the culture media to avoid overlapping cellular effects evoked by selection and experiment-specific treatments.
11. The cell culture support might differ from the μclear microplates employed here. The utilized μclear plates are flatbottom plates with physical surface treatment, lid with condensation rings. They are sterile and display a plate bottom thickness of 190 μm (with a batch dependent variation of maximal 10 μm) and plate bottom flatness variation max. 0.6 μm. Using plates with higher plate bottom thickness will necessarily require changing toward a long working distance objective and would therefore influence the properties of the acquired image.
12. The environmental control should be launched well ahead of the start of the experiment to allow the machine to adapt to 37°C. A change in temperature during the imaging process might cause a drift in the *z*-dimension and has to be avoided.
13. The objective employed in our study was a CFI PlanApo Lambda objective (Nikon, Tokio, Japan) with a numerical aperture of 0.75.
14. The number of view fields should be adapted to the confluence of cells at the time of imaging. Thus, a number of view fields that allows for a statistical evaluation should be chosen.
15. In our case the GFP filter set that was used for the acquisition was the GFP-3035D-NTE-ZERO filter set (Semrock, Rochester, NY, USA; 472/30 nm BrightLine single-band bandpass excitation filter/ 520/35 nm BrightLine single-band bandpass emission filter and a 495 nm edge BrightLine single-edge dichroic beamsplitter).

16. Here, we used the TxRed-4040C-NTE-ZERO filter set (Semrock, Rochester, NY, USA; 562/40–25 nm BrightLine single-band bandpass excitation filter/624/40–25 nm BrightLine single-band bandpass emission filter and a 593 nm edge BrightLine single-edge dichroic beam splitter).
17. In the present set up the focus was automatically determined for each acquisition using a laser autofocus that determines the bottom of the well based on the diffraction that causes reflections when the beam passes from gaseous to solid and solid to liquid phase. An additional *z*-offset in order to identify the focal plane with the highest signal-to-noise ratio was defined based on visual inspection and remains unalter during the timeframe of the experiment.
18. The interval for image acquisition has to be chosen based on several considerations: temporal resolution of the observed phenotype, phototoxicity, and fluorochrome exhaustion. Pilot experiments on a limited number of control conditions should yield reproducible results before considering full-blown live cell screening.
19. The software is using multithreading meaning that the number of cores available to the computing unit determines the speed of analysis. In our case a computer equipped with two Intel Xeon E5-2630v2 processors (Santa Clara, CA, USA) each having 6 cores (24 threads) and accessing to 32 GB of RAM was routinely used for the study and allowed to conduct the colocalization analysis within few hours.
20. The R source code can be downloaded at https://www.r-project.org.
21. In our case we used R version 3.3.1 for Windows (Microsoft, Redmond, WA, USA)
22. R precompiled binary versions are provided for various operating systems.
23. Depending on the platform other IDEs are available.
24. In our case we used the RStudio version 0.98.90 for Windows
25. Most modern image-based HTS systems like the one from Molecular Devices employed for this study are furnished together with a dedicated database and database management system. The database requests are designed to work with a SQL database (Microsoft, Redmond, CA, USA) but might be adapted by an experienced user to a given database structure.
26. The algorithm is optimized and has extensively been tested using a 20 × PlanApo objective.

27. Optimal assessment of colocalization depends on several properties of the raw images. Results thus have to be validated by visual inspection prior to HTS approaches.
28. Please consult the documentation available at the download platform for details on the utilized algorithm.

ACKNOWLEDGMENTS

The plasmid coding for mCherry-Parkin was a gift from Richard Youle (Addgene plasmid # 23956). H.Z. and P.L. are supported by the China Scholarship Council, L.Z. is supported by the Ligue contre le Cancer, G.K. is supported by the Ligue contre le Cancer (équipes labelisées); Agence National de la Recherche (ANR)—Projets blancs; ANR under the frame of E-Rare-2, the ERA-Net for Research on Rare Diseases; Association pour la recherche sur le cancer (ARC); Cancéropôle Ile-de-France; Institut National du Cancer (INCa); Fondation Bettencourt-Schueller; Fondation de France; Fondation pour la Recherche Médicale (FRM); the European Commission (ArtForce); the European Research Council (ERC); the LabEx Immuno-Oncology; the SIRIC Stratified Oncology Cell DNA Repair and Tumor Immune Elimination (SOCRATE); the SIRIC Cancer Research and Personalized Medicine (CARPEM); the Swiss Bridge Foundation, ISREC and the Paris Alliance of Cancer Research Institutes (PACRI).

Legal disclaimer: The described algorithm is for noncommercial purposes only. The authors are not responsible for any consequences resulting from the use of the algorithm or the R environment.

REFERENCES

Adler, J., & Parmryd, I. (2010). Quantifying colocalization by correlation: The Pearson correlation coefficient is superior to the Mander's overlap coefficient. *Cytometry Part A*, 77(8), 733–742. http://dx.doi.org/10.1002/cyto.a.20896.

Amadoro, G., Corsetti, V., Florenzano, F., Atlante, A., Ciotti, M. T., Mongiardi, M. P., … Calissano, P. (2014). AD-linked, toxic NH2 human tau affects the quality control of mitochondria in neurons. *Neurobiology of Disease*, *62*, 489–507. http://dx.doi.org/10.1016/j.nbd.2013.10.018.

Buhlman, L., Damiano, M., Bertolin, G., Ferrando-Miguel, R., Lombes, A., Brice, A., & Corti, O. (2014). Functional interplay between Parkin and Drp1 in mitochondrial fission and clearance. *Biochimica et Biophysica Acta*, *1843*(9), 2012–2026. http://dx.doi.org/10.1016/j.bbamcr.2014.05.012.

Codogno, P., Mehrpour, M., & Proikas-Cezanne, T. (2012). Canonical and non-canonical autophagy: Variations on a common theme of self-eating? *Nature Reviews Molecular Cell Biology*, *13*(1), 7–12. http://dx.doi.org/10.1038/nrm3249.

Deng, H., Dodson, M. W., Huang, H., & Guo, M. (2008). The Parkinson's disease genes pink1 and parkin promote mitochondrial fission and/or inhibit fusion in Drosophila. *Proceedings of the National Academy of Sciences of the United States of America*, *105*(38), 14503–14508. http://dx.doi.org/10.1073/pnas.0803998105.

Eid, N., Ito, Y., Maemura, K., & Otsuki, Y. (2013). Elevated autophagic sequestration of mitochondria and lipid droplets in steatotic hepatocytes of chronic ethanol-treated rats: An immunohistochemical and electron microscopic study. *Journal of Molecular Histology*, *44*(3), 311–326. http://dx.doi.org/10.1007/s10735-013-9483-x.

Feng, Y., He, D., Yao, Z., & Klionsky, D. J. (2014). The machinery of macroautophagy. *Cell Research*, *24*(1), 24–41. http://dx.doi.org/10.1038/cr.2013.168.
Galluzzi, L., Pietrocola, F., Bravo-San Pedro, J. M., Amaravadi, R. K., Baehrecke, E. H., Cecconi, F., … Kroemer, G. (2015). Autophagy in malignant transformation and cancer progression. *The EMBO Journal*, *34*((7), 856–880. http://dx.doi.org/10.15252/embj.201490784.
Galluzzi, L., Pietrocola, F., Levine, B., & Kroemer, G. (2014). Metabolic control of autophagy. *Cell*, *159*(6), 1263–1276. http://dx.doi.org/10.1016/j.cell.2014.11.006.
Klionsky, D. J., Abdelmohsen, K., Abe, A., Abedin, M. J., Abeliovich, H., Acevedo Arozena, A., … Zughaier, S. M. (2016). Guidelines for the use and interpretation of assays for monitoring autophagy (3rd edition). *Autophagy*, *12*(1), 1–222. http://dx.doi.org/10.1080/15548627.2015.1100356.
Klionsky, D. J., Baehrecke, E. H., Brumell, J. H., Chu, C. T., Codogno, P., Cuervo, A. M., … Tooze, S. A. (2011). A comprehensive glossary of autophagy-related molecules and processes (2nd edition). *Autophagy*, 7(11), 1273–1294. http://dx.doi.org/10.4161/auto.7.11.17661.
Lamb, C. A., Yoshimori, T., & Tooze, S. A. (2013). The autophagosome: Origins unknown, biogenesis complex. *Nature Reviews. Molecular Cell Biology*, *14*(12), 759–774. http://dx.doi.org/10.1038/nrm3696.
Madeo, F., Eisenberg, T., & Kroemer, G. (2009). Autophagy for the avoidance of neurodegeneration. *Genes & Development*, *23*(19), 2253–2259. http://dx.doi.org/10.1101/gad.1858009.
Marino, G., Madeo, F., & Kroemer, G. (2011). Autophagy for tissue homeostasis and neuroprotection. *Current Opinion in Cell Biology*, *23*(2), 198–206. http://dx.doi.org/10.1016/j.ceb.2010.10.001.
Narendra, D., Tanaka, A., Suen, D. F., & Youle, R. J. (2008). Parkin is recruited selectively to impaired mitochondria and promotes their autophagy. *The Journal of Cell Biology*, *183*(5), 795–803. http://dx.doi.org/10.1083/jcb.200809125.
Narendra, D. P., & Youle, R. J. (2011). Targeting mitochondrial dysfunction: Role for PINK1 and Parkin in mitochondrial quality control. *Antioxidants & Redox Signaling*, *14*(10), 1929–1938. http://dx.doi.org/10.1089/ars.2010.3799.
Narendra, D. P., & Youle, R. J. (2012). Neurodegeneration: Trouble in the cell's powerhouse. *Nature*, *483*(7390), 418–419. http://dx.doi.org/10.1038/nature10952.
Poole, A. C., Thomas, R. E., Andrews, L. A., McBride, H. M., Whitworth, A. J., & Pallanck, L. J. (2008). The PINK1/Parkin pathway regulates mitochondrial morphology. *Proceedings of the National Academy of Sciences of the United States of America*, *105*(5), 1638–1643. http://dx.doi.org/10.1073/pnas.0709336105.
Sica, V., Galluzzi, L., Bravo-San Pedro, J. M., Izzo, V., Maiuri, M. C., & Kroemer, G. (2015). Organelle-specific initiation of autophagy. *Molecular Cell*, *59*(4), 522–539. http://dx.doi.org/10.1016/j.molcel.2015.07.021.
Twig, G., Elorza, A., Molina, A. J., Mohamed, H., Wikstrom, J. D., Walzer, G., … Shirihai, O. S. (2008). Fission and selective fusion govern mitochondrial segregation and elimination by autophagy. *The EMBO Journal*, *27*(2), 433–446. http://dx.doi.org/10.1038/sj.emboj.7601963.
Yamamoto, A., Tagawa, Y., Yoshimori, T., Moriyama, Y., Masaki, R., & Tashiro, Y. (1998). Bafilomycin A1 prevents maturation of autophagic vacuoles by inhibiting fusion between autophagosomes and lysosomes in rat hepatoma cell line, H-4-II-E cells. *Cell Structure and Function*, *23*(1), 33–42.
Youle, R. J., & Narendra, D. P. (2011). Mechanisms of mitophagy. *Nature Reviews. Molecular Cell Biology*, *12*(1), 9–14. http://dx.doi.org/10.1038/nrm3028.
Zhou, H., Forveille, S., Sauvat, A., Yamazaki, T., Senovilla, L., Ma, Y., … Kroemer, G. (2016). The oncolytic peptide LTX-315 triggers immunogenic cell death. *Cell Death & Disease*. 7, e2134. http://dx.doi.org/10.1038/cddis.2016.47.

CHAPTER FOURTEEN

Assays to Monitor Lysophagy

Y.-P. Chu*, Y.-H. Hung*,†, H.-Y. Chang*,‡, W.Y. Yang*,†,‡,1
*Institute of Biological Chemistry, Academia Sinica, Taipei, Taiwan
†Institute of Biochemical Sciences, College of Life Sciences, National Taiwan University, Taipei, Taiwan
‡Genome and Systems Biology Degree Program, National Taiwan University, Taipei, Taiwan
[1]Corresponding author: e-mail address: weiyang@gate.sinica.edu.tw

Contents

Abstract

Lysophagy belongs to one of the many pathways cells activate in response to lysosomal damage. Damaged lysosomes attract glycan-binding galectins, become ubiquitinated, and are later on targeted for engulfment and degradation through lysophagy. Many triggers that are known to cause lysosomal membrane permeabilization have all been shown to induce lysophagy and can therefore be used to construct platforms for further molecular-level characterization of this process. In this chapter, we describe experimental parameters for triggering lysophagy through combined use of lysosome-specific dyes and light illumination. Within single cells, this optogenetic scheme allows easy manipulation on the amount of lysosomes to be impaired, the degree of damage desired, as well as when and where this should happen. On the other hand it can also be used to target all lysosomes within the entire cell population of a culture, allowing screening or bulk biochemical analyses to be carried out. The methodology will find use not only in monitoring lysophagy but also in probing lysosome damage responses in general.

Methods in Enzymology, Volume 588
ISSN 0076-6879
http://dx.doi.org/10.1016/bs.mie.2016.09.083

1. INTRODUCTION

By forming a double-layered membrane to capture substrates the autophagic machinery can take on virtually all cellular materials for degradation. Cells therefore rely on autophagy to control the number and quality of organelles (Okamoto, 2014; Sica et al., 2015). A particularly good example of this is the role autophagy plays in maintaining mitochondria. During fertilization paternal mitochondria are selectively removed by autophagy (Al Rawi et al., 2011; Sato & Sato, 2011). Cardiomyocytes utilize autophagy to completely turnover/replace their mitochondrial pool for proper heart development (Gong et al., 2015). Macrophages keep their inflammatory responses in check by targeting damaged mitochondria for elimination through autophagy (Zhong et al., 2016). The lack of autophagy-mediated mitochondrial quality control has been implicated in Parkinson's disease and cancer (Chourasia, Boland, & Macleod, 2015; Narendra, Tanaka, Suen, & Youle, 2008; Sumpter et al., 2016; Youle & Narendra, 2011).

The lysosome, initially known for its function in materials degradation, is now regarded as a signaling hub that regulates cell metabolism (Settembre, Fraldi, Medina, & Ballabio, 2013). It has been found that cells can also selectively target defective lysosomes for clearance through autophagy, a process termed lysophagy (Aits et al., 2015; Chen et al., 2014; Hung, Chen, Yang, & Yang, 2013; Maejima et al., 2013). Furthering the molecular details of this pathway will allow better understanding on when and how these events occur and the importance of this pathway in maintaining normal physiology. Progress along this direction will be greatly accelerated with robust experimental platforms for eliciting and monitoring lysophagy in place.

2. KNOWN EXPERIMENTAL TRIGGERS FOR LYSOPHAGY

Lysophagy belongs to one of the many pathways cells activate in response to lysosomal damage. It is therefore possible to experimentally probe lysophagy through robust induction of lysosomal dysfunction. A large number of small molecules are known to disrupt lysosome functions. For example, chloroquine, ammonium chloride, concanamycin A, and bafilomycin A1 can all be used to perturb lysosomal pH. E64D and pepstatin A can suppress cathepsins to interfere with protein degradation within the lysosomal lumen (Tanida, Minematsu-Ikeguchi, Ueno, & Kominami, 2005). To date, most reports demonstrating lysophagy have relied on compounds that can activate lysosomal membrane permeabilization (LMP). This list includes L-leucyl-L-leucine methyl ester (LLOMe), silica, monosodium urate, calcium phosphate precipitates, and lysosome-localized photosensitizers (Aits et al., 2015; Chen et al., 2014; Hung et al., 2013; Maejima et al., 2013).

2.1 L-Leucyl-L-Leucine Methyl Ester

LLOMe accumulates within lysosomes through endocytosis, and its membrane rupturing activity is activated when it encounters the lysosomal dipeptide peptidase I (DPPI). DPPI converts LLOMe into metabolites that possess detergent-like properties, which in turn disrupts lysosomal membranes to cause damage (Thiele & Lipsky, 1990). The effectiveness of LLOMe in eliciting lysosomal dysfunction depends on cellular DPPI activities, which are the highest in immune cells. Cultured cells can therefore be supplemented with LLOMe in the medium to trigger lysophagy (Aits et al., 2015; Maejima et al., 2013). Protocol detailing the use of LLOMe has been reported elsewhere (Hasegawa, Maejima, Iwamoto, & Yoshimori, 2015).

2.2 Calcium Phosphate Precipitates

Calcium phosphate precipitates (CPP) is commonly used in the delivery of DNA into cultured cells: LMP elicited by CPP allows DNA(s) to trespass into the cytoplasm, resulting in efficient transfection. It has been found that CPP treatments induced cellular formation of Lamp1 and Rab7 vesicles positive for Galectin 3, and these damaged late endosomes/lysosomes became selectively engulfed into autophagosomes (Chen et al., 2014). Suppressing cellular autophagy can therefore enhance CPP-mediated DNA delivery (Ege, Reisbig, & Rogne, 1984). Since many DNA transfection reagents also work by disrupting lysosomal integrity, one therefore expects them to also represent robust inducers of lysophagy.

2.3 Lysosome-Localized Photosensitizers

Biomolecules are sensitive to reactive oxygen species (ROS)—therefore, one can easily impair the functions of individual proteins, organelles, cells, and tissues through chromophore-assisted light inactivation (Hsieh, Chu, Lee, & Yuan Yang, 2015; Kobayashi et al., 2013; Lin et al., 2013; Qi, Garren, Shu, Tsien, & Jin, 2012; Surrey et al., 1998; Wang, Nartiss, Steipe, McQuibban, & Kim, 2012; Williams et al., 2013; Yang, 2015; Yang & Yang, 2011, 2013). Many chromophores, when excited, can efficiently mediate ROS formation, and these ROS locally attack nearby biomolecules to elicit damage (Jay, 1988). LMP can therefore be triggered in stained lysosomes (either by small molecule dyes or genetically encoded photosensitizers) through light illumination. Light-activated LMP has been utilized in the controlled delivery of small molecules into cells, a process termed photochemical internalization (Berg et al., 1999). It has also been shown to be able to elicit lysophagy (Hung et al., 2013). This scheme offers the opportunity to precisely tune the number of defective lysosomes per cell in an experiment. A detailed protocol for triggering LMP using photosensitizers is described in the following section.

3. EXPERIMENTAL PROTOCOL FOR TRIGGERING LYSOPHAGY WITH LIGHT

In this protocol, we describe the use of a far-red fluorescent dye Al(III) phthalocyanine chloride disulfonic acid ($AlPcS_{2a}$) to induce lysophagy. Other lysosome-specific dyes that mediate robust ROS generation can also be used in place of $AlPcS_{2a}$. One can therefore pick different colored dyes to induce lysophagy in accordance to the experimental needs (by being far-red fluorescent $AlPcS_{2a}$ is particularly compatible with downstream fluorescence imaging applications since it does not interfere with the typical blue, green, and red emission channels used). The described method allows one to selectively impair a subpopulation of lysosomes within single cells with spatiotemporal precision. It can also be used to conveniently damage lysosomes within the entire cell population (e.g., all cells on a culture dish) for image-based screening or biochemical assays (Fig. 1). Tuning of the $AlPcS_{2a}$ concentration or illumination conditions will allow one to alter the extent and degree of lysosome damage within the treated cells.

Materials

- Al(III) phthalocyanine chloride disulfonic acid ($AlPcS_{2a}$), (Frontier Scientific, Logan, UT, USA, P40632).
- FITC-dextran (Thermo Fisher Scientific, Waltham, MA, USA, D1821).
- TMR-dextran (Thermo Fisher Scientific, Waltham, MA, USA, D1816).
- Cell line of interest: e.g., HeLa cells (ATCC, Manassas, VA, USA, CCL-2).
- Glass bottom dishes (MatTek Corporation, Ashland, MA, USA, P35G-1.5-20-C).
- Poly-L-lysine (Sigma, St. Louis, MO, USA, P4707).
- Dulbecco's modified eagle's medium (DMEM) (Thermo Fisher Scientific, Waltham, MA, USA, 11965–092).

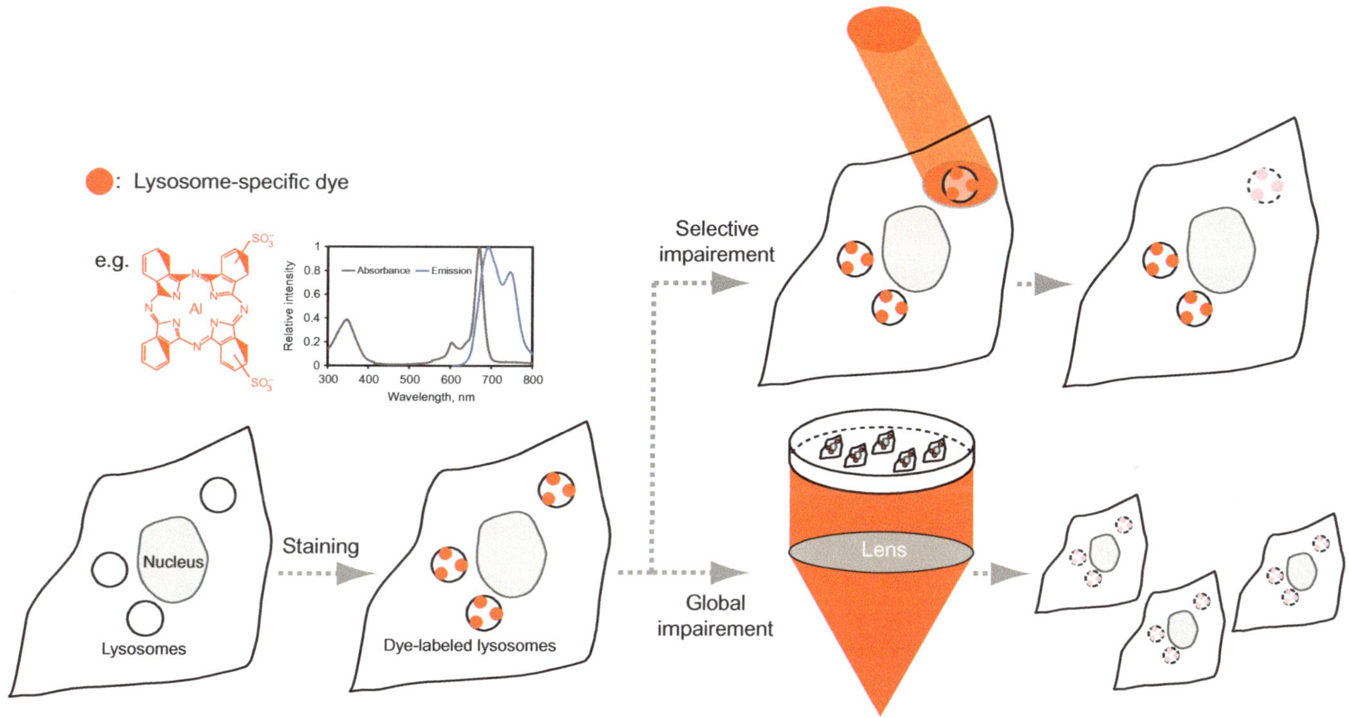

Fig. 1 Photochemical induction of LMP. Cells stained with lysosome-specific dyes (e.g., $AlPcS_{2a}$) can be illuminated to elicit LMP. By altering the light illumination patterns, one can choose either to damage small fractions of lysosomes within single cells or target lysosomes across the entire cell population.

- Phenol red-free DMEM (Thermo Fisher Scientific, Waltham, MA, USA, 31053–028).
- L-Glutamine (Thermo Fisher Scientific, Waltham, MA, USA, 15030–081).
- FBS (Gibco, 26140–079).
- Penicillin and streptomycin (P/S) liquid (Thermo Fisher Scientific, Waltham, MA, USA, 15140–122).
- 0.5% Trypsin–EGTA (10X) (Thermo Fisher Scientific, Waltham, MA, USA, 15400–054).
- Lipofectamine 2000 (Thermo Fisher Scientific, Waltham, MA, USA, 11668).
- Phosphate-buffered saline (PBS): 0.137 m*M* NaCl, 2.7 m*M* KCl, 8.0 m*M* Na_2HPO_4, 1.76 m*M* KH_2PO_4, pH 7.5.
- Fluorescent reporter constructs: TagRFP-Galectin 3; EGFP-Galectin 3, 8, and 9.
- Antibodies: anti-Lamp1 (Cell Signaling Technology, Danvers, MA, USA, 9091) and (Santa Cruz Biotechnology, Dallas, TX, USA, sc-8099); anti-LC3B (Novus Biologicals, Littleton, CO, USA, NB100-2220); anti-Ub (Merck Millipore, Billerica, MA, USA, 04–263).
- Lysotracker Blue (Thermo Fisher Scientific, Waltham, MA, USA, L7525).

Equipment

- Syringe (Terumo, Somerset, NJ, USA, SS-01T).
- 0.22-μm filters (Sartorius, Goettingen, Germany, 16534).
- Sonicator.
- Heat bath.
- UV–vis spectrophotometer.
- Confocal fluorescence microscope.
- Microscope stage top incubator (LCI, Seoul, South Korea, Chamlide IC).
- Collimated LED light source (Thorlabs, Newton, NJ, USA, M660L3-C1 and DC2100).

3.1 Staining Lysosomes With Appropriate Photosensitizers

3.1.1 Preparing the $AlPcS_{2a}$ Stock Solution

1. Add 5 mg $AlPcS_{2a}$ into 0.1–0.2 mL of 0.1 *M* NaOH. Dissolve by heating up the solution to 50°C followed by sonication at 40 kHz for 1 min.
2. Add PBS to make a 1 mL solution.
3. Filter the solution with a 0.22 μm filter to remove insoluble precipitates.

4. Measure the concentration of the filtered solution through UV–vis (extinction coefficient of $AlPcS_{2a}$ at 672 nm is 4×10^4 $cm^{-1}M^{-1}$)
5. Dilute the filtered solution with PBS to make a 125 μ*M* stock solution. Make small aliquots, wrap in aluminum foil to protect from light, and store at −20°C.

3.1.2 Staining Cells With $AlPcS_{2a}$

1. Culture HeLa cells in DMEM supplemented with 10% FBS and 1% P/S on 35 mm poly-L-lysine-coated glass bottom dishes. Allow cells to grow to 60–80% confluent before staining.
2. Dilute the $AlPcS_{2a}$ stock solution to 125 n*M* in DMEM supplemented with 10% FBS and 1% P/S for cell staining.
3. Stain cells at 37°C, 5% CO_2 for 16 h. This allows cells to internalize $AlPcS_{2a}$ through endocytosis.
4. Wash the cells twice with PBS.
5. Place cells into phenol red-free DMEM supplemented with 10% FBS, 1% P/S, and 1% L-G.
6. Incubate cells at 37°C, 5% CO_2 for 8 h. This allows trace amounts $AlPcS_{2a}$ to be removed from all cellular structures but the lysosomes.
7. Replace medium with fresh phenol red-free DMEM supplemented with 10% FBS, 1% P/S, and 1% L-G for downstream assays. Proper staining can be confirmed through imaging (there should be good colocalization between $AlPcS_{2a}$ and Lysotracker Blue). $AlPcS_{2a}$ incubation times can be adjusted according to the cell types used.

3.2 Activation of LMP

One can identify proper light illumination conditions for triggering lysophagy through the use of LMP indicators (e.g., fluorescently labeled dextrans or fluorescent protein-tagged galectins). Fluorescently labeled dextrans accumulate in lysosomes by endocytosis. Upon LMP, smaller dextrans become released into the cytoplasm, while larger dextrans remain trapped. High-molecular weight dextrans conjugated to pH-sensitive fluorescent dyes such as fluorescein isothiocyanate (FITC) can therefore be used to monitor LMP. FITC fluorescence is quenched within intact lysosomes (due to low pH), and display a large increase upon disruption of the acidic lysosomal pH following LMP (Hung et al., 2013). Galectins (β-galactoside-binding lectin), on the other hand, have also been found to be sensitive reporters for LMP (Fig. 2). Lysosomal lumens are heavily glycosylated. For example, both Lamp1 and 2 are unique transmembrane proteins of

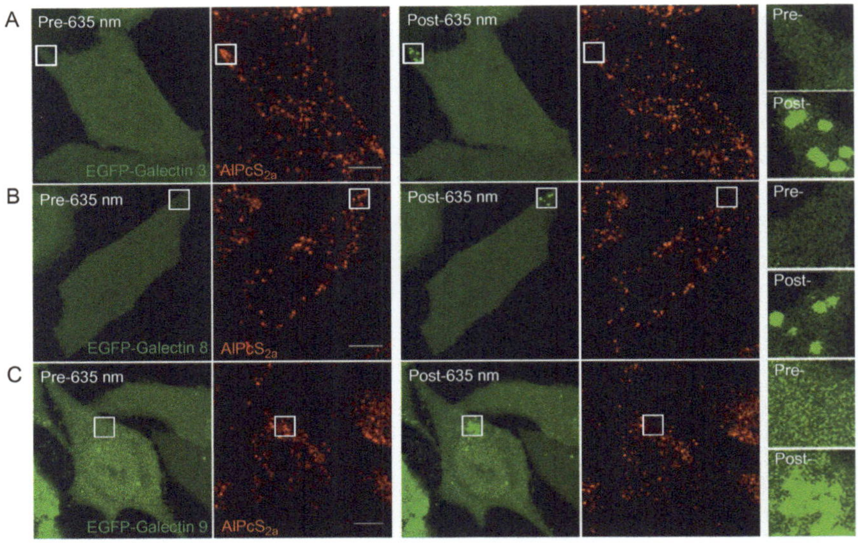

Fig. 2 Galectins are robust LMP reporters. HeLa cells expressing either (A) EGFP-Galectin 3, (B) EGFP-Galectin 8, and (C) EGFP-Galectin 9 were labeled with $AlPcS_{2a}$ and then 635 nm illuminated (30 s) within the respective *white square* regions (*left panels*). This led to the recruitment of Galectin 3, 8, and 9 to the photo-damaged areas (*middle panels*; $t=1$ min after illumination). *Right panels*: magnified view of EGFP-Galectin 3 (A), 8 (B), and 9 (C) within the 635 nm illuminated regions pre- and postlight activation. Scale bars, 10 μm.

the lysosome and are extensively modified by *N*-acetyllactosamines (Andrejewski et al., 1999; Do, Smith, & Cummings, 1990). Upon LMP glycans that line the inside of lysosomes become exposed to the cytoplasm, allowing for galectins to bind (Aits et al., 2015; Dupont et al., 2009; Maejima et al., 2013; Thurston, Wandel, von Muhlinen, Foeglein, & Randow, 2012). Both FITC-dextran and galectin display rapid signal changes (within minutes) following LMP, making them good indicators for efficient screening of appropriate light illumination conditions (Fig. 3). Various dyes have all been found to be capable of eliciting LMP when illuminated (Fig. 4).

3.2.1 Activating LMP Within Single Cells

1. Place cells grown on glass bottom dishes onto the confocal microscope equipped with a stage top CO_2 incubator.
2. Search for desired cells using low light (to minimize the activation $AlPcS_{2a}$).
3. Choose a 7×7 μm^2 ROI covering to-be-damaged lysosomes (lysosomes viewed through protein markers or directly through imaging $AlPcS_{2a}$), and scan the ROI with 50 μW of 635 nm light for 30 s to elicit damage.

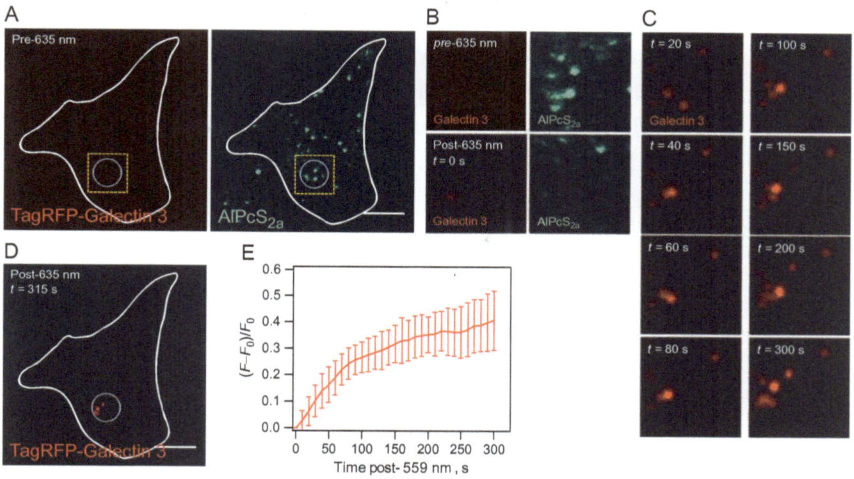

Fig. 3 Rapid recruitment of Galectin 3 to damaged lysosomes. (A) A TagRFP-Galectin 3 expressing, $AlPcS_{2a}$ stained HeLa cell was illuminated with 635 nm light (260 μW, 5 s) at the *blue circle* region. (B–C) Time-lapse view of the *orange dotted square* region in (A). Galectin 3 puncta rapidly formed following lysosome damage. (D) The HeLa cell in **A** 315 s following lysosome damage. (E) The kinetics of TagRFP-Galectin 3 fluorescence increase within the *orange dotted square* region over time ($n = 7$ cells; error bars represent S.D.). F_0, total TagRFP-Galectin 3 fluorescence within the *orange dotted square* region before lysosome damage; F, total TagRFP-Galectin 3 fluorescence within the *orange dotted square* region at a given time following lysosome damage. Scale bars, 10 μm.

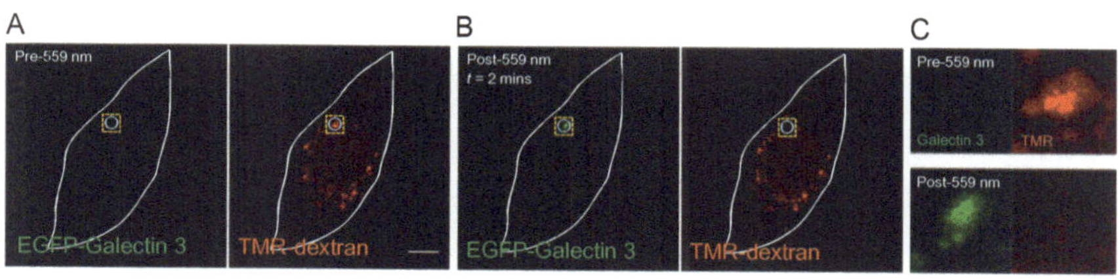

Fig. 4 Activating LMP through the use of TMR-dextran. Many lysosome-localized dyes can be used in place of $AlPcS_{2a}$ to achieve light-activated LMP. (A) An EGFP-Galectin 3 transfected HeLa cell was stained with 250 μ*M* TMR-dextran (10 kDa) for 15 h, then further incubated in dye-free culture medium for 4 h. *White solid line* marks the outline of the cell. (A–B) We used 559 nm light (190 μW, 5 s) to illuminate TMR-dextran puncta within the *blue circle* region, leading to the recruitment of EGFP-Galectin 3 onto damaged lysosomes after 2 min. (C) Magnified view of the *orange dotted square* regions in (A) and (B). Scale bar, 10 μm.

Note: $AlPcS_{2a}$ concentration, ROI size, 635 nm intensity used, as well as the illumination duration, can all be adjusted proportionally. To damage a large number of lysosomes one can also repeat the above-mentioned procedure at different locations of a cell. For long-term time-lapse imaging applications, it is desirable to use low $AlPcS_{2a}$ staining concentrations to minimize unwanted damage due to prolonged light exposure.

4. Acquire images (Since $AlPcS_{2a}$ shows low yet discernable absorbance across the visible spectrum, the imaging power should be kept low to avoid unwanted LMP induction.).

3.2.2 Activating LMP in all Cells on a Dish

Below is a list of parameters for inducing LMP in all cells on a culture plate using a collimated 440 mW, 660 nm LED light source from Thorlabs. The output beam width is wide enough to cover the culture area (1.4-mm diameter circle) of a 35-mm glass bottom dish. On the other hand, to do this for a 10-cm culture plate, we use a convex lens to slightly diverge the beam so that it is expanded to cover the entire dish. The illumination conditions used are listed in the following table:

	35-mm Glass Bottom Dishes	35-mm Glass Bottom Dishes	10 cm Culture Dishes
$AlPcS_2$ staining concentration	125 n*M*	625 n*M*	1.25 μ*M*
Illumination power density	97.3 mW/cm^2	97.3 mW/cm^2	3.8 mW/cm^2
Illumination duration	2.5 min	30 s	8 min

4. DETECTION OF LYSOPHAGY

Two simple strategies can be used to monitor the occurrence of lysophagy. One is to use a tandem fluorescent protein (e.g., EGFP-mRFP or TagRFP-mWasabi)-tagged fusion construct that localizes on damaged lysosomes (Kimura, Noda, & Yoshimori, 2007; Zhou et al., 2012). During lysophagy, this fusion protein will eventually enter low pH environments, resulting in the loss of EGFP (or mWasabi) fluorescence. An example of this has been previously described (Hasegawa et al., 2015; Maejima et al., 2013). Another is the use of immunofluorescence to monitor the earlier steps during lysophagy, namely, ubiquitination and autophagic engulfment of damaged lysosomes (Fig. 5), detailed as follows:

1. Induce LMP and incubate cells at 37°C, 5% CO_2 for desired periods of time.
2. Fix cells in PBS containing 4% paraformaldehyde for 15 min at room temperature.
3. Wash cells three times with PBS.
4. Permeabilize cells using 0.5% Triton X-100 in PBS for 10 min.

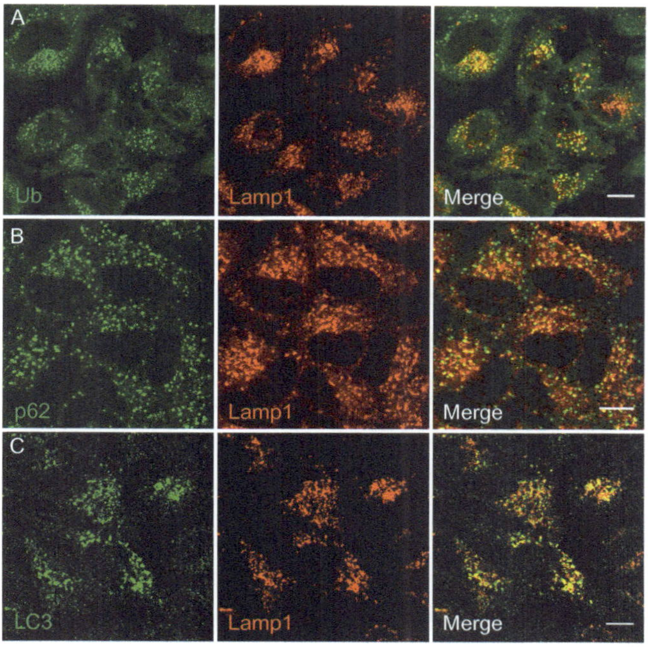

Fig. 5 Activating lysophagy in all cells on a culture dish. All $AlPcS_{2a}$ stained HeLa cells on a 35-mm glass bottom dish (following the protocol from Section 3.2.2) were simultaneously 660 nm illuminated, resulting in the appearance of ubiquitin (A), p62 (B), and LC3 (C) on damaged lysosomes. The images were obtained using the protocol provided in Section 4. Scale bars, 14 μm.

5. Wash cells three times with PBS.
6. Block cells with 3% BSA in PBS for 30 min at room temperature.
7. Dilute primary antibodies to desired concentrations in 100 μL PBS with 1% BSA. To probe ubiquitination of lysosomes: costain cells with mouse monoclonal antiubiquitin antibody (1: 500; Merck Millipore, 04–263) and rabbit monoclonal anti-Lamp1 antibody (1: 500; Cell Signaling Technology, 9091) for 1 h at room temperature. To probe LC3 on lysosomes: sequentially stain cells with rabbit polyclonal anti-LC3B antibody (1:500; Novus Biologicals, NB100-2220) and goat polyclonal anti-Lamp1 antibody (1: 200; Santa Cruz Biotechnology, sc-8099) for 1 h each.
8. Wash cells three times with PBS.
9. Incubate cells in dark with fluorochrome-conjugated secondary antibodies diluted in PBS with 1% BSA at room temperature for 45 min.
10. Wash cells five times with PBS.
11. Image cells on a confocal microscope.

5. CONCLUSIONS

With the range of lysosome-specific damaging schemes now available, one can easily monitor lysophagy within their systems of choice. In particular, the light-based damaging scheme allows precise control of the placement as well as the amount of damages lysosomes within a cell, thereby bringing about unique opportunities for the probing of lysophagy. For example, it will be interesting to use the strategy to investigate how lysophagy is differentially regulated in response to varying numbers of damaged lysosomes within a cell. Previously in Parkin-mediated mitophagy, it has been found that autophagosome initiation sites increase proportionally to the number of dysfunctional mitochondria within a cell (Hsieh et al., 2015).

It is now known that a number of mitochondrial maintenance mechanisms coexist within cells. Cells can utilize mitochondrial matrix localized proteases, mitoptosis, mitochondrial vesicular budding, as well as mitophagy to repair and eliminate defective mitochondria (Baker, Tatsuta, & Langer, 2011; Liu, Sakakibara, Chen, & Okamoto, 2014; Pickrell & Youle, 2015; Sugiura, McLelland, Fon, & McBride, 2014; Tinari, Garofalo, Sorice, Esposti, & Malorni, 2007). Mitochondrial biogenesis is also activated in response mitochondrial dysfunction (Ivankovic, Chau, Schapira, & Gegg, 2016). What is the corresponding scenario in the case of lysosome maintenance? The lysosome damaging schemes described here will help answer these questions by allowing one to generally probe the cellular repertoire of lysosome damage responses.

REFERENCES

Aits, S., Kricker, J., Liu, B., Ellegaard, A. M., Hamalisto, S., Tvingsholm, S., et al. (2015). Sensitive detection of lysosomal membrane permeabilization by lysosomal galectin puncta assay. *Autophagy*, *11*, 1408–1424.

Al Rawi, S., Louvet-Vallee, S., Djeddi, A., Sachse, M., Culetto, E., Hajjar, C., et al. (2011). Postfertilization autophagy of sperm organelles prevents paternal mitochondrial DNA transmission. *Science (New York, N. Y.)*, *334*, 1144–1147.

Andrejewski, N., Punnonen, E. L., Guhde, G., Tanaka, Y., Lullmann-Rauch, R., Hartmann, D., et al. (1999). Normal lysosomal morphology and function in LAMP-1-deficient mice. *The Journal of Biological Chemistry*, *274*, 12692–12701.

Baker, M. J., Tatsuta, T., & Langer, T. (2011). Quality control of mitochondrial proteostasis. *Cold Spring Harbor Perspectives in Biology*, *3*, a007559.

Berg, K., Selbo, P. K., Prasmickaite, L., Tjelle, T. E., Sandvig, K., Moan, J., et al. (1999). Photochemical internalization: A novel technology for delivery of macromolecules into cytosol. *Cancer Research*, *59*, 1180–1183.

Chen, X., Khambu, B., Zhang, H., Gao, W., Li, M., Chen, X., et al. (2014). Autophagy induced by calcium phosphate precipitates targets damaged endosomes. *The Journal of Biological Chemistry*, *289*, 11162–11174.
Chourasia, A. H., Boland, M. L., & Macleod, K. F. (2015). Mitophagy and cancer. *Cancer & Metabolism*, *3*, 4.
Do, K. Y., Smith, D. F., & Cummings, R. D. (1990). LAMP-1 in CHO cells is a primary carrier of poly-*N*-acetyllactosamine chains and is bound preferentially by a mammalian S-type lectin. *Biochemical and Biophysical Research Communications*, *173*, 1123–1128.
Dupont, N., Lacas-Gervais, S., Bertout, J., Paz, I., Freche, B., Van Nhieu, G. T., et al. (2009). Shigella phagocytic vacuolar membrane remnants participate in the cellular response to pathogen invasion and are regulated by autophagy. *Cell Host & Microbe*, *6*, 137–149.
Ege, T., Reisbig, R. R., & Rogne, S. (1984). Enhancement of DNA-mediated gene transfer by inhibitors of autophagic-lysosomal function. *Experimental Cell Research*, *155*, 9–16.
Gong, G., Song, M., Csordas, G., Kelly, D. P., Matkovich, S. J., & Dorn, G. W., 2nd. (2015). Parkin-mediated mitophagy directs perinatal cardiac metabolic maturation in mice. *Science (New York, N.Y.)*, *350*. aad2459.
Hasegawa, J., Maejima, I., Iwamoto, R., & Yoshimori, T. (2015). Selective autophagy: Lysophagy. *Methods*, *75*, 128–132.
Hsieh, C. W., Chu, C. H., Lee, H. M., & Yuan Yang, W. (2015). Triggering mitophagy with far-red fluorescent photosensitizers. *Scientific Reports*, *5*, 10376.
Hung, Y. H., Chen, L. M., Yang, J. Y., & Yang, W. Y. (2013). Spatiotemporally controlled induction of autophagy-mediated lysosome turnover. *Nature Communications*, *4*, 2111.
Ivankovic, D., Chau, K. Y., Schapira, A. H., & Gegg, M. E. (2016). Mitochondrial and lysosomal biogenesis are activated following PINK1/parkin-mediated mitophagy. *Journal of Neurochemistry*, *136*, 388–402.
Jay, D. G. (1988). Selective destruction of protein function by chromophore-assisted laser inactivation. *Proceedings of the National Academy of Sciences of the United States of America*, *85*, 5454–5458.
Kimura, S., Noda, T., & Yoshimori, T. (2007). Dissection of the autophagosome maturation process by a novel reporter protein, tandem fluorescent-tagged LC3. *Autophagy*, *3*, 452–460.
Kobayashi, J., Shidara, H., Morisawa, Y., Kawakami, M., Tanahashi, Y., Hotta, K., et al. (2013). A method for selective ablation of neurons in *C. elegans* using the phototoxic fluorescent protein, KillerRed. *Neuroscience Letters*, *548*, 261–264.
Lin, J. Y., Sann, S. B., Zhou, K., Nabavi, S., Proulx, C. D., Malinow, R., et al. (2013). Optogenetic inhibition of synaptic release with chromophore-assisted light inactivation (CALI). *Neuron*, *79*, 241–253.
Liu, L., Sakakibara, K., Chen, Q., & Okamoto, K. (2014). Receptor-mediated mitophagy in yeast and mammalian systems. *Cell Research*, *24*, 787–795.
Maejima, I., Takahashi, A., Omori, H., Kimura, T., Takabatake, Y., Saitoh, T., et al. (2013). Autophagy sequesters damaged lysosomes to control lysosomal biogenesis and kidney injury. *The EMBO Journal*, *32*, 2336–2347.
Narendra, D., Tanaka, A., Suen, D. F., & Youle, R. J. (2008). Parkin is recruited selectively to impaired mitochondria and promotes their autophagy. *The Journal of Cell Biology*, *183*, 795–803.
Okamoto, K. (2014). Organellophagy: Eliminating cellular building blocks via selective autophagy. *The Journal of Cell Biology*, *205*, 435–445.
Pickrell, A. M., & Youle, R. J. (2015). The roles of PINK1, parkin, and mitochondrial fidelity in Parkinson's disease. *Neuron*, *85*, 257–273.
Qi, Y. B., Garren, E. J., Shu, X., Tsien, R. Y., & Jin, Y. (2012). Photo-inducible cell ablation in *Caenorhabditis elegans* using the genetically encoded singlet oxygen generating protein miniSOG. *Proceedings of the National Academy of Sciences of the United States of America*, *109*, 7499–7504.

Sato, M., & Sato, K. (2011). Degradation of paternal mitochondria by fertilization-triggered autophagy in *C. elegans* embryos. *Science (New York, N.Y.)*, *334*, 1141–1144.

Settembre, C., Fraldi, A., Medina, D. L., & Ballabio, A. (2013). Signals from the lysosome: A control centre for cellular clearance and energy metabolism. *Nature Reviews. Molecular Cell Biology*, *14*, 283–296.

Sica, V., Galluzzi, L., Bravo-San Pedro, J. M., Izzo, V., Maiuri, M. C., & Kroemer, G. (2015). Organelle-specific initiation of autophagy. *Molecular Cell*, *59*, 522–539.

Sugiura, A., McLelland, G. L., Fon, E. A., & McBride, H. M. (2014). A new pathway for mitochondrial quality control: Mitochondrial-derived vesicles. *The EMBO Journal*, *33*, 2142–2156.

Sumpter, R., Jr., Sirasanagandla, S., Fernandez, A. F., Wei, Y., Dong, X., Franco, L., et al. (2016). Fanconi anemia proteins function in mitophagy and immunity. *Cell*, *165*, 867–881.

Surrey, T., Elowitz, M. B., Wolf, P. E., Yang, F., Nedelec, F., Shokat, K., et al. (1998). Chromophore-assisted light inactivation and self-organization of microtubules and motors. *Proceedings of the National Academy of Sciences of the United States of America*, *95*, 4293–4298.

Tanida, I., Minematsu-Ikeguchi, N., Ueno, T., & Kominami, E. (2005). Lysosomal turnover, but not a cellular level, of endogenous LC3 is a marker for autophagy. *Autophagy*, *1*, 84–91.

Thiele, D. L., & Lipsky, P. E. (1990). Mechanism of L-leucyl-L-leucine methyl ester-mediated killing of cytotoxic lymphocytes: Dependence on a lysosomal thiol protease, dipeptidyl peptidase I, that is enriched in these cells. *Proceedings of the National Academy of Sciences of the United States of America*, *87*, 83–87.

Thurston, T. L., Wandel, M. P., von Muhlinen, N., Foeglein, A., & Randow, F. (2012). Galectin 8 targets damaged vesicles for autophagy to defend cells against bacterial invasion. *Nature*, *482*, 414–418.

Tinari, A., Garofalo, T., Sorice, M., Esposti, M. D., & Malorni, W. (2007). Mitoptosis: Different pathways for mitochondrial execution. *Autophagy*, *3*, 282–284.

Wang, Y., Nartiss, Y., Steipe, B., McQuibban, G. A., & Kim, P. K. (2012). ROS-induced mitochondrial depolarization initiates PARK2/PARKIN-dependent mitochondrial degradation by autophagy. *Autophagy*, *8*, 1462–1476.

Williams, D. C., Bejjani, R. E., Ramirez, P. M., Coakley, S., Kim, S. A., Lee, H., et al. (2013). Rapid and permanent neuronal inactivation in vivo via subcellular generation of reactive oxygen with the use of KillerRed. *Cell Reports*, *5*, 553–563.

Yang, W. Y. (2015). Optogenetic probing of mitochondrial damage responses. *Annals of the New York Academy of Sciences*, *1350*, 48–51.

Yang, J. Y., & Yang, W. Y. (2011). Spatiotemporally controlled initiation of Parkin-mediated mitophagy within single cells. *Autophagy*, *7*, 1230–1238.

Yang, J. Y., & Yang, W. Y. (2013). Bit-by-bit autophagic removal of parkin-labelled mitochondria. *Nature Communications*, *4*, 2428.

Youle, R. J., & Narendra, D. P. (2011). Mechanisms of mitophagy. *Nature Reviews. Molecular Cell Biology*, *12*, 9–14.

Zhong, Z., Umemura, A., Sanchez-Lopez, E., Liang, S., Shalapour, S., Wong, J., et al. (2016). NF-kappaB restricts inflammasome activation via elimination of damaged mitochondria. *Cell*, *164*, 896–910.

Zhou, C., Zhong, W., Zhou, J., Sheng, F., Fang, Z., Wei, Y., et al. (2012). Monitoring autophagic flux by an improved tandem fluorescent-tagged LC3 (mTagRFP-mWasabi-LC3) reveals that high-dose rapamycin impairs autophagic flux in cancer cells. *Autophagy*, *8*, 1215–1226.

CHAPTER FIFTEEN

Kinetics of Protein Aggregates Disposal by Aggrephagy

S. Tan, E. Wong[1]

School of Biological Sciences, Nanyang Technological University, Singapore, Singapore
[1]Corresponding author: e-mail address: estherwong@ntu.edu.sg

Contents

Abstract

Macroautophagy has generated considerable interest because of its ability to recognize and target protein inclusions found prevalent in neurodegenerative diseases to

Methods in Enzymology, Volume 588
ISSN 0076-6879
http://dx.doi.org/10.1016/bs.mie.2016.09.084

lysosomes for elimination. Pharmacological upregulation of macroautophagy to remove disease-linked aggresomes has been reported to slow down disease progression and to improve survivability in many neurodegenerative models. Despite its bulk removal mechanism, we now know that there are selective forms of autophagy. Defect in selective autophagy represents a prominent cause of autophagic failures in many disorders. Aggrephagy refers to the selective degradation of protein aggregates by macroautophagy. In this chapter, we will review the mechanisms and factors that govern the selectivity of protein aggregates for disposal by macroautophagy. This understanding is paramount for us to develop therapeutic approaches to enhance specific cargo targeting in addition to general upregulation of macroautophagy for mitigating disease progression. In addition, we will provide an overview of the different experimental models of protein inclusions, as well as the microscopy and biochemical techniques used to quantitatively analyze the abundance of protein inclusions and the kinetics of their turnovers by macroautophagy.

1. INTRODUCTION

Formation of pathological protein lesions caused by aggregation of mutant or misfolded proteins is a hallmark of many neurodegenerative diseases (Ross & Poirier, 2004; Tan, Wong, & Lim, 2009). Generation of these damaged, misfolded proteins can be attributed to various factors, including (A) mutations, aberrant translation, and posttranslational modifications (PTMs) that disrupt the native conformations of the proteins; (B) deleterious environmental conditions including changes in pH, temperature, and oxidative stress that affect proper protein folding dynamics; and (C) compromised protein quality control systems in the cell including changes in levels of molecular chaperones, and decreasing cellular proteasomal and autophagic functions (Chiti & Dobson, 2006; Dobson, 2003; Meredith, 2005; Ross & Poirier, 2004; Rubinsztein, 2006; Soto, 2003). These faulty proteins often have elevated hydrophobicity, making them aggregation-prone with high propensity to organize into complex structures such as oligomers, protofibrils, small aggregates, and larger inclusion structures with decreasing solubility (Fig. 1). Many studies have pinpointed the soluble oligomeric and protofibrillar intermediates as the primary pathogenic species responsible for cytotoxicity in the disease states (Arrasate, Mitra, Schweitzer, Segal, & Finkbeiner, 2004; Baglioni et al., 2006; Campioni et al., 2010; Caughey & Lansbury, 2003; Cremades et al., 2012; Haass & Selkoe, 2007; Walsh & Selkoe, 2004; Winner et al., 2011), while formation of the insoluble inert protein inclusions is protective.

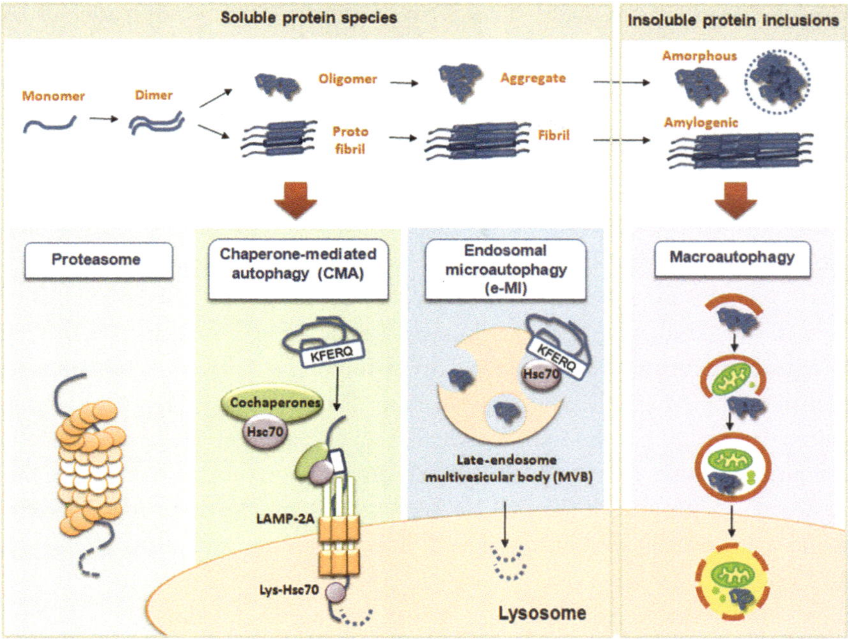

Fig. 1 Divergent roles of proteasome and different autophagy variants in proteostasis. Aggregation-prone polypeptides in the unfolded, monomeric soluble state can dimerize and further aggregate to form oligomer, aggregate and amorphous protein inclusions or β-sheet-enriched protofibril, fibril and amylogenic-ordered protein inclusions. Different types of proteolytic systems are responsible for clearance of these different protein species as indicated.

This is supported by recent evidence that misfolded proteins are actively sequestered into distinct deposition sites (different types of protein inclusions) during proteotoxic stress in an organized manner to differentially engage the proteolytic systems for degradation (Ben-Gedalya et al., 2011; Kaganovich, Kopito, & Frydman, 2008; Miller, Mogk, & Bukau, 2015; Tyedmers, Mogk, & Bukau, 2010). It is likely that these protein inclusions are protective only when they are efficiently removed from the cells by the degradative pathways.

Clear division of labor in maintaining proteostasis is seen among the ubiquitin proteasome system (UPS) and the different types of autophagy. The protein clients of the UPS are mainly short-lived monomeric or dimeric soluble proteins (Cuervo & Wong, 2014; Wong & Cuervo, 2010b). Recently, UPS has been shown to dissolve protein deposits by degrading soluble proteins disaggregated and extracted from the protein inclusions

(Miller, Mogk, et al., 2015). Both chaperone-mediated autophagy (CMA) and endosomal microautophagy (e-MI) degrade soluble pool of proteins containing the "KFERQ"-like targeting motif recognizable by Hsc70 (Fig. 1). In CMA, Hsc70–substrate complex binds to the LAMP-2A receptor in CMA active lysosomes, where the substrate will unfold and translocate into lysosome for degradation (Cuervo & Wong, 2014; Kaushik & Cuervo, 2012). Thus, CMA only handles highly specific soluble proteins, while e-MI has been shown to degrade soluble and small oligomeric proteins targeted to multivesicular body and internalized via vesicle formation (Sahu et al., 2011). Macroautophagy forms isolating membrane to engulf its cargoes in autophagosomes for delivery to the lysosomes for destruction (Mizushima & Komatsu, 2011) (Fig. 1). Thus, only macroautophagy has the bulk removal capacity to target the large, insoluble protein inclusions for degradation. The importance of macroautophagy for turnover of protein inclusions is highlighted in the neurodegenerative pathologies observed in autophagic deficient systems. Ablations of essential macroautophagic genes *atg5* and *atg7* led to motor impairment and accumulation of ubiquitinated cytosolic inclusions in mice (Hara et al., 2006; Komatsu, Wang, et al., 2007), and aberrant macroautophagy are linked to numerous neurodegenerative disorders (Martinez-Vicente et al., 2010; Wang et al., 2009; Wong & Cuervo, 2010a). In contrast, pharmacological activation of autophagy can alleviate aggregate-induced neurotoxicity in neurodegeneration (Rubinsztein, Gestwicki, Murphy, & Klionsky, 2007; Sarkar & Rubinsztein, 2008). Such evidence highlights that autophagy is an important pathway to degrade insoluble protein deposits and to protect the cell from proteotoxicity.

Once thought as a nondiscriminating bulk degradative pathway, macroautophagy pathway is now classified into selective and nonselective types (Khaminets, Behl, & Dikic, 2016; Kraft, Peter, & Hofmann, 2010). While random bulk processing of materials was initially observed under basal and starvation-induced macroautophagy (Kuma et al., 2004), new evidence supports the existence of specific selection of organelles or protein complexes for targeted degradation by macroautophagy even under these conditions. A recent global analysis of basal cellular protein flux indicates that macroautophagy contributes preferentially to the basal turnover of proteasome and CCT/TRiC chaperone complexes while ribosome is protected (Zhang, Shen, Qu, & Ghaemmaghami, 2016). On the other hand, starvation-induced autophagy has been shown to selectively remove organelles like endoplasmic reticulum (ER; Hamasaki, Noda, Baba, & Ohsumi,

2005), ribosomes (Kraft, Deplazes, Sohrmann, & Peter, 2008), and lipid droplets (Singh et al., 2009), as well as specific cytosolic proteins (Kageyama, Suzuki, & Ohsumi, 2009; Onodera & Ohsumi, 2004) and protein inclusions formed from aggregation-prone proteins implicated in neurodegenerative diseases (Wong et al., 2008).

Other cellular stressors have also shown to induce selective autophagy. For example, in response to proteotoxic stress, cell actively transports and sequesters aggregated proteins to transitionary storage compartment (e.g., aggresome (AGM)) to shield the intracellular constituents from their toxicity, as well as to repackage and flag the protein aggregates for selective disposal by macroautophagy (Tan, Wong, Dawson, Dawson, & Lim, 2007; Tan et al., 2009). The selective clearance of protein aggregates by macroautophagy is termed as aggrephagy (*Aggre*gates + Auto*phagy*). In this chapter, we will provide an overview on the topic of protein aggregates clearance by macroautophagy, with particular focus on the different types of protein aggregate sequestration sites and their connections to aggrephagy. Lastly, we will provide a series of methodologies and assays that have been established and widely accepted to study the autophagic clearance of protein inclusions in vivo and in vitro.

2. EXPERIMENTAL MODELS OF PROTEIN INCLUSIONS AND THEIR CHARACTERISTICS

Under normal cellular physiology, protein aggregation is infrequent and rare as quality control mechanisms guided by the molecular chaperones and proteasome help to refold or degrade misfolded and damaged proteins to maintain proteostasis (Hartl, Bracher, & Hayer-Hartl, 2011). From bacteria to human, studies have shown that spatial sequestration of misfolded proteins into subcellular deposition sites is an evolutionary conserved response in cells experiencing proteotoxic stress (Miller, Mogk, et al., 2015). Thus, the phenomenon of protein aggregation is closely studied in the context of cellular stress and aberrant accumulation of pathogenic mutant proteins associated with neurodegenerative diseases. Experimental models of protein inclusions and their susceptibility to aggrephagy can be studied in cultured cells and model organisms through overexpression of disease-related aggregate-prone mutant proteins (Section 2.1) or stress treatments that induce large-scale protein misfolding (Section 2.2).

Recent evidence suggests that cells are able to direct the deposition of misfolded proteins into various spatially distinct cellular compartments according to the type of stressors and the nature of the misfolded protein to better protect against proteotoxicity (Tyedmers, Mogk, et al., 2010). Three major protein deposition sites have been identified in yeast, including IPOD (insoluble protein deposit), INQ (intranuclear quality control compartment), and CytoQ (cytoplasmic quality control compartment) (Fig. 2). Protein sequestration compartments have also been reported in mammalian cells, which include the AGM structures localized perinuclearly (Cyto AGM) or intranuclearly (Nu AGM), and protein aggregates randomly dispersed through the cells (Fig. 2). Although these protein compartments serve to sequester aggregation-prone proteins, it is now clear that these protein inclusions are not uniform and perform different functions in yeast and mammals (Fig. 2) (Miller, Mogk, et al., 2015; Tyedmers, Mogk, et al., 2010; Wong et al., 2008, 2012). Different types of defective proteins, PTM,

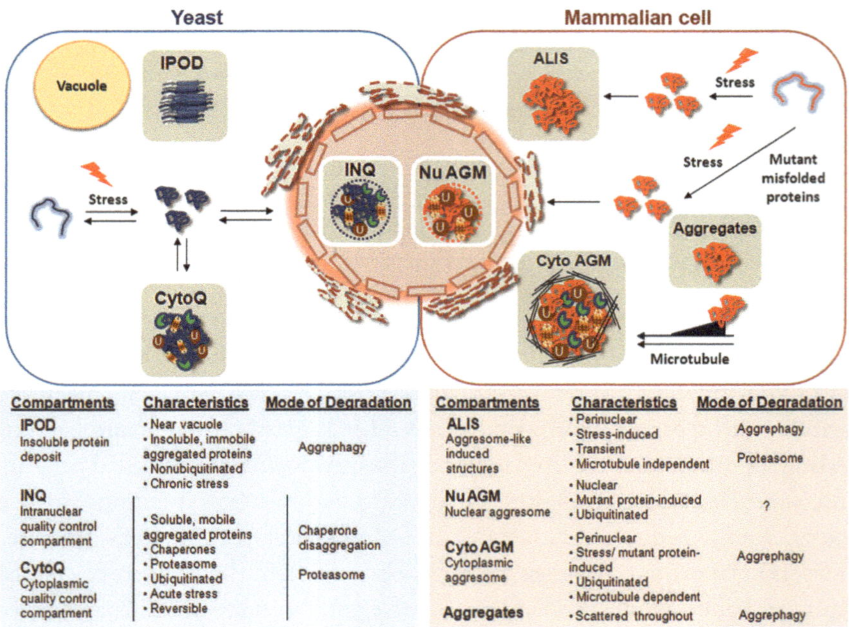

Fig. 2 Different types of inclusion compartments observed in yeast and mammalian cells. In yeast, three distinct types of inclusion bodies have been reported: (1) insoluble protein deposit (IPOD), (2) intranuclear quality control compartment (INQ), and (3) cytoplasmic quality control compartment (CytoQ). In mammals, nuclear and cytosolic AGM as well as AGM-like induced structures (ALIS) have been reported. Aggregates are precursors that can be sequestered into the AGM which arise due to accumulation of misfolded and damaged proteins. The corresponding characteristics and mode of degradation of the different inclusion compartments are summarized here.

and cellular components like chaperones and proteasomes are enriched in each of these compartments. The unique characteristics of each compartment suggest that each of them may serve different purposes and interact differently with the proteolytic pathways (Ben-Gedalya & Cohen, 2012; Miller, Mogk, et al., 2015). In yeast, IPOD harbors mainly terminally misfolded amyloidogenic proteins that are insoluble and do not interact with the environment. In addition, it is free of ubiquitins, proteasomes, and chaperones. Its appearance is associated with chronic stress and overexpression of mutant amyloidogenic proteins (Kaganovich et al., 2008; Miller, Mogk, et al., 2015; Tyedmers, Treusch, et al., 2010). IPOD is found next to the vacuole in yeast and shown to be degraded by adaptor Cue5-dependent aggrephagy (Lu, Psakhye, & Jentsch, 2014). In contrast, INQ and CytoQ are dynamic deposition sites formed during acute stress to sequester soluble aggregated proteins. These structures are enriched with chaperones (disaggregases) and proteasomes that can extract soluble proteins from the aggregates for refolding or proteasomal degradation to dissolve the structure once the condition is favorable (Miller, Ho, et al., 2015; Specht, Miller, Mogk, & Bukau, 2011). Sorting to INQ and CytoQ appears to be mediated by ubiquitination (Kaganovich et al., 2008; Miller, Ho, et al., 2015; Shiber, Breuer, Brandeis, & Ravid, 2013). In yeast, chaperone-assisted proteasome degradation seems to take a central role in the removal of aggregates formed upon acute stress such as heat shock, while selective aggrephagy is more prominent under chronic stress. Biophysical properties of the protein compartments also seem to influence the route of degradation of these inclusions (Miller, Mogk, et al., 2015).

In mammalian cells, the most well-characterized and widely studied inclusion model is the AGM (Johnston, Ward, & Kopito, 1998; Kopito, 2000). A variety of neurodegenerative disease-linked aggregation-prone proteins have been shown to form morphologically similar, perinuclear AGM-like inclusions in vivo and in vitro (Wong et al., 2008). Formation of AGM also represents a major cellular response to proteasomal inhibition where cells activate HDAC6, a microtubule-associated deacetylase, to facilitate retrograde transport of cytosolic protein aggregates along the microtubule network to form the AGM inclusion near microtubule-organizing center (MTOC). Thus, AGM is located close to nuclear envelope and is surrounded by the vimentin cage, which may promote stability of the structure and also to shield the intracellular environment from aberrant interactions with the sequestered protein aggregates (Kopito, 2000). Proteins deposited in AGM are normally ubiquitinated. To confirm the identity of AGM, inclusions can be immunostained for markers like γ-tubulin, ubiquitin, and vimentin (Kopito, 2000). Spatial localization of the AGM is important. Nuclear AGM induced by overexpression of polyglutamine repeat mutant proteins is refractory to macroautophagy degradation while perinuclearly localized cytosolic AGM serves as a staging ground for bulk removal of aggregated proteins by macroautophagy (Iwata et al., 2005). Different types of AGM-like induced structures (ALIS) have also been observed in mammalian cells. ALIS are distinct from AGM as its formation tends to be transient, mainly stress induced and does not depend on transport along the microtubules (Lelouard et al., 2004, 2002). In addition, ALIS are amenable to degradation by both aggrephagy and proteasome.

It is obvious that there is functional diversification among various types of protein inclusions in both yeast and mammalian cells where these aggregate depositories interact differently with the autophagy and proteasomal pathways. Hence, it is helpful to first determine the characteristics of the protein inclusion of interest to have an oversight on potential amenability for autophagic and/or proteasomal removal. In this chapter, we will mainly describe cell-based model systems for studying the removal of disease-associated AGM and stress-induced inclusions in mammalian system by autophagy.

2.1 Disease-Specific Protein Inclusions

We and others have shown that AGM formed upon overexpression of disease-related aggregate-prone proteins in cell culture share striking similarities to clinical protein lesions observed in various proteinopathies (Tan et al., 2008; Wong et al., 2008, 2012). These overexpression systems include transgenic model organisms and transient or stable inducible expression of the mutant proteins as well as their wild-type counterparts in cell culture system. Often, the aggregation-prone proteins are expressed with a fluorescence tag to facilitate the detection and monitoring of their behavior by microscopy. The inducible stable cell line system is particularly useful as it allows de novo protein synthesis to cease and clearance of existing aggregates to be quantified. Examples of neuronal cell lines with inducible expression of human wild-type and mutant α-synuclein implicated in Parkinson's disease as well as huntingtin (HTT) implicated in Huntington's disease (HD) have been well established (Lystad & Simonsen, 2015). For transiently expressed inclusion model, degradation of preformed inclusions can be traced using photoactivatable fluorescence probe. Examples of aggregation-prone

Table 1 Pathological Aggregate-Prone Proteins Susceptible to Autophagy Removal

Disease	Aggregating Protein	Brain Area Affected	Clinical Features
Alzheimer's disease	Tau Amyloid-β peptide	Temporal, parietal lobes	Decline in memory, cognitive function
Frontotemporal dementia	Tau	Frontal, temporal lobes	Decline in cognitive function
Amyloid lateral sclerosis	Superoxide dismutase 1	Motor cortex, brain stem, spinal cord	Muscle weakness, loss of muscle control
Parkinson's disease	α-Synuclein	Midbrain	Loss of muscle control
Dementia with Lewy bodies		Frontal, temporal lobes	Memory, cognitive function decline
Huntington's disease	Huntingtin with PolyQ expansion	Basal ganglia	Decline in cognitive function, loss of muscle control
Spinocerebellar ataxia	Ataxin with PolyQ expansion	Cerebellum	Loss of fine coordination of muscle movement
Spinal and bulbar muscular atrophy	Androgen receptor with PolyQ expansion	Brain stem, spinal cord	Muscle weakness, loss of muscle control

proteins associated with different types of neurodegenerative diseases that are susceptible to autophagic clearance are listed in Table 1. Interestingly, many of these disease-linked proteins also directly perturb the autophagy activity. This is observed in the HD model where mutant HTT interferes with cargo recognition by autophagosome leading to reduced lysosomal degradation (Martinez-Vicente et al., 2010). In this case, intact or even enhanced autophagic flux is accompanied by defective turnover of protein cargoes due to recognition failure. This highlights that functional autophagy activity does not necessarily equate to intact protein clearance. Hence, it is important to check for both the global autophagy activity (i.e., autophagosome or LC3-II flux) and protein cargo receptor turnover (i.e., p62 or NBR1 flux) to access the amenability of disease-linked inclusions to aggrephagy (Loos, du Toit, & Hofmeyr, 2014).

2.2 Stress-Induced Protein Inclusions

Besides managing disease-specific protein inclusions, autophagy also represents a major stress response pathway to handle generalized stress-induced protein inclusions that have more pleomorphic composition and size. Most stressors induce general accumulation of ubiquitinated proteins, and detection of ubiquitin-positive aggregated structures serves as a way to analyze the fate and turnover of protein inclusions by autophagy in stress paradigm. Proteasomal stress is well documented to elicit formation of ubiquitin-positive AGM in most cells to divert the ubiquitinated protein load from proteasome to autophagy for degradation in a HDAC6- and microtubule-dependent manner (Kawaguchi et al., 2003; Tan et al., 2009). Ubiquitinated aggregation-prone proteins are also found in AGM-like structures known as ALIS in several cell types in response to stressors such as oxidative stress, transfection, starvation, and puromycin (Lelouard et al., 2004, 2002). Besides being positive for ubiquitin, ALIS are largely made up of p62 (autophagy receptor) and hence are indistinguishable from p62 bodies which refer to aggregates formed by p62 in response to stressors. ALIS and p62 bodies constitute transient stress-induced aggregates formed independently of microtubule transport and their metabolism is regulated by both proteasome and selective autophagy (Fig. 2). Proteasomal stress-induced AGM can be generated by treating cells with proteasome inhibitors such as MG132 (reversible inhibitor) or lactacystin (irreversible inhibitor), while ALIS/p62 bodies can be generated in cells by exposure to puromycin. Clearance of such stress-induced protein inclusions can be analyzed at various time points after drug wash-out under control basal and autophagic conditions, by immunostaining against ubiquitin or p62 for inclusion number quantification or changes in insoluble species by Western blotting.

3. AGGREPHAGY: SELECTIVE CLEARANCE OF PROTEIN AGGREGATES

Emerging evidence now suggests that the cell copes with different types of protein inclusions by targeting them for clearance by different autophagy pathways (Wong et al., 2012). This specificity in autophagic targeting is observed in both yeast and mammalian systems (Escusa-Toret, Vonk, & Frydman, 2013; Wong et al., 2012). In aggrephagy, selectivity is conferred by the ability of the cell to differentiate between normal cytoplasmic proteins and deleterious aggregated proteins to only target the latter for autophagic disposal. This selectivity requires the presence of molecular determinants and players to mediate recognition of protein inclusions by the autophagy pathway. Here, we will focus on the mammalian system and discuss the machineries underlying the aggrephagy process and at the same time, highlight the key markers and players characterizing each step of the process.

3.1 Different Stages in the Aggrephagy Cascade

Aggrephagy involves the decision to compartmentalize proteins into AGM structures and degradation of the latter via macroautophagy (Fig. 3). Emerging evidence has shown that both processes are intermittently linked, as they share the same regulators. Autophagy cargo receptors (p62 and NBR1), the α-tubulin deacetylase HDAC6, and K63-linked polyubiquitination (K63U) PTM can drive both the AGM assembly and autophagic clearance. The cross talk between the AGM biogenesis pathway and aggrephagy provides timely disposal of the former to prevent accumulation of cytoprotective inclusion compartments to harmful pathological levels observed in late-stage neurodegeneration.

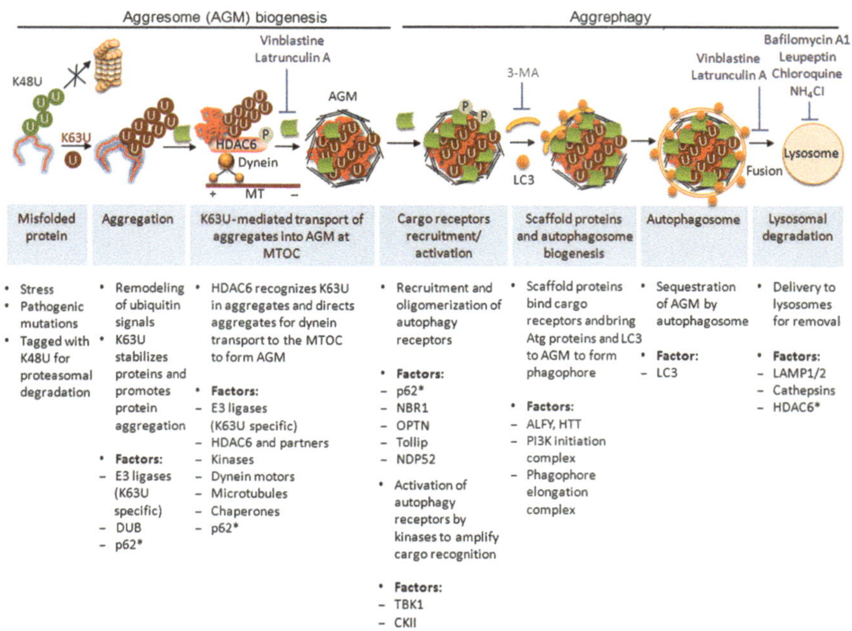

Fig. 3 Aggrephagy cascade. Schematic shows different steps of the aggrephagy process. Key factors modulating each step of the aggresome biogenesis and aggrephagy process are summarized below. *Factors that play multiple roles in the aggresome–aggrephagy pathway.

3.2 AGM Biogenesis

Formation of AGM is an active process in response to deteriorating proteasomal function. This process involves retrograde transportation of ubiquitinated protein aggregates along the microtubules to be sequestered into an inert structure at the MTOC region (Johnston et al., 1998). Two independent pathways have been shown to promote AGM formation. The first is a HDAC6 and ubiquitination-dependent mechanism (Fig. 3), which is more extensively characterized than the second, a BAG3-mediated process. The latter pathway does not depend on modification by ubiquitination. These two proteins, HDAC6 and BAG3, are also important for the downstream fate of AGM to be degraded by classical aggrephagy pathway and chaperone-assisted selective autophagy, respectively.

3.2.1 HDAC6-Mediated AGM Formation

HDAC6 interacts with the ubiquitinated protein aggregates and dynein of the dynein–dynactin motor complex and promotes loading of protein aggregates onto the microtubules for retrograde transport to the MTOC region (Kawaguchi et al., 2003; Kopito, 2000). K63U in protein aggregates serves as a sorting signal for engaging the machineries of both the AGM biogenesis and aggrephagy pathways, which include HDAC6 and cargo receptors like p62 and NBR1 (Kirkin, Lamark, et al., 2009; Olzmann et al., 2007). On the other hand, the K48-linked polyubiquitination (K48U), another degradative signal, preferentially targets misfolded proteins to the proteasome for degradation (Pickart & Fushman, 2004). It appears that the interplay between E3 ligases and deubiquitinating enzyme (DUB) is important for editing ubiquitin codes on protein substrates, to favor one type of degradation machinery over another (Tan et al., 2008, 2009; Yao, 2010). In aggrephagy, E3 ligases Parkin and TRAF6 are implicated in catalyzing K63U on protein aggregates to influence formation and clearance of AGM (Lim, Dawson, & Dawson, 2006; Moscat, Diaz-Meco, & Wooten, 2007). Ataxin-3, an aggregation-associated deubiquitining enzyme, cleaves away K63U in protein aggregates and inhibits AGM formation (Burnett & Pittman, 2005). It was also recently reported that ataxin-3 generates unanchored C-terminal ubiquitin chains from polyubiquitin linkages on protein aggregates which is recognized by HDAC6, suggesting an alternative pathway for selection of ubiquitinated protein aggregates for AGM formation (Ouyang et al., 2012).

HDAC6 is regulated at multiple levels. These include phosphorylation by casein kinase II (CKII) (Watabe & Nakaki, 2011) and protein kinase C (PKC) zeta (Du, Seibenhener, Yan, Jiang, & Wooten, 2015), interaction with cargo receptor protein p62 (Yan et al., 2013), E3 ligase Tripartite motif-containing 50 (TRIM50) (Fusco et al., 2012), and ubiquitin-like modifier FAT10 (Kalveram, Schmidtke, & Groettrup, 2008). Thus, PTM and interacting partners of HDAC6 play a role in modifying its deacetylase activity and thereby influencing AGM formation and clearance. The cargo receptor p62 also helps to promote formation of AGM as a response to autophagy failure to help sequester ubiquitinated proteins into inert inclusion structure (Komatsu, Waguri, et al., 2007). Molecular chaperone p97/VCP is also important for AGM formation in response to proteasomal stress, by delivering ubiquitinated proteins to HDAC6 (Boyault et al., 2006).

3.2.2 BAG3-Mediated AGM Formation

Under proteasomal inhibition, stress-induced BAG3 recognizes Hsp70-bound misfolded proteins and interacts with the dynein motor to transport the Hsp70 clients along microtubules to the MTOC for sequestration into AGM (Gamerdinger, Kaya, Wolfrum, Clement, & Behl, 2011). The interaction between BAG3 and dynein motor requires the presence of 14-3-3 regulatory protein, which can interact with both the BAG3 and dynein intermediate chain. This form of dimeric binding makes 14-3-3 an intermediate adaptor to bridge the Hsp70 associated proteins to the dynein motor complex for trafficking of misfolded proteins (Xu et al., 2013). Unlike the HDAC6-mediated transport, ubiquitination of proteins is not a criterion for substrate recognition and delivery to the MTOC region. This route of trafficking provides an alternative mechanism to sequester non-ubiquitinated, chaperoned misfolded protein substrates into the AGM (Gamerdinger et al., 2011).

3.3 Linking Protein Inclusions to Autophagic Compartments

Selective autophagy depends on special adaptors known as cargo receptor proteins (Zaffagnini & Martens, 2016). These autophagy cargo receptors are themselves autophagic substrates and at the same time, facilitate degradation of protein aggregates by binding and tethering them to the autophagosome membrane (Kirkin, McEwan, Novak, & Dikic, 2009; Pankiv et al., 2007). Cargo receptor proteins contain a short, linear LIR (LC3-interacting region) motif that interacts with LC3. Currently, the consensus sequence is D/E − D − W − x − x − L (x = any amino acid)

(Johansen & Lamark, 2011). Following the initial discovery in receptor protein p62, this motif has been characterized in other proteins, increasing the list of cargo receptors (Fig. 3). Thus far, p62 and NBR1 are the most well-characterized aggrephagy receptors in selective degradation of protein aggregates in mammalian models (Lamark, Kirkin, Dikic, & Johansen, 2009). Both p62 and NBR1 contain an N-terminal PB1, intermediate LIR motifs, and a C-terminal ubiquitin-binding domain (UBA). The UBA domain recognizes and binds to polyubiquitinated protein substrates and targets them to the autophagosome. Phosphorylation can also alter aggregation and clearance propensity of p62- and NBR1-associated proteins. CKII-mediated phosphorylation of the UBA domain of p62 increases its binding affinity and promotes formation and clearance of p62 inclusions (Matsumoto, Wada, Okuno, Kurosawa, & Nukina, 2011). Conversely, GSK3-mediated phosphorylation of NBR1 prevents formation and selective degradation of ubiquitinated protein aggregates (Nicot et al., 2014). Despite the preference for ubiquitinated substrates, autophagic degradation of nonubiquitinated protein substrates by p62 has also been reported and is mediated by the PB1 domain. The PB1 domain in p62 and NBR1 is important for oligomerization, and these two adaptors can associate with each other via the domain. p62 and NBR1 have been shown to act cooperatively to promote aggregation and selective degradation of ubiquitinated protein aggregates (Lamark et al., 2009). By itself, p62 self-oligomerization also promotes selective degradation of p62 by autophagy. In *Drosophila* cells, self-oligomerized p62 is recruited to the ER-associated autophagosome formation site called omegasome (Itakura & Mizushima, 2011), and phagophores may preferentially form at p62 aggregates near lysosomes. In addition, p62 colocalizes with Unc-51-like autophagy activating kinase 1 (ULK1) at the omegasome (Itakura & Mizushima, 2011) and associates with mammalian target of rapamycin complex 1 (mTORC1) (Duran et al., 2011), suggesting a role of p62 in initiating autophagosome membrane formation around protein aggregates by promoting recruitment and assembly of upstream Atg proteins to the initiation site. Thus, p62 functions not only as a bridge between protein aggregates and autophagosome but it also regulates initiation of autophagosome membrane formation around the sequestered aggregates. Phosphorylation also regulates cargo receptor activity where TANK-binding kinase 1 (TBK1)-mediated phosphorylation of p62 and Optineurin (OPTN) amplifies cargo recognition and commits it to degradation by prolonging the residence time of cargo receptors on the surface of cargoes (Herhaus & Dikic, 2015).

3.4 Recruitment of Scaffolding Proteins

Scaffold proteins interact with cargo receptors in protein inclusions to recruit autophagosome initiation and elongation complexes to the inclusions to facilitate formation of autophagosome membrane around the protein cargo to be degraded. Scaffold proteins identified so far in mammalian cells include autophagy-linked FYVE protein (Alfy; Isakson, Holland, & Simonsen, 2013) and HTT proteins (Rui et al., 2015). Alfy contains a PH-BEACH (Pleckstrin homology-Beige and Chediak-Higashi) domain that interacts with p62 (Clausen et al., 2010), five WD40 repeats (beta-transducin repeats) that interact with Atg5 (Filimonenko et al., 2010), and a FYVE domain that binds to phosphatidylinositol 3-phosphate (PI3P) found in autophagic membranes. This triad interaction promotes the recruitment of the Atg5–Atg12–Atg16L complex important for autophagosome biogenesis around the p62 protein cargo (Filimonenko et al., 2010). HTT performs dual roles of acting as a scaffold to promote efficient interactions between p62, LC3, and the K63U signals in protein cargoes for recognition and also regulating the activation of ULK1, a kinase that initiates autophagy (Rui et al., 2015).

3.5 Trafficking of Autophagosomes Along Microtubule Network for Fusion With Lysosomes

Autophagosomes can directly fuse with the lysosomes to form the autophagolysosomes, or can fuse with late endosomes or multivesicular bodies to form amphisomes prior to lysosome fusion (Larsen & Sulzer, 2002). Fusion between autophagosome and lysosome is dependent on dynein–dynactin trafficking along the microtubule network for autophagosome maturation at the perinuclear region where lysosomes are abundant. During proteasomal stress, the formation of autophagolysosome involves remodeling of actin–cortactin network by HDAC6, where assembly of the F-actin network promotes autophagosome–lysosome fusion and clearance of ubiquitinated protein aggregates (Lee et al., 2010). p62 in protein aggregates plays a role in this step by recruiting the F-actin network assemblies to the perinuclear region, ensuring effective autophagic clearance of protein aggregates (Yan et al., 2013). Enrichment and positioning of the lysosomes at the MTOC region is also important. This is achieved by reduced movement of lysosomes along microtubules at the perinuclear region which has been observed upon misfolded protein stress induced by proteasomal failure (Zaarur et al., 2014). Wild-type HTT was also reported to play a role in endosomal and lysosomal trafficking to the perinuclear region by facilitating loading of the organelles to the dynein motor on the microtubules (Caviston, Zajac, Tokito, & Holzbaur, 2011). Tethering of autophagosome membrane with the lysosomal membrane prior to fusion is mediated by late-endosome protein Rab7A and the homotypic fusion and protein sorting (HOPS) complex. Briefly, Rab7A is converted to its active GTP-bound form by HOPS complex to facilitate membrane tethering. Meanwhile, another group of late-endosomal proteins, soluble NSF attachment protein receptor (SNARE) proteins, will increase membrane permeability and induce membrane opening to facilitate fusion (Itakura, Kishi-Itakura, & Mizushima, 2012; Zucchi & Zick, 2011).

3.6 Disassembly of AGM Structure Prior to Lysosomal Degradation

Recent evidence shows that the AGM structure is disassembled prior to degradation by the autophagy pathway. Proteasome-associated deubiquitinase Poh1 cleaves off K63U chains from protein aggregates, releasing unbound and free K63U chains that can activate HDAC6 and promote cortactin-dependent F-actin remodeling important for the formation of autophagolysosomes (Hao et al., 2013). This process requires the dissociation of the 20S proteasome subunit from the AGM, and is mediated by molecular chaperone Hsp90. Inhibition of Hsp90 suppresses 26S proteasome remodeling, unanchored K63U ubiquitin chain production, and AGM clearance (Nanduri, Hao, Fitzpatrick, & Yao, 2015).

4. METHODS TO MEASURE AGGREPHAGY

The study of protein aggregate removal by macroautophagy is most informative in cell-based models using a combination of microscopy and biochemical techniques. Microscopy methods are useful for determining the biophysical properties, cellular localization and morphology of protein inclusions, and their spatial interaction with the machinery of aggrephagy. Microscopy may offer a more sensitive tracking of the abundance and structural changes of protein inclusions in response to various macroautophagy stimulation and inhibition. On the other hand, biochemical analysis gives the advantage of measuring changes in the total pool of insoluble proteins in response to autophagy modulation. The success of biochemical studies relies heavily on the efficacy of solubilization and extraction of aggregated proteins, and may not be a sensitive method for studying aggrephagy in low transfection efficiency models.

5. MODULATION OF AUTOPHAGIC RESPONSES

Often the first hint that a protein inclusion of interest interacts with the macroautophagy pathway is the presence of autophagic components in association with the inclusions. While this is encouraging, it is important to determine the aggrephagy flux (rate of protein inclusion turnover) with respect to macroautophagy stimulation and inhibition through drug treatments and/or genetic manipulations. Commonly used chemical activators of macroautophagy include rapamycin, torin, and lithium chloride. Starvation stress-induced macroautophagy can also be used (Wong et al., 2008, 2012). Macroautophagy inhibition can be achieved either by targeting the early stage or by late stage of the pathway. 3-Methyladenine (3-MA), an inhibitor of phosphatidylinositol 3-kinase (PI3K), inhibits the initiation of autophagosome formation (Fig. 3). However, 3-MA has recently been shown to have opposing effects under certain culture conditions and length of treatment which may be confusing. Alternatively, blockade of macroautophagy can be achieved by inhibiting fusion of autophagosomes with lysosomes using drugs to perturb the cellular cytoskeleton network such as vinblastine and latrunculin A. Since formation of AGM is also dependent on trafficking along the cytoskeleton network, application of these cytoskeleton-disrupting drugs may also perturb AGM biogenesis. Biophysical and intrinsic properties of protein inclusions have shown to influence the mechanism of autophagic disposal (Kaganovich et al., 2008; Wong et al., 2012). Hence, it may be necessary to optimize the drug dosage and length of treatment to capture a window where maturation of autophagosomes is prevented without affecting AGM formation. Lysosomal inhibition with bafilomycin A1 that inhibits vacuolar/lysosomal ATPase, leupeptin that inhibits lysosomal hydrolytic enzymes, and ammonium chloride or chloroquine that neutralizes the intrinsic lysosomal acidic pH can also be employed to study the accumulation of protein inclusions caused by dysfunctional lysosomes (Fig. 3). Recent advancements have identified additional modulators that can activate global macroautophagy or targeting specific steps in selective aggrephagy as seen in Table 2. For genetic controls, ablation of essential autophagy genes such as Beclin 1, Atg5, Atg7, and ULK1 through siRNA/shRNA interference or CRISPR/Cas9 knockout system can be carried out. More and more autophagy proteins have shown to perform

Table 2 List of Recent Modulators of Autophagy Tested in Various Disease Models

Step in Autophagy Cascade	Compounds/Drugs	Target(s)	Positive or Negative Regulation	Disease Model(s)
Autophagy initiation	Rapamycin	mTORC1	–	AD, HD, PD, FTD, SCA Type III, Prion disease
	Lithium	IMPase	–	HD, PD
	Sodium valproate, carbamazepine	Inositol	–	HD, PD
	Rilmenidine, clonidine	α2 adrenoreceptor, I_1 receptor	–	HD
	Trehalose	mTOR-independent signaling	+	HD, PD, ALS, DLB
	Metformin	mTORC1, AMPK	–, +	AD, Taupathies
	AICAR	AMPK	+	Taupathies
Selective sequestration and clearance of protein aggregates	Deprenyl	Nrf2	+	p62-associated diseases
	Trichostatin A	HDAC6	–	HD
	Tubacin			AD(?)
	Mercaptoacetamides			Neuronal cell injury
	Sodium valproate, sodium butyrate, vorinostat	HDAC1	–	AD
Lysosomal degradation	Cystatin B antagonist	Cystatin B	–	LSD
	Valproate lithium	GSK-3β	–	
	Calpain inhibitor	Calpain	–	

nonautophagic functions. Hence, it is important to consider possible autophagy-independent roles of these proteins in the evaluation of aggregates removal. To ascertain that an aggregate-prone protein is degraded by cargo receptor-dependent aggrephagy (Fig. 3), the flux of the aggregates can be checked under conditions of specific autophagy receptor depletion.

6. MICROSCOPIC ANALYSIS OF PROTEIN INCLUSION CLEARANCE BY AGGREPHAGY

6.1 Fluorescence Microscopy

Fluorescence microscopy is a common and useful technique for monitoring aggrephagy. By overexpressing fluorescence-tagged proteins or immunostaining protein aggregates, one can study the association of these inclusions with the autophagic machinery in live and fixed cells, and quantify the protein inclusion levels under autophagic inhibition or induction to measure the turnover. However, it is important to take note of the nature of the inclusions that are being examined when studying aggrephagy. Unfortunately, unless ultrastructural or high-resolution microscopy techniques like transmission electron microscopy (TEM), atomic force microscopy, or X-ray diffraction are employed, protein inclusions are generally only distinguishable by their intracellular localization and size under fluorescence microscope visualization. Nonetheless, the use of 63 × or 100 × magnification in fluorescence microscopy is sufficient to generate both qualitative and quantitative data when studying aggrephagy.

Generally, inclusion bodies observed under fluorescence microscopy can be broadly classified into two different types based on their size and localization: (1) the small cytosolic aggregates or (2) large perinuclear, AGM-like inclusion (Quinlan, Brenner, Goldman, & Messing, 2007; Zaarur, Meriin, Gabai, & Sherman, 2008). However for the latter, it is important to stain the inclusion body for AGM markers like γ-tubulin, ubiquitin, and vimentin to verify the AGM identity (Kopito, 2000). Due to variability in the sizes of protein aggregates that can be formed, it is important to set parameters for sizing of the aggregates in order to classify and standardize the structures that are to be tracked and monitored. For example, PD-linked aggregation-prone protein Synphilin-1 (Sph1) can form cytosolic aggregates as well as perinuclear AGM when overexpressed in cell culture model. The typical size of an AGM formed by Sph1 is 5–6 μm in diameter, and protein aggregates that are smaller than this measurement range will be classified as the cytosolic form (Fig. 4) (Wong et al., 2012). Based on the size and

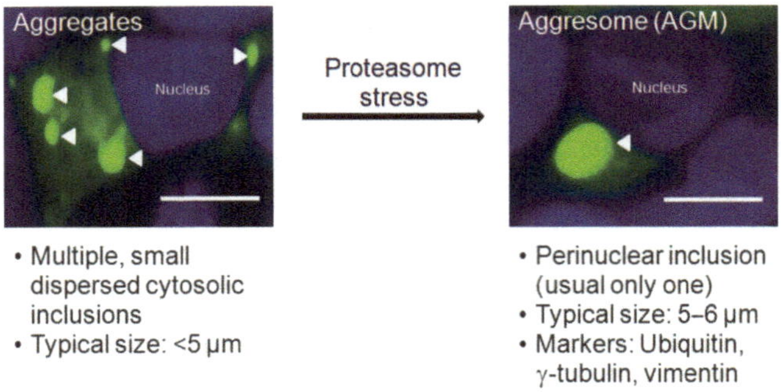

Fig. 4 Fluorescence images of Parkinson's disease-linked GFP-synphilin-1 protein inclusions in neuronal cells. GFP-synphilin-1 (*green*) forms small cytosolic protein aggregates when overexpressed (*left*) basally and large perinuclear AGM inclusion (*right*) in the presence of proteasomal inhibition. These inclusion structures are distinguishable under 63 × magnification using fluorescence microscopy. Characteristics and size ranges of the aggregates and AGM are summarized. The nucleus can be counterstained with DAPI or Hoechst 33342 to confirm the perinuclear localization. To validate the AGM identity, the inclusion body has to localize with AGM markers, including ubiquitin, γ-tubulin, and vimentin. Scale bar is 10 μm.

localization, one can track specific pools of protein aggregates and investigate its autophagic susceptibility.

6.2 Interaction of Inclusion Bodies With Components of the Aggrephagy Machinery

Associations between aggrephagy markers and protein inclusions are indications of susceptibility to autophagy. Recruitments of lysosome and autophagosome to inclusion bodies could be monitored real time in live cells using LysoTracker® and overexpression with GFP- or RFP-tagged aggregate-prone proteins, respectively. In fixed cells, protein inclusions can be immunostained with lysosome marker LAMP1 or autophagosome marker LC3. In this case, a colocalization between the protein inclusion and the autophagic markers would suggest the former is amenable to degradation by autophagy. To pinpoint that the association is a response to selective autophagy rather than nonselective bulk degradation, it is important to establish that the enhanced interaction between the protein inclusion and autophagosomes or lysosomes is also accompanied by higher association with cargo receptor proteins for selective targeting. Colocalization between inclusion bodies and receptors like p62 and NBR1 is an indication for selective autophagy. Under physiological condition, colocalization between protein inclusion and components of the aggrephagy pathway might be low.

If a protein inclusion is predisposed for clearance, activation of autophagy will lead to enhanced presence of autophagic makers in the inclusion. Hence, by comparing the percentages of cells exhibiting colocalization under basal and autophagic induction conditions, it can indicate whether the protein inclusions recruit the autophagic structures during autophagy activation to facilitate removal by aggrephagy.

For quantitative analysis, a colocalization is scored when there is overlap of fluorescence signals between both markers. It is important to adjust the exposure timing of the individual fluorescence channel to ensure that both channel signals are of equal intensity during image acquisition to display a merged image that shares the individual signals uniformly. Overexposure of one channel over the other will discount a positive colocalization, leading to underscoring of colocalization percentage. It is also critical to avoid pixel saturation of both channels to rule out false colocalization. A variety of free online colocalization analysis plugins (e.g., JaCoP, WCIF) are available for use with ImageJ processing software for quantification of colocalization. Briefly, the region of interest is outlined and "thresholding" of fluorescence signals is applied to limit the range of intensity values that are picked up for colocalization analysis. This helps to discriminate the background noise from the actual fluorescence signals to be analyzed. The same thresholding parameters should be apply to all experimental samples to avoid biasness. A comprehensive guide for various statistical approaches commonly employed to analyze colocalization, including Pearson's correlation coefficient, Manders' colocalization or overlap coefficient, and Costes' method are reviewed by Dunn, Kamocka, and McDonald (2011).

6.3 Protein Inclusions Turnover by Aggrephagy

Monitoring changes in the levels of protein inclusions under autophagic induction and/or inhibition will demonstrate the amenability of the inclusion bodies toward aggrephagy turnover. One can either quantify the changes in the number or size of the protein inclusions within each cell or analyze in a population of cells, the changes in the percentage of cells harboring protein inclusions. For the latter study, typically large sample size of more than 100 cells is analyzed. Inhibition of proteasomal activity is a way to artificially induce formation of protein inclusions to assess their turnover (Fig. 5). Varying concentrations of MG132 (2–10 μ*M*), bortezomib (1 n*M*–1 μ*M*), and lactacystin (5 μ*M*) are common proteasome inhibitors used to induce AGM formation.

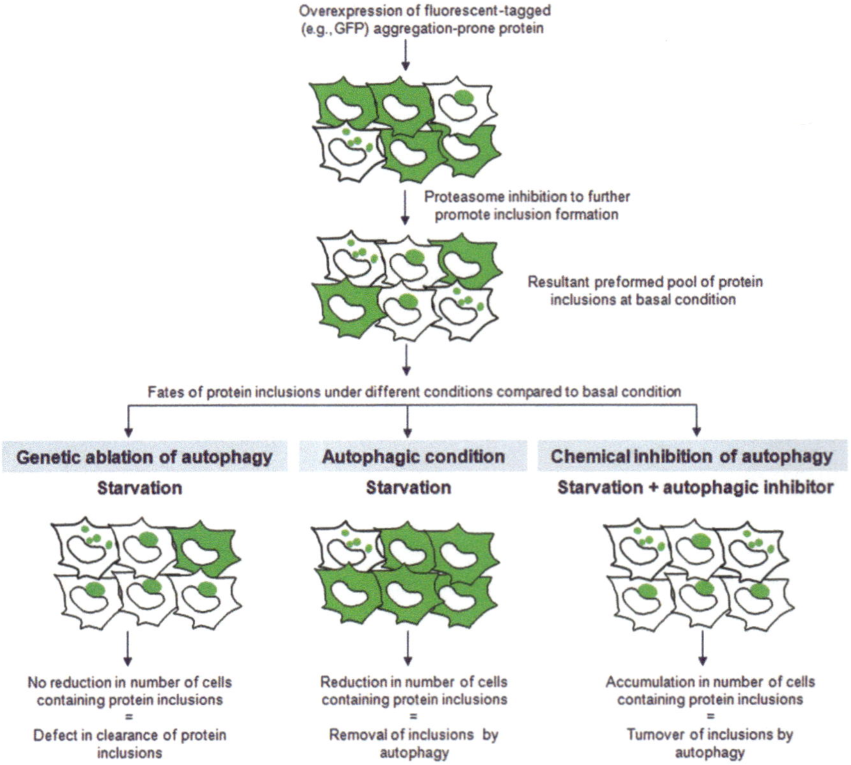

Fig. 5 Schematic outline for monitoring turnover of protein inclusions by aggrephagy and the readouts. Formation of protein inclusions is promoted by proteasome inhibition to give rise to an adequate pool of inclusions to monitor autophagic removal. Starvation (*S*–) reduces the number of cells containing protein inclusions, which indicates the clearance of inclusions by autophagy. Addition of autophagic inhibitor under starvation will lead to accumulation of protein inclusions due to blockage in their autophagic turnover. Ablation of autophagic essential genes and cargo receptor proteins via knockout or knockdown systems will perturb clearance of protein inclusions and indicates selective removal of the inclusions by autophagy.

With the presence of an adequate preexisting pool of protein inclusions, one can monitor changes in the percentage of protein inclusions in response to autophagic activation or inhibition to measure the turnover (if any) by aggrephagy (Fig. 5). Upregulation of autophagy in response to starvation is an adaptive mechanism by the cell to cope with energy demand and has been shown to promote inclusion bodies turnover (Mortimore & Poso, 1987; Wong et al., 2008). If there is aggrephagy flux, autophagic upregulation will reduce the percentage of protein inclusions as compared to untreated control condition, and the reduction represents the protein inclusions turnover by autophagy.

Accumulation of protein inclusions upon autophagy inhibition is another readout for autophagy turnover. The accumulation under autophagy inhibition would represent the pool of inclusion bodies that would have been cleared if the autophagic process is intact (Fig. 5). This method of looking at accumulation of protein inclusions in the presence of autophagy inhibitor is particularly useful for studying basal aggrephagy (Wong et al., 2012).

Sometimes, it is not clear if the observed decline in the levels of protein inclusions is due to a defect in formation, or the clearance of the preexisting pool of inclusion bodies. To bypass this issue, a tetracycline-inducible system can be adopted to temporally regulate the stable expression of the aggregation-prone protein (Yamamoto, Cremona, & Rothman, 2006). Briefly, stress-induced synthesis of proteins can be switched on for a period of time by treating with tetracycline (*tet-on* system), or removal of tetracycline (*tet-off* system). After this, clearance of protein inclusions can be monitored at various time points after antibiotic addition or wash-out. This method allows de novo synthesis to be halted, and the subsequent clearance process to be monitored in the absence of synthesis variations.

Recently, a new optical pulse-chase method that uses photoswitchable Dendra2 fluorescence protein tagged to the aggregation-prone protein of interest has been used to monitor turnover of aggregated protein. Briefly, cells are transfected with Dendra2-fusion protein reporter and the latter is photoactivated by short-wavelength visible light (blue light). This causes Dendra2 to undergo conformational change and switches from green to red fluorescence emission where the signals can be subsequently tracked (Tsvetkov et al., 2013). This method has been used to assess the turnover kinetics of wild-type and mutant HTT proteins by monitoring the decline in red fluorescence intensity overtime and determining the half-life of the HTT proteins. A shorter half-life would indicate accelerated clearance phenomenon (Koga, Martinez-Vicente, Macian, Verkhusha, & Cuervo, 2011; Tsvetkov et al., 2013).

To show that macroautophagy is the pathway responsible for any clearance profiles observed, it is important to determine the readouts under conditions where autophagy genes are silenced to genetically ablate autophagic functions. This also serves to eliminate potential off-target effects of drugs used to chemically inhibit the autophagy pathways (Wu et al., 2010). Genetic ablation of key components of the autophagic machinery like

Beclin 1, Atg5, Atg7, or ULK1 using knockout or siRNA/shRNA-mediated knockdown in cell or animal models can be used to create an autophagy-deficient environment. Perturbed turnover of protein aggregates under macroautophagy inhibition will suggest that the clearance is a macroautophagy-dependent process. To pinpoint that protein clearance is due to selective targeting of substrates to macroautophagy and not due to bulk degradation, it is essential to perform experiments in cargo receptor-deficient conditions via genetic knockout. Perturbation in turnover of protein inclusions under this condition will indicate presence of selectivity in targeting the inclusions for clearance (Fig. 5).

6.4 TEM and Correlative Light Electron Microscopy

TEM is routinely employed to visualize autophagic structures in mammalian cells and tissue samples (Martinez-Vicente et al., 2010; Yla-Anttila, Vihinen, Jokitalo, & Eskelinen, 2009). TEM analysis of autophagic components has been covered in other chapter, hence not discussed here. Correlative Light Electron Microscopy (CLEM) is an advanced electron microscopy technique that allows specific visualization of the ultrastructural details of the cellular environment directly surrounding a fluorescent-tagged protein inclusion identified through fluorescence microscopy (Wong et al., 2012). Enrichment of autophagic vacuoles and lysosomes in the vicinity of the protein inclusion will indicate enhanced recruitment of the autophagosome–lysosomal machinery and autophagic susceptibility of the protein inclusion (Hanson et al., 2010). Despite TEM being a powerful imaging to observe autophagic structures at high-resolution nanometer scale, it is limited to visualization of fixed samples. In view of this limitation, super resolution techniques like photoactivation localization microscopy (PALM) and stochastic optical reconstruction microscopy (STORM) have allowed applications in live cells to provide information on protein structures as well as imaging of protein aggregation at increased resolution compared to TEM (Huang, Babcock, & Zhuang, 2010).

6.5 Fluorescence Recovery After Photobleaching

The biophysical property of a protein inclusion may play a role in influencing its ability to engage the autophagy components (e.g., cargo receptors, autophagosomes, and lysosomes) for aggrephagy. In addition, kinetics of these autophagy components can also influence the amenability of the inclusion bodies toward autophagic clearance (Wang, Chen, Yang, & Zhang,

2013; Wong et al., 2012). Using fluorescence recovery after photobleaching (FRAP) analysis, a technique that involves photobleaching a proportion of protein pool and examining the recovery or migration of fluorescence signals back into the bleached region, the propensity for a protein inclusion to recruit and retain components of the aggrephagy pathway can be examined as an indication of its autophagic susceptibility.

To quantify the molecular dynamics of protein inclusions or autophagic components associated with the inclusion bodies, analysis of the FRAP recovery curve is helpful by plotting the fluorescence intensity of the protein species-of-interest in the photobleached region (*x*-axis) over time (*y*-axis). From the recovery curve, the proportion of mobile and immobile protein pools in the bleached region can be measured by determining the plateau level of the fluorescence intensity. By setting the fluorescence intensity at time = 0 (prebleached) as 100%, the percentage mobile fraction is represented by the level of FRAP recovery at the plateau (Fig. 6). Additionally, the

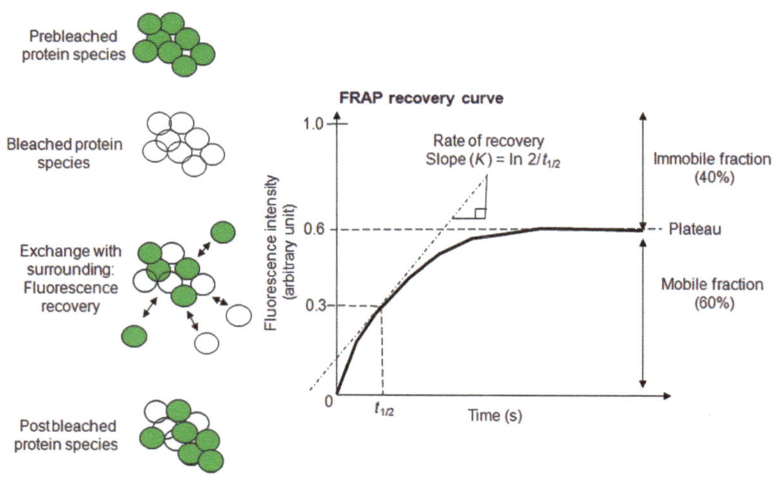

Fig. 6 Interpretation of FRAP recovery curve. A FRAP recovery curve provides useful information regarding the molecular dynamics of the protein species. A region in the pool of proteins is irradiated and photobleached, and the fluorescence signal in the region (recovery of signal due to exchange of protein species with the surrounding) is monitored over time. To plot the recovery curve, the fluorescence intensity of the prebleached protein species is set to arbitrary unit 1, and the fluorescence intensity following bleaching is recorded at different time intervals. In aggrephagy, the percentage of mobile protein species can be interpreted from the recovery curve by determining the fraction of fluorescence intensity recovered at the plateau. The gradient at the half-life ($t_{1/2}$) of fluorescence recovery will give a quantitative value to the kinetics of the proteins' mobility.

rate of exchange or recovery (association and dissociation kinetics) between the mobile protein species and surrounding could be determined by measuring the gradient of the slope ($K = \text{In } 2/t_{1/2}$) of the curve at the half-life ($t_{1/2}$) of fluorescence recovery (Fig. 6).

Generally, low mobile protein fraction and slow recovery of aggregate-prone protein species in the inclusion body after photobleaching (low K value) is indicative of a stable inclusion structure and has been associated with enhanced autophagic susceptibility (Scotter et al., 2014; Wong et al., 2012). Reduced mobility of protein species in AGM has been shown to provide a stable platform for robust assembly of the autophagic components on the surface of the inclusion to facilitate clearance (Wong et al., 2012). FRAP approach has also been used to access the dynamic properties of LC3, p62, and ubiquitin in protein aggregates (Lelouard et al., 2004; Matsumoto et al., 2011; Wang et al., 2013). Generally, high mobility and rapid association and dissociation kinetics (high K value) of autophagic components on the inclusion body suggests transient interactions between the autophagic components and protein inclusions. On the other hand, a slow recovery of the autophagic components on the protein inclusions would most likely indicate formation of a stable complex between the components and inclusion that may favor autophagic clearance.

7. BIOCHEMICAL ANALYSIS OF PROTEIN INCLUSION TURNOVER BY AGGREPHAGY

7.1 Sequential Extraction of Soluble and Insoluble Proteins for Sodium Dodecyl Sulfate Polyacrylamide Gel Electrophoresis

Turnover of protein inclusions by autophagy can also be monitored biochemically using sodium dodecyl sulfate polyacrylamide gel electrophoresis (SDS-PAGE) and Western blotting to examine changes in the levels of insoluble proteins upon autophagy activation or inhibition. This technique provides a nonbias analysis to complement protein turnover analysis by microscopy techniques. Differential detergent extraction method can be used to separate out the soluble and the insoluble protein fractions for analysis of changes in each fraction. Soluble proteins are extracted under mild detergent conditions (e.g., NP-40, Triton X-100), while the insoluble protein aggregates can only be extracted under harsh detergent conditions (e.g., 1% SDS; Fig. 7).

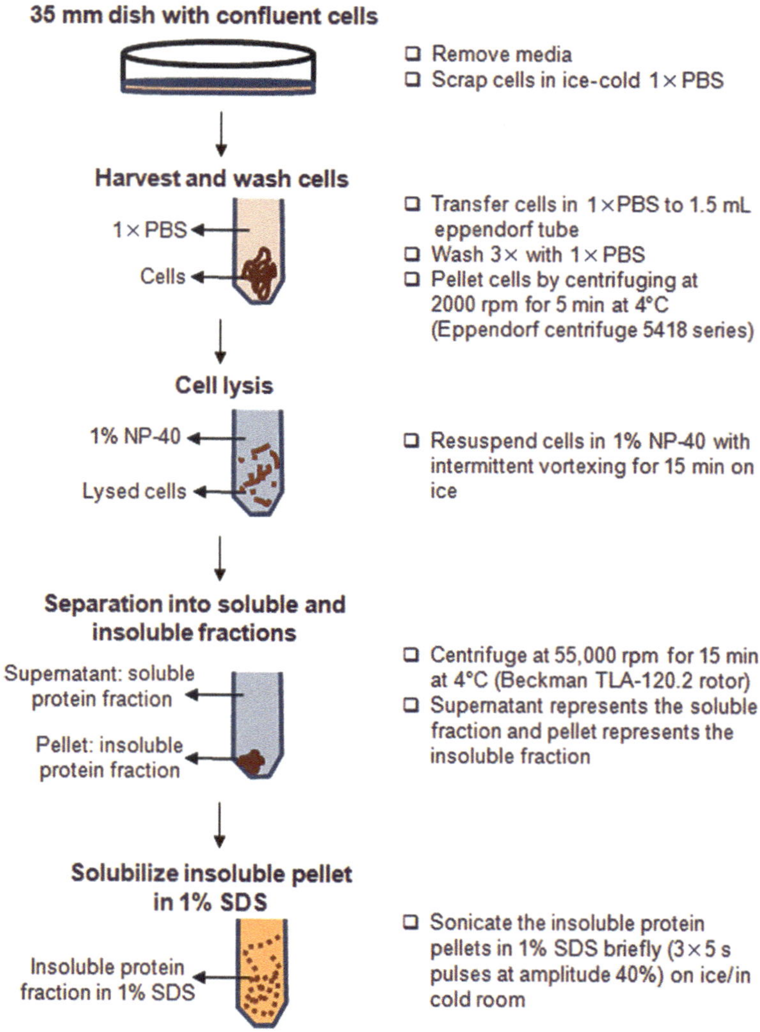

Fig. 7 Schematic outline of the sequential extraction of soluble and insoluble protein fractions. The stepwise protocol for sequential protein extractions to separate the soluble and insoluble protein fractions of the aggregation-prone proteins is illustrated here.

For aggrephagy study, it is relevant to monitor changes in the levels of the aggregate-prone proteins in the insoluble protein fractions by analyzing the intensities of the respective protein bands in Western blots using ImageJ software. Briefly, scans of the protein blots are converted into 8-bit images and then the background is subtracted ("Process" > "Subtract background"). The protein band is then outlined and the "Relative Intensity" value is determined (Measure). Fig. 8 illustrates the correlation between changes in the insoluble protein fractions under various autophagic conditions and the susceptibility of the protein inclusions toward autophagic turnover. Briefly, a reduction in the level of aggregate-prone protein in the insoluble protein fraction (lower protein band intensity compared to basal control) upon autophagy activation (e.g., starvation (*S*−) for >12 h) suggests that the aggregate-prone proteins are cleared by autophagy (Fig. 8). The extent of the accumulation of the aggregate-prone proteins in the insoluble fraction (higher protein band intensity compared to starvation (*S*−)) upon autophagy activation in the presence of autophagy inhibitory drugs (e.g., 10 m*M* 3-MA treatment for 12 h) is indicative of the magnitude of the aggrephagy flux (Fig. 8). The higher the accumulation, the greater the aggrephagy flux, suggesting faster turnover of the aggregate-prone proteins by the autophagy pathway (see inlet in Fig. 8).

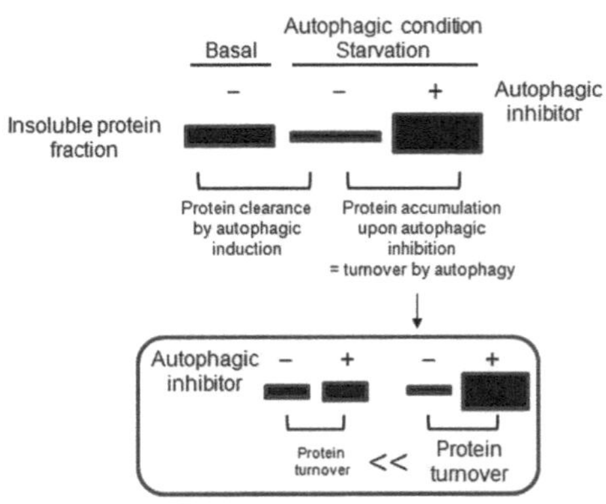

Fig. 8 Schematic showing changes in the insoluble protein fraction of an aggregate-prone protein under different autophagic conditions by Western blotting and the associated interpretations. The insoluble protein fraction is of interest when it comes to monitoring aggrephagy. Reduction in intensity of the insoluble protein band under starvation as compared to basal condition suggests removal of aggregation-prone protein by autophagy. To accurately pinpoint that the reduction is due to clearance by autophagy, it is important to assess the band intensity of the fraction under autophagy inhibition (e.g., 3-MA). An increase in the insoluble protein band intensity upon addition of 3-MA under starvation represents accumulation of the fraction of aggregate-prone protein targeted for turnover by autophagy. Inlet: The greater the accumulation of the insoluble protein fraction in the presence of autophagic inhibition, the faster the rate of aggregate-prone protein turnover by autophagy.

7.1.1 Protocol for Differential Detergent Extraction of Soluble and Insoluble Protein Fractions for a Single 35 mm Culture Dish (Fig. 7)

1. Prepare mild lysis buffer containing 1% NP-40 in 1 × PBS, and harsh lysis buffer containing 1% SDS in 1 × PBS buffer*. Both are supplemented with protease and phosphatase inhibitors.
2. Remove culture media and scrape cells in ice-cold 1 × PBS and transfer to eppendorf tubes. Wash three times with 1 × PBS by centrifugation at 2000 rpm** for 5 min at 4°C.
3. Add 100–150 μL of 1% NP-40 lysis buffer to resuspend the cell pellet and incubate for 20 min on ice with intermittent vortexing during lysis.
4. Transfer the lysed cells to high-speed centrifuge tubes, and centrifuge at 55,000 rpm*** for 15 min at 4°C to separate the total protein lysate into soluble (supernatant) and the insoluble (pellet) fractions.
5. Transfer the supernatant into a new eppendorf tube and store the soluble fraction at −20°C until further use.
6. Add 100–150 μL of 1% SDS lysis buffer to the remaining cell pellet, and sonicated briefly (three times of 5 s pulses at amplitude 40%) until pellet is no longer seen. This fraction represents the insoluble fraction, and is to be stored at room temperature* until further use.
7. Optional: If pellet particles are still seen after sonication, the debris can be removed by centrifuging at 55,000 rpm*** for 5 min.

*Note 1% SDS precipitates on ice and should be kept at room temperature.
**Centrifugation is done with Eppendorf Centrifuge 5418 series.
***Centrifugation is done with Beckman High Speed Ultracentrifuge using TLA-120.2 rotor.

7.2 Filter Trap Retardation Assay

The filter-trap assay is an alternative method to separate soluble and insoluble protein fractions to quantify the levels of protein aggregates. This method is based on the principle that SDS-insoluble mutant HTT protein inclusions are retained on a cellulose acetate filter but not the cleaved forms of the SDS-soluble HTT (Wanker et al., 1999). By immunoblotting the filter paper with antibody against the protein of interest, the levels of SDS-insoluble protein inclusions can be determined. Since then, this technique has been used to quantify levels of other neurodegenerative disorder-linked aggregate-prone proteins in tissues, including Aβ amyloid, tau, and α-synuclein (Xu, Gonzales, & Borchelt, 2002). A comprehensive protocol for the filter-trap assay has been reviewed by Lystad and Simonsen (2015).

8. CONCLUDING REMARKS

Protein aggregation underlies the pathogenesis of neurodegenerative disorders and upregulation of selective aggrephagy is a rational therapeutic strategy to slow down the progression of the diseases. Currently, small-molecule HDAC modulators in preclinical trials have shown beneficial effects in protection against neurological disorders, making modulation of HDAC6 activity an appealing therapeutic solution (Dallavalle, Pisano, & Zunino, 2012; Li, Jiang, Chang, Xie, & Hu, 2011). However, HDAC6 plays multiple roles in the selective aggrephagy cascade, and caution has to be exercised when attempting to modulate HDAC6 deacetylase activity as a solution for protein aggregation-associated diseases. In light of this, the search for other therapeutic targets is important. Recent advancement in super resolution microscopy techniques will allow more precise monitoring of the protein aggregation and autophagy processes which will facilitate greater in-depth elucidation and dissection of the molecular mechanisms involved in various steps of aggrephagy. This information will be useful for development of more suitable therapeutic targets for management of protein aggregate stress in proteinopathies.

ACKNOWLEDGMENTS

Work in our laboratory is supported by Grants from MOE Tier 2 M4020161.080 (ARC 25/13), MOE Tier 1 M4011565.080 (RG139/15), and SUG M4080753.080.

REFERENCES

Arrasate, M., Mitra, S., Schweitzer, E. S., Segal, M. R., & Finkbeiner, S. (2004). Inclusion body formation reduces levels of mutant huntingtin and the risk of neuronal death. *Nature*, *431*(7010), 805–810. http://dx.doi.org/10.1038/nature02998.

Baglioni, S., Casamenti, F., Bucciantini, M., Luheshi, L. M., Taddei, N., Chiti, F., … Stefani, M. (2006). Prefibrillar amyloid aggregates could be generic toxins in higher organisms. *The Journal of Neuroscience*, *26*(31), 8160–8167. http://dx.doi.org/10.1523/jneurosci.4809-05.2006.

Ben-Gedalya, T., & Cohen, E. (2012). Quality control compartments coming of age. *Traffic*, *13*(5), 635–642. http://dx.doi.org/10.1111/j.1600-0854.2012.01330.x.

Ben-Gedalya, T., Lyakhovetsky, R., Yedidia, Y., Bejerano-Sagie, M., Kogan, N. M., Karpuj, M. V., … Cohen, E. (2011). Cyclosporin-A-induced prion protein aggresomes are dynamic quality-control cellular compartments. *Journal of Cell Science*, *124*(Pt. 11), 1891–1902. http://dx.doi.org/10.1242/jcs.077693.

Boyault, C., Gilquin, B., Zhang, Y., Rybin, V., Garman, E., Meyer-Klaucke, W., … Khochbin, S. (2006). HDAC6-p97/VCP controlled polyubiquitin chain turnover. *The EMBO Journal*, *25*(14), 3357–3366. http://dx.doi.org/10.1038/sj.emboj.7601210.

Burnett, B. G., & Pittman, R. N. (2005). The polyglutamine neurodegenerative protein ataxin 3 regulates aggresome formation. *Proceedings of the National Academy of Sciences of the United States of America*, *102*(12), 4330–4335. http://dx.doi.org/10.1073/pnas.0407252102.

Campioni, S., Mannini, B., Zampagni, M., Pensalfini, A., Parrini, C., Evangelisti, E., ... Chiti, F. (2010). A causative link between the structure of aberrant protein oligomers and their toxicity. *Nature Chemical Biology*, *6*(2), 140–147. http://dx.doi.org/10.1038/nchembio.283.

Caughey, B., & Lansbury, P. T. (2003). Protofibrils, pores, fibrils, and neurodegeneration: Separating the responsible protein aggregates from the innocent bystanders. *Annual Review of Neuroscience*, *26*, 267–298. http://dx.doi.org/10.1146/annurev.neuro.26.010302.081142.

Caviston, J. P., Zajac, A. L., Tokito, M., & Holzbaur, E. L. (2011). Huntingtin coordinates the dynein-mediated dynamic positioning of endosomes and lysosomes. *Molecular Biology of the Cell*, *22*(4), 478–492. http://dx.doi.org/10.1091/mbc.E10-03-0233.

Chiti, F., & Dobson, C. M. (2006). Protein misfolding, functional amyloid, and human disease. *Annual Review of Biochemistry*, *75*, 333–366. http://dx.doi.org/10.1146/annurev.biochem.75.101304.123901.

Clausen, T. H., Lamark, T., Isakson, P., Finley, K., Larsen, K. B., Brech, A., ... Johansen, T. (2010). p62/SQSTM1 and ALFY interact to facilitate the formation of p62 bodies/ALIS and their degradation by autophagy. *Autophagy*, *6*(3), 330–344.

Cremades, N., Cohen, S. I., Deas, E., Abramov, A. Y., Chen, A. Y., Orte, A., ... Klenerman, D. (2012). Direct observation of the interconversion of normal and toxic forms of alpha-synuclein. *Cell*, *149*(5), 1048–1059. http://dx.doi.org/10.1016/j.cell.2012.03.037.

Cuervo, A. M., & Wong, E. (2014). Chaperone-mediated autophagy: Roles in disease and aging. *Cell Research*, *24*, 92–104.

Dallavalle, S., Pisano, C., & Zunino, F. (2012). Development and therapeutic impact of HDAC6-selective inhibitors. *Biochemical Pharmacology*, *84*(6), 756–765. http://dx.doi.org/10.1016/j.bcp.2012.06.014.

Dobson, C. M. (2003). Protein folding and misfolding. *Nature*, *426*(6968), 884–890. http://dx.doi.org/10.1038/nature02261.

Du, Y., Seibenhener, M. L., Yan, J., Jiang, J., & Wooten, M. C. (2015). aPKC phosphorylation of HDAC6 results in increased deacetylation activity. *PLoS One*, *10*(4), e0123191. http://dx.doi.org/10.1371/journal.pone.0123191.

Dunn, K. W., Kamocka, M. M., & McDonald, J. H. (2011). A practical guide to evaluating colocalization in biological microscopy. *American Journal of Physiology Cell Physiology*, *300*(4), C723–C742. http://dx.doi.org/10.1152/ajpcell.00462.2010.

Duran, A., Amanchy, R., Linares, J. F., Joshi, J., Abu-Baker, S., Porollo, A., ... Diaz-Meco, M. T. (2011). p62 is a key regulator of nutrient sensing in the mTORC1 pathway. *Molecular Cell*, *44*(1), 134–146. http://dx.doi.org/10.1016/j.molcel.2011.06.038.

Escusa-Toret, S., Vonk, W. I., & Frydman, J. (2013). Spatial sequestration of misfolded proteins by a dynamic chaperone pathway enhances cellular fitness during stress. *Nature Cell Biology*, *15*(10), 1231–1243. http://dx.doi.org/10.1038/ncb2838.

Filimonenko, M., Isakson, P., Finley, K. D., Anderson, M., Jeong, H., Melia, T. J., ... Yamamoto, A. (2010). The selective macroautophagic degradation of aggregated proteins requires the PI3P-binding protein Alfy. *Molecular Cell*, *38*(2), 265–279. http://dx.doi.org/10.1016/j.molcel.2010.04.007.

Fusco, C., Micale, L., Egorov, M., Monti, M., D'Addetta, E. V., Augello, B., ... Merla, G. (2012). The E3-ubiquitin ligase TRIM50 interacts with HDAC6 and p62, and promotes the sequestration and clearance of ubiquitinated proteins into the aggresome. *PLoS One*, *7*(7), e40440. http://dx.doi.org/10.1371/journal.pone.0040440.

Gamerdinger, M., Kaya, A. M., Wolfrum, U., Clement, A. M., & Behl, C. (2011). BAG3 mediates chaperone-based aggresome-targeting and selective autophagy of misfolded proteins. *EMBO Reports*, *12*(2), 149–156. http://dx.doi.org/10.1038/embor.2010.203.

Haass, C., & Selkoe, D. J. (2007). Soluble protein oligomers in neurodegeneration: Lessons from the Alzheimer's amyloid beta-peptide. *Nature Reviews. Molecular Cell Biology*, *8*(2), 101–112. http://dx.doi.org/10.1038/nrm2101.

Hamasaki, M., Noda, T., Baba, M., & Ohsumi, Y. (2005). Starvation triggers the delivery of the endoplasmic reticulum to the vacuole via autophagy in yeast. *Traffic*, *6*(1), 56–65. http://dx.doi.org/10.1111/j.1600-0854.2004.00245.x.

Hanson, H. H., Kang, S., Fernandez-Monreal, M., Oung, T., Yildirim, M., Lee, R., … Phillips, G. R. (2010). LC3-dependent intracellular membrane tubules induced by gamma-protocadherins A3 and B2: A role for intraluminal interactions. *The Journal of Biological Chemistry*, *285*(27), 20982–20992. http://dx.doi.org/10.1074/jbc.M109.092031.

Hao, R., Nanduri, P., Rao, Y., Panichelli, R. S., Ito, A., Yoshida, M., & Yao, T. P. (2013). Proteasomes activate aggresome disassembly and clearance by producing unanchored ubiquitin chains. *Molecular Cell*, *51*(6), 819–828. http://dx.doi.org/10.1016/j.molcel.2013.08.016.

Hara, T., Nakamura, K., Matsui, M., Yamamoto, A., Nakahara, Y., Suzuki-Migishima, R., … Mizushima, N. (2006). Suppression of basal autophagy in neural cells causes neurodegenerative disease in mice. *Nature*, *441*(7095), 885–889. http://dx.doi.org/10.1038/nature04724.

Hartl, F. U., Bracher, A., & Hayer-Hartl, M. (2011). Molecular chaperones in protein folding and proteostasis. *Nature*, *475*(7356), 324–332. http://dx.doi.org/10.1038/nature10317.

Herhaus, L., & Dikic, I. (2015). Expanding the ubiquitin code through post-translational modification. *EMBO Reports*, *16*(9), 1071–1083. http://dx.doi.org/10.15252/embr.201540891.

Huang, B., Babcock, H., & Zhuang, X. (2010). Breaking the diffraction barrier: Super-resolution imaging of cells. *Cell*, *143*(7), 1047–1058. http://dx.doi.org/10.1016/j.cell.2010.12.002.

Isakson, P., Holland, P., & Simonsen, A. (2013). The role of ALFY in selective autophagy. *Cell Death and Differentiation*, *20*(1), 12–20. http://dx.doi.org/10.1038/cdd.2012.66.

Itakura, E., Kishi-Itakura, C., & Mizushima, N. (2012). The hairpin-type tail-anchored SNARE syntaxin 17 targets to autophagosomes for fusion with endosomes/lysosomes. *Cell*, *151*(6), 1256–1269. http://dx.doi.org/10.1016/j.cell.2012.11.001.

Itakura, E., & Mizushima, N. (2011). p62 Targeting to the autophagosome formation site requires self-oligomerization but not LC3 binding. *The Journal of Cell Biology*, *192*(1), 17–27. http://dx.doi.org/10.1083/jcb.201009067.

Iwata, A., Christianson, J. C., Bucci, M., Ellerby, L. M., Nukina, N., Forno, L. S., & Kopito, R. R. (2005). Increased susceptibility of cytoplasmic over nuclear polyglutamine aggregates to autophagic degradation. *Proceedings of the National Academy of Sciences of the United States of America*, *102*(37), 13135–13140. http://dx.doi.org/10.1073/pnas.0505801102. 0505801102 [pii].

Johansen, T., & Lamark, T. (2011). Selective autophagy mediated by autophagic adapter proteins. *Autophagy*, 7(3), 279–296.

Johnston, J. A., Ward, C. L., & Kopito, R. R. (1998). Aggresomes: A cellular response to misfolded proteins. *The Journal of Cell Biology*, *143*(7), 1883–1898.

Kaganovich, D., Kopito, R., & Frydman, J. (2008). Misfolded proteins partition between two distinct quality control compartments. *Nature*, *454*(7208), 1088–1095. http://dx.doi.org/10.1038/nature07195.

Kageyama, T., Suzuki, K., & Ohsumi, Y. (2009). Lap3 is a selective target of autophagy in yeast, Saccharomyces cerevisiae. *Biochemical and Biophysical Research Communications*, *378*(3), 551–557. http://dx.doi.org/10.1016/j.bbrc.2008.11.084.

Kalveram, B., Schmidtke, G., & Groettrup, M. (2008). The ubiquitin-like modifier FAT10 interacts with HDAC6 and localizes to aggresomes under proteasome inhibition. *Journal of Cell Science*, *121*(Pt. 24), 4079–4088. http://dx.doi.org/10.1242/jcs.035006.

Kaushik, S., & Cuervo, A. M. (2012). Chaperone-mediated autophagy: A unique way to enter the lysosome world. *Trends in Cell Biology*, *22*(8), 407–417. http://dx.doi.org/10.1016/j.tcb.2012.05.006.

Kawaguchi, Y., Kovacs, J. J., McLaurin, A., Vance, J. M., Ito, A., & Yao, T. P. (2003). The deacetylase HDAC6 regulates aggresome formation and cell viability in response to misfolded protein stress. *Cell*, *115*(6), 727–738.

Khaminets, A., Behl, C., & Dikic, I. (2016). Ubiquitin-dependent and independent signals in selective autophagy. *Trends in Cell Biology*, *26*(1), 6–16. http://dx.doi.org/10.1016/j.tcb.2015.08.010.

Kirkin, V., Lamark, T., Sou, Y. S., Bjorkoy, G., Nunn, J. L., Bruun, J. A., … Johansen, T. (2009). A role for NBR1 in autophagosomal degradation of ubiquitinated substrates. *Molecular Cell*, *33*(4), 505–516. http://dx.doi.org/10.1016/j.molcel.2009.01.020.

Kirkin, V., McEwan, D. G., Novak, I., & Dikic, I. (2009). A role for ubiquitin in selective autophagy. *Molecular Cell*, *34*(3), 259–269. http://dx.doi.org/10.1016/j.molcel.2009.04.026.

Koga, H., Martinez-Vicente, M., Macian, F., Verkhusha, V. V., & Cuervo, A. M. (2011). A photoconvertible fluorescent reporter to track chaperone-mediated autophagy. *Nature Communications*, *2*(386). http://dx.doi.org/10.1038/ncomms1393.

Komatsu, M., Waguri, S., Koike, M., Sou, Y. S., Ueno, T., Hara, T., … Tanaka, K. (2007). Homeostatic levels of p62 control cytoplasmic inclusion body formation in autophagy-deficient mice. *Cell*, *131*(6), 1149–1163. http://dx.doi.org/10.1016/j.cell.2007.10.035.

Komatsu, M., Wang, Q. J., Holstein, G. R., Friedrich, V. L., Jr., Iwata, J., Kominami, E., … Yue, Z. (2007). Essential role for autophagy protein Atg7 in the maintenance of axonal homeostasis and the prevention of axonal degeneration. *Proceedings of the National Academy of Sciences of the United States of America*, *104*(36), 14489–14494. http://dx.doi.org/10.1073/pnas.0701311104.

Kopito, R. R. (2000). Aggresomes, inclusion bodies and protein aggregation. *Trends in Cell Biology*, *10*(12), 524–530.

Kraft, C., Deplazes, A., Sohrmann, M., & Peter, M. (2008). Mature ribosomes are selectively degraded upon starvation by an autophagy pathway requiring the Ubp3p/Bre5p ubiquitin protease. *Nature Cell Biology*, *10*(5), 602–610. http://dx.doi.org/10.1038/ncb1723.

Kraft, C., Peter, M., & Hofmann, K. (2010). Selective autophagy: Ubiquitin-mediated recognition and beyond. *Nature Cell Biology*, *12*(9), 836–841. http://dx.doi.org/10.1038/ncb0910-836.

Kuma, A., Hatano, M., Matsui, M., Yamamoto, A., Nakaya, H., Yoshimori, T., … Mizushima, N. (2004). The role of autophagy during the early neonatal starvation period. *Nature*, *432*(7020), 1032–1036. http://dx.doi.org/10.1038/nature03029.

Lamark, T., Kirkin, V., Dikic, I., & Johansen, T. (2009). NBR1 and p62 as cargo receptors for selective autophagy of ubiquitinated targets. *Cell Cycle*, *8*(13), 1986–1990.

Larsen, K. E., & Sulzer, D. (2002). Autophagy in neurons: A review. *Histology and Histopathology*, *17*(3), 897–908.

Lee, J. Y., Koga, H., Kawaguchi, Y., Tang, W., Wong, E., Gao, Y. S., … Yao, T. P. (2010). HDAC6 controls autophagosome maturation essential for ubiquitin-selective quality-control autophagy. *The EMBO Journal*, *29*(5), 969–980. http://dx.doi.org/10.1038/emboj.2009.405.

Lelouard, H., Ferrand, V., Marguet, D., Bania, J., Camosseto, V., David, A., … Pierre, P. (2004). Dendritic cell aggresome-like induced structures are dedicated areas for ubiquitination and storage of newly synthesized defective proteins. *The Journal of Cell Biology*, *164*(5), 667–675. http://dx.doi.org/10.1083/jcb.200312073.

Lelouard, H., Gatti, E., Cappello, F., Gresser, O., Camosseto, V., & Pierre, P. (2002). Transient aggregation of ubiquitinated proteins during dendritic cell maturation. *Nature*, *417*(6885), 177–182. http://dx.doi.org/10.1038/417177a.

Li, G., Jiang, H., Chang, M., Xie, H., & Hu, L. (2011). HDAC6 alpha-tubulin deacetylase: A potential therapeutic target in neurodegenerative diseases. *Journal of the Neurological Sciences*, *304*(1–2), 1–8. http://dx.doi.org/10.1016/j.jns.2011.02.017.

Lim, K. L., Dawson, V. L., & Dawson, T. M. (2006). Parkin-mediated lysine 63-linked polyubiquitination: A link to protein inclusions formation in Parkinson's and other conformational diseases? *Neurobiology of Aging*, *27*(4), 524–529. http://dx.doi.org/10.1016/j.neurobiolaging.2005.07.023.

Loos, B., du Toit, A., & Hofmeyr, J. H. (2014). Defining and measuring autophagosome flux-concept and reality. *Autophagy*, *10*(11), 2087–2096. http://dx.doi.org/10.4161/15548627.2014.973338.

Lu, K., Psakhye, I., & Jentsch, S. (2014). Autophagic clearance of polyQ proteins mediated by ubiquitin-Atg8 adaptors of the conserved CUET protein family. *Cell*, *158*(3), 549–563. http://dx.doi.org/10.1016/j.cell.2014.05.048.

Lystad, A. H., & Simonsen, A. (2015). Assays to monitor aggrephagy. *Methods*, *75*, 112–119. http://dx.doi.org/10.1016/j.ymeth.2014.12.019.

Martinez-Vicente, M., Talloczy, Z., Wong, E., Tang, G., Koga, H., Kaushik, S., … Cuervo, A. M. (2010). Cargo recognition failure is responsible for inefficient autophagy in Huntington's disease. *Nature Neuroscience*, *13*(5), 567–576. http://dx.doi.org/10.1038/nn.2528.

Matsumoto, G., Wada, K., Okuno, M., Kurosawa, M., & Nukina, N. (2011). Serine 403 phosphorylation of p62/SQSTM1 regulates selective autophagic clearance of ubiquitinated proteins. *Molecular Cell*, *44*(2), 279–289. http://dx.doi.org/10.1016/j.molcel.2011.07.039.

Meredith, S. C. (2005). Protein denaturation and aggregation: Cellular responses to denatured and aggregated proteins. *The Annals of the New York Academy of Sciences*, *1066*, 181–221. http://dx.doi.org/10.1196/annals.1363.030.

Miller, S. B., Ho, C. T., Winkler, J., Khokhrina, M., Neuner, A., Mohamed, M. Y., Bukau, B. … (2015). Compartment-specific aggregases direct distinct nuclear and cytoplasmic aggregate deposition. *The EMBO Journal*, *34*(6), 778–797. http://dx.doi.org/10.15252/embj.201489524.

Miller, S. B., Mogk, A., & Bukau, B. (2015). Spatially organized aggregation of misfolded proteins as cellular stress defense strategy. *Journal of Molecular Biology*, *427*(7), 1564–1574. http://dx.doi.org/10.1016/j.jmb.2015.02.006.

Mizushima, N., & Komatsu, M. (2011). Autophagy: Renovation of cells and tissues. *Cell*, *147*, 728–741.

Mortimore, G. E., & Poso, A. R. (1987). Intracellular protein catabolism and its control during nutrient deprivation and supply. *Annual Review of Nutrition*, 7, 539–564. http://dx.doi.org/10.1146/annurev.nu.07.070187.002543.

Moscat, J., Diaz-Meco, M. T., & Wooten, M. W. (2007). Signal integration and diversification through the p62 scaffold protein. *Trends in Biochemical Sciences*, *32*(2), 95–100. http://dx.doi.org/10.1016/j.tibs.2006.12.002.

Nanduri, P., Hao, R., Fitzpatrick, T., & Yao, T. P. (2015). Chaperone-mediated 26S proteasome remodeling facilitates free K63 ubiquitin chain production and aggresome clearance. *The Journal of Biological Chemistry*, *290*(15), 9455–9464. http://dx.doi.org/10.1074/jbc.M114.627950.

Nicot, A. S., Lo Verso, F., Ratti, F., Pilot-Storck, F., Streichenberger, N., Sandri, M., … Goillot, E. (2014). Phosphorylation of NBR1 by GSK3 modulates protein aggregation. *Autophagy*, *10*(6), 1036–1053. http://dx.doi.org/10.4161/auto.28479.

Olzmann, J. A., Li, L., Chudaev, M. V., Chen, J., Perez, F. A., Palmiter, R. D., & Chin, L. S. (2007). Parkin-mediated K63-linked polyubiquitination targets misfolded DJ-1 to aggresomes via binding to HDAC6. *The Journal of Cell Biology*, *178*(6), 1025–1038. http://dx.doi.org/10.1083/jcb.200611128.

Onodera, J., & Ohsumi, Y. (2004). Ald6p is a preferred target for autophagy in yeast, Saccharomyces cerevisiae. *The Journal of Biological Chemistry*, *279*(16), 16071–16076. http://dx.doi.org/10.1074/jbc.M312706200.

Ouyang, H., Ali, Y. O., Ravichandran, M., Dong, A., Qiu, W., MacKenzie, F., … Zhai, R. G. (2012). Protein aggregates are recruited to aggresome by histone deacetylase 6 via unanchored ubiquitin C termini. *The Journal of Biological Chemistry*, *287*(4), 2317–2327. http://dx.doi.org/10.1074/jbc.M111.273730.

Pankiv, S., Clausen, T. H., Lamark, T., Brech, A., Bruun, J. A., Outzen, H., … Johansen, T. (2007). p62/SQSTM1 binds directly to Atg8/LC3 to facilitate degradation of ubiquitinated protein aggregates by autophagy. *The Journal of Biological Chemistry*, *282*(33), 24131–24145. http://dx.doi.org/10.1074/jbc.M702824200.

Pickart, C. M., & Fushman, D. (2004). Polyubiquitin chains: Polymeric protein signals. *Current Opinion in Chemical Biology*, *8*(6), 610–616. http://dx.doi.org/10.1016/j.cbpa.2004.09.009.

Quinlan, R. A., Brenner, M., Goldman, J. E., & Messing, A. (2007). GFAP and its role in Alexander disease. *Experimental Cell Research*, *313*(10), 2077–2087. http://dx.doi.org/10.1016/j.yexcr.2007.04.004.

Ross, C. A., & Poirier, M. A. (2004). Protein aggregation and neurodegenerative disease. *Nature Medicine*, *10*(Suppl.), S10–S17. http://dx.doi.org/10.1038/nm1066.

Rubinsztein, D. C. (2006). The roles of intracellular protein-degradation pathways in neurodegeneration. *Nature*, *443*(7113), 780–786. http://dx.doi.org/10.1038/nature05291.

Rubinsztein, D. C., Gestwicki, J. E., Murphy, L. O., & Klionsky, D. J. (2007). Potential therapeutic applications of autophagy. *Nature Reviews. Drug Discovery*, *6*(4), 304–312. http://dx.doi.org/10.1038/nrd2272.

Rui, Y. N., Xu, Z., Patel, B., Chen, Z., Chen, D., Tito, A., … Zhang, S. (2015). Huntingtin functions as a scaffold for selective macroautophagy. *Nature Cell Biology*, *17*(3), 262–275. http://dx.doi.org/10.1038/ncb3101.

Sahu, R., Kaushik, S., Cannizzo, E. S., Scharf, B., Follenzi, A., Clement, C., … Santambrogio, L. (2011). Microautohagy of cytosolic proteins by late endosomes. *Developmental Cell*, *20*(1), 131–139. http://dx.doi.org/10.1016/j.devcel.2010.12.003.

Sarkar, S., & Rubinsztein, D. C. (2008). Small molecule enhancers of autophagy for neurodegenerative diseases. *Molecular BioSystems*, *4*(9), 895–901. http://dx.doi.org/10.1039/b804606a.

Scotter, E. L., Vance, C., Nishimura, A. L., Lee, Y. B., Chen, H. J., Urwin, H., … Shaw, C. E. (2014). Differential roles of the ubiquitin proteasome system and autophagy in the clearance of soluble and aggregated TDP-43 species. *Journal of Cell Science*, *127*(Pt. 6), 1263–1278. http://dx.doi.org/10.1242/jcs.140087.

Shiber, A., Breuer, W., Brandeis, M., & Ravid, T. (2013). Ubiquitin conjugation triggers misfolded protein sequestration into quality control foci when Hsp70 chaperone levels are limiting. *Molecular Biology of the Cell*, *24*(13), 2076–2087. http://dx.doi.org/10.1091/mbc.E13-01-0010.

Singh, R., Kaushik, S., Wang, Y., Xiang, Y., Novak, I., Komatsu, M., … Czaja, M. J. (2009). Autophagy regulates lipid metabolism. *Nature*, *458*(7242), 1131–1135. http://dx.doi.org/10.1038/nature07976. nature07976 [pii].

Soto, C. (2003). Unfolding the role of protein misfolding in neurodegenerative diseases. *Nature Reviews. Neuroscience*, *4*(1), 49–60. http://dx.doi.org/10.1038/nrn1007.
Specht, S., Miller, S. B., Mogk, A., & Bukau, B. (2011). Hsp42 is required for sequestration of protein aggregates into deposition sites in Saccharomyces cerevisiae. *The Journal of Cell Biology*, *195*(4), 617–629. http://dx.doi.org/10.1083/jcb.201106037.
Tan, J. M., Wong, E. S., Dawson, V. L., Dawson, T. M., & Lim, K. L. (2007). Lysine 63-linked polyubiquitin potentially partners with p62 to promote the clearance of protein inclusions by autophagy. *Autophagy*, *4*(2), 251–253. 5444 [pii].
Tan, J. M., Wong, E. S., Kirkpatrick, D. S., Pletnikova, O., Ko, H. S., Tay, S. P., … Lim, K. L. (2008). Lysine 63-linked ubiquitination promotes the formation and autophagic clearance of protein inclusions associated with neurodegenerative diseases. *Human Molecular Genetics*, *17*(3), 431–439. http://dx.doi.org/10.1093/hmg/ddm320.
Tan, J. M., Wong, E. S., & Lim, K. L. (2009). Protein misfolding and aggregation in Parkinson's disease. *Antioxidants and Redox Signaling*, *11*(9), 2119–2134. http://dx.doi.org/10.1089/ars.2009.2490.
Tsvetkov, A. S., Arrasate, M., Barmada, S., Ando, D. M., Sharma, P., Shaby, B. A., & Finkbeiner, S. (2013). Proteostasis of polyglutamine varies among neurons and predicts neurodegeneration. *Nature Chemical Biology*, *9*(9), 586–592. http://dx.doi.org/10.1038/nchembio.1308.
Tyedmers, J., Mogk, A., & Bukau, B. (2010). Cellular strategies for controlling protein aggregation. *Nature Reviews. Molecular Cell Biology*, *11*(11), 777–788. http://dx.doi.org/10.1038/nrm2993.
Tyedmers, J., Treusch, S., Dong, J., McCaffery, J. M., Bevis, B., & Lindquist, S. (2010). Prion induction involves an ancient system for the sequestration of aggregated proteins and heritable changes in prion fragmentation. *Proceedings of the National Academy of Sciences of the United States of America*, *107*(19), 8633–8638. http://dx.doi.org/10.1073/pnas.1003895107.
Walsh, D. M., & Selkoe, D. J. (2004). Oligomers on the brain: The emerging role of soluble protein aggregates in neurodegeneration. *Protein and Peptide Letters*, *11*(3), 213–228.
Wang, L., Chen, M., Yang, J., & Zhang, Z. (2013). LC3 fluorescent puncta in autophagosomes or in protein aggregates can be distinguished by FRAP analysis in living cells. *Autophagy*, *9*(5), 756–769. http://dx.doi.org/10.4161/auto.23814.
Wang, Y., Martinez-Vicente, M., Kruger, U., Kaushik, S., Wong, E., Mandelkow, E. M., … Mandelkow, E. (2009). Tau fragmentation, aggregation and clearance: the dual role of lysosomal processing. *Human Molecular Genetics*, *18*(21), 4153–4170. http://dx.doi.org/10.1093/hmg/ddp367.
Wanker, E. E., Scherzinger, E., Heiser, V., Sittler, A., Eickhoff, H., & Lehrach, H. (1999). Membrane filter assay for detection of amyloid-like polyglutamine-containing protein aggregates. *Methods in Enzymology*, *309*, 375–386.
Watabe, M., & Nakaki, T. (2011). Protein kinase CK2 regulates the formation and clearance of aggresomes in response to stress. *Journal of Cell Science*, *124*(Pt. 9), 1519–1532. http://dx.doi.org/10.1242/jcs.081778.
Winner, B., Jappelli, R., Maji, S. K., Desplats, P. A., Boyer, L., Aigner, S., … Riek, R. (2011). In vivo demonstration that alpha-synuclein oligomers are toxic. *Proceedings of the National Academy of Sciences of the United States of America*, *108*(10), 4194–4199. http://dx.doi.org/10.1073/pnas.1100976108.
Wong, E., Bejarano, E., Rakshit, M., Lee, K., Hanson, H. H., Zaarur, N., … Cuervo, A. M. (2012). Molecular determinants of selective clearance of protein inclusions by autophagy. *Nature Communications*, *3*, 1240. http://dx.doi.org/10.1038/ncomms2244.
Wong, E., & Cuervo, A. M. (2010a). Autophagy gone awry in neurodegenerative diseases. *Nature Neuroscience*, *13*(7), 805–811. http://dx.doi.org/10.1038/nn.2575.

Wong, E., & Cuervo, A. M. (2010b). Integration of clearance mechanisms: the proteasome and autophagy. *Cold Spring Harbor Perspective Biology*, *2*(12). http://dx.doi.org/10.1101/cshperspect.a006734.

Wong, E., Tan, J. M., Soong, W. E., Hussein, K., Nukina, N., Dawson, V. L., … Lim, K. L. (2008). Autophagy-mediated clearance of aggresomes is not a universal phenomenon. *Human Molecular Genetics*, *17*(16), 2570–2582. http://dx.doi.org/10.1093/hmg/ddn157.

Wu, Y. T., Tan, H. L., Shui, G., Bauvy, C., Huang, Q., Wenk, M. R., … Shen, H. M. (2010). Dual role of 3-methyladenine in modulation of autophagy via different temporal patterns of inhibition on class I and III phosphoinositide 3-kinase. *The Journal of Biological Chemistry*, *285*(14), 10850–10861. http://dx.doi.org/10.1074/jbc.M109.080796.

Xu, G., Gonzales, V., & Borchelt, D. R. (2002). Rapid detection of protein aggregates in the brains of Alzheimer patients and transgenic mouse models of amyloidosis. *Alzheimer Disease and Associated Disorders*, *16*(3), 191–195.

Xu, Z., Graham, K., Foote, M., Liang, F., Rizkallah, R., Hurt, M., … Zhou, Y. (2013). 14-3-3 protein targets misfolded chaperone-associated proteins to aggresomes. *Journal of Cell Science*, *126*(Pt 18), 4173–4186. http://dx.doi.org/10.1242/jcs.126102.

Yamamoto, A., Cremona, M. L., & Rothman, J. E. (2006). Autophagy-mediated clearance of huntingtin aggregates triggered by the insulin-signaling pathway. *The Journal of Cell Biology*, *172*(5), 719–731. http://dx.doi.org/10.1083/jcb.200510065.

Yan, J., Seibenhener, M. L., Calderilla-Barbosa, L., Diaz-Meco, M. T., Moscat, J., Jiang, J., … Wooten, M. C. (2013). SQSTM1/p62 interacts with HDAC6 and regulates deacetylase activity. *PLoS One*, *8*(9), e76016. http://dx.doi.org/10.1371/journal.pone.0076016.

Yao, T. P. (2010). The role of ubiquitin in autophagy-dependent protein aggregate processing. *Genes & Cancer*, *1*(7), 779–786. http://dx.doi.org/10.1177/1947601910383277.

Yla-Anttila, P., Vihinen, H., Jokitalo, E., & Eskelinen, E. L. (2009). Monitoring autophagy by electron microscopy in mammalian cells. *Methods in Enzymology*, *452*, 143–164. http://dx.doi.org/10.1016/s0076-6879(08)03610-0.

Zaarur, N., Meriin, A. B., Bejarano, E., Xu, X., Gabai, V. L., Cuervo, A. M., & Sherman, M. Y. (2014). Proteasome failure promotes positioning of lysosomes around the aggresome via local block of microtubule-dependent transport. *Molecular Cell Biology*, *34*(7), 1336–1348. http://dx.doi.org/10.1128/mcb.00103-14.

Zaarur, N., Meriin, A. B., Gabai, V. L., & Sherman, M. Y. (2008). Triggering aggresome formation. Dissecting aggresome-targeting and aggregation signals in synphilin 1. *The Journal of Biological Chemistry*, *283*(41), 27575–27584. http://dx.doi.org/10.1074/jbc.M802216200.

Zaffagnini, G., & Martens, S. (2016). Mechanisms of selective autophagy. *Journal of Molecular Biology*, *428*(9 Pt. A), 1714–1724. http://dx.doi.org/10.1016/j.jmb.2016.02.004.

Zhang, T., Shen, S., Qu, J., & Ghaemmaghami, S. (2016). Global analysis of cellular protein flux quantifies the selectivity of basal autophagy. *Cell Reports*, *14*(10), 2426–2439. http://dx.doi.org/10.1016/j.celrep.2016.02.040.

Zucchi, P. C., & Zick, M. (2011). Membrane fusion catalyzed by a Rab, SNAREs, and SNARE chaperones is accompanied by enhanced permeability to small molecules and by lysis. *Molecular Biology of the Cell*, *22*(23), 4635–4646. http://dx.doi.org/10.1091/mbc.E11-08-0680.

CHAPTER SIXTEEN

Methods to Study Chaperone-Mediated Autophagy

E. Arias[1]
Institute for Aging Studies, Albert Einstein College of Medicine, Bronx, NY, United States
[1]Corresponding author: e-mail address: esperanza.arias-perez@einstein.yu.edu

Contents

Abstract

Chaperone-mediated autophagy (CMA), a selective form of degradation of cytosolic proteins in lysosomes, contributes to maintenance of proteostasis and to the cellular adaptation to stress. CMA substrates are selectively recognized and delivered by a cytosolic chaperone to the lysosomal surface, where, upon unfolding, they are internalized through a membrane translocation complex. Defective or dysfunctional CMA has been associated with human pathologies such as neurodegeneration, cancer, immunodeficiency, or diabetes, increasing the overall interest in methods to monitor this selective autophagic pathway. In this chapter, we review the different experimental approaches used to evaluate CMA activity in different organs from animals or in cell cultures in vitro.

ABBREVIATIONS

3MA 3-methyladenine
ATG autophagy-related protein
CMA chaperone-mediated autophagy
e-MI endosomal microautophagy

Methods in Enzymology, Volume 588
ISSN 0076-6879
http://dx.doi.org/10.1016/bs.mie.2016.10.009

hsc heat shock cognate protein
LAMP lysosome-associated membrane protein
MVB multivesicular body

1. INTRODUCTION

Three different types of autophagy coexist in almost all mammalian cells: macroautophagy, microautophagy, and chaperone-mediated autophagy (CMA). CMA is a very selective form. In basal conditions, there are some levels of CMA activity in almost all cell types (Koga, Martinez-Vicente, Macian, Verkhusha, & Cuervo, 2011); nevertheless, maximal CMA activation is attained in response to cellular stressors such as starvation (Cuervo, Knecht, Terlecky, & Dice, 1995), oxidative stress (Kiffin, Christian, Knecht, & Cuervo, 2004), hypoxia (Hubbi et al., 2013), lipotoxicity (Rodriguez-Navarro et al., 2012), or DNA damage (Park, Suh, & Cuervo, 2015).

Under stress conditions, the selective removal of proteins by CMA contributes to the elimination of altered proteins, recycling of amino acids resulting from proteolysis, and adjustment of the cellular proteome composition. In aging, CMA activity is decreased in almost all tissues analyzed, which may contribute to aggravate the course of age-related disorders related to proteotoxicity (Cuervo & Dice, 2000a; Kiffin et al., 2007; Valdor et al., 2014). Actually, reduced CMA activity has been described in neurodegenerative diseases including Parkinson's disease (Cuervo, Stefanis, Fredenburg, Lansbury, & Sulzer, 2004) and tauopathies (Wang et al., 2009), but also in metabolic disorders, such as diabetes (Sooparb, Price, Shaoguang, & Franch, 2004) and diet-induced obesity (Rodriguez-Navarro et al., 2012). Reduced CMA also underlies the basis of defective T-cell function with age and of immunosenescence (Valdor et al., 2014). Contrarily, upregulated CMA activity is common in many types of cancer (Kon et al., 2011) and has been proposed to contribute to the pathogenesis of immune disorders such as lupus (Macri et al., 2014). This well-established connection between CMA and different human diseases has motivated a growing interest in understanding this fundamental cellular process and manipulating it with therapeutic purposes.

There are specific characteristics that distinguish CMA from other forms of autophagy: substrate proteins for this pathway are not sequestered in vesicles, but reach the lysosomal lumen after directly crossing the lysosomal membrane (Arias & Cuervo, 2011). The molecular components that contribute to targeting of CMA substrate proteins toward the lysosomal compartment and their translocation through the lysosomal membrane includes: integral lysosomal membrane proteins, cytosolic proteins that associate transiently with the lysosomal membrane, and luminal lysosomal proteins. Fig. 1 represents the sequential steps that mediate degradation of cytosolic proteins through CMA. Selectivity is determined by the presence on the substrate proteins of a pentapeptide amino acid sequence (the CMA targeting motif) that is biochemically similar to the pentapeptide KFERQ (Dice, 1990). This motif in the substrate proteins is identified by a cytosolic chaperone, the constitutively expressed hsc70, that guides them to the lysosomal membrane surface (Chiang, Terlecky, Plant, & Dice, 1989). Binding of substrates to the cytosolic tail of the lysosome-associated membrane

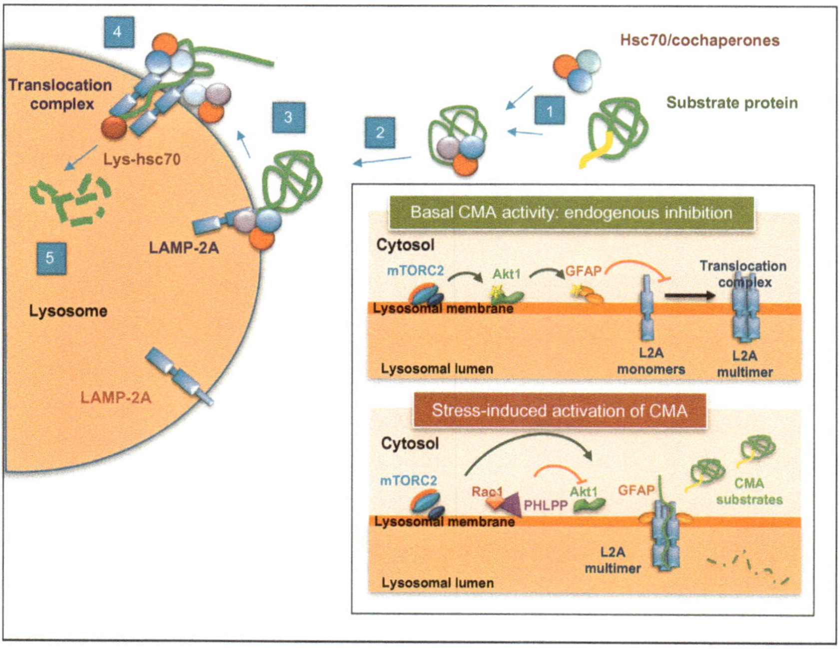

Fig. 1 Scheme of the sequential steps that mediate degradation of cytosolic proteins through chaperone-mediated autophagy: 1. Recognition of cytosolic proteins by hsc70 and its cochaperones. 2. Binding of the chaperone/substrate complex to the receptor at the lysosomal membrane. 3. Unfolding of the substrate protein. 4. Translocation across the lysosomal membrane. 5. Degradation in the lysosomal lumen. *Inset*: Phosphorylation of lysosome-associated Akt by mTORC2 represses CMA activation by negatively regulating the assembly of LAMP-2A into the CMA translocation complex. Activation of CMA in response to stress is attained by the recruitment to lysosomes of the phosphatase PHLPP1 (that dephosphorylates Akt) and the GTPase Rac1 (that stabilizes PHLPP1 at the membrane). Reduced Akt activity enhances CMA by increasing the stability of the CMA translocation complex at the lysosomal membrane.

protein type 2A (LAMP-2A) (Cuervo & Dice, 1996) induces the organization of this single-span membrane protein into a multimeric complex that facilitates translocation of proteins across the membrane (Bandyopadhyay, Kaushik, Varticovski, & Cuervo, 2008). A lysosomal form of hsp90 assures LAMP-2A stability at the lysosomal membrane during these multimerization events (Bandyopadhyay et al., 2008). Chaperones sited at the membrane of the lysosome mediate substrate unfolding and those in the luminal side of the lysosomal compartment help with substrate translocation (Agarraberes, Terlecky, & Dice, 1997). To allow for a new cycle of substrate binding and translocation, the translocation complex disassembles into monomeric forms of LAMP-2A once the substrate protein crosses the lysosomal membrane (Bandyopadhyay et al., 2008). Two proteins, glial fibrillary acidic protein (GFAP) and elongation factor 1α (EF1α), modulate the dynamics of LAMP-2A assembly and disassembly (Bandyopadhyay et al., 2008). When LAMP-2A is in the multimeric status, GFAP binds to it and contributes to the stabilization. In contrast, phosphorylated GFAP (pGFAP) has low binding affinity for LAMP-2A, and instead associates at the lysosomal membrane with EF1α in a complex. Once the substrate is translocated, EF1α is released, and the higher affinity of GFAP for pGFAP when compared to LAMP-2A facilitates disassembly of LAMP-2A as GFAP leaves the translocation complex (Bandyopadhyay et al., 2008).

Although all lysosomes contain in their membrane LAMP-2A, two populations of lysosomes with different CMA activity can be distinguished and biochemically isolated on the basis of their density and size. A major difference between these populations is that only one of them contains hsc70 within the lysosomal lumen and consequently only these lysosomes can perform CMA. The increased stability of hsc70 in this population of lysosomes is primarily a consequence of this lysosomal population's more acidic pH. This lysosomal fraction with higher CMA activity is referred to throughout the text as CMA+ or CMA active lysosomal population (Cuervo, Dice, & Knecht, 1997) (Fig. 2).

A lysosomal multiprotein complex comprised of the Pleckstrin homology (PH) domain and leucine-rich repeat phosphatase 1 (PHLPP1), the mammalian target of rapamycin complex 2 (mTORC2), and their common downstream target Akt has been recently show to modulate basal CMA activity and its activation in response to cellular stressors (Arias et al., 2015). The two kinases, mTORC2 and Akt, exert an inhibitory effect on CMA directly at the membrane of the subgroup of lysosomes committed to CMA. From there, they negatively modulate the dynamics of the CMA translocation complex, through—at least in part—phosphorylation

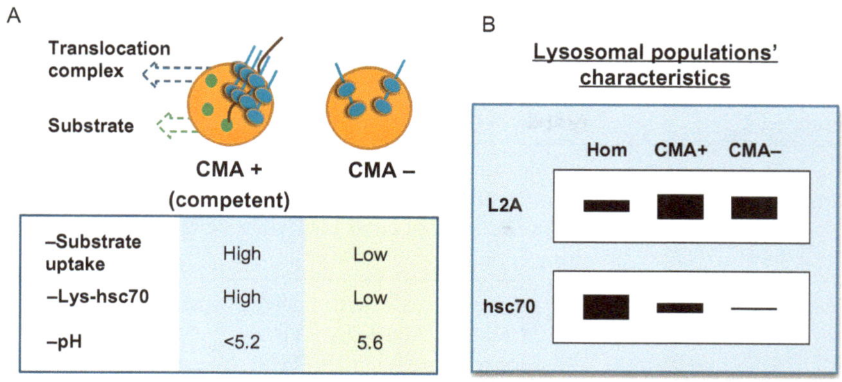

Fig. 2 Lysosomal populations' characteristics. (A) Details of the most relevant characteristics to distinguish between the two different populations of lysosomes involved in the autophagic processes. (B) Representative immunoblot for L2A and hsc70 in *Hom*, homogenate; *CMA+*, CMA competent; *CMA−*, CMA lower activity lysosomes.

of GFAP (Arias et al., 2015). This inhibitory effect is neutralized during stress conditions by the recruitment of PHLPP1 to the lysosomal membrane (Arias et al., 2015).

Treatment with chemical compounds that behave as atypical antagonists of the nuclear retinoic acid receptor alpha (RARα) induces CMA, revealing this signaling pathway as an endogenous inhibitor of CMA (Anguiano et al., 2013). In the case of T-cells, CMA activation is mainly attained through calcineurin/NFAT signaling in an ROS-dependent manner (Yang & Klionsky, 2010). Although there is a big effort in studying the signaling cascades responsible for CMA activation/inhibition, the chemical modulation of this pathway for therapeutic purposes is still challenging.

This review summarizes two types of CMA-related methods: (a) methods to determine if a specific protein is a CMA substrate, which will be important for understanding the consequences of malfunctioning of this autophagic pathway, and (b) those that assess CMA activity in a specific condition (physiological or pathological) or in response to possible therapeutic interventions.

2. METHODS TO TEST A PROTEIN AS POSSIBLE CMA SUBSTRATE

Considering that around 30% of total cytosolic proteins are known to bear KFERQ-like CMA targeting motifs (Dice, 1990), it becomes of great interest to elucidate the possible contribution of CMA to their degradation

Table 1 Requirements to Consider a Protein as a CMA Substrate

Specifications	**Checking**
KFERQ-like pentapeptide	–Protein sequencing (Dice, 1990)
Presence in lysosomal compartment	–Colocalization with lysosomal markers by immunofluorescence –Detection in isolated CMA-active lysosomes by immunoblot
CMA modulation	–Substrate increase with CMA blockage (lysosomal inhibitors or LAMP-2A KD) –Substrate decrease with CMA activation (LAMP-2A overexpression or chemical activators)
Interaction to CMA components	–Coimmunoprecipitation with hsc70 in cytosolic fractions (Cuervo, Hildebrand, Bomhard, & Dice, 1999; Cuervo, Hu, Lim, & Dice, 1998) –Coimmunoprecipitation with LAMP-2A in lysosomal fractions (Cuervo et al., 2004)
Translocation into lysosomes	–In vitro binding and uptake using isolated CMA+ lysosomes (Cuervo et al., 1999, 1998, 1995) –Degradation by intact isolated CMA+ lysosomes (Cuervo et al., 1999, 1998, 1995)
Lys-LAMP-2A multimer changes	–Blue native electrophoresis of isolated CMA-active lysosomes (Bandyopadhyay et al., 2008)

when studying specific proteins and their role in different physiological or pathological conditions. Although the presence of a CMA targeting motif in a protein allows discriminating its potential to undergo degradation by CMA, it does not imply that this pathway continuously degrades the protein in question. Frequently, the CMA targeting motif is not accessible to the chaperone when the protein is folded, or if the motif is covered by other interacting proteins or bears posttranslational modifications. Consequently, more direct experimental proof is required to validate a protein as a CMA substrate. Here, we detail different approaches that can be used to determine if a protein is a CMA substrate (Table 1).

2.1 Presence at the Lysosomal Compartment

Immunofluorescence methods can be used to quantify percentage of colocalization of the protein of interest with the lysosomal compartment since if a protein is a CMA substrate it will be delivered to lysosomes for degradation. Some of the lysosomal markers that can be used for this approach include cathepsins (D or B) or LAMPs (LAMP1, LAMP2). Given that both groups of proteins can be also found in endosomes (the compartment utilized for endosomal microautophagy (e-MI)), it is crucial to exclude colocalization of the protein of interest with late endosomal/MVB markers (such as rab9 or ESCRT proteins) (Sahu et al., 2011). This approach is suitable for cultured cellular models, but a different approach is needed in the case of studies in intact tissues. In this case, physical isolation of lysosomes from the tissues of interest followed by immunoblot analysis of the extracted material is the preferred approach. Enrichment of a specific protein in CMA-active lysosomes is a good indicator of its possible targeting and degradation through CMA (Cuervo et al., 1998; Orenstein et al., 2013; Schneider, Suh, & Cuervo, 2014) (Fig. 2). However, confirming its degradation in lysosomes is always important to discard that the protein of our interest associates with lysosomes for purposes other than its degradation. A good way to confirm degradation is to show the increase in lysosomal association in the presence of lysosomal inhibitors (Orenstein et al., 2013). Full proteomics of CMA-active lysosomes isolated from untreated and leupeptin-treated mice have been recently used as an unbiased approach to efficiently determine the proteins undergoing degradation by CMA at a given time or condition (Schneider et al., 2014). In summary, a CMA substrate will be defined as a protein whose levels increase in CMA-active lysosomes in the presence of leupeptin but not if LAMP-2A is absent (since other proteins may end in this specific subtype of lysosomes by endocytosis or by other types of autophagy).

2.2 Modulation of CMA Activity

The use of protease inhibitors to block lysosomal degradation usually increases cellular levels of CMA substrates, but the support for CMA degradation requires additional analysis because blockage of lysosomal proteolysis will also increase cellular levels of proteins degraded by other autophagic pathways and by endocytosis (Kaushik & Cuervo, 2009). A reduction in cellular levels of the protein of interest under conditions of CMA activation can also be used as an indication of its possible degradation through this pathway. Cellular stress, including nutrient deprivation, oxidative stress, and proteo- and lipotoxicity, results in CMA upregulation. However, direct modulation of CMA by genetic or chemic approaches is required as ultimate proof of degradation through this pathway (Table 1).

2.2.1 How to Activate CMA

Upregulation of CMA can be attained easily by overexpressing the CMA limiting protein LAMP-2A (Cuervo & Dice, 1996; Zhang & Cuervo, 2008). In contrast, for a long time the lack of selective chemical activators has been a major limitation in the modulation of this pathway. Part of the problem has been that the proteins involved in this pathway, for the most part, also participate in other cellular processes.

Early studies proposed 6-aminonicotinamide (glucose-6-phosphate dehydrogenase inhibitor) and geldanamycin (hsp90 inhibitor) as possible activators of CMA (Finn, Mesires, Vine, & Dice, 2005). However, the use of these molecules as CMA activators has been recently discouraged because of their lack of selectivity and variable effects. Thus, 6-aminonicotinamide affects other forms of lysosomal degradation, and more direct assays to measure CMA have revealed a minimal effect in this pathway. Similarly, the effect of geldanamycin on CMA seems to be highly cell type dependent and biphasic (early activation followed by inhibition of CMA) (Bandyopadhyay et al., 2008).

Posterior studies have identified that signaling through the RARα is inhibitory for CMA (Anguiano et al., 2013). Three different genes encode the three types of RARs identified in mammals; in contrast to the complex tissue-dependent expression of RARβ and RARγ, RARα is ubiquitously expressed. RARs are attractive druggable targets because their natural substrate all-*trans*-retinoic acid (ATRA) and similar retinoids have been well characterized, and because they efficiently traffic across lipid bilayers due to their small size and hydrophobicity. This together with growing understanding of the chemical modifications that retinoid derivatives undergo in cells made ATRA derivatives good drug candidates to target intracellular processes. Once identified that the RARα was an endogenous CMA inhibitory signaling pathway, the use of structure-based chemical design on ATRA has rendered selective inhibitors for particular RARα downstream pathways, including CMA. The leading final molecules in these studies have been shown capable of selective activation of CMA in the absence of detectable changes in other autophagic pathways (Anguiano et al., 2013). The antagonistic effect of the retinoid derivatives most likely results from a combination of tight binding to RARα (favored by the multiplicity of contact sites within the binding pocket) and high stability (obtained by protecting the sites in retinoids usually amenable to intracellular modification). Furthermore, the provided enhancement of CMA confers cells resistance to oxidative stress and proteotoxicity (Anguiano et al., 2013). These are, so far, the first CMA-specific activators with no effect at least in other protein degradation pathways.

In the case of CMA substrate proteins, we expect a decrease in its cellular levels after treatment with the CMA activators. It is important to note that due to the generally long half-life of CMA substrates, if the protein of interest is abundant in the cell, it becomes difficult to observe a decrease in overall cellular levels by immunoblot. In this case, radiolabeling of the whole protein content (by incubation with a radiolabeled amino acid as described in later sections in this review), followed by immunoprecipitation of the protein of interest at different times, provides more accurate assessment of accelerated degradation upon CMA activation (Cuervo et al., 2004).

2.2.2 How to Inhibit CMA

Currently, the most selective and efficient way to inhibit CMA is by knocking down or out LAMP-2A which prevents substrate binding/translocation into lysosomes (Massey, Kaushik, Sovak, Kiffin, & Cuervo, 2006; Schneider et al., 2014). Blockage of other CMA components (i.e., hsc70) is more problematic due to their involvement in other important cellular functions. For example, blocking hsc70 will also affect e-MI, macroautophagy, and endocytosis (by inhibiting clathrin disassembly), as well as processes such as protein folding or aggregation.

There are three isoforms of LAMP-2 that bear identical luminal regions. Functions associated to this common luminal region could be compensating by LAMP-2B and LAMP-2C upon reduction of LAMP-2A levels. However, the reduction of CMA after LAMP-2A blockage cannot be compensated for, because this function relies on the distinctive cytosolic tail of LAMP-2A different from other LAMP-2 protein variants (Massey et al., 2006; Schneider et al., 2014). Genetic reduction of LAMP-2A levels does not affect lysosomal stability and functions beyond CMA (Massey et al., 2006; Schneider et al., 2014; Valdor et al., 2014), and it is particularly useful to discriminate between CMA and e-MI degradation. Both autophagic pathways rely on hsc70 for substrate targeting, but only CMA depends on LAMP-2A for substrate translocation, and consequently, knockdown of LAMP-2A does not reduce degradation by e-MI (Sahu et al., 2011). An important limitation of the genetic reduction of LAMP-2A is the long half-life of this protein, which makes necessary a waiting period of at least 5 days before cellular levels of LAMP-2A are significantly reduced. This long time makes difficult the study of the effect of acute CMA inhibition effect, making thus of high priority the search for specific CMA chemical inhibitors (Table 2).

Table 2 CMA and e-Micro Substrate Differences

CMA Characteristics	**e-Micro Characteristics**
Specific pool of CMA-active lysosomes	–Late endosomes (and multivesicular bodies)
Interaction to LAMP-2A	–No interaction to LAMP-2A
Substrate unfolding required	–Substrate unfolding not required
Inhibition by LAMP-2A KD	–Modulation by ESCRT proteins KD

There are no available chemical compounds to selectively block CMA. Although previous studies demonstrate inhibition of CMA protein degradation by using inhibitors of protein synthesis (Finn et al., 2005), the fact that it also associate with many other cellular changes makes this approach unsuitable to specifically study the effect of CMA inhibition.

Due to the cross talk between the different degradation pathways, when CMA is compromised, it is also important to check the status of the other forms of autophagy and whether the protein of interest could have switch toward degradation through macroautophagy and/or e-MI. Contribution of these pathways to the degradation of the protein of interest can be estimated by knocking down the essential components for macroautophagy (i.e., ATG7 or ATG5) or e-MI (i.e., Vps4).

2.3 Interaction With CMA Components

As described before, any substrate protein degraded by CMA requires interaction with several CMA components (Table 1). Cytosolic interaction with the hsc70 chaperone is the first step in CMA (Chiang et al., 1989), and consequently, coimmunoprecipitating of the protein of interest with this chaperone is indicative of possible degradation by CMA. However, it is important to remember that this interaction between hsc70 and the substrate is not proportional to substrate uptake rates, and in fact binding to hsc70 could be even higher if lysosomal binding/uptake are blocked (Schneider et al., 2014). In addition, binding to hsc70 can also happen as a part of protein degradation via e-MI (Sahu et al., 2011), making necessary the use of additional studies to support CMA degradation such as testing interaction of the candidate protein with LAMP-2A.

2.4 Modification of the Targeting Motif

The absolute requirement for all CMA substrates of the pentapeptide motif recognized by hsc70 makes mutation of this pentapeptide sequence in the protein of interest a useful approach to support degradation through this pathway. The motif contains a glutamine residue (Q) flanked on either side by four amino acid residues consisting of only one acidic residue (D, E), and one or two basic (K, R) or bulky hydrophobic residues (F, I, L, V) (Dice, 1990). Mutation of Q and the residue next to it in the motif is usually sufficient to disrupt the hsc70-substrate interaction (Cuervo et al., 2004; Orenstein et al., 2013; Valdor et al., 2014). However, this approach alone still does not discriminate between degradation through CMA and e-MI, making necessary additional confirmatory experiments. Organelle colocalization can be of assistance in this case since elimination or alteration of the targeting motif will lead to decreased association of the substrate with lysosomes in the case of CMA or with late endosomes/MVB in the case of e-MI.

There are instances in which the interaction of a protein with hsc70 persists even upon elimination of the KFERQ-like motif due to the ability of this chaperone to also bind hydrophobic residues. Contrary to binding to the KFERQ-like motif, association of hsc70 through hydrophobic residues cannot be compete with other proteins carrying the CMA targeting motif such as RNaseA (Orenstein et al., 2013).

3. METHODS TO ASSAY CMA ACTIVITY

In this section we include useful cell culture and tissue assays as well as experimental models commonly used to analyze CMA activity.

3.1 Useful Approaches to Monitor CMA Activity

Changes in the abundance of the main molecular components of CMA and in the amount and distribution of CMA-active lysosomes can be used as an indirect way to infer changes in CMA activity. These analyses are only correlative and should be complemented, when possible, with some of the functional assays described in the following sections.

Immunoblot for different CMA components is an easy starting point to detect changes in CMA activity. Since the levels of LAMP-2A are limiting for CMA (Cuervo & Dice, 1996), very often, changes in the abundance of this protein in lysosomes correlate well with changes in CMA activity.

Nevertheless, the presence of LAMP-2A in lysosomes is necessary but not enough to conclude that lysosomes are active for CMA. Although all lysosomes contain LAMP-2A, only those that also contain lys-hsc70 in the luminal compartment are CMA competent (Cuervo et al., 1997). As mention before, LAMP-2A is one of the three spliced variants of a single gene, *lamp2*. Because the three LAMP-2 proteins have identical luminal regions, it is very important to use antibodies that can specifically recognize the 12 amino acid residues that constitute the cytosolic tail of LAMP-2A and that are different from those in LAMP-2B and LAMP-2C tails (Eskelinen et al., 2005). The most commonly used commercial antibodies have been generated against the luminal region of LAMP-2 and, consequently, cannot be used to obtain specific information about the CMA status. In fact, the other LAMP-2 variants often show similar cellular distribution because of their participation in vesicular trafficking, macroautophagy, lysosomal maturation, and even nucleotide transport into lysosomes (Eskelinen et al., 2005). When tracking changes in LAMP-2A levels is also important to remember that, it is the LAMP-2A present at the lysosomal membrane the one that directly correlates with CMA activity (Cuervo & Dice, 2000b). Thus, analysis of LAMP-2A levels in total cell lysates is of less value than measurement of LAMP-2A levels in lysosome-enriched fractions. Changes in LAMP-2A mRNA levels can be proportional to changes in CMA activity as previously demonstrated in certain conditions that activate CMA such as oxidative stress (Kiffin et al., 2004). However, the absence of mRNA changes cannot be interpreted as unchanged CMA since CMA activation often occurs without de novo synthesis of LAMP-2A (i.e., during starvation).

The levels of lysosomal-hsc70 are also proportional to CMA activity (Agarraberes et al., 1997; Cuervo et al., 1997), but the immunoblot for hsc70 in total cellular lysates or tissue homogenates is not informative since this is one of the most abundant cellular chaperones, and the fraction located in lysosomes is small. For the same reason, mRNA levels of this constitutive chaperone do not have any predictive value for CMA activity. When performing immunoblot of lysosome-enriched fractions or immunofluorescence to look for lysosomal association of the chaperone, it is important to utilize antibodies that recognize only hsc70 but not those that recognize both the constitutive (hsc70) and the inducible (hsp70) chaperones that only differ in a very small number of amino acid residues (Agarraberes et al., 1997).

The amount of CMA-positive lysosomes in proportion to the whole lysosomal pool increases under conditions that activate CMA (Kiffin et al., 2004). Image-based procedures can be used to estimate the abundance of this subset of lysosomes such as colocalization of hsc70 with lysosomal markers (LAMP-2, LAMP-1, etc.). In this case, methanol fixation is necessary to extract the cytosolic hsc70 and retain only the vesicle-associated hsc70 (Kaushik & Cuervo, 2009) (Fig. 2). Immunogold staining for hsc70 and electron microscopy can also provide useful information about the pool of CMA-active lysosomes (Cuervo et al., 1997). Still unknown why, but in many cell types CMA activation leads to subcellular mobilization of hsc70-positive lysosomes toward the perinuclear region making changes in the cellular distribution of these lysosomes a useful indirect indication of CMA activation (Kaushik & Cuervo, 2009).

Changes in lysosomal levels of standard CMA markers such as LAMP-2A and lys-hsc70 (Fig. 2) are cell type dependent: different cells have different basal levels of these proteins and of CMA activity. Consequently, studies using these markers should be done comparatively rather than in absolute values. Contrary to most CMA components that are constitutively present in lysosomes, association of the positive CMA regulator PHLPP with the lysosomal membrane preferentially occurs under conditions of CMA activation (Arias et al., 2015). This makes costaining for PHLPP and LAMP-2A a possibly useful way to detect CMA activation. There are other proteins present in the lysosomal compartment that have an active role in CMA, such as lysosomal hsp90, lysosomal rictor, and glial fibrillary acidic protein. However, they are not used regularly to determine CMA activity because of its participation in other cellular functions and/or their lack of changes in abundance or location during CMA activation.

Lastly, when working with isolated lysosomes from cells or tissues, an increase in the levels of well-known CMA substrates is also a good indication of CMA failure. Because degradation of CMA substrates occurs rapidly after translocation, comparison of levels of CMA substrates in lysosomes from cells or animals untreated or treated with inhibitors of lysosomal proteases (i.e., leupeptin) allows measuring flux through CMA (Schneider et al., 2014). Confirmation of CMA involvement in degradation can be attained, showing that LAMP-2A knockdown reduces the leupeptin effect.

3.2 Experimental Models

Although chaperones are conserved through evolution, the limiting CMA component LAMP-2A is not conserved in invertebrates, explaining why this pathway has only been described in mammals. The LAMP-2A spliced variant is not present in yeast, worms, flies, or even zebrafish, but appears in mammals and birds (Gough, Hatem, & Fambrough, 1995). This explains why all the methods for the study of CMA have been developed in mammals.

CMA activity has been reported in many different mammalian cell types, including fibroblasts (Massey et al., 2006), hepatocytes (Schneider et al., 2014), astrocytes (Martin, Joseph, & Cuervo, 2002), primary neurons (Orenstein et al., 2013), macrophages (Cannizzo et al., 2012), dendritic cells (Cannizzo et al., 2012), T-cells (Valdor et al., 2014), retinal cells (Rodriguez-Muela et al., 2013), and in an important subset of cancer cells of different origins (Kon et al., 2011). CMA has been also studied using lysosomes isolated from liver (Zhang & Cuervo, 2008), spleen (Cuervo et al., 1998), different brain regions (Orenstein et al., 2013), and kidney (Cuervo et al., 1999). Among these tissues, liver is the most studied and best-characterized organ, where most of the molecular dissection of CMA has been performed, making methods to measure CMA activity in this organ the most developed.

3.3 Functional Assays

To track CMA activity over time in cells, tissues, and isolated organelles we can use several functional assays.

3.3.1 Photoconvertible CMA Reporters

The Cuervo group has developed a series of photoconvertible fluorescent reporters, either photoswitchable (KFERQ-ps-CFP; fluorescence switches from CFP to GFP upon exposure to 405 nm light) or photoactivable (KFERQ-PA-mCherry1; becomes fluorescent upon exposure to 405 nm light) (Koga et al., 2011), that allow for tracking and measuring of CMA. By inducing photoconversion at a given time, it is possible to track the association of the photoconverted protein with lysosomes in a different fluorescence channel than the newly synthesized protein (that fluoresces in the original wavelength because it is not photoconverted) (Koga et al., 2011). Changes in the number of fluorescent puncta per cell are an easy readout of CMA activity. Although the magnitude of CMA activation depends on the cell type and basal CMA levels, in general, a two- to fourfold increase in puncta/cell is a good indication of CMA activation (Fig. 3). These reporters have limitations and need to be complemented with other methods to measure CMA activity, but are useful to get an overall estimate of CMA changes. One of these limitations is the fact that changes in degradation—once inside the lysosomal compartment—cannot be detected by direct fluorescence, since the fluorescent protein requires unfolding before translocation, and that disrupts its ability to fluoresce. However, degradation of reporters could still be measured, as for any other substrate, using

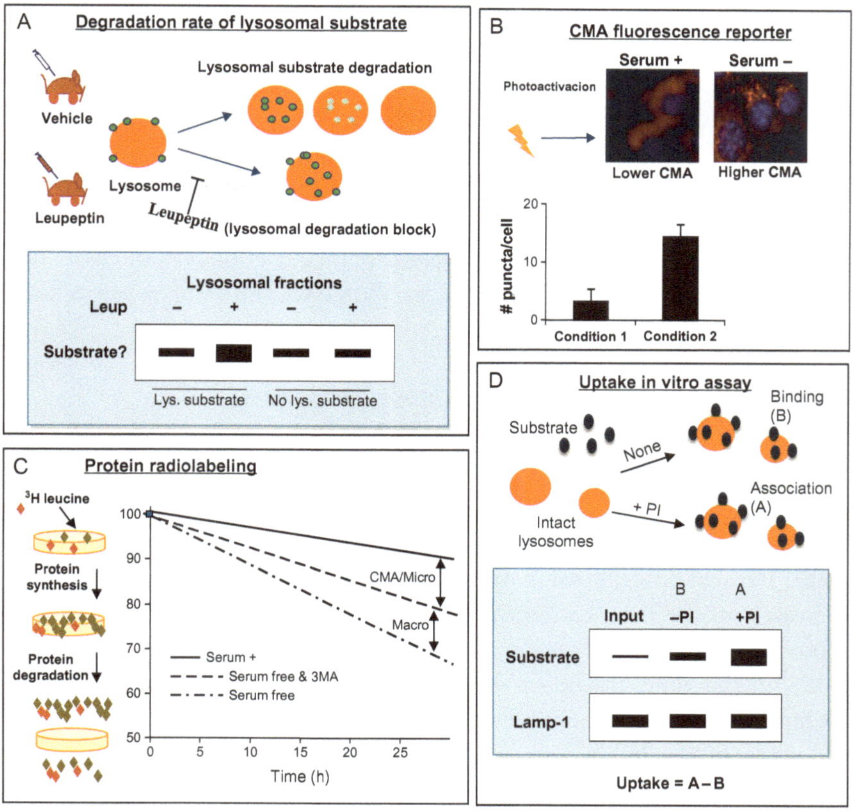

Fig. 3 Scheme of the most common assays to measure CMA activity; complete assessment of CMA requires the use of functional assays. (A) Representative scheme to confirm degradation in lysosomes, by showing the increase in lysosomal association in the presence of lysosomal inhibitors. A CMA substrate will be defined as a protein that increases in CMA-active lysosomes in the presence of leupeptin (Orenstein et al., 2013). (B) Representative images of mouse fibroblasts stably expressing KFERQ-PA-mCherry1 were photoactivated and maintained in media supplemented or not with serum for 16 h. Nuclei are staining with DAPI. *Bottom*: Representative quantification of the number of red fluorescent puncta per cell (Koga et al., 2011). (C) Measurement of long-lived protein degradation. Confluent cells in culture are incubated with a radiolabeled amino acid for 48 h and after extensive washing the amount of acid-soluble radioactivity (amino acids and small peptides) released into the medium at different times is determined. *Right*: Example of rates of degradation of long-lived proteins in cultured cells due to CMA, micro-, or macroautophagy. CMA activity is calculated as the increase in protein degradation during serum deprivation sensitive to lysosomal protease inhibition and insensitive to the effect of PI3-K type III inhibitors (Auteri, Okada, Bochaki, & Dice, 1983; Fuertes, Martin De Llano, Villarroya, Rivett, & Knecht, 2003). (D) Uptake in vitro assay. Incubation of CMA substrates with intact lysosomes treated or not with protease inhibitors allows the determination of the amount of substrate bound to the lysosomal membrane via immunoblot against the substrate after collecting the lysosomes by centrifugation. The amount of substrate translocated into the lysosomal lumen can be calculated by subtracting the amount of "bound" substrate (B) from the total amount of substrate associated with the protease-inhibited lysosomes (A) (Orenstein & Cuervo, 2010).

antibodies against the fluorescence protein in samples treated or not with leupeptin. Also, in cases with compromised translocation of substrates across the lysosomal membrane, the reporter will remind bound to lysosomes, giving a false impression of enhanced CMA. Once differences are detected with these reporters, more in-depth studies with specific endogenous CMA substrates need to be carried out (i.e., analysis of flux of endogenous CMA substrates or reconstitution of CMA in vitro with lysosomes isolated from those cells or tissues as described in later sections in this review).

3.3.2 Intracellular Protein Degradation Assessment

Levels of cytosolic proteins degraded at a specific time by CMA depend on the cellular type and the cellular conditions. Based on the fact that majority of proteins degraded by lysosomes through any autophagic pathway have long half-lives (Dice, 1987), pulse and chase experiments using radiolabeled amino acids and inhibitors of either lysosomal proteases or other autophagic pathways can be used to discriminate proteins undergoing degradation in lysosomes through CMA from other pathways (Kaushik & Cuervo, 2009). In this type of studies, de novo synthesized proteins are first radiolabeled by addition to the media of ^{3}H-leucine or ^{3}H-valine (this last amino acid is recommended in cells where leucine has been shown to have inhibitory effect on macroautophagy). To assure labeling of long-lived proteins a pulse duration of 48 h is necessary. Breakdown of the radiolabeled proteins is monitored upon removal of the radiolabeled amino acid as the release of free radiolabeled amino acid into the cell culture media. Reincorporation of the released amino acids into proteins is prevented by supplementing the chase media with an excess of the nonlabeled amino acid (e.g., leucine or valine) (Auteri et al., 1983). The fraction of long-lived proteins degraded in lysosomes is calculated by comparing proteolysis values in untreated cells to those in cells treated with general inhibitors of lysosomal proteolysis. Due to the long-time incubation with inhibitors (up to 20 h of chase) required in these studies, the combination of leupeptin, a selective inhibitor of cysteine and serine peptidases, with ammonium chloride (NH_4Cl), a weak base that neutralizes the lysosomal acidic pH necessary for lysosomal protease activity (Fuertes et al., 2003), is preferred to the use of bafilomycin A_1, also used to dissipate lysosomal acidification but that when present in cells for long period of times can result in compensatory upregulation of macroautophagy or even pronounced cellular toxicity. To estimate lysosomal degradation through macroautophagy in these assays, the most commonly used inhibitor is 3-methyladenine (3MA), an inhibitor of phosphatidylinositol-3-kinases (Seglen & Gordon, 1982). CMA-dependent degradation is calculated as the percentage of protein degradation sensitive to lysosomal inhibitors but insensitive to 3MA inhibition (Fig. 3). Due to the fact that degradation by e-MI can still be part of the fraction of lysosomal proteolysis insensitive to 3MA, this approach alone cannot be used to conclude that any observed change is solely due to changes in CMA activity. In addition, prolonged blockage of macroautophagy can lead to CMA upregulation and subsequent underestimation of the fraction of cellular proteins degraded through CMA (Massey et al., 2006).

3.3.3 In Vitro Analysis of CMA Using Isolated Lysosomes

The existence of cross talk and sharing of molecular components among the different autophagic pathways makes difficult to accurately assess CMA activity in intact cells. Consequently, the most reliable approach to estimate CMA is obtained by in vitro reconstituting this process with isolated lysosomes. Since CMA occurs at the lysosomal compartment, functional assays with isolated lysosomes are the ultimate way to measure functional CMA differences. The preferred protocols for isolation of lysosomes are those that allow for the separation of lysosomes exclusively contributing to CMA (hsc70-positive (Cuervo et al., 1997)) rather than methods that isolate a pool of mixed lysosomes. To be able to reproduce each of the steps of CMA in vitro, it is crucial to preserve the integrity of the lysosomal membrane throughout the whole incubation time; consequently, protocols that include detergents extraction or physical cell disruption methods such as sonication should be avoided. Nitrogen cavitation for cultured cells (Storrie & Madden, 1990), and speed-controlled homogenization in the case of tissues are the best approaches for cell disruption, preserving the integrity of the lysosomal membrane (more details can be found at Kaushik and Cuervo (2009)). Isolation of lysosomes is based on ultra-high-speed centrifugation and flotation in metrizamide discontinuous density gradient [modified from Wattiaux, Wattiaux-De Coninck, Ronveaux-Dupal, and Dubois (1978)]. CMA-active lysosomes in these gradients can be differentiated based on their density and size from those lysosomes more active for other autophagic processes or for endocytosis (Cuervo et al., 1997). Analysis of purity and enrichment of lysosomal enzymes in the isolated lysosomes, as described before (Storrie & Madden, 1990), is a mandatory requirement, especially when interested in comparatively checking activity in fractions isolated from different cells and conditions.

The amount of cells or tissue required to get enough CMA-active lysosomes is usually a limiting point to the above proposed in vitro *CMA studies*. In those instances, it is still possible using an enriched lysosomal fraction to test for changes in CMA components, as it will be more accurate than similar measurements in the whole homogenate or cellular lysate. When compared with purified lysosomes, this enriched fraction has a considerably larger amount of mitochondria, making necessary normalization of any observed change in CMA substrate levels to the amount of mitochondrial content (using mitochondrial markers) in the fractions.

Isolation of the specific fraction of lysosomes active for CMA allows also analyzing the content of endogenous CMA substrates in these compartments right after isolation. Blockage of lysosomal proteolysis before isolation (using ammonium chloride/leupeptin in cultured cells, or injection of animals with leupeptin for in vivo studies) provides information of flux of endogenous substrates through CMA (Orenstein et al., 2013; Schneider et al., 2014) (Fig. 3). Isolated lysosomes also permit reconstituting CMA in vitro to follow the steps involved in this process—substrate binding, lysosomal uptake, and lysosomal degradation (Kaushik & Cuervo, 2009). To that effect, [^{14}C]-radiolabeled CMA substrates (i.e., RNaseA) or a pool of radiolabeled long-lived cytosolic proteins (enriched in CMA substrates) can be incubated with isolated lysosomes, and their internalization and degradation in these compartments can be monitored as the amount of free [^{14}C]-radiolabeled amino acids released in the incubation media (Terlecky, Chiang, Olson, & Dice, 1992). Substrate uptake can be distinguished from degradation by comparing degradation of these substrates upon incubation with intact or broken lysosomes (Cuervo et al., 1997).

To distinguish lysosomal binding of CMA substrates from their lysosomal uptake another immunoblot-based assay with isolated lysosomes can be used (Aniento, Roche, Cuervo, & Knecht, 1993; Orenstein & Cuervo, 2010). Upon incubation of isolated lysosomes with purified CMA substrates, the internalized substrate can be detected if its degradation by lysosomal proteases is prevented with lysosomal inhibitors, as otherwise the substrate is rapidly degraded as soon as it reaches the lysosomal lumen and only the fraction bound at the cytosolic side of the lysosomal membrane remains. Binding can be calculated as the amount of substrate recovered in the lysosomal fraction after its incubation with intact lysosomes not treated with protease inhibitors. Incubation of the CMA substrate with lysosomes pretreated with protease inhibitors allows measuring both the substrate bound and translocated into lysosomes (binding + uptake).

Lastly, uptake is calculated by discounting from this last group the amount of CMA substrate associated with lysosomes in which proteolysis has not been inhibited (Bandyopadhyay & Cuervo, 2008) (Fig. 3).

Isolated lysosomal fractions can also be subjected to immunoblot for essential CMA components to directly detect changes in their lysosomal content and in their organization at the lysosomal membrane. A decrease in lysosomal abundance of LAMP-2A or hsc70 is indicative of decreased CMA activity (Sahu et al., 2011; Salvador, Aguado, Horst, & Knecht, 2000), whereas increased LAMP-2A levels can be used to support possible CMA upregulation (Eskelinen et al., 2005). However, in some instances, an increase in LAMP-2A levels can be observed without an increase in CMA activity. For example, although LAMP-2A is initially limiting for CMA, if the levels of lys-hsc70 in the lumen are not sufficient to accommodate increased binding to LAMP-2A, uptake does not increase proportionally because lys-hsc70 becomes limiting (Cuervo et al., 1995). Moreover, substrate uptake depends on the organization of LAMP into the CMA translocation complex (Orenstein & Cuervo, 2010). Lysosomal LAMP-2A multimerization is driven by the presence of CMA substrates, making this a dynamic process. The ratio of lysosomal LAMP-2A assembled into multimeric complex at a given time can be determined using blue native electrophoresis of isolated lysosomes and immunoblot for LAMP-2A (Bandyopadhyay & Cuervo, 2008).

4. CONCLUDING REMARKS

The different methods now available to study CMA make possible to quantify changes in the activity of this selective form of autophagy under different conditions. Combination of several of these experimental approaches allows now differentiating degradation through CMA from other forms of autophagy, such as macroautophagy and e-MI. Some of the new approaches also allow now gathering preliminary information on the state of CMA even when the amount of available biological material is limited. As in the case of other protein degradation pathways, use of functional assays is essential to conclusively identify changes in CMA activity. The array of methods described in this review have been successfully used to gather information on CMA malfunctioning in different models of disease such as neurological disorders, metabolic diseases, or cancer, and in physiological states such as aging.

The dynamic nature of the CMA process and the fact that translocation of substrate proteins into lysosomes is the rate-limiting step in this type of autophagy make in vitro experiments using isolated lysosomes the most precise approach for analysis of CMA activity.

Remaining challenges in the study of CMA include the need for the development of chemical approaches to selectively inhibit this process. Selective acute inhibition of CMA should help identifying additional new physiological functions of CMA, and may have application with therapeutic purposes in those conditions with abnormally enhanced CMA activity such as cancer. Another important challenge is the developing of methods for analysis of CMA function in vivo in a whole organism.

ACKNOWLEDGMENTS

Our work is supported by the Nathan Shock Center of Excellence for the Biology of Aging, Pilot Grant Program (NIH/NIA P30AG038072-P&F) and past funding from the Ellison Medical Foundation/American Federation for Aging Research (AFAR), and the Micinn/Fulbright program. I thank Dr. Cuervo and members of the Cuervo lab for insightful discussions and feedback.

REFERENCES

Agarraberes, F., Terlecky, S., & Dice, J. (1997). An intralysosomal hsp70 is required for a selective pathway of lysosomal protein degradation. *The Journal of Cell Biology*, *137*, 825–834.

Anguiano, J., Garner, T. P., Mahalingam, M., Bas, B. C., Gavathiotis, E., & Cuervo, A. M. (2013). Chemical modulation of chaperone-mediated autophagy by retinoic acid derivatives. *Nature Chemical Biology*, *9*, 374–382.

Aniento, F., Roche, E., Cuervo, A. M., & Knecht, E. (1993). Uptake and degradation of glyceraldehyde-3-phosphate dehydrogenase by rat liver lysosomes. *The Journal of Biological Chemistry*, *268*, 10463–10470.

Arias, E., & Cuervo, A. M. (2011). Chaperone-mediated autophagy in protein quality control. *Current Opinion in Cell Biology*, *23*, 184–189.

Arias, E., Koga, H., Diaz, A., Mocholi, E., Patel, B., & Cuervo, A. M. (2015). Lysosomal mTORC2/PHLPP1/Akt regulate chaperone-mediated autophagy. *Molecular Cell*, *59*, 270–284.

Auteri, J. S., Okada, A., Bochaki, V., & Dice, J. F. (1983). Regulation of intracellular protein degradation in IMR-90 human diploid fibroblasts. *Journal of Cellular Physiology*, *115*, 159–166.

Bandyopadhyay, U., & Cuervo, A. M. (2008). Entering the lysosome through a transient gate by chaperone-mediated autophagy. *Autophagy*, *4*, 1101–1103.

Bandyopadhyay, U., Kaushik, S., Varticovski, L., & Cuervo, A. M. (2008). The chaperone-mediated autophagy receptor organizes in dynamic protein complexes at the lysosomal membrane. *Molecular and Cellular Biology*, *28*, 5747–5763.

Cannizzo, E. S., Clement, C. C., Morozova, K., Valdor, R., Kaushik, S., Almeida, L. N., et al. (2012). Age-related oxidative stress compromises endosomal proteostasis. *Cell Reports*, *2*, 136–149.

Chiang, H., Terlecky, S., Plant, C., & Dice, J. F. (1989). A role for a 70 kDa heat shock protein in lysosomal degradation of intracellular protein. *Science*, *246*, 382–385.

Cuervo, A. M., & Dice, J. F. (1996). A receptor for the selective uptake and degradation of proteins by lysosomes. *Science*, *273*, 501–503.

Cuervo, A. M., & Dice, J. F. (2000a). Age-related decline in chaperone-mediated autophagy. *The Journal of Biological Chemistry*, *275*, 31505–31513.

Cuervo, A. M., & Dice, J. F. (2000b). Regulation of lamp2a levels in the lysosomal membrane. *Traffic*, *1*, 570–583.

Cuervo, A. M., Dice, J. F., & Knecht, E. (1997). A population of rat liver lysosomes responsible for the hsc73-mediated degradation of cytosolic proteins in lysosomes. *The Journal of Biological Chemistry*, *272*, 5606–5615.

Cuervo, A. M., Hildebrand, H., Bomhard, E. M., & Dice, J. F. (1999). Direct lysosomal uptake of α2-microglobulin contributes to chemically induced nephropathy. *Kidney International*, *55*, 529–545.

Cuervo, A. M., Hu, W., Lim, B., & Dice, J. F. (1998). IκB is a substrate for a selective pathway of lysosomal proteolysis. *Molecular Biology of the Cell*, *9*, 1995–2010.

Cuervo, A. M., Knecht, E., Terlecky, S. R., & Dice, J. F. (1995). Activation of a selective pathway of lysosomal proteolysis in rat liver by prolonged starvation. *The American Journal of Physiology*, *269*, C1200–C1208.

Cuervo, A. M., Stefanis, L., Fredenburg, R., Lansbury, P. T., & Sulzer, D. (2004). Impaired degradation of mutant α-synuclein by chaperone-mediated autophagy. *Science*, *305*, 1292–1295.

Dice, J. F. (1987). Molecular determinants of protein half-lives in eukaryotic cells. *The FASEB Journal*, *1*, 349–357.

Dice, J. F. (1990). Peptide sequences that target cytosolic proteins for lysosomal proteolysis. *Trends in Biochemical Sciences*, *15*, 305–309.

Eskelinen, E. L., Cuervo, A. M., Taylor, M. R., Nishino, I., Blum, J. S., Dice, J. F., et al. (2005). Unifying nomenclature for the isoforms of the lysosomal membrane protein LAMP-2. *Traffic*, *6*, 1058–1061.

Finn, P., Mesires, N., Vine, M., & Dice, J. F. (2005). Effects of small molecules on chaperone-mediated autophagy. *Autophagy*, *1*, 141–145.

Fuertes, G., Martin De Llano, J., Villarroya, A., Rivett, A., & Knecht, E. (2003). Changes in the proteolytic activities of proteasomes and lysosomes in human fibroblasts produced by serum withdrawal, amino-acid deprivation and confluent conditions. *The Biochemical Journal*, *375*, 75–86.

Gough, N. R., Hatem, C. L., & Fambrough, D. M. (1995). The family of LAMP-2 proteins arises by alternative splicing from a single gene: Characterization of the avian LAMP-2 gene and identification of mammalian homologs of LAMP-2b and LAMP-2c. *DNA and Cell Biology*, *14*, 863–867.

Hubbi, M. E., Hu, H., Kshitiz, Ahmed, I., Levchenko, A., & Semenza, G. L. (2013). Chaperone-mediated autophagy targets hypoxia-inducible factor-1α (HIF-1α) for lysosomal degradation. *The Journal of Biological Chemistry*, *288*, 10703–10714.

Kaushik, S., & Cuervo, A. M. (2009). Methods to monitor chaperone-mediated autophagy. *Methods in Enzymology*, *452*, 297–324.

Kiffin, R., Christian, C., Knecht, E., & Cuervo, A. M. (2004). Activation of chaperone-mediated autophagy during oxidative stress. *Molecular Biology of the Cell*, *15*, 4829–4840.

Kiffin, R., Kaushik, S., Zeng, M., Bandyopadhyay, U., Zhang, C., Massey, A. C., et al. (2007). Altered dynamics of the lysosomal receptor for chaperone-mediated autophagy with age. *Journal of Cell Science*, *120*, 782–791.

Koga, H., Martinez-Vicente, M., Macian, F., Verkhusha, V. V., & Cuervo, A. M. (2011). A photoconvertible fluorescent reporter to track chaperone-mediated autophagy. *Nature Communications*, *2*, 386.

Kon, M., Kiffin, R., Koga, H., Chapochnick, J., Macian, F., Varticovski, L., et al. (2011). Chaperone-mediated autophagy is required for tumor growth. *Science Translational Medicine, 3*, 109ra117.

Macri, C., Wang, F., Tasset, I., Schall, N., Page, N., Briand, J.-P., et al. (2014). Modulation of deregulated chaperone-mediated autophagy by a phosphopeptide. *Autophagy, 11*, 472–486.

Martin, A., Joseph, J. A., & Cuervo, A. M. (2002). Stimulatory effect of vitamin C on autophagy in glial cells. *Journal of Neurochemistry, 82*, 538–549.

Massey, A. C., Kaushik, S., Sovak, G., Kiffin, R., & Cuervo, A. M. (2006). Consequences of the selective blockage of chaperone-mediated autophagy. *Proceedings of the National Academy of Sciences of the United States of America, 103*, 5905–5910.

Orenstein, S. J., & Cuervo, A. M. (2010). Chaperone-mediated autophagy: Molecular mechanisms and physiological relevance. *Seminars in Cell & Developmental Biology, 21*, 719–726.

Orenstein, S. J., Kuo, S. H., Tasset, I., Arias, E., Koga, H., Fernandez-Carasa, I., et al. (2013). Interplay of LRRK2 with chaperone-mediated autophagy. *Nature Neuroscience, 16*, 394–406.

Park, C., Suh, Y., & Cuervo, A. M. (2015). Regulated degradation of Chk1 by chaperone-mediated autophagy in response to DNA damage. *Nature Communications, 6*, 6823.

Rodriguez-Muela, N., Koga, H., Garcia-Ledo, L., de la Villa, P., de la Rosa, E. J., Cuervo, A. M., et al. (2013). Balance between autophagic pathways preserves retinal homeostasis. *Aging Cell, 12*, 478–488.

Rodriguez-Navarro, J. A., Kaushik, S., Koga, H., Dall'armi, C., Shui, G., Wenk, M. R., et al. (2012). Inhibitory effect of dietary lipids on chaperone-mediated autophagy. *Proceedings of the National Academy of Sciences of the United States of America, 109*, E705–E714.

Sahu, R., Kaushik, S., Cannizzo, E., Scharf, B., Follenzi, A., Clement, C., et al. (2011). Microautophagy of cytosolic proteins by late endosomes. *Developmental Cell, 20*, 131–139.

Salvador, N., Aguado, C., Horst, M., & Knecht, E. (2000). Import of a cytosolic protein into lysosomes by chaperone-mediated autophagy depends on its folding state. *The Journal of Biological Chemistry, 275*, 27447–27456.

Schneider, J. L., Suh, Y., & Cuervo, A. M. (2014). Deficient chaperone-mediated autophagy in liver leads to metabolic dysregulation. *Cell Metabolism, 20*, 417–432.

Seglen, P., & Gordon, P. (1982). 3-Methyladenine: Specific inhibitor of autophagic/lysosomal protein degradation in isolated rat hepatocytes. *Proceedings of the National Academy of Sciences of the United States of America, 79*, 1889–1892.

Sooparb, S., Price, S. R., Shaoguang, J., & Franch, H. A. (2004). Suppression of chaperone-mediated autophagy in the renal cortex during acute diabetes mellitus. *Kidney International, 65*, 2135–2144.

Storrie, B., & Madden, E. (1990). Isolation of subcellular organelles. *Methods in Enzymology, 182*, 203–225.

Terlecky, S. R., Chiang, H.-L., Olson, T. S., & Dice, J. F. (1992). Protein and peptide binding and stimulation of in vitro lysosomal proteolysis by the 73-kDa heat shock cognate protein. *The Journal of Biological Chemistry, 267*, 9202–9209.

Valdor, R., Mocholi, E., Botbol, Y., Guerrero-Ros, I., Chandra, D., Koga, H., et al. (2014). Chaperone-mediated autophagy regulates T cell responses through targeted degradation of negative regulators of T cell activation. *Nature Immunology, 15*, 1046–1054.

Wang, Y., Martinez-Vicente, M., Kruger, U., Kaushik, S., Wong, E., Mandelkow, E. M., et al. (2009). Tau fragmentation, aggregation and clearance: The dual role of lysosomal processing. *Human Molecular Genetics, 18*, 4153–4170.

Wattiaux, R., Wattiaux-De Coninck, S., Ronveaux-Dupal, M., & Dubois, F. (1978). Isolation of rat liver lysosomes by isopycnic centrifugation in a metrizamide gradient. *The Journal of Cell Biology*, *78*, 349–368.

Yang, Z., & Klionsky, D. J. (2010). Eaten alive: A history of macroautophagy. *Nature Cell Biology*, *12*, 814–822.

Zhang, C., & Cuervo, A. M. (2008). Restoration of chaperone-mediated autophagy in aging liver improves cellular maintenance and hepatic function. *Nature Medicine*, *14*, 959–965.

CHAPTER SEVENTEEN

Quantitative Assay of Macroautophagy Using Pho8Δ60 Assay and GFP-Cleavage Assay in Yeast

Y. Araki*, S. Kira*, T. Noda*,†,1

*Center for Frontier Oral Science, Graduate School of Dentistry, Osaka University, Osaka, Japan
†Graduate School of Frontier BioSciences, Osaka University, Osaka, Japan
[1]Corresponding author: e-mail address: takenoda@dent.osaka-u.ac.jp

Contents

Abstract

It is intrinsically difficult to directly measure a specific protein reduction that is mediated by nonselective autophagy, because the degradation proceeds at a relatively slow pace, and the residual nondegraded part becomes the background. Several methods for measuring nonselective autophagy have been reported in the past. One classical simple method is called bulk degradation assay, which measures the release of degraded amino acids after radioisotope labeling of the total cellular proteins. In 1995, we developed a quantitative Pho8Δ60 assay in the yeast, *Saccharomyces cerevisiae*, for studying autophagy, which is widely used for its advantages that are described in the following sections. Another method used in recent times is the GFP-based processing assay in yeast. We will describe these two methods in this chapter.

Methods in Enzymology, Volume 588
ISSN 0076-6879
http://dx.doi.org/10.1016/bs.mie.2016.10.027

1. INTRODUCTION TO PHO8Δ60 ASSAY

Pho8Δ60 assay employs the yeast endogenous enzyme, vacuolar alkaline phosphatase (ALP), encoded by the *PHO8* gene (Kaneko, Toh-e, & Oshima, 1982). ALP has long been considered a nonspecific ALP. However, recently, it has been found to function as a nucleotidase, generating nucleosides derived from autophagy-induced RNA degradation. Its null mutant leads to accumulation of 3′-NMPs (Huang et al., 2015). In most cases, the physiological relevance of the absence of this enzyme would be quite minor.

ALP/Pho8 is a type II single membrane spanning protein, and it is synthesized as an enzymatically inactive proform. After insertion into the ER membrane, it transits through the Golgi and reaches the vacuole. In the vacuole, the proregion of the C-terminal proenzyme is cleaved by vacuolar proteases (Pep4/Prb1) to generate the enzymatically active form (Klionsky & Emr, 1989). By deleting the N-terminal 60 amino acid fragment of Pho8, which forms the cytosolic and transmembrane region, Pho8Δ60 cannot enter the ER and instead, resides in the cytosol as a soluble protein. When nonselective autophagy is induced, Pho8Δ60 is incorporated into the autophagosome as a nonselective component together with the other cytosolic proteins and transported to a vacuole by autophagy, following which, it immediately becomes the enzymatically mature form. Therefore, ALP activity is proportional to the amount of cytosol delivered to the vacuole by autophagy. Because ALP is an endogenous vacuolar enzyme, it is inherently resistant to degradation by vacuolar proteases. Therefore, it is possible to estimate the total amount of the enzyme delivered for autophagic activity. As it is an enzymatic assay using fluorophotometry, it is very sensitive compared to the other methods such as Western blotting-based green fluorescent protein (GFP)-cleavage assay. Therefore, Pho8Δ60 assay is suitable for analyzing the subtle effects of experimental perturbation.

This method could be further extended to analyze selective autophagy. For monitoring the degradation of mitochondria, a chimera of mitochondrial protein COXIV-Pho8Δ60 was used (Campbell & Thorsness, 1998). Degradation of ER by autophagy is analyzed by chimeric forms of ER membrane proteins such as Sec63, fused with Pho8Δ60 (Schuck, Gallagher, & Walter, 2014).

2. STANDARD METHOD OF PHO8Δ60 ASSAY

2.1 Construction of a Yeast Strain Expressing Pho8Δ60

Budding yeast strains expressing Pho8Δ60 protein from endogenous locus can be generated by transforming cells with PCR-amplified DNA. The two oligonucleotide primers used for PCR amplification contain approximately 50 bases homologous to the target locus; thus, the amplified DNA has a 50 bp homology with the target locus at its ends, which serves as sites for chromosomal integration mediated by homologous recombination. To delete 60 amino acids at the N-terminus of Pho8, PCR-amplified DNA has to be integrated at the 5′ end of *PHO8*, which would destroy its own promoter. To drive the expression of the N-terminally deleted protein, we use a GPD promoter derived from *TDH3* gene.

Additionally, it is possible to use a 2 μm ori, *URA3* plasmid expressing Pho8Δ60, under the control of GPD promoter (pTN3) (Noda, Matsuura, Wada, & Ohsumi, 1995). Although plasmids containing the 2 μm ori are maintained at a high copy number up to 200 copies per cell, these types of plasmids require continuous selection; otherwise, they will be lost from growing populations. In contrast, chromosomal integration of the expression construct using PCR-based gene targeting enables stable, selection-free culturing of the modified yeast strains.

1. Procurement of plasmids (PCR-TOOLBOX Knop) through EUROSCARF (http://www.euroscarf.de/search.php?search=&project=Projects&gene=PCR-TOOLBOX%20Knop&selectedProject=Projects). GPD promoter derived from *TDH3* gene is cloned into cassettes with *kanMX4* (pYM-N14) and *natNT2* (pYM-N15 and pYM-N16) selection markers. pYM-N16 allows the substitution of the promoter upstream of *PHO8* and the introduction of an N-terminal 3-HA tag to the gene simultaneously (Janke et al., 2004).
2. To design primers for PCR, the sequences of the 500 bp upstream and *PHO8* are retrieved from the *Saccharomyces* Genome Database (http://www.yeastgenome.org/). Two oligonucleotide primers are synthesized. The first one is the PHO8-S1 with 45–55 bases upstream of the ATG, including the start codon of the *PHO8*, followed by 5′-cgt acg ctg cag gtc gac-3′. The second is the PHO8-S4 designed by adding 45–55 bases downstream of 60th amino acid (excluding the 60th codon) to the end of the following sequence, 5′-cat cga tga att ctc tgt cg-3′. The

oligonucleotide primers are ordered antiparallel to this sequence. We routinely use PHO8-S1 (CAT ATC AGC ATA CGG GAC ATT ATT TGA ACG CGC ATT AGC AGC ATG cgt acg ctg cag gtc gac) and PHO8-S4 (ATC CGT CAC GAA GAA TAT GAC ATT CTT CTT CTT GTG TGA TGC AGA cat cga tga att ctc tgt cg).

3. PCR reaction is set up in a total volume of 50 μL, using 32 μL water, 1 μL of 100 ng/μL template DNA, 1.5 μL of 10 μ*M* PHO8-S1, 1.5 μL of 10 μ*M* PHO8-S4, 5 μL of 10 × Buffer for KOD-plus Ver. 2, 5 μL of 2 m*M* dNTPs, 3 μL of 25 m*M* $MgSO_4$, and 1 μL of KOD-plus.
4. PCR reaction is performed using the following program on a thermocycler: (1) 94°C for 2 min, (2) 98°C for 10 s, (3) 53°C for 30 s, (4) 68°C for 1 min/kb, (5) repeat steps 2–4 30 times, (6) 4°C, until the samples are removed. To check whether amplification yielded PCR products, 3–5 μL of each reaction mixture is electrophoresed on a 1% agarose gel containing 0.5 μg/mL ethidium bromide (size of product pYM-N14: 2143 bp, pYM-N15: 1983 bp, pYM-N16: 2133 bp).
5. Yeast strain suitable for the selection marker is chosen and cultured to obtain 400 μL competent cells per 50 mL of cell culture. Yeast transformation is based on the lithium acetate method as described previously (Knop et al., 1999). Transformation reaction should include 50 μL of competent cells and 5 μL of the unpurified PCR product. After heat shock treatment, the cells are resuspended in 1 mL of YPD (1% yeast extract, 2% peptone, and 2% glucose) and incubated at 30°C for 3 h to allow the expression of antibiotic resistance genes. Cells are plated on a selection plate and incubated at 30°C. For antibiotic selection, G418 (200 μg/mL) for *kanMX4* or Nourseothricin (100 μg/mL) for *natNT2* is added to YPD plates. If the background for transiently antibiotic-resistant cells is high, the cells need to be transferred by replica plating onto a fresh selective plate. Three transformed colonies are picked with a sterile toothpick and streaked on a fresh selective plate to isolate a single colony.
6. Analytical colony PCR is used to check whether the cassette has integrated into the correct chromosomal location using PHO8 check primer-F (CAT TGA CGA TAG CGA TAA GC) and PHO8 check primer-R (TGG GCC TCC ATG TCG CTG G). Correct integration should give an approximately 500-bp band for pYM-N14, approximately 250-bp band for pYM-N15 and pYM-N16. To check for expression of 3HA-Pho8Δ60 (pYM-N16), whole yeast cell extract with

the NaOH/TCA method is analyzed by SDS-PAGE, followed by Western blotting using anti-HA antibody (Kaufmann & Knop, 2011).

2.2 Activity Measurement

Two methods that differ in substrate and instrumentation can be used to measure the activity of vacuolar ALP. Here, we describe a spectrofluorometric assay that we have developed (Noda et al., 1995). For spectrophotometric assay, please refer to Noda and Klionsky (2008) or Klionsky, Cuervo, and Seglen (2007). *Saccharomyces cerevisiae* has the other ALP, Pho13, which is localized in the cytoplasm. Tuleva et al. showed that Pho13 has a strong preference for *p*-nitrophenyl phosphate, which is used as a substrate for spectrophotometric assay. However, the activity of Pho13 on α-naphthyl phosphate, a substrate for spectrofluorometric assay, is very low (Tuleva, Vasileva-Tonkova, & Galabova, 1998). Therefore, it is not necessary to delete the *PHO13* gene for the assay described later.

1. Strains of interest harboring *Pho8Δ60* gene, together with appropriate positive (*pho8Δ60*) and negative (*pho8Δ60*, with a deletion of one of the *ATG* genes essential for macroautophagy) control strains, are inoculated in a liquid growth medium such as YPD or SD/CA medium (0.17% yeast nitrogen base w/o amino acids and ammonium sulfate, 0.5% ammonium sulfate, 0.5% casamino acids, 2% glucose, and auxotrophic amino acids, and Uracil as needed) and grown to mid-log phase (OD_{600} ~ 1.5). For better induction of macroautophagy, the OD_{600} should be over 1.0.
2. Cells are collected by centrifugation (3000 × *g*) at room temperature for 2 min and supernatant discarded. The cells are washed twice with an equivalent volume of SD-N medium (0.17% yeast nitrogen base w/o amino acids and ammonium sulfate, and 2% glucose). For induction of macroautophagy, the cells are resuspended in SD-N medium and incubated at 30°C with shaking. At each point, 4–6 OD_{600} units of cells are harvested by centrifugation at 3000 × *g* for 2 min at 4°C, and the medium discarded. We usually take samples at 0, 2, and 4 h after starvation treatment. The cell pellets can be kept at −80°C until they are ready for processing. Alternatively, macroautophagy may be induced by the addition of rapamycin (1 mg/mL stock in 90% ethanol and 10% Tween-20). However, we have found that ALP activity induced by rapamycin (0.2 μg/mL final concentration) is lower than that induced by SD-N medium, for unknown reasons.

3. The cell pellet is suspended in 0.1 mL chilled assay buffer (250 m*M* Tris–HCl, pH 9.0; 10 m*M* $MgSO_4$, and 10 μ*M* $ZnSO_4$), transferred to a 1.5-mL microcentrifuge tube, centrifuged at 3000 × *g* for 2 min at 4°C, and the supernatant removed completely by aspiration. The cell pellet is resuspended in 100 μL of chilled assay buffer containing 1 m*M* PMSF. It is important that PMSF (100 m*M* stock in ethanol or 2-propanol) be added last, because it is rapidly degraded in water. Acid-washed glass beads (425–600 μm) are added to a level just below the meniscus. The beads should not be more than the cell suspension volume, because lysis is insufficient, if the beads are above the liquid level in the tube. The tubes are placed on ice for at least 5 min. The cells are then lysed by mixing the tubes vigorously on a vortex for 15 min at 4°C, either in a cold room or in a refrigerated chamber. Additionally, 50 μL of chilled assay buffer containing 1 m*M* PMSF is added and mixed well. The lysed cells are centrifuged at 15,000 × *g* for 5 min at 4°C to remove unlysed cells and debris. The supernatant is used for subsequent analysis.
4. Ten microliters of the cell extract is added to a tube containing 0.5 mL of assay buffer. Setting up the assay in triplicate is recommended. The tubes are placed at 30°C for at least 5 min.
5. The assay reaction is initiated by the addition of 50 μL of 55 m*M* α-naphthyl phosphate disodium salt dissolved in assay buffer and mixing well. If there are multiple tubes for the assay, the initiation should be staggered at 10–15 s intervals to achieve exactly the same performance. The tubes are incubated at 30°C for 10 min or an appropriate time.
6. The reaction is stopped by adding 0.5 mL of stop buffer (2 *M* glycine–NaOH (pH 11.0)) at 10–15 s intervals and mixing well. The tubes are kept on ice until all the samples are ready.
7. Fluorescence in the reaction tube is measured at a wavelength of 345 nm for excitation and 472 nm for emission. Special care should be taken to ensure that all the measurements are in the linear range of the spectrofluorometer.
8. The protein concentrations of the cell extracts are determined by the BCA assay using the Pierce BCA Protein Assay Kit (Thermo Fisher Scientific, USA) according to the manufacturer's protocol. A standard curve with BSA (0–20 μg) is used as reference.
9. ALP activity is calculated as emission per unit protein in the reaction (mg) and the reaction time (min) (Fig. 1).

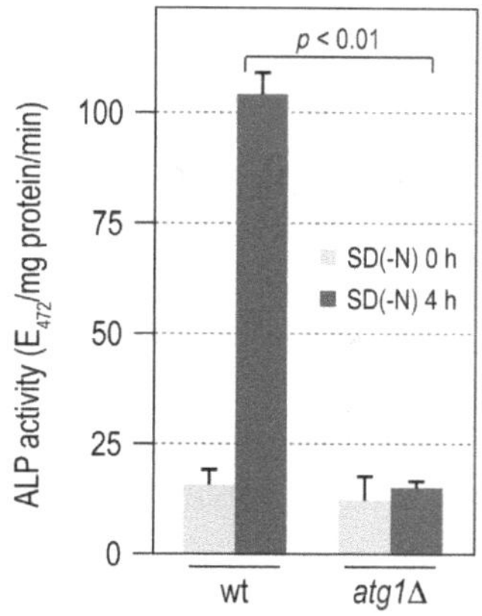

Fig. 1 Pho8Δ60 assay to follow and quantify macroautophagy. Wild-type and *atg1*Δ cells expressing pho8Δ60 were grown in growth medium (YPD) and shifted to starvation medium (SD-N) for 4 h. Samples were collected and lysates from each group of cells were tested for alkaline phosphatase (ALP) activity to monitor the maturation of Pho8Δ60 proform. ALP activity is shown as mean ± SD of three-independent experiments. Significance of differences was determined using an unpaired two-tailed Student's *t* test. The inactive proform is transported into the vacuole in an autophagy-dependent manner and processed to an active form, the activity of which quantitatively reflects the progress of autophagy. In wild-type cells, the ALP activity increased in response to starvation, whereas in autophagy-deficient *atg1*Δ cells no elevation of ALP activity was observed.

Note: It is recommended that the Pho8Δ60 protein amount in each experiment be evaluated by Western blot using anti-Pho8 antibody (e.g., 1D3A10 available from Abcam plc, UK) for pYM-N14 and pYM-N15 or anti-HA antibody (e.g., 12CA5 commercially available) for pYM-N16 to avoid misinterpretation of results. The ALP activity is largely influenced by changes in the expression level of the Pho8Δ60 protein in a given genetic background or specific culture conditions. One reason for these changes may be the use of GPD promoter derived from *TDH3*. If there is a difference between samples, one way to circumvent the problem is to use a different promoter, such as a TEF promoter instead of a GPD promoter. TEF promoter cloned into cassettes with *kanMX4* (pYM-N18) and *natNT2* (pYM-N19 and pYM-N20) are available from EUROSCARF. Another issue could be that mutants of genes involved in the regulation of membrane trafficking processes may have some defects in the delivery or processing of vacuolar enzymes such as Pep4 and Prb1 (Huang et al., 2015; Kraft, Deplazes, Sohrmann, & Peter, 2008). It is recommended that the maturation of these enzymes under the experimental conditions be evaluated.

3. GENOME WIDE PHO8Δ60 ASSAY

In addition to the test tube-based method, a method utilizing 96-well plate is also available (Kira et al., 2014). For genome-wide evaluation of autophagic activity by Pho8Δ60 assay, we utilized synthetic genetic array (SGA) method for genomic insertion of Pho8Δ60 allele into mutant library. Originally, SGA method was developed for the systematic construction of a double mutant library by mating a particular deletion mutant with yeast knock-out library (Tong et al., 2001). In this section, we describe the method of library construction for Pho8Δ60 assay and 96-well plate-based Pho8Δ60 assay.

3.1 Library Construction

1. For making SGA parental strain, Pho8Δ60 allele is introduced into SGA host stain harboring *Δcan1::MFA1pr-HIS3-MFα1pr-LEU2* allele (Tong et al., 2001), following the earlier section, ensuring that the mating type of host strain is the opposite of the objective library.
2. Parental Pho8Δ60 strain and the objective library strains are grown separately on rectangular YPD plate (omnitray, Thermo Fischer Scientific). Plating position is marked with a 96-pin replicator prior to plating. The plates are incubated at 30°C for 1–2 days.
3. Parental Pho8Δ60 and the objective library on the YPD plate are mated using a 96-pin replicator and incubated at 30°C for 1 day.
4. The cultures are passaged to diploid selection plate. In case the library contains kan resistance marker, the YPD selection plate should contain 100 μg/mL nourseothricin and 200 μg/mLG418. The plates are incubated at 30°C for 2 days.
5. For induction of sporulation, the cells are passaged with 96-well pin onto sporulation plate and incubated at 22°C for 5 days.
6. For selection of haploids, cells are passaged onto haploid selection plate and incubated at 30°C for 2 days.
7. Haploid selection is repeated one more time and incubated at 30°C for 1 day.
8. For selection of haploids containing Pho8Δ60 allele, cells are passaged onto haploid selection medium containing 100 μg/mL nourseothricin and incubated at 30°C for 2 days.
9. For selection of haploids containing both Pho8Δ60 allele and alleles derived from the objective library, cells are passaged onto haploid selection plate containing 100 μg/mL nourseothricin and 200 mg/mL G418 and incubated at 30°C for 1–2 days.

3.2 Medium Recipe

Diploid selection plate:

Yeast extract	10 g
Peptone	20 g
Agar	20 g

Dissolve in 950 mL dH_2O. After autoclaving, add 50 mL of 40% glucose, 1 mL of 100 mg/mL Nourseothricin, and 200 mg/mL G418.

Sporulation plate:

Potassium acetate	10 g
Yeast extract	1.0 g
Glucose	0.5 g
Amino acids add-back	0.1 g

(mixture of 2 g uracil, 2 g histidine, and 10 g leucine)

Agar	20 g

Dissolve in 1000 mL dH_2O.

Haploid selection plate:

Yeast nitrogen base w/o amino acids and ammonium sulfate	1.7 g
Monosodium glutamic acid	1.0 g
Glucose	0.5 g
Dropout mix w/o histidine and arginine (see later)	2.0 g

Dissolve in 950 mL dH_2O.

Agar	20 g

After autoclaving, add 50 mL of 40% glucose and 0.5 mL of 100 mg/mL canavanine.

Dropout mix w/o histidine and arginine:

Adenine	2.4 g
Uracil	2.4 g
Tryptophan	2.4 g
Methionine	2.4 g
Tyrosine	3.6 g
Leucine	7.2 g
Lysine	3.6 g
Phenylalanine	6.0 g
Asparagine	12.0 g
Threonine	24.0 g

3.3 Procedure for 96-Well Plate-Based Pho8Δ60 Assay

1. The strains are grown to early- to mid-log phase(OD_{600} 1.0–1.8) in YPD 0.5–1.2 mL media in 96-deep well plate with well volume of 2.5 mL (37001-518, VWR international) sealed with air-exchangeable seal (AB-0718, Thermo Fischer Scientific). The plate is incubated with shaking at 1200 rpm on a rotary shaker. To prevent sedimentation of cells during incubation, the rotary rate should not be less than 1200 rpm.
2. Cultures (300 μL) are transferred to flat bottom type 96-well plate.
3. The plate is centrifuged at 490 × *g* at 4°C for 5 min and the supernatant aspirated out with a multichannel micropipette. The cells are washed twice with 200 μL of SD-N medium (0.17% yeast nitrogen base without amino acids and ammonium sulfate and 2% glucose). For induction of autophagy, the cells are resuspended in 200 μL SD-N medium and incubated for 3–6 h at 30°C with shaking at 1200 rpm on a rotary shaker.
4. The cells are washed twice with 200 μL of ice-cold ALP lysis buffer (50 m*M* Tris–Cl, pH 9.0, 10 m*M* $MgSO_4$, 10 μ*M* $ZnSO_4$).
5. The cells are resuspended in 50 μL of ALP lysis buffer containing 1 m*M* PMSF and roughly 10 μL of zirconia–silica beads (φ=0.6 mm), and

the plate is sealed with parafilm. Using glass beads did not yield sufficient cell lysis.

6. The cells are lysed by mixing vigorously on a microplate rotary shaker at 2500 rpm for 10 min at 4°C.
7. The parafilm is removed after brief centrifugation; 150 μL of ALP lysis buffer containing 1 m*M* PMSF is added and centrifuged at 490 × *g* for 15 min at 4°C.
8. Twenty-five microliters of the lysate supernatant is transferred to a new 96-well plate, 200 μL of BCA assay working solution is added and incubated for 30 min at 37°C. The protein concentration is determined spectrophotometrically at 572 nm wavelength. For constructing the standard curve, 1–10 μL of 2 mg/mL BSA standard solution is dispensed and made upto 25 μL with the addition of ALP lysis buffer.
9. Fifty microliters of the lysate supernatant is transferred to another 96-well plate. The reaction is initiated by adding 50 μL of substrate solution (55 m*M* of α-naphthyl phosphate dissolved in 500 m*M* Tris–Cl, pH 9.0, 10 m*M* $MgSO_4$, and 10 μ*M* $ZnSO_4$) and allowed to proceed for 10 min at 30°C. For the blank solution, 50 μL of ALP buffer is aliquoted into an empty well. 100 μL of STOP buffer (2 *M* glycine–NaOH, pH 11.0) is added after 10 min. The fluorescence is measured at 345 nm for excitation and 472 nm for emission.
10. ALP activity is calculated as units of emission/mg protein/min.

4. GFP-CLEAVAGE ASSAY FOR AUTOPHAGY

In this assay, three proteins, Atg8, Atg1, and Pgk1, are tagged with the GFP. Atg8 is an ubiquitin-like protein required for autophagy. Atg8 is conjugated to phosphatidylethanolamine (PE) at the C-terminus and recruited to the isolation membrane, where it is involved in cargo recruitment and biogenesis of autophagosomes. A part of Atg8-PE is anchored to the inner membrane of the autophagosome and transported to the vacuole, where the autophagic bodies containing Atg8 are released into the vacuolar lumen. The breakdown of the autophagic bodies by vacuolar lipases leads to the degradation of Atg8. Atg1 is a serine–threonine kinase whose activity is indispensable for autophagy. Atg1 interacts with Atg8, followed by incorporation into the vacuole (Kraft et al., 2012; Nakatogawa et al., 2012). Pgk1 is an abundantly expressed enzyme involved in glycolysis and gluconeogenesis, and is not essential for autophagy. Pgk1 is shown to be distributed in the cytosol and be delivered randomly to the vacuole for degradation, primarily through macroautophagy (Welter, Thumm, & Krick, 2010).

Once GFP-Atg8, Atg1-GFP, or Pgk1-GFP protein is transported to the vacuole, proteolytic degradation of the fusion protein occurs. However, GFP is released from the GFP-fusion protein by vacuolar proteases and preserved intact in the vacuole, because of its high resistance to proteolysis. This free GFP level can be monitored and quantified by Western blot analysis using antibodies against GFP. Therefore, increasing GFP levels indicate higher macroautophagy rates.

Although this is a less precise quantitative method compared to the ALP assay, GFP-cleavage assay is an easy method to follow and quantify nonselective macroautophagy in yeast, because this does not require special devices such as a spectrofluorometer. GFP-cleavage assay can be recommended as a second option.

4.1 Construction of a Yeast Strain for GFP-Cleavage Assay

1. One of the PCR cassettes, pYM12 and pYM25-29 (PCR-TOOLBOX Knop) (Janke et al., 2004; Knop et al., 1999) is procured through EUROSCARF as described earlier. They contain a GFP derivative in combination with *kanMX6* or *hphNT1* marker or a *klTPR1* gene.
2. Two primers for Atg1 or Pgk1 are synthesized. One is the S2 primer designed by adding 45 bases downstream of the STOP codon (including the STOP) to the end of the following sequence, 5′-cga gct cga att cat cga t-3′. Oligonucleotides antiparallel to this sequence (ATG1-S2: CAG GTC ATT TGT ACT TAA TAA GAA AAC CAT ATT ATG CAT CAC TTA atc gat gaa ttc gag ctc g, PGK1-S2: GAA AAG AAA AAA ATT GAT CTA TCG ATT TCA ATT CAA TTC AAT TTA atc gat gaa ttc gag ctc g) are procured. The other is the S3 primer with 45 bases upstream of the STOP codon excluding STOP of the gene, followed by 5′-cgt acg ctg cag gtc gac-3′ (ATG1-S3: GCA AAC AGG TTG AAA ATA TTG AGG CAG AAG ATG AAC CAC CAA AAT cgt acg ctg cag gtc gac, PGK1-S3: GAA GGT AAG GAA TTG CCA GGT GTT GCT TTC TTA TCC GAA AAG AAA cgt acg ctg cag gtc gac). PCR reaction is performed, as described in the previous section.
3. In the case of Atg8, C-terminal tagging cannot be used due to its conjugation to PE. Therefore, either pRS316-GFP-ATG8 (CEN plasmid-borne GFP-Atg8 from the endogenous promoter) (Suzuki et al., 2001) or pRS306-GFP-ATG8 (integrating plasmid-borne GFP-Atg8 from the

endogenous promoter) is prepared (Suzuki, Kubota, Sekito, & Ohsumi, 2007). pRS306-GFP-ATG8 has to be digested with *Sna*BI for integration into *ATG8* locus.

4. A yeast strain suitable for the selection marker is chosen. Yeast transformation is based on the standard lithium acetate/ssDNA/polyethylene glycol method as described in the previous section.
5. Expression of GFP-Atg8, Atg1-GFP, and Pgk1-GFP is confirmed by Western blotting using anti-GFP antibody. The expected sizes of GFP-Atg8, Atg1-GFP, and Pgk1-GFP are 37, 126, and 75 kD, respectively.

4.2 GFP-Cleavage Assay

1. Strains of interest expressing either GFP-Atg8, Atg1-GFP, or Pgk1-GFP, together with appropriate positive (wild-type expressing GFP-fusion proteins) and negative (deletion strains of one of the *ATG* genes essential for bulk autophagy expressing GFP-fusion proteins) control strains, are inoculated into liquid growth medium such as YPD or SD/CA medium. In the case of pRS316-GFP-ATG8, it is necessary to use SD/CA lacking Uracil. Cells are grown to mid-log phase ($OD_{600}\sim 1.5$) on rotator at 30°C. For better induction of autophagy, the OD_{600} should be over 1.0.
2. Macroautophagy is induced by culturing cells in SD-N medium or adding rapamycin (0.2 μg/mL final concentration) to the culture. Two to three OD_{600} units of cells are collected by centrifugation at 5000 × *g* for 1 min at 4°C, and the supernatant discarded. We usually draw samples at intervals of 2 h, for 8 h after starvation treatment. Cell pellets can be frozen in liquid nitrogen and stored at −80°C.
3. The cell pellet is resuspended in 500 μL of ice-cold water with 75 μL of 1.85 *M* NaOH and mixed on a vortex. The samples are incubated for 10 min on ice, and 43 μL 100% TCA (stored in the dark) is added, followed by vortexing and an additional 10 min incubation on ice. The precipitated protein is pelleted by centrifugation at 15,000 × *g* for 15 min at 4°C. The supernatant is discarded completely by aspiration, and the protein pellets are resuspended in 100 μL of HU-buffer (8 *M* urea, 200 m*M* sodium phosphate buffer pH 6.8, 5% SDS, 1 m*M* EDTA, 100 m*M* DTT, traces of bromophenol blue as coloring, and pH indicator). If the pH of the samples becomes particularly low due to residual traces of TCA (yellow color), 1–3 μL of 2 *M* Tris-base should be added

for neutralization of the samples. Otherwise, some proteins are degraded in acidic conditions.

4. Starved cells are extremely difficult to lyse compared to nontreated cells; hence, lysis by beads is required. Zirconia beads (Yasui kikai, Japan), 0.5 mm in size, are added to a level below the meniscus. Cells are lysed in FastPrep (speed 5.5, 30 s), with subsequent heating at 65°C for 15 min, followed by vortexing for another 2 min at room temperature. Lysate is separated from the beads by hot needle puncture method. The tube is inverted and the bottom punctured with heated needle (19G). Then, the lid is punctured. The tube is placed on top of a new 1.5 mL tube and centrifuged at 2000 × g for 1 min at room temperature. Then the extract is centrifuged at 15,000 × *g* for 10 min at room temperature to pelletize the insoluble material. 10–20 μL of the samples are loaded on a 10–12% acrylamide gel, and proteins are separated by SDS-PAGE, followed by Western blotting. The expected size of the free GFP fragment, which is a product of vacuolar degradation, is 25 kD.
5. Bands of GFP-Atg8, Atg1-GFP, Pgk1-GFP, and free GFP are detected using a digital imaging system for quantitative imaging of blots by chemiluminescence and fluorescence. The intensities of free GFP and GFP-fused full-length protein bands at each time point are quantified using ImageJ or an equivalent software. The degradation of GFP-fused proteins is expressed as the ratio of the intensity of the free GFP band to the sum of free GFP and GFP-fused full-length protein band intensities and normalized to that of wild-type cells at the final time point (set to 100%). Care should be exercised to ensure that all the measurements are in the linear range of the digital imaging system.

REFERENCES

Campbell, C. L., & Thorsness, P. E. (1998). Escape of mitochondrial DNA to the nucleus in yme1 yeast is mediated by vacuolar-dependent turnover of abnormal mitochondrial compartments. *Journal of Cell Science*, *111*, 2455–2464.

Huang, H., Kawamata, T., Horie, T., Tsugawa, H., Nakayama, Y., Ohsumi, Y., et al. (2015). Bulk RNA degradation by nitrogen starvation-induced autophagy in yeast. *The EMBO Journal*, *34*, 154–168.

Janke, C., Magiera, M. M., Rathfelder, N., Taxis, C., Reber, S., Maekawa, H., et al. (2004). A versatile toolbox for PCR-based tagging of yeast genes: New fluorescent proteins, more markers and promoter substitution cassettes. *Yeast*, *21*, 947–962.

Kaneko, Y., Toh-e, A., & Oshima, Y. (1982). Identification of the genetic locus for the structural gene and a new regulatory gene for the synthesis of repressible alkaline phosphatase in *Saccharomyces cerevisiae*. *Molecular and Cellular Biology*, *2*, 127–137.

Kaufmann, A., & Knop, M. (2011). Genomic promoter replacement cassettes to alter gene expression in the yeast *Saccharomyces cerevisiae*. *Methods in Molecular Biology*, *765*, 275–294.

Kira, S., Tabata, K., Shirahama-Noda, K., Nozoe, A., Yoshimori, T., & Noda, T. (2014). Reciprocal conversion of Gtr1 and Gtr2 nucleotide-binding states by Npr2-Npr3 inactivates TORC1 and induces autophagy. *Autophagy*, *10*, 1–14.

Klionsky, D. J., Cuervo, A. M., & Seglen, P. O. (2007). Methods for monitoring autophagy from yeast to human. *Autophagy*, *3*, 181–206.

Klionsky, D. J., & Emr, S. D. (1989). Membrane protein sorting: Biosynthesis, transport and processing of yeast vacuolar alkaline phosphatase. *The EMBO Journal*, *8*, 2241–2250.

Knop, M., Siegers, K., Pereira, G., Zachariae, W., Winsor, B., Nasmyth, K., et al. (1999). Epitope tagging of yeast genes using a PCR-based strategy: More tags and improved practical routines. *Yeast*, *15*, 963–972.

Kraft, C., Deplazes, A., Sohrmann, M., & Peter, M. (2008). Mature ribosomes are selectively degraded upon starvation by an autophagy pathway requiring the Ubp3p/Bre5p ubiquitin protease. *Nature Cell Biology*, *10*, 602–610.

Kraft, C., Kijanska, M., Kalie, E., Siergiejuk, E., Lee, S. S., Semplicio, G., et al. (2012). Binding of the Atg1/ULK1 kinase to the ubiquitin-like protein Atg8 regulates autophagy. *The EMBO Journal*, *31*, 3691–3703.

Nakatogawa, H., Ohbayashi, S., Sakoh-Nakatogawa, M., Kakuta, S., Suzuki, S. W., Kirisako, H., et al. (2012). The autophagy-related protein kinase Atg1 interacts with the ubiquitin-like protein Atg8 via the Atg8 family interacting motif to facilitate autophagosome formation. *The Journal of Biological Chemistry*, *287*, 28503–28507.

Noda, T., & Klionsky, D. J. (2008). The quantitative Pho8Delta60 assay of nonspecific autophagy. *Methods in Enzymology*, *451*, 33–42.

Noda, T., Matsuura, A., Wada, Y., & Ohsumi, Y. (1995). Novel system for monitoring autophagy in the yeast *Saccharomyces cerevisiae*. *Biochemical and Biophysical Research Communications*, *210*, 126–132.

Schuck, S., Gallagher, C. M., & Walter, P. (2014). ER-phagy mediates selective degradation of endoplasmic reticulum independently of the core autophagy machinery. *Journal of Cell Science*, *127*, 4078–4088.

Suzuki, K., Kirisako, T., Kamada, Y., Mizushima, N., Noda, T., & Ohsumi, Y. (2001). The pre-autophagosomal structure organized by concerted functions of APG genes is essential for autophagosome formation. *The EMBO Journal*, *20*, 5971–5981.

Suzuki, K., Kubota, Y., Sekito, T., & Ohsumi, Y. (2007). Hierarchy of Atg proteins in pre-autophagosomal structure organization. *Genes to Cells*, *12*, 209–218.

Tong, A. H., Evangelista, M., Parsons, A. B., Xu, H., Bader, G. D., Page, N., et al. (2001). Systematic genetic analysis with ordered arrays of yeast deletion mutants. *Science*, *294*, 2364–2368.

Tuleva, B., Vasileva-Tonkova, E., & Galabova, D. (1998). A specific alkaline phosphatase from *Saccharomyces cerevisiae* with protein phosphatase activity. *FEMS Microbiology Letters*, *161*, 139–144.

Welter, E., Thumm, M., & Krick, R. (2010). Quantification of nonselective bulk autophagy in *S. cerevisiae* using Pgk1-GFP. *Autophagy*, *6*, 794–797.

CHAPTER EIGHTEEN

Monitoring the Formation of Autophagosomal Precursor Structures in Yeast *Saccharomyces cerevisiae*

R. Gómez-Sánchez, J. Sánchez-Wandelmer, F. Reggiori[1]
University of Groningen, University Medical Center Groningen, Groningen, The Netherlands
[1]Corresponding author: e-mail address: f.m.reggiori@umcg.nl

Contents

Abstract

The budding yeast *Saccharomyces cerevisiae* is a powerful and versatile model organism for studying multiple aspects of the biology of eukaryotic cells, including the molecular principles underlying autophagy. One of the unique advantages of this unicellular system is its amenability to genetic and biochemical approaches, which had a pivotal role in the discovery and characterization of most of the autophagy-related (Atg) proteins, the central players of autophagy. The relevance of investigating autophagy in this cell model lies in the high conservation of this pathway among eukaryotes, i.e., most of the yeast Atg proteins possess one or more mammalian orthologs. In addition to the experimental advantages, a very large collection of reagents keeps *S. cerevisiae* in a leading position for the study of the molecular mechanism and regulation of autophagy. In this chapter, we describe fluorescence microscopy and biochemical methods that allow to monitor in vivo the assembly the of Atg machinery, a key step of autophagy. These

Methods in Enzymology, Volume 588
ISSN 0076-6879
http://dx.doi.org/10.1016/bs.mie.2016.09.085

approaches can be very useful to those researchers that would like to assess the progression of the autophagosomal precursor structure formation under various conditions, in the presence of specific Atg protein mutants or in the absence of other factors.

1. INTRODUCTION

Autophagy is a highly conserved degradative pathway that allows cells to rapidly eliminate unwanted structures, including organelles and invading pathogens (Karanasios & Ktistakis, 2016). The basic mechanism of autophagy is the sequestration of the structures targeted to destruction by large double-membrane vesicles called autophagosomes (Fig. 1). Autophagosomes are formed by expansion and sealing of a small membrane cistern termed the phagophore or isolation membrane, which is generated at a location adjacent to the vacuole known in yeast as the phagophore assembly site or

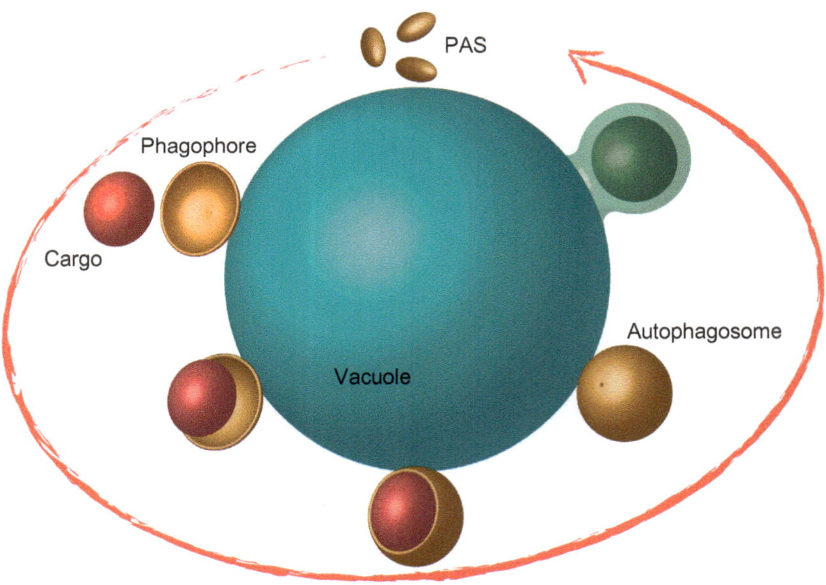

Fig. 1 Schematic overview of autophagosome biogenesis in yeast. The basic mechanism of autophagy is the sequestration of the structure(s) targeted to destruction by autophagosomes. These large double-membrane vesicles are formed adjacently to the vacuole by nucleation, expansion, and sealing of the phagophore. These events take places at the so-called PAS. Upon completion, autophagosome fuses with the vacuole releasing their internal vesicles and the cargo in the interior of this organelle, where they are turned over by resident hydrolases.

preautophagosomal structure (PAS) (Kim, Huang, Stromhaug, & Klionsky, 2002; Suzuki et al., 2001). Once completed, autophagosomes fuse with the vacuole (the yeast counterpart of the mammalian lysosome) delivering the cargo in the interior of this organelle, where it is degraded into metabolites that are recycled by the cell (Fig. 1).

Autophagy is subdivided into selective or nonselective types depending on the cargo. During the selective types of autophagy, specific structures, such as mitochondria (mitophagy), peroxisomes (pexophagy), and endoplasmic reticulum (ER-phagy), are exclusively sequestered into autophagosomes. Yeast also possesses a biosynthetic selective type of autophagy called the cytoplasm-to-vacuole targeting (Cvt) pathway, in which a large oligomer mostly composed by the Ape1 peptidase is transported into the vacuole (Scott et al., 1996). In contrast, there is what appears to be a random engulfment of cytosolic components during nonselective or bulk autophagy.

Autophagosome biogenesis is regulated through an orderly recruitment and action, and subsequent release of autophagy-related (Atg) proteins at the PAS (Suzuki, Kubota, Sekito, & Ohsumi, 2007). The core molecular machinery essential for autophagosome formation can be subdivided into five functional clusters: the Atg1 complex (composed in yeast by Atg1, Atg13, Atg17, Atg29, and Atg31), the Atg9 trafficking system (Atg9, Atg2, Atg18, Atg23, and Atg27), the phosphatidylinositol 3-kinase complex (Vps15, Vps34, Atg6, Atg14, and Atg38), and the two ubiquitin-like protein conjugation systems, the one centered around Atg8 (Atg8, Atg3, Atg4, Atg7) and the other connected to Atg12 (Atg5, Atg7, Atg10, Atg12, and Atg16) (Feng, He, Yao, & Klionsky, 2014). In addition, there are several additional Atg proteins essential to guarantee the specificity of the different selective types of autophagy, such as, for example, Atg11, an adaptor protein required for all these processes, but also autophagy receptors like Atg19 (for the Cvt pathway), Atg32 (for mitophagy), Atg36 (for pexophagy), and Atg39 and Atg40 (for ER-phagy). As a result, the analysis of the localization and the interactions of these Atg proteins permit to mechanistically assess the progression of bulk autophagy, but also specific selective types of autophagy. In this context, yeast presents a unique advantage compared to mammalian cells because autophagosomes are formed one after the other, and not several simultaneously (Geng, Baba, Nair, & Klionsky, 2008). Thus, autophagosome biogenesis is somehow synchronized at the cell level facilitating in particular imaging approaches.

After the discovery of Atg proteins and trafficking pathways linked to autophagy, such as the endosomal one and the early steps of secretion, an intense effort has started to be invested in unveiling the dynamic interrelationship and interactions between the different players in order to unveil and acquire molecular details into the mechanism of autophagosome biogenesis. Although approaches such as the yeast two-hybrid (Y2H) system (Fields & Song, 1989) have been pivotal in providing insights into certain interactions between Atg proteins, they do not provide important mechanistic information about Atg proteins including in vivo interrelationships, localization or the site of interaction. In this chapter, we describe a series of standard methods used to monitor PAS formation in yeast. In particular, we present fluorescence microscopy-based approaches to analyze the recruitment of Atg proteins to the PAS, their distribution on the phagophore, and the localization of their interaction. Moreover, we also describe a general pull-down protocol to biochemically study bindings between them.

2. STANDARD PROCEDURES

2.1 Yeast Culture

Yeast is mostly grown in shaking liquid precultures at 30°C overnight before being diluted in the same medium and regrown to a logarithmic phase (OD_{600} 0.8–1.2) the next day. Depending on the strain and the auxotrophic requirements, cells are typically grown in rich medium (YPD; 1% yeast extract, 2% peptone, 2% glucose) or synthetic minimal media (SMD; 0.67% yeast nitrogen base, 2% glucose, amino acids, and vitamins as needed).

2.1.1 Yeast Preparation for Transformation or Cvt Pathway Analyses

Cells can be directly transformed or imaged after they have reached a logarithmic growth phase.

2.1.2 Sample Preparation for Bulk Autophagy Analysis

Nitrogen starvation is the most commonly used condition to induce bulk autophagy in yeast. To this aim, cells in a logarithmic growth phase are transferred into a synthetic medium lacking nitrogen (SD-N; 0.17% yeast nitrogen base without amino acids and ammonium sulfate, 2% glucose).

1. Centrifuge 10 mL of cell culture in 15-mL conical centrifuge tubes at 3750 × *g* for 3 min.
2. Discard the supernatant and wash the cells with 10 mL of sterile distilled water to completely remove the rich medium.

3. Centrifuge the suspension at $3750 \times g$ for 3 min and discard the supernatant.
4. Resuspend the cell pellet in 10 mL of SD-N and transfer the suspension in a culture tube.
5. Incubate the cell culture in a shaking incubator at 30°C for the time determined by the experiment design.

Alternatively to the SD-N medium treatment, autophagy can be induced by directly adding 200 ng/mL of rapamycin (dissolved in ethanol–Tween 40 (9:1) or dimethyl sulfoxide) to rich culture media to mimic nitrogen starvation conditions (Kamada et al., 2000). Rapamycin is a specific inhibitor of the TOR kinase (Kamada et al., 2000). For analysis of selective types of autophagy, such as mitophagy, pexophagy, reticulophagy, or ribophagy, specific media are required for culture and induction of the degradative process (Hutchins, Veenhuis, & Klionsky, 1999; Kissova, Deffieu, Manon, & Camougrand, 2004; Kraft, Deplazes, Sohrmann, & Peter, 2008; Mochida et al., 2015).

2.2 Strain and Plasmid Generation

2.2.1 Strain Construction

Although plasmids overexpressing fluorescently-tagged Atg proteins have been used in the past to analyze their subcellular localization, the most rapid approach to generate this type of reporter fusion proteins is by polymerase chain reaction (PCR)-based integration at the 3′end of *ATG* genes of the sequence encoding for the different fluorescent proteins (Tables 1 and 2) (Longtine et al., 1998). This is possible when the C-terminal tagging does not affect the functionality of the Atg protein of interest. The major advantage of this approach is that the resulting chimeras are under the control of the endogenous promoter. Since its discovery, green fluorescent protein (GFP) (fluorescence excitation wavelength (λ_{ex}): 488 nm; fluorescence emission wavelength (λ_{em}): 509 nm) has been the most used protein to generate fluorescent fusion proteins (Klionsky et al., 2016). However, generation of new GFP variants, such as blue fluorescent protein (BFP; λ_{ex}: 383 nm; λ_{em}: 448 nm), cyan fluorescent protein (CFP) (λ_{ex}: 433 nm; λ_{em}: 475 nm), yellow fluorescent protein (YFP) (λ_{ex}: 513 nm; λ_{em}: 527 nm), monomeric red fluorescent protein (mRFP) (λ_{ex}: 584 nm; λ_{em}: 607 nm), or mCherry (λ_{ex}: 587 nm; λ_{em}: 610 nm), has strongly promoted live-cell imaging experiments simultaneously analyzing more than one protein (Day & Davidson, 2009). Other protein tags, including hemagglutinin (HA), myc, FLAG, glutathione S-transferase (GST), or tandem affinity

Table 1 Functional N- and C-Terminal Taggings of Atg Proteins That Are Part of the Core Machinery

Atg Protein	N-Terminal Tagging	C-Terminal Tagging
Atg1	Cheong, Nair, Geng, and Klionsky (2008), He, Baba, Cao, and Klionsky (2008), Kim et al. (2002), Straub, Bredschneider, and Thumm (1997), and Suzuki et al. (2001)	Cheong et al. (2005), Kawamata, Kamada, Kabeya, Sekito, and Ohsumi (2008), Reggiori, Tucker, Stromhaug, and Klionsky (2004), Stromhaug, Reggiori, Guan, Wang, and Klionsky (2004), and Suzuki et al. (2007)
Atg2	Guan et al. (2001), He et al. (2008), Shintani, Suzuki, Kamada, Noda, and Ohsumi (2001), van der Vaart, Griffith, and Reggiori (2010), and Wang et al. (2001)	Kawamata et al. (2008), Kobayashi, Suzuki, and Ohsumi (2012), Ma et al. (2008), Suzuki et al. (2007), and Watanabe et al. (2012)
Atg3	Tag inserted within amino acids 117–118 (Ngu, Hirata, and Suzuki, 2015) or 265–266 (Sakoh-Nakatogawa, Kirisako, Nakatogawa, and Ohsumi, 2015)	
Atg4	Lang et al. (1998) and Perez-Perez, Zaffagnini, Marchand, Crespo, and Lemaire (2014)	Kim, Huang, et al. (2001), Kim, Kamada, et al. (2001), Lang et al. (1998), Ma et al. (2008), and Nakatogawa, Ishii, Asai, and Ohsumi (2012)
Atg5	Chang and Huang (2007), Mizushima et al. (1998), and Mizushima, Noda, and Ohsumi (1999)	George et al. (2000), Guan et al. (2001), Kim, Huang, et al. (2001), Kim, Kamada, et al. (2001), Kim et al. (2002), Stromhaug et al. (2004), and Suzuki et al. (2001)
Atg6	Kihara, Noda, Ishihara, and Ohsumi (2001)	Araki et al. (2013), Noda et al. (2012), Obara, Sekito, and Ohsumi (2006), Suzuki et al. (2007), and Suzuki, Akioka, Kondo-Kakuta, Yamamoto, and Ohsumi (2013)
Atg7	—	Hanada and Ohsumi (2005), Ichimura et al. (2000), Kim, Dalton, Eggerton, Scott, and Klionsky (1999), Ma et al. (2008), and Yamazaki-Sato, Tanida, Ueno, and Kominami (2003)

Table 1 Functional N- and C-Terminal Taggings of Atg Proteins That Are Part of the Core Machinery—cont'd

Atg Protein	N-Terminal Tagging	C-Terminal Tagging
Atg8	Ichimura et al. (2000), Kim, Huang, et al. (2001), Kim, Kamada, et al. (2001), Kirisako et al. (1999), Kirisako et al. (2000), and Lang et al. (1998)	—
Atg9	Chang and Huang (2007), Kim et al. (2002), Lang, Reiche, Straub, Bredschneider, and Thumm (2000), and Wang et al. (2001)	Noda et al. (2000), Reggiori et al. (2004), Shintani and Klionsky (2004), Suzuki et al. (2001), and Tucker, Reggiori, Dunn, and Klionsky (2003)
Atg10	Shintani et al. (1999)	Hanada and Ohsumi (2005)
Atg12	Komatsu et al. (2001), Mizushima et al. (1998), Mizushima et al. (1999), Shintani et al. (1999), and Tanida et al. (1999)	—
Atg13	Chew et al. (2015) and He et al. (2008)	Graef, Friedman, Graham, Babu, and Nunnari (2013), Kawamata et al. (2008), Kraft et al. (2012) Stephan, Yeh, Ramachandran, Deminoff, and Herman (2009), and Suzuki et al. (2007)
Atg14	Kametaka, Okano, Ohsumi, and Ohsumi (1998), Kihara et al. (2001), and Kim et al. (2002)	Cheong et al. (2008), Obara et al. (2006), Reggiori et al. (2004), Stromhaug et al. (2004), and Suzuki et al. (2007)
Atg16	Mizushima et al. (1999)	Cheong et al. (2008), Matsushita et al. (2007), Nair, Cao, Xie, and Klionsky (2010), Suzuki et al. (2001), and Suzuki et al. (2007)
Atg17	Cheong et al. (2005), Kabeya et al. (2009), Kamada et al. (2010), Nice, Sato, Stromhaug, Emr, and Klionsky (2002), and Sekito, Kawamata, Ichikawa, Suzuki, and Ohsumi (2009)	Cao, Cheong, Song, and Klionsky (2008), Cheong et al. (2008), Kabeya, Kawamata, Suzuki, and Ohsumi (2007), Kawamata et al. (2008), and Suzuki et al. (2007)

Continued

Table 1 Functional N- and C-Terminal Taggings of Atg Proteins That Are Part of the Core Machinery—cont'd

Atg Protein	N-Terminal Tagging	C-Terminal Tagging
Atg18	Dove et al. (2004), Efe, Botelho, and Emr (2007), Krick, Tolstrup, Appelles, Henke, and Thumm (2006), Krick, Henke, Tolstrup, and Thumm (2008)	Barth, Meiling-Wesse, Epple, and Thumm (2001), Guan et al. (2001), Reggiori et al. (2004), Stromhaug et al. (2004), and Suzuki et al. (2007)
Atg23	—	Legakis, Yen, and Klionsky (2007), Ma et al. (2008), Reggiori et al. (2004), and Tucker et al. (2003)
Atg27	—	Legakis et al. (2007), Segarra, Boettner, and Lemmon (2015), Wang et al. (2012), and Yen, Legakis, Nair, and Klionsky (2007)
Atg29	—	Cao, Nair, Yasumura-Yorimitsu, and Klionsky (2009), Kawamata et al. (2005), Kawamata et al. (2008), Kraft et al. (2012), Mao, Chew, et al. (2013), Mao, Wang, et al. (2013), and Wang, Menon, et al. (2013)
Atg31	Feng et al. (2015)	Cao et al. (2009), Kabeya et al. (2007), Kabeya et al. (2009), Kawamata et al. (2008), and Ma et al. (2008)
Atg38	—	Araki et al. (2013)
Vps15	—	Araki et al. (2013), Budovskaya, Hama, DeWald, and Herman (2002), Heenan et al. (2009), Obara et al. (2006), and Suzuki et al. (2013)
Vps34	Grunau et al. (2011)	Araki et al. (2013), Grunau et al. (2011), Obara et al. (2006), and Slessareva, Routt, Temple, Bankaitis, and Dohlman (2006)

Up to five references demonstrating the N- and/or C-terminal taggings that permit to have a functional fusion protein are indicated.

Table 2 Functional N- and C-Terminal Taggings of Autophagy Adaptors and Receptors

Atg Protein (Function)	N-Terminal Tagging	C-Terminal Tagging
Atg11 (Cvt pathway)	Guan et al. (2001), Kim, Huang, et al. (2001), Kim, Kamada, et al. (2001), Kim et al. (2002), Nice et al. (2002), and Tucker et al. (2003)	Mochida et al. (2015) and Tanaka et al. (2014)
Atg19 (Cvt pathway)	Cao et al. (2008), Ho, Chang, and Huang (2009), Leber, Silles, Sandoval, and Mazon (2001), Pfaffenwimmer et al. (2014), Shintani, Huang, Stromhaug, and Klionsky (2002), and Yorimitsu and Klionsky (2005)	Guan et al. (2001), Kim et al. (2002), Nice et al. (2002), Scott, Guan, Hutchins, Kim, and Klionsky (2001), and Tucker et al. (2003)
Atg32 (Mitophagy)	Aoki et al. (2011), Bockler and Westermann (2014), Kanki, Wang, Cao, Baba, and Klionsky (2009), Mao, Wang, Zhao, Xu, and Klionsky (2011), and Wang, Jin, Liu, and Klionsky (2013)	Kondo-Okamoto et al. (2012) and Sakakibara et al. (2015)
Atg33 (Mitophagy)	—	Kanki, Wang, Baba, et al. (2009)
Atg34 (Mitophagy)	—	Mochida, Ohsumi, and Nakatogawa (2014) and Suzuki, Kondo, Morimoto, and Ohsumi (2010)
Atg36 (Pexophagy)	Mao, Liu, Feng, and Klionsky (2014)	Motley, Nuttall, and Hettema (2012) and Tanaka et al. (2014)
Atg39 (ER-phagy)	—	Mochida et al. (2015)
Atg40 (ER-phagy)	—	Mochida et al. (2015)

Up to five references demonstrating the N- and/or C-terminal taggings that permit to have a functional fusion protein are indicated.

purification (TAP), which allow protein detection by Western blot when specific antibodies are not available or facilitate biochemical experiments, can be fused at the C-terminus of proteins using the same PCR-based strategy (Longtine et al., 1998).

2.2.1.1 Integration of Sequences Encoding for Fluorescent and Tag Proteins at 3′End of Genes

To generate strains endogenously expressing C-terminally tagged fluorescent Atg chimeras, the sequence coding for the protein to be fused to the Atg protein with either an auxotrophic marker such as *HIS3, LEU2, TRP1, LYS2,* or *URA3*, or an antibiotic resistance gene like *kanMX6, natMX6, hphMX6,* or *bleMX6*, is integrated at the 3′end of the *ATG* gene open reading frame (ORF). These sequences are generated by PCR from standard template vectors using primers containing at least 50–60 bases of identity to the regions flanking the gene stop codon (Gueldener, Heinisch, Koehler, Voss, & Hegemann, 2002; Hentges, Van Driessche, Tafforeau, Vandenhaute, & Carr, 2005; Longtine et al., 1998).

PCR products are then directly transformed into yeast using the lithium acetate procedure or any other standard protocol (Gietz, St Jean, Woods, & Schiestl, 1992). Positive clones grown on the appropriate selection plates have to be subsequently verified for the correct insertion of the tag. One approach is to analyze either the *ATG* locus by PCR (Gueldener et al., 2002; Longtine et al., 1998) or the expression of C-terminally tagged Atg protein by Western blot, using antibodies against the protein or the tag. In the case of fluorescent fusion proteins, examination of the cells with a fluorescence microscope can also help to confirm the correct integration in the chromosomal locus.

The resulting strains express the fusion proteins under the control of their native promoters. A critical step is to assess the functionality of the Atg chimeras (Tables 1 and 2), and this can be done by either analyzing prApe1 maturation by Western blot (Kim, Scott, Oda, & Klionsky, 1997; Noda et al., 2000) or assessing autophagy using standard methods, such as Pho8Δ60 and GFP-Atg8 processing assays (Guimaraes, Delorme-Axford, Klionsky, & Reggiori, 2015).

2.2.1.2 Deletions of Genes

Knockout strains are also generated using the same approach as the one for PCR-based integration of proteins at the 3′end of the *ATG* genes. The entire ORF is disrupted by homologous recombination through replacement with either an auxotrophic marker or an antibiotic resistance gene generated by PCR using primers containing at least 50–60 bases of identity to the regions flanking the ORF (Gueldener et al., 2002; Hentges et al., 2005; Longtine et al., 1998).

The correct knocking out of the *ATG* genes of interest can be verified in three different ways: (1) Examination of the *ATG* locus by PCR (Gueldener et al., 2002; Longtine et al., 1998), (2) Western blot analysis by using specific antibodies against the deleted Atg protein, or (3) assessment of prApe1 maturation by Western blot (Kim et al., 1997; or alternatively measurement of autophagy progression; Guimaraes et al., 2015).

Other possible approaches to study the role of a specific Atg protein in the formation of autophagosomes are by either using strains carrying a thermosensitive allele of the gene of interest (Suzuki et al., 2001) or employing auxin-inducible degron systems (Dohmen, Wu, & Varshavsky, 1994; Nishimura, Fukagawa, Takisawa, Kakimoto, & Kanemaki, 2009). The advantage of these methods is that they permit to examine the direct consequences upon a rapid and relatively brief inactivation of an Atg protein.

2.2.2 Plasmids for the Expression of N-Terminally Tagged Atg Proteins

One of the Atg proteins that cannot be tagged C-terminally is Atg8 (Table 1). Its C-terminus is proteolytically processed by Atg4 before being linked to phosphatidyletaholamine (PE) through the action of two ubiquitin-like-conjugation systems (Fig. 2) (Ichimura et al., 2000; Kirisako et al., 1999). Nonetheless, the C-terminal tagging of Atg8 with a large protein tag like GFP can be employed to assess Atg4 activity as the liberation of the fused protein indicates that this protease is functional (Kim, Huang, & Klionsky, 2001).

To detect Atg8 by fluorescence microscopy or Western blot, when specific antibodies are not available, this protein has to be N-terminally-tagged. *ATG8* has thus often been cloned into plasmids at the 3′end of the sequence coding for the tag of choice and under the control of either the endogenous or the stronger *CUP1* promoter (Kim, Huang, et al., 2001; Suzuki et al., 2001). Fluorescent Atg8 chimeras driven by the authentic promoter generate weak signals when cells carrying them are grown in rich media, but the intensity of the signal becomes more pronounced under starvation conditions because of an augmentation in *ATG8* expression (Kirisako et al., 1999). Using a stronger promoter, such as the cupper-inducible *CUP1*, allows a better visualization of Atg8 protein under growing conditions without the need of adding extra cupper in the medium (Kim, Huang, et al., 2001; Kim, Kamada, et al., 2001). The *CUP1* promoter, however, does not allow measuring induction of Atg8 expression by Western

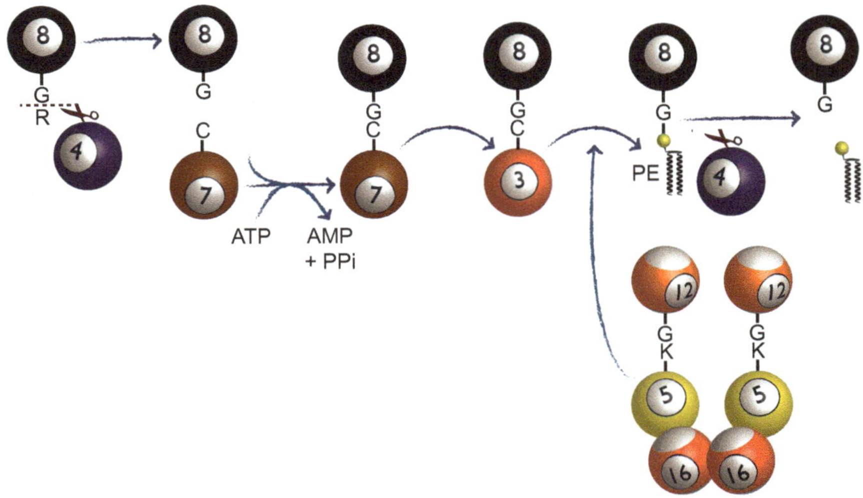

Fig. 2 The mechanism of Atg8 conjugation to phosphatidylethanolamine. In yeast, ubiquitin-like Atg8 is initially synthesized with a C-terminal arginine (i.e., R117). The Atg4 protease constitutively removes this residue allowing the exposure of a glycine (i.e., G116). Upon autophagosome induction, Atg8 is activated by the E1-like enzyme Atg7 and then transferred from Atg7 to the E2-like Atg3. Finally, Atg8 is conjugated to the amino group of phosphatidyletholamine (PE) present on PAS/phagophore membranes, guided by a homodimeric protein complex composed by Atg5, Atg12, and Atg16. Atg8 can be recycled from autophagosomal membranes by the cleavage of the bond between its C-terminal G116 and PE, which is mediated by Atg4.

blot during autophagy-inducing conditions. Two considerations have to be kept in mind when using expression plasmids. First, if those are centromeric (Sikorski & Hieter, 1989), the expression of Atg8 throughout a cell population is heterogeneous because there will be 1–4 plasmids per cell. This could hamper statistical evaluations. Therefore, it is strongly recommended to employ integration vectors (Sikorski & Hieter, 1989). Second, it is recommended to delete the *ATG8* gene before transforming plasmids expressing tagged versions of this protein. In fluorescence microscopy experiments, in particular, this will improve imaging because punctuate signals will not be decreased by the presence of a nondetectable Atg8.

Another marker protein that is frequently used to monitor the progression of the Cvt pathway and autophagy by either Western blot or fluorescence microscopy is Ape1. Expression of fluorescent chimeras under the control of its endogenous promoter is sufficient to obtain good signals because this peptidase is abundant and form oligomers (Kim et al., 1997). Ape1 has been very often tagged on its N-terminus

(Shintani et al., 2002). However, this cannot be done on its extremity because the profragment is essential for oligomerization and transport (Oda, Scott, Hefner-Gravink, Caffarelli, & Klionsky, 1996). As a result, sequence encoding for tags has to be inserted after this part of the *APE1* gene (Shintani et al., 2002). Nonetheless, it appears that Ape1 can also be C-terminally tagged (Dziedzic & Caplan, 2011; Obara et al., 2006; Papinski et al., 2014) (Section 2.2.1.1).

3. RECRUITMENT OF ATG MACHINERY TO THE PAS

In order to properly assess the recruitment of specific proteins to the PAS, it is crucial to accurately identify this subcellular location. One of the most frequently employed marker proteins for the PAS is Atg8. This protein is associated with both the inner and the outer surface of autophagosomal membranes and, as a result, the Atg8 pool localized within the complete autophagosome is delivered and degraded together with the cargo in the vacuolar lumen (Kirisako et al., 1999). The fluorescent signal from Atg8 chimeras is mostly dispersed uniformly in the cytosol in wild-type strains growing in a nutrient-rich medium, but in approximately 1/3 of the cells, there is also the presence of a single perivacuolar punctum, which corresponds to the PAS (Shintani et al., 2002) (Fig. 3). Under starvation conditions, the increase in autophagic flux due to a high frequency in the generation and subsequent fusion of double-membrane vesicles with the vacuole leads to the delivery of a major amount of Atg8 in the vacuolar lumen (Fig. 3). Because of this peculiar distribution, Atg8 has been very often employed in live-cell imaging experiments as a fluorescent chimera because it allows localizing all the autophagosomal biogenesis intermediates but also autophagy functionality through the appearance of the fluorescent signal in the vacuolar interior (Suzuki et al., 2001).

Through the analysis of Atg8 kinetic association to the PAS, it has been measured that one autophagy cycle, defined as the period from the phagophore formation until the autophagosome fusion with the vacuole, takes approximately 7–9 min under starvation conditions, leading to the estimation that autophagosome formation requires around 4–5 min (Geng et al., 2008). As a result, it is difficult and time-consuming to analyze the dynamics association of a specific protein to the PAS in wild-type cells. A common approach to overcome this difficulty has been to disrupt one *ATG* gene in order to halt the process of autophagosome biogenesis. This also permits visualizing more PAS in any given cell population. Using this approach, it

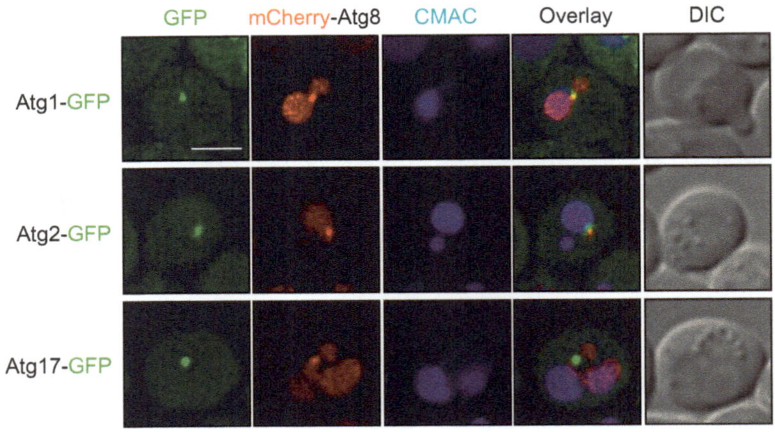

Fig. 3 Visualization of subcellular distribution of Atg proteins in nitrogen starvation conditions by fluorescence microscopy. Cells expressing mCherry-V5-Atg8 and either Atg1-GFP (RGY153, *MATα leu2-3,112 ura3-52 his3-Δ200 trp1-Δ901 suc2-Δ9 lys2-801; GAL atg8Δ::loxP-kanMX-loxP pCUP1-mCheV5-ATG8::URA3 ATG1-GFP::NatMX6*), Atg2-GFP (RGY154, *MATα leu2-3,112 ura3-52 his3-Δ200 trp1-Δ901 suc2-Δ9 lys2-801; GAL atg8Δ:: loxP-kanMX-loxP pCUP1-mCheV5-ATG8::URA3 ATG2-GFP::NatMX6*), or Atg17-GFP (RGY155, *MATα leu2-3,112 ura3-52 his3-Δ200 trp1-Δ901 suc2-Δ9 lys2-801; GAL atg8Δ:: loxP-kanMX-loxP pCUP1-mCheV5-ATG8::URA3 ATG17-GFP::NatMX6*) were grown in SMD medium at 30°C to a logarithmic phase before being after nitrogen starved in SD-N medium for 1 h to induce autophagy. Cells were also incubated with the CMAC dye 10 min prior to imaging in order to label the vacuolar lumen. Z-stacks of 22 pictures, with focal planes 0.2 μm apart, were acquired and deconvolved with a DeltaVision RT fluorescence microscope system (Applied Precision, Issaquah, WA). A single focal plane is shown. The green fluorescent signal of Atg1-GFP, Atg2-GFP, and Atg17-GFP appears as a single perivacuolar punctum. This structure represents the PAS, and therefore, it is also positive for mCherry-V5-Atg8. Note that the signal of mCherry-V5-Atg8 is also present within the vacuole highlighting the fusion of autophagosomes with this organelle, i.e., normal progression of autophagy. *DIC*, differential interference contrast. Scale bar: 2 μm.

has been established which Atg proteins localize to the PAS in most of the *atg* knockout mutants, leading to a model for the hierarchical assembly of the Atg machinery (Suzuki et al., 2001, 2007).

In specific mutants, in particular in those that affect Atg8 conjugation to PE such as *atg3Δ* or *atg7Δ* (Fig. 2), Atg8 cannot be used to visualize the PAS because it is exclusively cytoplasmic (Kim, Huang, et al., 2001; Kim, Kamada, et al., 2001). Therefore, a different Atg protein has to be used to highlight the PAS (Suzuki et al., 2007). Nonetheless, it is recommended to always use a second reliable PAS marker protein to ensure that results obtained using Atg8 are consistent, and not connected with a peculiar localization of this protein. If the experiment design requires to distinguish between the PAS and autophagosomes, two marker proteins have to be employed (Cebollero et al., 2012). One has to be Atg8, because localizing to both the PAS and autophagosomes, and the other can be another Atg protein only presents at the PAS. All the core Atg proteins can serve for this purpose except Atg1 and Atg9, which are also found on autophagosomes (Kraft et al., 2012; Yamamoto et al., 2012).

It must be noted that the prApe1 oligomer has often been used as a PAS marker protein but it presents a caveat. Although Atg8 concentration into a punctum requires the induction of PAS formation, Ape1 always appears as a punctuate structure because of its constitutive assembly into a large oligomer (Kim et al., 1997; Shintani et al., 2002). As a matter of fact, Ape1 localizes in a single spot also in mutants such as *atg11Δ* and *atg19Δ*, where no PAS is formed in growing conditions (Shintani et al., 2002).

Because the PAS is specifically localized in near proximity of the vacuole (Suzuki et al., 2001), determining the spatial distribution of this structure can add accuracy to the monitoring of the autophagosome formation by live-cell imaging. The vacuole is the largest organelle of yeast, and therefore, it is easily detectable by differential interference contract (DIC). Alternatively, vacuolar staining by either labeling of the membrane or the lumen of the vacuole using dyes such as 20 μ*M* FM 4-64 (Vida & Emr, 1995) and 100 n*M* CellTracker™ Blue CMAC (both from Molecular Probes™, Thermo Fisher Scientific, Waltham, MA) (Stefan & Blumer, 1999), respectively, permits the visualization of this organelle by fluorescence microscopy (Fig. 3). Another approach to image the vacuole is to fuse a vacuole membrane protein, like Vph1, with a fluorescent protein (Urbanowski & Piper, 1999).

To study the localization of the Atg machinery by fluorescence microscopy, sample preparation has to be performed as described in Section 3.1. Typically, Atg proteins appear as fluorescent punctuate structures by live-cell imaging; therefore, appropriate image acquisition settings have to be established to optimize spatial and temporal resolution. In this regard, it is suggested to acquire images from single focal planes (0.2–0.4 μm). When possible, it is highly recommended to analyze three-dimensional reconstructions from series of Z-stack images covering the whole cell (4–6 μm in diameter) to allow monitoring with high accuracy through different focal planes, the fast moving organelles (Ahmad, Nakajima, Kojima, Homma, & Fukuda, 2008). To this aim, it is warmly suggested to determine the minimal excitation required to highlight the structure of interest before starting with image acquisition. This permits to collect a maximum of pictures before the fluorescence signal gets photobleached. Moreover, when detecting more than one fluorescent signal, images for each wavelength have to be collected before changing focal plane. This is to avoid to miss a colocalization when the structure of interest moves fast and could thus change position over the time when collecting a Z-stack of images using a single wavelength before passing to the other one. Finally, image analyses have not to rely on maximum-intensity projections because those will superpose several pictures and lead to colocalization of structures on different focal planes.

3.1 Visualization of Fluorescence Signals

The fastest and easiest method to check the subcellular localization of fluorescent fusion proteins in *Saccharomyces cerevisiae* is by live-cell imaging (Kim, Dalton, Eggerton, Scott, & Klionsky, 1999; Suzuki et al., 2001).

1. Collect 1 mL of culture with exponentially growing or starved yeast.
2. Centrifuge at 5000 rpm for 1 min and remove almost all the supernatant leaving 50–100 μL of liquid. If cells are grown in YPD medium, they have to be washed with 1 mL of SMD medium to completely remove the YPD medium, which is autofluorescent and thus affects the quality of the acquired images.
3. Resuspend homogenously the cell pellet in the remaining medium by vortexing.
4. Spot 2–3 μL of the resuspension on a glass slide before placing a coverslip onto the drop and immobilizing the sample with finger pressure.
5. Once the sample is prepared, acquisition of fluorescence images can immediately begin using either a confocal or a fluorescence microscope.

This type of preparations last approximately 10–15 min before they dry. This time can be extended by sealing the coverslip on the glass slide using nail polish. One important consideration to keep in mind is that nutrients and oxygen are consumed over time, and therefore, the physiological status of the imaged cells could change. It is thus recommended to prepare a new sample for imaging rather than using one for a long time. If live-cell imaging has to be performed over a long period, i.e., hours, the use of microfluidic devices is a better option (Hansen, Hao, & O'Shea, 2015).

Generally, yeast cells display very low levels of autofluorescence and this does not interfere with the imaging of most of the fluorescent chimeras. In strains that have a mutation in either *ADE1* or *ADE2* genes, which are involved in adenine biosynthesis, a red fluorescent metabolic intermediate is accumulated in the vacuole, and this can make difficult to distinguish red fluorescent protein signals (Stotz & Linder, 1990). Addition of 20 μg/mL adenine (or more) in the medium is sufficient to minimize this type of autofluorescence.

Mild fixation of the sample (Reggiori, Shintani, Nair, & Klionsky, 2005) or immunofluorescence (Holthuis, Nichols, & Pelham, 1998) can be employed as an alternative to live-cell imaging. Their advantages are that (1) organelles with an intracellular mobility stay in place and they are thus easily imaged, and (2) there are no time limitations in picture acquisition. However, the disadvantages of these two approaches are (1) the longer processing time of the sample and (2) cell fixation reduces either the fluorescence of the proteins (mild fixation approach) or the available epitopes recognized by the employed antibodies (immunofluorescence). As a result of this latter limitation, the best procedure for imaging protein of low abundance is by live-cell imaging.

4. LOCALIZATION OF ATG PROTEINS ON FORMING AUTOPHAGOSOMES

The diameter of autophagosomes is 300–900 nm (Baba, Takeshige, Baba, & Ohsumi, 1994), and therefore, they appear as a punctum in fluorescence microscopy in most yeast backgrounds. This disadvantage makes almost impossible to determine whether a specific Atg protein localizes to a discrete area of forming phagophores/autophagosomes. Although a recent study has revealed the localization of several Atg proteins onto the phagophore in wild-type cells (Graef et al., 2013), overexpression of Ape1 has been used as a tool to facilitate the accumulation and visualization of this membranous cistern and accurately analyze the precise distribution of the Atg machinery (Papinski et al., 2014; Pfaffenwimmer et al., 2014; Suzuki et al., 2013) (Fig. 4). This approach permits the generation of giant prApe1 oligomers, which cannot be completely enwrapped by the nascent autophagosomes and as a result expanding phagophores are accumulated around prApe1. For this purpose, Ape1 has to be under control of an inducible promoter such as the *CUP1* (Papinski et al., 2014; Suzuki et al., 2013), which permit a massive overexpression upon addition of 250 μ*M* $CuSO_4$ into the culture medium. Cloning of *APE1* into a 2 μ multicopy plasmid (Sikorski & Hieter, 1989) can additionally enhance its expression (Suzuki et al., 2013).

Fusion of Ape1 with a fluorescent protein allows an easier localization of the giant Ape1 oligomer (Papinski et al., 2014; Pfaffenwimmer et al., 2014; Suzuki et al., 2013) (Fig. 4). This also facilitates the localization of

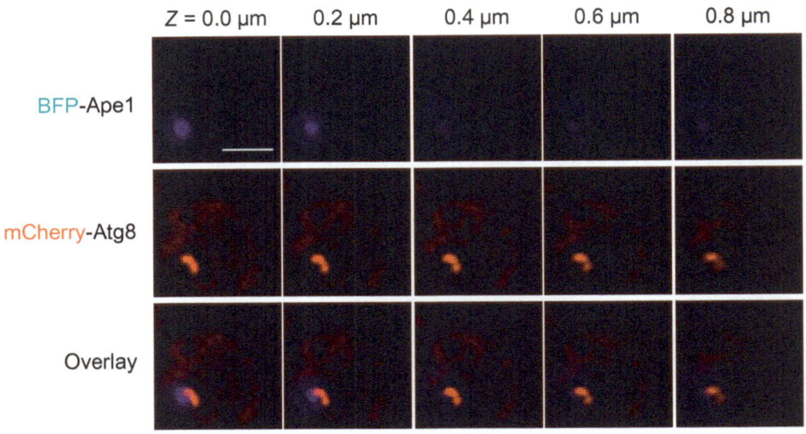

Fig. 4 Phagophore expansion around giant Ape1 oligomers. Wild-type (RGY114, *MATα leu2-3,112 ura3-52 his3-Δ200 trp1-Δ901 suc2-Δ9 lys2-801; GAL atg8Δ::loxP-kanMX-loxP pCUP1-mCheV5-ATG8::URA3*) cells transformed with a plasmid expressing BFP-Ape1 (pDP105; Pfaffenwimmer et al., 2014) were grown in SMD medium at 30°C to a logarithmic phase before inducing autophagy by nitrogen starvation in SD-N medium for 1 h. Overexpression of Ape1 was induced by adding 250 μ*M* $CuSO_4$ for 3 h. Z-stacks of 22 pictures, with focal planes 0.2 μm apart, were acquired and deconvolved with a DeltaVision RT fluorescence microscope system. Five consecutive focal planes are shown. Blue fluorescent giant Ape1 oligomer is surrounded by membranous structures positive for mCherry-V5-Atg8, which represents the phagophore. Scale bar: 2 μm.

different Atg proteins on the forming autophagosome and eventually clusters them depending on their distribution patterns (in one or both edges of the forming double-membrane vesicle, or colocalizing with the entire structure) (Suzuki et al., 2013) (Fig. 4). It must be noted, however, that this is an artificial system, and therefore, it should be kept in mind that the distribution of some of the Atg proteins could be altered due to their direct and indirect interaction with Ape1.

To visualize the different fluorescent signals, sample preparation and fluorescence microscopy image acquisition settings should be performed as mentioned in Section 3.

5. BIMOLECULAR FLUORESCENCE COMPLEMENTATION (BiFC)

The cellular levels of yeast Atg proteins are low, and autophagosome formation is a highly dynamic process governed by transient protein interactions. These two aspects often make difficult the detection of binding between Atg proteins. The BiFC-based assays allow monitoring weak interactions, and they have the additional big advantage of highlighting the principal subcellular location(s), where the association between the two studied proteins takes place (Kerppola, 2008).

The BiFC system is based on the tagging of two different proteins with the nonfluorescent N- and C-terminal fragments of GFP, respectively, and the interaction between the assayed proteins leads to the reformation of the functional GFP, which emits a fluorescent signal upon excitation. The generation of enhanced GFP, which exhibits a 35-fold increase in fluorescence compared to GFP, as well as the development of some spectral variants such as BFP, CFP, YFP, and the YFP variant Venus (λ_{ex}: 515 nm; λ_{em}: 528 nm), has broadened the use of the original BiFC assay because providing more sensitive readouts. Moreover, the development of the multicolor BiFC has made possible the simultaneous evaluation of pairwise interactions (Lipatova et al., 2012). Multicolor BiFC relies on the fact that the C-terminal fragments of CFP and YFP are identical and when fused with a specific protein, it is possible to determine in vivo whether it interacts with a protein tagged with the N-terminus of CFP or another one fused with the same fragment of YFP based on the wavelength of the emitted fluorescent signal.

One of the limitations of the BiFC assay is the irreversibility of the fluorophore assembly. Although this characteristic can help in detecting weak interactions, the fact that the two associated proteins barely separate could lead to their accumulation in a location different from the one where the interaction initially takes place. It must be also kept in mind that this method can also allow the detection of proteins that are in proximity without being in direct contact, such as, for example, subunits of a same protein complex. Moreover, BiFC does not reveal all bindings, and consequently if a signal is not detected, it does not indicate that a determined interaction does not take place. As a result, it is highly recommended to always use in parallel other approaches, such as coimmunoprecipitation (Co-IP) experiments, to explore an interaction between two proteins (Section 6).

The most frequently used split fluorescent protein for yeast BiFC experiments is Venus because of the bright fluorescent signal, its fast maturation time and insensitivity to changes in pH and ion concentrations (Nagai et al., 2002). Moreover, Venus has higher specificity compared to YFP (based on BiFC fluorescent signals obtained from known specific interactions), and consequently, it is better in detecting weak interaction (Shyu, Liu, Deng, & Hu, 2006). A series of plasmids that allow the C-terminal tagging of endogenous proteins with the N-terminal and C-terminal fragments of Venus, i.e., VN and VC, respectively (Section 2.2.1.1), as well as the N-terminal tagging of proteins under the control of an inducible (*pGAL1*) or constitutive promoter (*pCET1*) are available (Sung & Huh, 2007). It must be noted that as for the Y2H, fluorescent protein fragments have to be initially exchanged between the two tested interaction partners because it could be that only one of the two possible combinations works.

BiFC has been successfully used to study the interaction of different Atg proteins during autophagosome biogenesis, but also during selective types of autophagy (Tables 3 and 4). For example, this technique has helped to unveil the molecular determinants for Atg2–Atg18 interaction at the PAS upon autophagy induction (Rieter et al., 2013). Moreover, it has recently been shown using BiFC that phosphorylation of serine 122 of Atg9 is required for the interaction of this protein with Atg23 and Atg27 (Feng et al., 2016). Finally, BiFC has also been key in revealing the involvement of the mitochondrial fission machinery and the ER-mitochondria encounter structure during yeast mitophagy (Bockler & Westermann, 2014; Mao, Wang, Liu, & Klionsky, 2013). Finally, multicolor BiFC has permitted to reveal that Ypt1 interacts with both Atg11 and Trs85 in locations positive for Atg9 and Atg8 under nutrient-rich conditions (Lipatova et al., 2012).

When possible (Tables 1 and 2), tagging of the endogenous protein is the most appropriate strategy to optimally preserve the physiological conditions of binding. Protein overexpression is an alternative approach because of the low abundance of some Atg proteins and the weakness of specific protein–protein interactions. However, this has to be considered when interpreting the results, especially when the overexpressed protein is in complex with others. In order to detect of some weak and transient interactions without altering the stoichiometry of a protein complex, a possible solution is to perform the BiFC assay in a mutant background in which autophagosome formation is blocked at a precise step that permits an enrichment of the analyzed interaction (Rieter et al., 2013).

Although split Venus moieties do not emit a fluorescent signal, it is important to evaluate eventual background to prove the specificity of eventual interactions detected by BiFC. First, strains in which only one of the proteins is tagged with a Venus fragment have to be included as negative controls (Fig. 5). Second, it is recommendable to perform the BiFC assay

Table 3 Interactions Between Atg Proteins That Are Part of the Core Machinery

Protein-Protein Interactions

Atg Protein	Atg Protein	Methods: BiFC	Methods: Co-IP
Atg1	Atg1	—	Yeh, Wrasman, and Herman (2010) and Yeh, Shah, and Herman (2011)
	Atg8	—	Kraft et al. (2012) and Nakatogawa et al. (2012)
	Atg11	Lipatova et al. (2012)	Cheong et al. (2008), Kawamata et al. (2008), Kim, Huang, et al. (2001), Kim, Kamada, et al. (2001), Kraft et al. (2012), Mao, Chew, et al. (2013), and Mao, Wang, et al. (2013)
	Atg13	—	Cheong et al. (2008), He et al. (2008), Kabeya et al. (2005), Kamada et al. (2000), and Kawamata et al. (2008)
	Atg17	—	Cheong et al. (2005), Kabeya et al. (2005), Kamada et al. (2010), Kawamata et al. (2008), and Yeh et al. (2010)
	Atg29	—	Kawamata et al. (2008) and Kraft et al. (2012)
	Atg31	—	Kawamata et al. (2008)
Atg2	Atg9	—	Wang et al. (2001)
	Atg18	Rieter et al. (2013)	Obara, Sekito, Niimi, and Ohsumi (2008), Rieter et al. (2013), Suzuki et al. (2007), and Watanabe et al. (2012)
Atg3	Atg7	—	Komatsu et al. (2001)
	Atg8	—	Ichimura et al. (2000)
Atg5	Atg12	—	Kim, Huang, et al. (2001), Kim, Kamada, et al. (2001), Komatsu et al. (2001), Mizushima et al. (1998), Mizushima et al. (1999), and Shintani et al. (1999)
	Atg16	—	Mizushima et al. (1999)

Continued

Table 3 Interactions Between Atg Proteins That Are Part of the Core Machinery—cont'd

		Protein-Protein Interactions	
		Methods	
Atg Protein	**Atg Protein**	**BiFC**	**Co-IP**
Atg6	Atg14	—	Araki et al. (2013), Kametaka et al. (1998), Kihara et al. (2001), Noda et al. (2012), and Obara et al. (2006)
	Atg38	—	Araki et al. (2013)
	Vps15	—	Kihara et al. (2001) and Obara et al. (2006)
	Vps34	—	Kihara et al. (2001), Noda et al. (2012), and Obara et al. (2006)
Atg7	Atg3	—	Komatsu et al. (2001)
	Atg7	—	Komatsu et al. (2001)
	Atg8	—	Ho et al. (2009), Ichimura et al. (2000), and Komatsu et al. (2001)
	Atg12	—	Tanida et al. (1999)
Atg8	Atg1	—	Kraft et al. (2012) and Nakatogawa et al. (2012)
	Atg3	—	Ichimura et al. (2000)
	Atg7	—	Ho et al. (2009), Ichimura et al. (2000), and Komatsu et al. (2001)
	Atg19	—	Ho et al. (2009) and Kim et al. (2002)
	Atg27	—	Wurmser and Emr (2002)
	Atg32	—	Kondo-Okamoto et al. (2012) and Okamoto, Kondo-Okamoto, and Ohsumi (2009)
	Atg36	—	Motley et al. (2012)
	Atg39	—	Mochida et al. (2015)
	Atg40	—	Mochida et al. (2015)
Atg9	Atg2	—	Wang et al. (2001)
	Atg9	—	He et al. (2008)
	Atg11	—	Chang and Huang (2007), He et al. (2006), and Sekito et al. (2009)

Table 3 Interactions Between Atg Proteins That Are Part of the Core Machinery—cont'd
Protein-Protein Interactions

Atg Protein	Atg Protein	Methods	
		BiFC	Co-IP
	Atg13	—	Suzuki et al. (2015)
	Atg17	—	Sekito et al. (2009)
	Atg18	—	Papinski et al. (2014) and Reggiori et al. (2004)
	Atg23	Feng et al. (2016)	He et al. (2008), Legakis et al. (2007), and Tucker et al. (2003)
	Atg27	Feng et al. (2016)	—
Atg10	Atg12	—	Shintani et al. (1999)
Atg12	Atg5	—	Hanada and Ohsumi (2005), Kim, Huang, et al. (2001), Kim, Kamada, et al. (2001), Komatsu et al. (2001), Mizushima et al. (1998), Mizushima et al. (1999), and Shintani et al. (1999)
	Atg7	—	Tanida et al. (1999)
	Atg10	—	Shintani et al. (1999)
	Atg16	—	Hanada and Ohsumi (2005) and Mizushima et al. (1999)
Atg13	Atg1	—	Cheong et al. (2008), He et al. (2008), Kabeya et al. (2005), Kamada et al. (2000), and Kawamata et al. (2008)
	Atg9	—	Suzuki et al. (2015)
	Atg17	—	Cheong et al. (2005), Fujioka et al. (2014), Kabeya et al. (2005), Stephan et al. (2009), and Yeh et al. (2011)
Atg14	Atg6	—	Araki et al. (2013), Kametaka et al. (1998), Kihara et al. (2001), Noda et al. (2012), and Obara et al. (2006)
	Atg38	—	Araki et al. (2013)
	Vps15	—	Araki et al. (2013), Heenan et al. (2009), and Kihara et al. (2001)
	Vps34	—	Araki et al. (2013), Kihara et al. (2001), and Obara et al. (2006)

Continued

Table 3 Interactions Between Atg Proteins That Are Part of the Core Machinery—cont'd

		Protein-Protein Interactions	
		Methods	
Atg Protein	**Atg Protein**	**BiFC**	**Co-IP**
Atg16	Atg5	—	Mizushima et al. (1999)
	Atg12	—	Hanada and Ohsumi (2005) and Mizushima et al. (1999)
	Atg16	—	Mizushima et al. (1999)
Atg17	Atg1	—	Cheong et al. (2005), Kabeya et al. (2005), Kamada et al. (2010), Kawamata et al. (2008), and Yeh et al. (2010)
	Atg9	—	Sekito et al. (2009)
	Atg11	—	Mao, Chew, et al. (2013) and Mao, Wang, et al. (2013)
	Atg13	—	Cheong et al. (2005), Fujioka et al. (2014), Kabeya et al. (2005), Stephan et al. (2009), and Yeh et al. (2011)
	Atg29	—	Cao et al. (2009), Chew et al. (2015), Kabeya et al. (2009), and Kawamata et al. (2008)
	Atg31	—	Kabeya et al. (2007), Kabeya et al. (2009), and Kawamata et al. (2008)
Atg18	Atg2	Rieter et al. (2013)	Obara et al. (2008), Rieter et al. (2013), Suzuki et al. (2007), and Watanabe et al. (2012)
	Atg9	—	Papinski et al. (2014) and Reggiori et al. (2004)
Atg23	Atg9	Feng et al. (2016)	He et al. (2008), Legakis et al. (2007), and Tucker et al. (2003)
	Atg27	—	Legakis et al. (2007)
Atg27	Atg8	—	Wurmser and Emr (2002)
	Atg9	Feng et al. (2016)	—
	Atg23	—	Legakis et al. (2007)

Table 3 Interactions Between Atg Proteins That Are Part of the Core Machinery—cont'd

Atg Protein	Atg Protein	Protein-Protein Interactions: Methods	
		BiFC	Co-IP
Atg29	Atg1	—	Kawamata et al. (2008) and Kraft et al. (2012)
	Atg11	Mao, Chew, et al. (2013) and Mao, Wang, et al. (2013)	Mao, Chew, et al. (2013) and Mao, Wang, et al. (2013)
	Atg17	—	Cao et al. (2009), Chew et al. (2015), Kabeya et al. (2009), and Kawamata et al. (2008)
	Atg31	—	Cao et al. (2009), Mao, Chew, et al. (2013), and Mao, Wang, et al. (2013)
Atg31	Atg1	—	Kawamata et al. (2008)
	Atg17	—	Kabeya et al. (2007), Kabeya et al. (2009), and Kawamata et al. (2008)
	Atg29	—	Cao et al. (2009), Mao, Chew, et al. (2013), and Mao, Wang, et al. (2013)
Atg38	Atg6	—	Araki et al. (2013)
	Atg14	—	Araki et al. (2013)
	Vps15	—	Araki et al. (2013)
	Vps34	—	Araki et al. (2013)
Vps15	Atg6	—	Kihara et al. (2001) and Obara et al. (2006)
	Atg14	—	Araki et al. (2013), Heenan et al. (2009), and Kihara et al. (2001)
	Atg38	—	Araki et al. (2013)
	Vps34	—	Stack, Herman, Schu, and Emr (1993)
Vps34	Atg6	—	Kihara et al. (2001), Noda et al. (2012), and Obara et al. (2006)
	Atg14	—	Araki et al. (2013), Kihara et al. (2001), and Obara et al. (2006)
	Atg38	—	Araki et al. (2013)
	Vps15	—	Stack et al. (1993)

Up to five references demonstrating the interaction between a determined pair of proteins are indicated.

Table 4 Interactions Between Autophagy Adaptors and Receptors, With Other Atg Proteins

Atg Protein	Atg Protein	Protein-Protein Interactions: Methods – BiFC	Protein-Protein Interactions: Methods – Co-IP
Atg11	Atg1	Lipatova et al. (2012)	Cheong et al. (2008), Kawamata et al. (2008), Kim, Huang, et al. (2001), Kim, Kamada, et al. (2001), Kraft et al. (2012), Mao, Chew, et al. (2013), and Mao, Wang, et al. (2013)
	Atg9	—	Chang and Huang (2007), He et al. (2006), and Sekito et al. (2009)
	Atg11	—	Kim, Huang, et al. (2001), Kim, Kamada, et al. (2001), and Yorimitsu and Klionsky (2005)
	Atg17	—	Mao, Chew, et al. (2013) and Mao, Wang, et al. (2013)
	Atg19	Lipatova et al. (2012)	Kamber, Shoemaker, and Denic (2015), Pfaffenwimmer et al. (2014), Tanaka et al. (2014), and Yorimitsu and Klionsky (2005)
	Atg29	Mao, Chew, et al. (2013) and Mao, Wang, et al. (2013)	Mao, Chew, et al. (2013) and Mao, Wang, et al. (2013)
	Atg32	Mao, Chew, et al. (2013) and Mao, Wang, et al. (2013)	Aoki et al. (2011), Kanki, Wang, Baba, et al. (2009), Kanki, Wang, Cao, et al. (2009), Kondo-Okamoto et al. (2012), Okamoto et al. (2009), and Wang, Jin, et al. (2013)
	Atg34	—	Mochida et al. (2014)
	Atg36	—	Motley et al. (2012) and Tanaka et al. (2014)
	Atg39	—	Mochida et al. (2015)
	Atg40	—	Mochida et al. (2015)
Atg19	Atg8	—	Ho et al. (2009) and Kim et al. (2002)
	Atg11	Lipatova et al. (2012)	Kamber et al. (2015), Pfaffenwimmer et al. (2014), Tanaka et al. (2014), and Yorimitsu and Klionsky (2005)

Table 4 Interactions Between Autophagy Adaptors and Receptors, With Other Atg Proteins—cont'd

		Protein-Protein Interactions	
		Methods	
Atg Protein	Atg Protein	BiFC	Co-IP
Atg32	Atg8	—	Kondo-Okamoto et al. (2012) and Okamoto et al. (2009)
	Atg11	Mao, Chew, et al. (2013) and Mao, Wang, et al. (2013)	Aoki et al. (2011), Kanki, Wang, Baba, et al. (2009), Kanki, Wang, Cao, et al. (2009), Kondo-Okamoto et al. (2012), Okamoto et al. (2009), and Wang, Jin, et al. (2013)
Atg34	Atg11	—	Mochida et al. (2014)
Atg36	Atg8	—	Motley et al. (2012)
	Atg11	—	Motley et al. (2012) and Tanaka et al. (2014)
Atg39	Atg8	—	Mochida et al. (2015)
	Atg11	—	Mochida et al. (2015)
Atg40	Atg8	—	Mochida et al. (2015)
	Atg11	—	Mochida et al. (2015)

Up to five references demonstrating the interaction between a determined pair of proteins are indicated.

with each protein of interest and some unrelated proteins (Mao, Wang, et al., 2013). Alternatively, the examined interaction can be tested in a mutant background known to interfere or block it (Rieter et al., 2013).

Importantly, BiFC interactions can also be colocalized with protein markers fused with fluorescent proteins emitting in the blue (BFP, CFP, …) or the red (mRFP, mCherry, …) light spectrum, providing information where the studied interaction takes place in the cell. On the same line, BiFC signals are also compatible with fluorescent probes for specific organelles such as CMAC and FM 4-64 (Stefan & Blumer, 1999; Vida & Emr, 1995).

Once the corresponding yeast strains are generated (Section 2.2.1), sample preparation and fluorescence microscopy image acquisition settings have to be carried out as mentioned earlier to visualize the different fluorescent signals (Section 3).

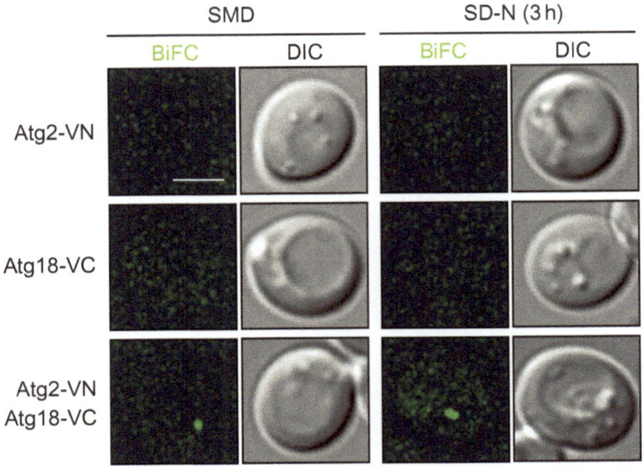

Fig. 5 In vivo interaction between Atg2 and Atg18 by BiFC. Wild-type cells expressing endogenous Atg2-VN (ERY115, *MATα leu2-3,112 ura3-52 his3-Δ200 trp1-Δ901 suc2-Δ9 lys2-801; GAL ATG2-VN::His3MX6*), Atg18-VC (ERY116, *MATα leu2-3,112 ura3-52 his3-Δ200 trp1-Δ901 suc2-Δ9 lys2-801; GAL ATG18-VC::TRP1*), and both Atg2-VN and Atg18-VC (ERY117, *MATα leu2-3,112 ura3-52 his3-Δ200 trp1-Δ901 suc2-Δ9 lys2-801; GAL ATG2-VN::His3MX6 ATG18-VC::TRP1*) were grown in SMD medium at 30°C to a logarithmic phase before being nitrogen starved in SD-N medium for 3 h. Fluorescence images were taken before and after nitrogen starvation. Z-stacks of 18 images, with focal planes 0.2 μm apart, were acquired and deconvolved with a DeltaVision RT fluorescence microscope system. A single focal plane is shown. In cells expressing either Atg2-VN or Atg18-VC alone, fluorescent signals are not observed. In contrast, a single perivacuolar fluorescent punctum is detected in cells expressing both chimeras, revealing the interaction between Atg2 and Atg18 at the PAS (Rieter et al., 2013). DIC, differential interference contrast. Scale bar: 2 μm.

6. COIMMUNOPRECIPITATION (Co-IP) EXPERIMENTS

Co-IP or pull-down experiments are routinely used to unveil protein–protein interactions. Co-IP experiments consist in capturing a protein under native conditions in order to isolate it together with the proteins that are associated. Two major strategies are often adopted to capture a protein: (1) Specific antibodies against the protein or an epitope, such as HA, GFP, myc or FLAG, fused to it or (2) a tag like, for example, GST, protein A, or a stretch of six histidine residues that specifically bind to Sepharose bead-conjugated glutathione, IgG, and nickel, respectively.

Co-IP experiments have been extensively used to gain a better understanding about the molecular mechanism of the autophagosome biogenesis and, in particular, to unveil the interactions between Atg proteins during their assembly at the PAS. One example is the dissection of the interactions in between the subunits of the pentameric Atg1 complex. Atg17, Atg29, and Atg31 would form a ternary subcomplex (Kabeya et al., 2009; Kawamata et al., 2008) that interacts with the Atg1–Atg13 subcomplex (Kamada et al., 2000) via a binding between Atg17 and Atg13 (Cheong et al., 2005; Fujioka et al., 2014; Kabeya et al., 2005). Moreover, Atg17 mediates the dimerization of the pentameric Atg1 complex to be fully functional (Fujioka et al., 2014; Yeh et al., 2011). Other interactions that have been characterized using Co-IP include the interactions between Atg5, Atg12, and Atg16 within the Atg12 ubiquitin-like conjugation system (Mizushima et al., 1998, 1999) or the subunits of the phosphatidylinositol 3-kinase complex (Kihara et al., 2001). A complete comprehensive list is presented in Tables 3 and 4.

6.1 Realization of the Co-IP Experiment

Given the low expression of numerous Atg proteins, a considerable amount of cells is usually required for Co-IP experiments. Although 50 OD_{600} equivalents of cells can often be sufficient for stable interactions, 5 or 10 times higher quantities of cells are generally required for the detection of weak and/or transient interactions.

Appropriate controls are essential to confirm the specificity of the detected bindings. To this aim, when using antibodies, the preimmunization serum and/or strains lacking the protein recognized by the antibodies have to be used as a control. If tagged proteins are employed for the Co-IP experiments, strains that are not expressing the corresponding fusion proteins have to be included to determine whether there is binding background (Fig. 6).

6.1.1 Cell Lysis

It is important to employ a lysis method that optimally preserves the integrity of the studied subcellular protein complex. The cryogenic grinding or cryo-milling, in which cells are mechanically lysed at −196°C, alters minimally the conditions found in living cells (Schultz, Hockman, Harkness, Garinther, & Altheim, 1997). However, other mechanical lysis methods, such as vortexing cells in the presence of small glass beads or Dounce homogenization of spheroplasts, are also efficient if a cryogenic grinding machine is not available (Rao, Perna, Hofmann, Beier, & Wollert, 2016; Reggiori et al., 2004). Due to their less effectiveness in cell lysis compared with the cryogenic grinding, bigger amount of cells might be needed when using these alternative methods. Independently from the cell lysis method, it is key to avoid protein degradation because this will reduce the chances of

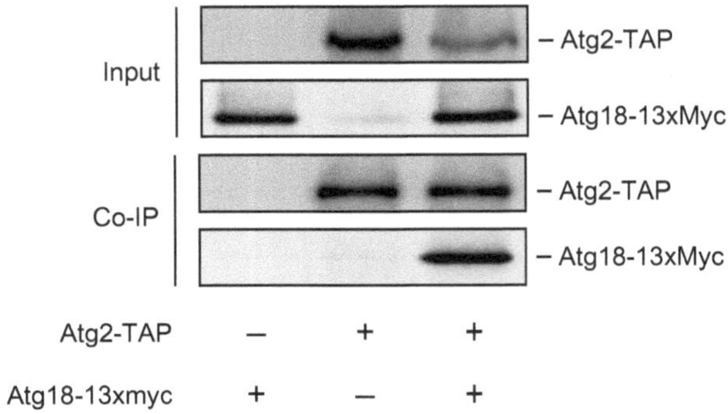

Fig. 6 Coimmunoprecipitation of Atg2 and Atg18. Wild-type cells expressing Atg18-13xmyc (ERY108, *MATα leu2-3,112 ura3-52 his3-Δ200 trp1-Δ901 suc2-Δ9 lys2-801; GAL ATG18-13xmyc::TRP1*), Atg2-TAP (RGY016, *MATα leu2-3,112 ura3-52 his3-Δ200 trp1-Δ901 suc2-Δ9 lys2-801; GAL atg2Δ::HIS3 pATG2WT-TAP::LEU2*), and both Atg18-13xmyc and Atg2-TAP (RGY012, *MATα leu2-3,112 ura3-52 his3-Δ200 trp1-Δ901 suc2-Δ9 lys2-801; GAL ATG18-13xmyc::TRP1 atg2Δ::HIS3 pATG2WT-TAP::LEU2*) were grown in SMD at 30°C to a logarithmic phase before starting the Co-IP procedure. Cell lysates, obtained by vortexing 100 OD_{600} equivalents of cells in the presence of small glass beads and cleared by subsequent contrifugations, were incubated with 50 μL of IgG-Sepharose beads (GE Healthcare, Waukesha, WI). After washing, immune-isolates were eluted with 80 μL of sample buffer, resolved by SDS-PAGE, and analyzed by Western blot using anti-TAP (Thermo Scientific, Waltham, MA) and anti-myc antibodies (Santa Cruz Biotechnology, Santa Cruz, CA). For the *Input*, 37.5% of the cell lysate was analyzed, whereas 37.5% of the isolated material was load on the SDS-PAGE gel (Co-IP). Atg18 is specifically pulled down by Atg2-TAP (Rieter et al., 2013).

detecting binding between proteins. It is therefore crucial to always add protease inhibitors to the employed lysis buffer. Moreover, samples have to be handled on ice in order to keep the solutions at low temperature. It is thus warmly recommend to work in a cold room, in this way tubes and pipette tips are chilled as well.

The composition of the lysis buffers must be optimized depending on the type of interaction under investigation. They have to mimic the intracellular pH and salt concentrations. The use of mild detergents, i.e., Tween 20 or octyl-β-D-glucopyranoside, at low concentrations helps to preserve weak interactions more than high concentrations and/or stronger detergents such as Triton X-100. Moreover, glycerol as well as dithiothretiol (DTT) stabilizes protein structures of specific proteins. A buffer successfully used for the study of Atg proteins is the one composed by 20 m*M* Tris–HCl pH 8.0, 150 m*M* KCl, 5 m*M* $MgCl_2$, 1% Triton X-100, 1 m*M* PMSF, and protease inhibitor cocktail (Rieter et al., 2013). Other ones are, for example: 50 m*M* HEPES pH 7.5, 150 m*M* NaCl, 1 m*M*

EDTA, 0.5% Tween 20, 1 m*M* PMSF, and protease inhibitor cocktail (Cheong et al., 2005), or 50 m*M* Tris–HCl pH 8.0, 150 m*M* NaCl, 50 m*M* NaF, 5 m*M* EDTA, 5 m*M* EGTA, 10% glycerol, 0.4% *n*-dodecyl-β-D-maltoside, 500 n*M* microcystin, 1 m*M* PMSF, 1X protease inhibitor cocktail for mammal (Sigma-Aldrich) and 1X PhosSTOP (Roche) (Fujioka et al., 2014).

Protocol for cryogenic grinding

1. Take 50–200 mL of cell culture and harvest the cells by centrifugation at 3000 × *g* for 5 min.
2. Wash the cell pellet twice with 10 mL ice-cold water by centrifugation at 3000 × *g* for 5 min at 4°C.
3. After the second wash, dry the pellet by an additional centrifugation step at 3500 rpm for 10 min and aspirate all the liquid, in this way the resulting pellet has a paste-like consistency.
4. Place the cell pellet into a syringe with a spatula and push it out into a 50-mL centrifugation tube containing liquid nitrogen to freeze it in droplets/noodles.
5. Discard all the liquid nitrogen from the tube. At this point, the cell droplets/noodles can be store at −80°C.
6. Disrupt the cells using a cryogenic grinding machine with four cycles of 3 min (25 Hz).
7. Resuspend the resulting powder in 2 mL of lysis buffer of choice by pipetting before clearing it from insoluble cell debris by consecutive centrifugations at 5000 × *g* for 5 min (4°C), and twice at 13,000 × *g* for 15 min (4°C).

6.1.2 Isolation of Complexes

Here, we described a basic standard procedure for Co-IP experiments using as a bait a protein fused with a tag that allow purification using a ligand conjugated to Sepharose beads.

1. If Sepharose beads without conjugated ligand are available, incubate the cell lysates (Section 6.1.1) with 30–50 μL of them on a rotatory wheel for 30 min at 4°C. The Sepharose beads have to be washed with the employed lysis buffer before use.
2. Centrifuge at 1500 × *g* for 3 min at 4°C and collect the supernatant.
3. Save up to 50 μL of supernatant, they will serve as the *Input* control.
4. Incubate the rest of the supernatant with 30–50 μL of Sepharose beads conjugated with the appropriate ligand on a rotatory wheel for 2 h at 4°C. Sepharose beads have to be washed with the employed lysis buffer before use.

5. Centrifuge at 1500 × *g* for 3 min at 4°C and discard the supernatant.
6. Wash the beads three times with 1 mL of lysis buffer on a rotatory wheel for 15 min at 4°C.
7. Once the last wash is discarded, centrifuge again at 1500 × *g* for 3 min at 4°C and remove completely the liquid.
8. Add 40–80 μL of sample buffer to the beads and boil to elute the protein complexes. Complexes can alternatively be eluted by competition through the addition of ligand in solution, i.e., not conjugated to the Sepharose beads. If a specific protease cleavage site is present between the bait protein and the tag, complexes can also be eluted by adding the appropriate protease.
9. Centrifuge at 13,000 × *g* for 5 min and collect the supernatant, which will contain the complexes isolated by Co-IP.

Finally, the *Input* and the isolated complexes are separated by SDS-PAGE followed by Western blot with antibodies recognizing the different proteins that are being examined (Fig. 6). The analysis of the *Input* permits to demonstrate the presence and distribution of the different proteins of interest in the different samples. Examination of the Co-IP material, taking in consideration the control samples, will reveal whether a protein specifically interacts with another one.

Few adaptations can be implemented if unspecific interactions are detected in the control samples. The easiest solution is to increase the time and/or number of washes (step 6 of the above protocol). Salt concentrations can also be increased to reduce ionic, and electrostatic attractions, e.g., 150–300 m*M* NaCl, can be included in the lysis buffer used for the washes. If unspecific binding is still not eliminated, one possibility is to increase the concentration of the detergent or change it for a stronger one. Low levels of reducing agents, such as DTT (1–2 m*M*) or β-mercaptoethanol, may help to decrease nonspecific interactions by nucleophilic or disulphide bridges. The addition of EDTA, in general at a concentration of 1 m*M*, is also an option, but it must be kept in mind that some interactions require divalent ions.

7. CONCLUDING REMARKS

Assembly of preautophagosomal precursor structures is a highly regulated process that involves the dynamic assembly of the Atg machinery, which also critically relies on protein–protein interactions. The standard methods presented in this chapter, especially when applied side-by-side, permit to precisely assess the progression of the initial steps of autophagosome biogenesis at the molecular level. Although a good starting point, they are not the only available experimental procedures. When combined with additional ones like the protease protection assays (Nair, Thumm, Klionsky, & Krick, 2011; Yen & Klionsky, 2012) or electron microscopy (Baba, 2008; Mari et al., 2010), they can also provide information about the membrane rearrangements taking place during autophagosome biogenesis. All together, these methods are very useful to study new genes potentially involved in autophagy but also, through mutations, domains or motifs of known Atg players. They can also help to determine whether under specific conditions, formation of an autophagosome proceeds in part differently like, for example, during nonselective and selective types of autophagy (Shintani & Klionsky, 2004).

ACKNOWLEDGMENTS

The authors thank Rodrigo Soares Guimaraes and Ester Rieter for the Co-IP and BiFC images. F.R. is supported by ALW Open Program (822.02.014), DFG-NWO cooperation (DN82-303), SNF Sinergia (CRSII3_154421), and ZonMW VICI (016.130.606) Grants. R.G.-S. is supported by a Marie Sklodowska-Curie Individual Fellowship (IF-EF) from the European Commission.

REFERENCES

Ahmad, M. R., Nakajima, M., Kojima, S., Homma, M., & Fukuda, T. (2008). The effects of cell sizes, environmental conditions, and growth phases on the strength of individual W303 yeast cells inside ESEM. *IEEE Transactions on Nanobioscience*, 7(3), 185–193. http://dx.doi.org/10.1109/TNB.2008.2002281.

Aoki, Y., Kanki, T., Hirota, Y., Kurihara, Y., Saigusa, T., Uchiumi, T., & Kang, D. (2011). Phosphorylation of serine 114 on Atg32 mediates mitophagy. *Molecular Biology of the Cell*, *22*(17), 3206–3217. http://dx.doi.org/10.1091/mbc.E11-02-0145.

Araki, Y., Ku, W. C., Akioka, M., May, A. I., Hayashi, Y., Arisaka, F., … Ohsumi, Y. (2013). Atg38 is required for autophagy-specific phosphatidylinositol 3-kinase complex integrity. *The Journal of Cell Biology*, *203*(2), 299–313. http://dx.doi.org/10.1083/jcb.201304123.

Baba, M. (2008). Electron microscopy in yeast. *Methods in Enzymology*, *451*, 133–149. http://dx.doi.org/10.1016/S0076-6879(08)03210-2.

Baba, M., Takeshige, K., Baba, N., & Ohsumi, Y. (1994). Ultrastructural analysis of the autophagic process in yeast: detection of autophagosomes and their characterization. *The Journal of Cell Biology*, *124*(6), 903–913.

Barth, H., Meiling-Wesse, K., Epple, U. D., & Thumm, M. (2001). Autophagy and the cytoplasm to vacuole targeting pathway both require Aut10p. *FEBS Letters*, *508*(1), 23–28. doi:S0014-5793(01)03016-2.

Bockler, S., & Westermann, B. (2014). Mitochondrial ER contacts are crucial for mitophagy in yeast. *Developmental Cell*, *28*(4), 450–458. http://dx.doi.org/10.1016/j.devcel.2014.01.012.

Budovskaya, Y. V., Hama, H., DeWald, D. B., & Herman, P. K. (2002). The C terminus of the Vps34p phosphoinositide 3-kinase is necessary and sufficient for the interaction with the Vps15p protein kinase. *The Journal of Biological Chemistry*, *277*(1), 287–294. http://dx.doi.org/10.1074/jbc.M109263200.

Cao, Y., Cheong, H., Song, H., & Klionsky, D. J. (2008). In vivo reconstitution of autophagy in Saccharomyces cerevisiae. *The Journal of Cell Biology*, *182*(4), 703–713. http://dx.doi.org/10.1083/jcb.200801035.

Cao, Y., Nair, U., Yasumura-Yorimitsu, K., & Klionsky, D. J. (2009). A multiple ATG gene knockout strain for yeast two-hybrid analysis. *Autophagy*, *5*(5), 699–705. http://dx.doi.org/10.4161/auto.5.5.8382.

Cebollero, E., van der Vaart, A., Zhao, M., Rieter, E., Klionsky, D. J., Helms, J. B., & Reggiori, F. (2012). Phosphatidylinositol-3-phosphate clearance plays a key role in autophagosome completion. *Current Biology: CB*, *22*(17), 1545–1553. http://dx.doi.org/10.1016/j.cub.2012.06.029.

Chang, C. Y., & Huang, W. P. (2007). Atg19 mediates a dual interaction cargo sorting mechanism in selective autophagy. *Molecular Biology of the Cell*, *18*(3), 919–929. http://dx.doi.org/10.1091/mbc.E06-08-0683.

Cheong, H., Nair, U., Geng, J., & Klionsky, D. J. (2008). The Atg1 kinase complex is involved in the regulation of protein recruitment to initiate sequestering vesicle formation for nonspecific autophagy in Saccharomyces cerevisiae. *Molecular Biology of the Cell*, *19*(2), 668–681. http://dx.doi.org/10.1091/mbc.E07-08-0826.

Cheong, H., Yorimitsu, T., Reggiori, F., Legakis, J. E., Wang, C. W., & Klionsky, D. J. (2005). Atg17 regulates the magnitude of the autophagic response. *Molecular Biology of the Cell*, *16*(7), 3438–3453. http://dx.doi.org/10.1091/mbc.E04-10-0894.

Chew, L. H., Lu, S., Liu, X., Li, F. K., Yu, A. Y., Klionsky, D. J., … Yip, C. K. (2015). Molecular interactions of the Saccharomyces cerevisiae Atg1 complex provide insights into assembly and regulatory mechanisms. *Autophagy*, *11*(6), 891–905. http://dx.doi.org/10.1080/15548627.2015.1040972.

Day, R. N., & Davidson, M. W. (2009). The fluorescent protein palette: tools for cellular imaging. *Chemical Society Reviews*, *38*(10), 2887–2921. http://dx.doi.org/10.1039/b901966a.

Dohmen, R. J., Wu, P., & Varshavsky, A. (1994). Heat-inducible degron: a method for constructing temperature-sensitive mutants. *Science (New York, N.Y.)*, *263*(5151), 1273–1276.

Dove, S. K., Piper, R. C., McEwen, R. K., Yu, J. W., King, M. C., Hughes, D. C., … Lemmon, M. A. (2004). Svp1p defines a family of phosphatidylinositol 3,5-bisphosphate effectors. *The EMBO Journal*, *23*(9), 1922–1933. http://dx.doi.org/10.1038/sj.emboj.7600203.

Dziedzic, S. A., & Caplan, A. B. (2011). Identification of autophagy genes participating in zinc-induced necrotic cell death in Saccharomyces cerevisiae. *Autophagy*, 7(5), 490–500. http://dx.doi.org/10.4161/auto.7.5.14872.

Efe, J. A., Botelho, R. J., & Emr, S. D. (2007). Atg18 regulates organelle morphology and Fab1 kinase activity independent of its membrane recruitment by phosphatidylinositol 3,5-bisphosphate. *Molecular Biology of the Cell*, *18*(11), 4232–4244. http://dx.doi.org/10.1091/mbc.E07-04-0301.

Feng, Y., Backues, S. K., Baba, M., Heo, J. M., Harper, J. W., & Klionsky, D. J. (2016). Phosphorylation of Atg9 regulates movement to the phagophore assembly site and the rate of autophagosome formation. *Autophagy*, *12*(4), 648–658. http://dx.doi.org/10.1080/15548627.2016.1157237.

Feng, Y., He, D., Yao, Z., & Klionsky, D. J. (2014). The machinery of macroautophagy. *Cell Research*, *24*(1), 24–41. http://dx.doi.org/10.1038/cr.2013.168.

Feng, W., Wu, T., Dan, X., Chen, Y., Li, L., Chen, S., … Yu, L. (2015). Phosphorylation of Atg31 is required for autophagy. *Protein & Cell*, *6*(4), 288–296. http://dx.doi.org/10.1007/s13238-015-0138-4.

Fields, S., & Song, O. (1989). A novel genetic system to detect protein-protein interactions. *Nature*, *340*(6230), 245–246. http://dx.doi.org/10.1038/340245a0.

Fujioka, Y., Suzuki, S. W., Yamamoto, H., Kondo-Kakuta, C., Kimura, Y., Hirano, H., … Noda, N. N. (2014). Structural basis of starvation-induced assembly of the autophagy initiation complex. *Nature Structural & Molecular Biology*, *21*(6), 513–521. http://dx.doi.org/10.1038/nsmb.2822.

Geng, J., Baba, M., Nair, U., & Klionsky, D. J. (2008). Quantitative analysis of autophagy-related protein stoichiometry by fluorescence microscopy. *The Journal of Cell Biology*, *182*(1), 129–140. http://dx.doi.org/10.1083/jcb.200711112.

George, M. D., Baba, M., Scott, S. V., Mizushima, N., Garrison, B. S., Ohsumi, Y., & Klionsky, D. J. (2000). Apg5p functions in the sequestration step in the cytoplasm-to-vacuole targeting and macroautophagy pathways. *Molecular Biology of the Cell*, *11*(3), 969–982.

Gietz, D., St Jean, A., Woods, R. A., & Schiestl, R. H. (1992). Improved method for high efficiency transformation of intact yeast cells. *Nucleic Acids Research*, *20*(6), 1425.

Graef, M., Friedman, J. R., Graham, C., Babu, M., & Nunnari, J. (2013). ER exit sites are physical and functional core autophagosome biogenesis components. *Molecular Biology of the Cell*, *24*(18), 2918–2931. http://dx.doi.org/10.1091/mbc.E13-07-0381.

Grunau, S., Lay, D., Mindthoff, S., Platta, H. W., Girzalsky, W., Just, W. W., & Erdmann, R. (2011). The phosphoinositide 3-kinase Vps34p is required for pexophagy in Saccharomyces cerevisiae. *The Biochemical Journal*, *434*(1), 161–170. http://dx.doi.org/10.1042/BJ20101115.

Guan, J., Stromhaug, P. E., George, M. D., Habibzadegah-Tari, P., Bevan, A., Dunn, W. A., Jr., & Klionsky, D. J. (2001). Cvt18/Gsa12 is required for cytoplasm-to-vacuole transport, pexophagy, and autophagy in Saccharomyces cerevisiae and Pichia pastoris. *Molecular Biology of the Cell*, *12*(12), 3821–3838.

Gueldener, U., Heinisch, J., Koehler, G. J., Voss, D., & Hegemann, J. H. (2002). A second set of loxP marker cassettes for cre-mediated multiple gene knockouts in budding yeast. *Nucleic Acids Research*, *30*(6), e23.

Guimaraes, R. S., Delorme-Axford, E., Klionsky, D. J., & Reggiori, F. (2015). Assays for the biochemical and ultrastructural measurement of selective and nonselective types of autophagy in the yeast Saccharomyces cerevisiae. *Methods (San Diego, Calif)*, *75*, 141–150. http://dx.doi.org/10.1016/j.ymeth.2014.11.023.

Hanada, T., & Ohsumi, Y. (2005). Structure-function relationship of Atg12, a ubiquitin-like modifier essential for autophagy. *Autophagy*, *1*(2), 110–118. http://dx.doi.org/10.4161/auto.1.2.1858.

Hansen, A. S., Hao, N., & O'Shea, E. K. (2015). High-throughput microfluidics to control and measure signaling dynamics in single yeast cells. *Nature Protocols*, *10*(8), 1181–1197. http://dx.doi.org/10.1038/nprot.2015.079.

He, C., Baba, M., Cao, Y., & Klionsky, D. J. (2008). Self-interaction is critical for Atg9 transport and function at the phagophore assembly site during autophagy. *Molecular Biology of the Cell*, *19*(12), 5506–5516. http://dx.doi.org/10.1091/mbc.E08-05-0544.

He, C., Song, H., Yorimitsu, T., Monastyrska, I., Yen, W. L., Legakis, J. E., & Klionsky, D. J. (2006). Recruitment of Atg9 to the preautophagosomal structure by Atg11 is essential for selective autophagy in budding yeast. *The Journal of Cell Biology*, *175*(6), 925–935. http://dx.doi.org/10.1083/jcb.200606084.

Heenan, E. J., Vanhooke, J. L., Temple, B. R., Betts, L., Sondek, J. E., & Dohlman, H. G. (2009). Structure and function of Vps15 in the endosomal G protein signaling pathway. *Biochemistry*, *48*(27), 6390–6401. http://dx.doi.org/10.1021/bi900621w.

Hentges, P., Van Driessche, B., Tafforeau, L., Vandenhaute, J., & Carr, A. M. (2005). Three novel antibiotic marker cassettes for gene disruption and marker switching in Schizosaccharomyces pombe. *Yeast (Chichester, England)*, *22*(13), 1013–1019. http://dx.doi.org/10.1002/yea.1291.

Ho, K. H., Chang, H. E., & Huang, W. P. (2009). Mutation at the cargo-receptor binding site of Atg8 also affects its general autophagy regulation function. *Autophagy*, *5*(4), 461–471. http://dx.doi.org/10.4161/auto.5.4.7696.

Holthuis, J. C., Nichols, B. J., & Pelham, H. R. (1998). The syntaxin Tlg1p mediates trafficking of chitin synthase III to polarized growth sites in yeast. *Molecular Biology of the Cell*, *9*(12), 3383–3397.

Hutchins, M. U., Veenhuis, M., & Klionsky, D. J. (1999). Peroxisome degradation in Saccharomyces cerevisiae is dependent on machinery of macroautophagy and the cvt pathway. *Journal of Cell Science*, *112*(Pt. 22), 4079–4087.

Ichimura, Y., Kirisako, T., Takao, T., Satomi, Y., Shimonishi, Y., Ishihara, N., … Ohsumi, Y. (2000). A ubiquitin-like system mediates protein lipidation. *Nature*, *408*(6811), 488–492. http://dx.doi.org/10.1038/35044114.

Kabeya, Y., Kamada, Y., Baba, M., Takikawa, H., Sasaki, M., & Ohsumi, Y. (2005). Atg17 functions in cooperation with Atg1 and Atg13 in yeast autophagy. *Molecular Biology of the Cell*, *16*(5), 2544–2553. http://dx.doi.org/10.1091/mbc.E04-08-0669.

Kabeya, Y., Kawamata, T., Suzuki, K., & Ohsumi, Y. (2007). Cis1/Atg31 is required for autophagosome formation in Saccharomyces cerevisiae. *Biochemical and Biophysical Research Communications*, *356*(2), 405–410. http://dx.doi.org/10.1016/j.bbrc.2007.02.150.

Kabeya, Y., Noda, N. N., Fujioka, Y., Suzuki, K., Inagaki, F., & Ohsumi, Y. (2009). Characterization of the Atg17-Atg29-Atg31 complex specifically required for starvation-induced autophagy in Saccharomyces cerevisiae. *Biochemical and Biophysical Research Communications*, *389*(4), 612–615. http://dx.doi.org/10.1016/j.bbrc.2009.09.034.

Kamada, Y., Funakoshi, T., Shintani, T., Nagano, K., Ohsumi, M., & Ohsumi, Y. (2000). Tor-mediated induction of autophagy via an Apg1 protein kinase complex. *The Journal of Cell Biology*, *150*(6), 1507–1513.

Kamada, Y., Yoshino, K., Kondo, C., Kawamata, T., Oshiro, N., Yonezawa, K., & Ohsumi, Y. (2010). Tor directly controls the Atg1 kinase complex to regulate autophagy. *Molecular and Cellular Biology*, *30*(4), 1049–1058. http://dx.doi.org/10.1128/MCB.01344-09.

Kamber, R. A., Shoemaker, C. J., & Denic, V. (2015). Receptor-bound targets of selective autophagy use a scaffold protein to activate the Atg1 kinase. *Molecular Cell*, *59*(3), 372–381. http://dx.doi.org/10.1016/j.molcel.2015.06.009.

Kametaka, S., Okano, T., Ohsumi, M., & Ohsumi, Y. (1998). Apg14p and Apg6/Vps30p form a protein complex essential for autophagy in the yeast, Saccharomyces cerevisiae. *The Journal of Biological Chemistry*, *273*(35), 22284–22291.

Kanki, T., Wang, K., Baba, M., Bartholomew, C. R., Lynch-Day, M. A., Du, Z., … Klionsky, D. J. (2009a). A genomic screen for yeast mutants defective in selective mitochondria autophagy. *Molecular Biology of the Cell*, *20*(22), 4730–4738. http://dx.doi.org/10.1091/mbc.E09-03-0225.

Kanki, T., Wang, K., Cao, Y., Baba, M., & Klionsky, D. J. (2009b). Atg32 is a mitochondrial protein that confers selectivity during mitophagy. *Developmental Cell*, *17*(1), 98–109. http://dx.doi.org/10.1016/j.devcel.2009.06.014.

Karanasios, E., & Ktistakis, N. T. (2016). *Autophagy at the cell, tissue and organismal level. SpringerBriefs in Cell Biology*. Switzerland: Springer International Publishing AG. http://dx.doi.org/10.1007/978-3-319-33145-4.

Kawamata, T., Kamada, Y., Kabeya, Y., Sekito, T., & Ohsumi, Y. (2008). Organization of the pre-autophagosomal structure responsible for autophagosome formation. *Molecular Biology of the Cell*, *19*(5), 2039–2050. http://dx.doi.org/10.1091/mbc.E07-10-1048.

Kawamata, T., Kamada, Y., Suzuki, K., Kuboshima, N., Akimatsu, H., Ota, S., … Ohsumi, Y. (2005). Characterization of a novel autophagy-specific gene, ATG29. *Biochemical and Biophysical Research Communications*, *338*(4), 1884–1889. http://dx.doi.org/10.1016/j.bbrc.2005.10.163.

Kerppola, T. K. (2008). Bimolecular fluorescence complementation (BiFC) analysis as a probe of protein interactions in living cells. *Annual Review of Biophysics*, *37*, 465–487. http://dx.doi.org/10.1146/annurev.biophys.37.032807.125842.

Kihara, A., Noda, T., Ishihara, N., & Ohsumi, Y. (2001). Two distinct Vps34 phosphatidylinositol 3-kinase complexes function in autophagy and carboxypeptidase Y sorting in Saccharomyces cerevisiae. *The Journal of Cell Biology*, *152*(3), 519–530.

Kim, J., Dalton, V. M., Eggerton, K. P., Scott, S. V., & Klionsky, D. J. (1999a). Apg7p/Cvt2p is required for the cytoplasm-to-vacuole targeting, macroautophagy, and peroxisome degradation pathways. *Molecular Biology of the Cell*, *10*(5), 1337–1351.

Kim, J., Huang, W. P., & Klionsky, D. J. (2001). Membrane recruitment of Aut7p in the autophagy and cytoplasm to vacuole targeting pathways requires Aut1p, Aut2p, and the autophagy conjugation complex. *The Journal of Cell Biology*, *152*(1), 51–64.

Kim, J., Huang, W. P., Stromhaug, P. E., & Klionsky, D. J. (2002). Convergence of multiple autophagy and cytoplasm to vacuole targeting components to a perivacuolar membrane compartment prior to de novo vesicle formation. *The Journal of Biological Chemistry*, *277*(1), 763–773. http://dx.doi.org/10.1074/jbc.M109134200.

Kim, J., Kamada, Y., Stromhaug, P. E., Guan, J., Hefner-Gravink, A., Baba, M., … Klionsky, D. J. (2001). Cvt9/Gsa9 functions in sequestering selective cytosolic cargo destined for the vacuole. *The Journal of Cell Biology*, *153*(2), 381–396.

Kim, J., Scott, S. V., Oda, M. N., & Klionsky, D. J. (1997). Transport of a large oligomeric protein by the cytoplasm to vacuole protein targeting pathway. *The Journal of Cell Biology*, *137*(3), 609–618.

Kirisako, T., Baba, M., Ishihara, N., Miyazawa, K., Ohsumi, M., Yoshimori, T., … Ohsumi, Y. (1999). Formation process of autophagosome is traced with Apg8/Aut7p in yeast. *The Journal of Cell Biology*, *147*(2), 435–446.

Kirisako, T., Ichimura, Y., Okada, H., Kabeya, Y., Mizushima, N., Yoshimori, T., … Ohsumi, Y. (2000). The reversible modification regulates the membrane-binding state of Apg8/Aut7 essential for autophagy and the cytoplasm to vacuole targeting pathway. *The Journal of Cell Biology*, *151*(2), 263–276.

Kissova, I., Deffieu, M., Manon, S., & Camougrand, N. (2004). Uth1p is involved in the autophagic degradation of mitochondria. *The Journal of Biological Chemistry*, *279*(37), 39068–39074. http://dx.doi.org/10.1074/jbc.M406960200.

Klionsky, D. J., Abdelmohsen, K., Abe, A., Abedin, M. J., Abeliovich, H., Acevedo Arozena, A., … Zughaier, S. M. (2016). Guidelines for the use and interpretation of assays for monitoring autophagy (3rd edition). *Autophagy*, *12*(1), 1–222. http://dx.doi.org/10.1080/15548627.2015.1100356.

Kobayashi, T., Suzuki, K., & Ohsumi, Y. (2012). Autophagosome formation can be achieved in the absence of Atg18 by expressing engineered PAS-targeted Atg2. *FEBS Letters*, *586*(16), 2473–2478. http://dx.doi.org/10.1016/j.febslet.2012.06.008.

Komatsu, M., Tanida, I., Ueno, T., Ohsumi, M., Ohsumi, Y., & Kominami, E. (2001). The C-terminal region of an Apg7p/Cvt2p is required for homodimerization and is essential for its E1 activity and E1-E2 complex formation. *The Journal of Biological Chemistry*, *276*(13), 9846–9854. http://dx.doi.org/10.1074/jbc.M007737200.

Kondo-Okamoto, N., Noda, N. N., Suzuki, S. W., Nakatogawa, H., Takahashi, I., Matsunami, M., … Okamoto, K. (2012). Autophagy-related protein 32 acts as autophagic degron and directly initiates mitophagy. *The Journal of Biological Chemistry*, *287*(13), 10631–10638. http://dx.doi.org/10.1074/jbc.M111.299917.

Kraft, C., Deplazes, A., Sohrmann, M., & Peter, M. (2008). Mature ribosomes are selectively degraded upon starvation by an autophagy pathway requiring the Ubp3p/Bre5p ubiquitin protease. *Nature Cell Biology*, *10*(5), 602–610. http://dx.doi.org/10.1038/ncb1723.

Kraft, C., Kijanska, M., Kalie, E., Siergiejuk, E., Lee, S. S., Semplicio, G., ... Peter, M. (2012). Binding of the Atg1/ULK1 kinase to the ubiquitin-like protein Atg8 regulates autophagy. *The EMBO Journal*, *31*(18), 3691–3703. http://dx.doi.org/10.1038/emboj.2012.225.

Krick, R., Henke, S., Tolstrup, J., & Thumm, M. (2008). Dissecting the localization and function of Atg18, Atg21 and Ygr223c. *Autophagy*, *4*(7), 896–910. http://dx.doi.org/10.4161/auto.6801.

Krick, R., Tolstrup, J., Appelles, A., Henke, S., & Thumm, M. (2006). The relevance of the phosphatidylinositolphosphat-binding motif FRRGT of Atg18 and Atg21 for the cvt pathway and autophagy. *FEBS Letters*, *580*(19), 4632–4638. http://dx.doi.org/10.1016/j.febslet.2006.07.041.

Lang, T., Reiche, S., Straub, M., Bredschneider, M., & Thumm, M. (2000). Autophagy and the cvt pathway both depend on AUT9. *Journal of Bacteriology*, *182*(8), 2125–2133.

Lang, T., Schaeffeler, E., Bernreuther, D., Bredschneider, M., Wolf, D. H., & Thumm, M. (1998). Aut2p and Aut7p, two novel microtubule-associated proteins are essential for delivery of autophagic vesicles to the vacuole. *The EMBO Journal*, *17*(13), 3597–3607. http://dx.doi.org/10.1093/emboj/17.13.3597.

Leber, R., Silles, E., Sandoval, I. V., & Mazon, M. J. (2001). Yol082p, a novel CVT protein involved in the selective targeting of aminopeptidase I to the yeast vacuole. *The Journal of Biological Chemistry*, *276*(31), 29210–29217. http://dx.doi.org/10.1074/jbc.M101438200.

Legakis, J. E., Yen, W. L., & Klionsky, D. J. (2007). A cycling protein complex required for selective autophagy. *Autophagy*, *3*(5), 422–432. http://dx.doi.org/10.4161/auto.4129.

Lipatova, Z., Belogortseva, N., Zhang, X. Q., Kim, J., Taussig, D., & Segev, N. (2012). Regulation of selective autophagy onset by a ypt/rab GTPase module. *Proceedings of the National Academy of Sciences of the United States of America*, *109*(18), 6981–6986. http://dx.doi.org/10.1073/pnas.1121299109.

Longtine, M. S., McKenzie, A., 3rd, Demarini, D. J., Shah, N. G., Wach, A., Brachat, A., ... Pringle, J. R. (1998). Additional modules for versatile and economical PCR-based gene deletion and modification in Saccharomyces cerevisiae. *Yeast (Chichester, England)*, *14*(10), 953–961. http://dx.doi.org/10.1002/(SICI)1097-0061(199807)14:10<953::AID-YEA293>3.0.CO;2-U.

Ma, J., Bharucha, N., Dobry, C. J., Frisch, R. L., Lawson, S., & Kumar, A. (2008). Localization of autophagy-related proteins in yeast using a versatile plasmid-based resource of fluorescent protein fusions. *Autophagy*, *4*(6), 792–800. http://dx.doi.org/10.4161/auto.6308.

Mao, K., Chew, L. H., Inoue-Aono, Y., Cheong, H., Nair, U., Popelka, H., ... Klionsky, D. J. (2013). Atg29 phosphorylation regulates coordination of the Atg17-Atg31-Atg29 complex with the Atg11 scaffold during autophagy initiation. *Proceedings of the National Academy of Sciences of the United States of America*, *110*(31), E2875–E2884. http://dx.doi.org/10.1073/pnas.1300064110.

Mao, K., Liu, X., Feng, Y., & Klionsky, D. J. (2014). The progression of peroxisomal degradation through autophagy requires peroxisomal division. *Autophagy*, *10*(4), 652–661. http://dx.doi.org/10.4161/auto.27852.

Mao, K., Wang, K., Liu, X., & Klionsky, D. J. (2013). The scaffold protein Atg11 recruits fission machinery to drive selective mitochondria degradation by autophagy. *Developmental Cell*, *26*(1), 9–18. http://dx.doi.org/10.1016/j.devcel.2013.05.024.

Mao, K., Wang, K., Zhao, M., Xu, T., & Klionsky, D. J. (2011). Two MAPK-signaling pathways are required for mitophagy in Saccharomyces cerevisiae. *The Journal of Cell Biology*, *193*(4), 755–767. http://dx.doi.org/10.1083/jcb.201102092.

Mari, M., Griffith, J., Rieter, E., Krishnappa, L., Klionsky, D. J., & Reggiori, F. (2010). An Atg9-containing compartment that functions in the early steps of autophagosome biogenesis. *The Journal of Cell Biology*, *190*(6), 1005–1022. http://dx.doi.org/10.1083/jcb.200912089.

Matsushita, M., Suzuki, N. N., Obara, K., Fujioka, Y., Ohsumi, Y., & Inagaki, F. (2007). Structure of Atg5.Atg16, a complex essential for autophagy. *The Journal of Biological Chemistry*, *282*(9), 6763–6772. http://dx.doi.org/10.1074/jbc.M609876200.

Mizushima, N., Noda, T., & Ohsumi, Y. (1999). Apg16p is required for the function of the Apg12p-Apg5p conjugate in the yeast autophagy pathway. *The EMBO Journal*, *18*(14), 3888–3896. http://dx.doi.org/10.1093/emboj/18.14.3888.

Mizushima, N., Noda, T., Yoshimori, T., Tanaka, Y., Ishii, T., George, M. D., … Ohsumi, Y. (1998). A protein conjugation system essential for autophagy. *Nature*, *395*(6700), 395–398. http://dx.doi.org/10.1038/26506.

Mochida, K., Ohsumi, Y., & Nakatogawa, H. (2014). Hrr25 phosphorylates the autophagic receptor Atg34 to promote vacuolar transport of alpha-mannosidase under nitrogen starvation conditions. *FEBS Letters*, *588*(21), 3862–3869. http://dx.doi.org/10.1016/j.febslet.2014.09.032.

Mochida, K., Oikawa, Y., Kimura, Y., Kirisako, H., Hirano, H., Ohsumi, Y., & Nakatogawa, H. (2015). Receptor-mediated selective autophagy degrades the endoplasmic reticulum and the nucleus. *Nature*, *522*(7556), 359–362. http://dx.doi.org/10.1038/nature14506.

Motley, A. M., Nuttall, J. M., & Hettema, E. H. (2012). Pex3-anchored Atg36 tags peroxisomes for degradation in Saccharomyces cerevisiae. *The EMBO Journal*, *31*(13), 2852–2868. http://dx.doi.org/10.1038/emboj.2012.151.

Nagai, T., Ibata, K., Park, E. S., Kubota, M., Mikoshiba, K., & Miyawaki, A. (2002). A variant of yellow fluorescent protein with fast and efficient maturation for cell-biological applications. *Nature Biotechnology*, *20*(1), 87–90. http://dx.doi.org/10.1038/nbt0102-87.

Nair, U., Cao, Y., Xie, Z., & Klionsky, D. J. (2010). Roles of the lipid-binding motifs of Atg18 and Atg21 in the cytoplasm to vacuole targeting pathway and autophagy. *The Journal of Biological Chemistry*, *285*(15), 11476–11488. http://dx.doi.org/10.1074/jbc.M109.080374.

Nair, U., Thumm, M., Klionsky, D. J., & Krick, R. (2011). GFP-Atg8 protease protection as a tool to monitor autophagosome biogenesis. *Autophagy*, 7(12), 1546–1550. http://dx.doi.org/10.4161/auto.7.12.18424.

Nakatogawa, H., Ishii, J., Asai, E., & Ohsumi, Y. (2012). Atg4 recycles inappropriately lipidated Atg8 to promote autophagosome biogenesis. *Autophagy*, *8*(2), 177–186. http://dx.doi.org/10.4161/auto.8.2.18373.

Ngu, M., Hirata, E., & Suzuki, K. (2015). Visualization of Atg3 during autophagosome formation in Saccharomyces cerevisiae. *The Journal of Biological Chemistry*, *290*(13), 8146–8153. http://dx.doi.org/10.1074/jbc.M114.626952.

Nice, D. C., Sato, T. K., Stromhaug, P. E., Emr, S. D., & Klionsky, D. J. (2002). Cooperative binding of the cytoplasm to vacuole targeting pathway proteins, Cvt13 and Cvt20, to phosphatidylinositol 3-phosphate at the pre-autophagosomal structure is required for selective autophagy. *The Journal of Biological Chemistry*, *277*(33), 30198–30207. http://dx.doi.org/10.1074/jbc.M204736200.

Nishimura, K., Fukagawa, T., Takisawa, H., Kakimoto, T., & Kanemaki, M. (2009). An auxin-based degron system for the rapid depletion of proteins in nonplant cells. *Nature Methods*, *6*(12), 917–922. http://dx.doi.org/10.1038/nmeth.1401.

Noda, T., Kim, J., Huang, W. P., Baba, M., Tokunaga, C., Ohsumi, Y., & Klionsky, D. J. (2000). Apg9p/Cvt7p is an integral membrane protein required for transport vesicle formation in the cvt and autophagy pathways. *The Journal of Cell Biology*, *148*(3), 465–480.

Noda, N. N., Kobayashi, T., Adachi, W., Fujioka, Y., Ohsumi, Y., & Inagaki, F. (2012). Structure of the novel C-terminal domain of vacuolar protein sorting 30/autophagy-related protein 6 and its specific role in autophagy. *The Journal of Biological Chemistry*, *287*(20), 16256–16266. http://dx.doi.org/10.1074/jbc.M112.348250.

Obara, K., Sekito, T., Niimi, K., & Ohsumi, Y. (2008). The Atg18-Atg2 complex is recruited to autophagic membranes via phosphatidylinositol 3-phosphate and exerts an essential function. *The Journal of Biological Chemistry*, *283*(35), 23972–23980. http://dx.doi.org/10.1074/jbc.M803180200.

Obara, K., Sekito, T., & Ohsumi, Y. (2006). Assortment of phosphatidylinositol 3-kinase complexes—Atg14p directs association of complex I to the pre-autophagosomal structure in Saccharomyces cerevisiae. *Molecular Biology of the Cell*, *17*(4), 1527–1539. http://dx.doi.org/10.1091/mbc.E05-09-0841.

Oda, M. N., Scott, S. V., Hefner-Gravink, A., Caffarelli, A. D., & Klionsky, D. J. (1996). Identification of a cytoplasm to vacuole targeting determinant in aminopeptidase I. *The Journal of Cell Biology*, *132*(6), 999–1010.

Okamoto, K., Kondo-Okamoto, N., & Ohsumi, Y. (2009). Mitochondria-anchored receptor Atg32 mediates degradation of mitochondria via selective autophagy. *Developmental Cell*, *17*(1), 87–97. http://dx.doi.org/10.1016/j.devcel.2009.06.013.

Papinski, D., Schuschnig, M., Reiter, W., Wilhelm, L., Barnes, C. A., Maiolica, A., … Kraft, C. (2014). Early steps in autophagy depend on direct phosphorylation of Atg9 by the Atg1 kinase. *Molecular Cell*, *53*(3), 471–483. http://dx.doi.org/10.1016/j.molcel.2013.12.011.

Perez-Perez, M. E., Zaffagnini, M., Marchand, C. H., Crespo, J. L., & Lemaire, S. D. (2014). The yeast autophagy protease Atg4 is regulated by thioredoxin. *Autophagy*, *10*(11), 1953–1964. http://dx.doi.org/10.4161/auto.34396.

Pfaffenwimmer, T., Reiter, W., Brach, T., Nogellova, V., Papinski, D., Schuschnig, M., … Kraft, C. (2014). Hrr25 kinase promotes selective autophagy by phosphorylating the cargo receptor Atg19. *EMBO Reports*, *15*(8), 862–870. http://dx.doi.org/10.15252/embr.201438932.

Rao, Y., Perna, M. G., Hofmann, B., Beier, V., & Wollert, T. (2016). The Atg1-kinase complex tethers Atg9-vesicles to initiate autophagy. *Nature Communications*, 7, 10338. http://dx.doi.org/10.1038/ncomms10338.

Reggiori, F., Shintani, T., Nair, U., & Klionsky, D. J. (2005). Atg9 cycles between mitochondria and the pre-autophagosomal structure in yeasts. *Autophagy*, *1*(2), 101–109. http://dx.doi.org/10.4161/auto.1.2.1840.

Reggiori, F., Tucker, K. A., Stromhaug, P. E., & Klionsky, D. J. (2004). The Atg1-Atg13 complex regulates Atg9 and Atg23 retrieval transport from the pre-autophagosomal structure. *Developmental Cell*, *6*(1), 79–90. http://dx.doi.org/10.1016/S1534-5807(03)00402-7.

Rieter, E., Vinke, F., Bakula, D., Cebollero, E., Ungermann, C., Proikas-Cezanne, T., & Reggiori, F. (2013). Atg18 function in autophagy is regulated by specific sites within its beta-propeller. *Journal of Cell Science*, *126*(Pt. 2), 593–604. http://dx.doi.org/10.1242/jcs.115725.

Sakakibara, K., Eiyama, A., Suzuki, S. W., Sakoh-Nakatogawa, M., Okumura, N., Tani, M., … Okamoto, K. (2015). Phospholipid methylation controls Atg32-mediated mitophagy and Atg8 recycling. *The EMBO Journal*, *34*(21), 2703–2719. http://dx.doi.org/10.15252/embj.201591440.

Sakoh-Nakatogawa, M., Kirisako, H., Nakatogawa, H., & Ohsumi, Y. (2015). Localization of Atg3 to autophagy-related membranes and its enhancement by the Atg8-family interacting motif to promote expansion of the membranes. *FEBS Letters*, *589*(6), 744–749. http://dx.doi.org/10.1016/j.febslet.2015.02.003.

Schultz, M. C., Hockman, D. J., Harkness, T. A., Garinther, W. I., & Altheim, B. A. (1997). Chromatin assembly in a yeast whole-cell extract. *Proceedings of the National Academy of Sciences of the United States of America*, *94*(17), 9034–9039.

Scott, S. V., Guan, J., Hutchins, M. U., Kim, J., & Klionsky, D. J. (2001). Cvt19 is a receptor for the cytoplasm-to-vacuole targeting pathway. *Molecular Cell*, 7(6), 1131–1141. http://dx.doi.org/10.1016/S1097-2765(01)00263-5.

Scott, S. V., Hefner-Gravink, A., Morano, K. A., Noda, T., Ohsumi, Y., & Klionsky, D. J. (1996). Cytoplasm-to-vacuole targeting and autophagy employ the same machinery to deliver proteins to the yeast vacuole. *Proceedings of the National Academy of Sciences of the United States of America*, *93*(22), 12304–12308.

Segarra, V. A., Boettner, D. R., & Lemmon, S. K. (2015). Atg27 tyrosine sorting motif is important for its trafficking and Atg9 localization. *Traffic (Copenhagen, Denmark)*, *16*(4), 365–378. http://dx.doi.org/10.1111/tra.12253.

Sekito, T., Kawamata, T., Ichikawa, R., Suzuki, K., & Ohsumi, Y. (2009). Atg17 recruits Atg9 to organize the pre-autophagosomal structure. *Genes to Cells: Devoted to Molecular & Cellular Mechanisms*, *14*(5), 525–538. http://dx.doi.org/10.1111/j.1365-2443.2009.01299.x.

Shintani, T., Huang, W. P., Stromhaug, P. E., & Klionsky, D. J. (2002). Mechanism of cargo selection in the cytoplasm to vacuole targeting pathway. *Developmental Cell*, *3*(6), 825–837. http://dx.doi.org/10.1016/S1534-5807(02)00373-8.

Shintani, T., & Klionsky, D. J. (2004). Cargo proteins facilitate the formation of transport vesicles in the cytoplasm to vacuole targeting pathway. *The Journal of Biological Chemistry*, *279*(29), 29889–29894. http://dx.doi.org/10.1074/jbc.M404399200.

Shintani, T., Mizushima, N., Ogawa, Y., Matsuura, A., Noda, T., & Ohsumi, Y. (1999). Apg10p, a novel protein-conjugating enzyme essential for autophagy in yeast. *The EMBO Journal*, *18*(19), 5234–5241. http://dx.doi.org/10.1093/emboj/18.19.5234.

Shintani, T., Suzuki, K., Kamada, Y., Noda, T., & Ohsumi, Y. (2001). Apg2p functions in autophagosome formation on the perivacuolar structure. *The Journal of Biological Chemistry*, *276*(32), 30452–30460. http://dx.doi.org/10.1074/jbc.M102346200.

Shyu, Y. J., Liu, H., Deng, X., & Hu, C. D. (2006). Identification of new fluorescent protein fragments for bimolecular fluorescence complementation analysis under physiological conditions. *BioTechniques*, *40*(1), 61–66. http://dx.doi.org/10.2144/000112036.

Sikorski, R. S., & Hieter, P. (1989). A system of shuttle vectors and yeast host strains designed for efficient manipulation of DNA in Saccharomyces cerevisiae. *Genetics*, *122*(1), 19–27.

Slessareva, J. E., Routt, S. M., Temple, B., Bankaitis, V. A., & Dohlman, H. G. (2006). Activation of the phosphatidylinositol 3-kinase Vps34 by a G protein alpha subunit at the endosome. *Cell*, *126*(1), 191–203. http://dx.doi.org/10.1016/j.cell.2006.04.045.

Stack, J. H., Herman, P. K., Schu, P. V., & Emr, S. D. (1993). A membrane-associated complex containing the Vps15 protein kinase and the Vps34 PI 3-kinase is essential for protein sorting to the yeast lysosome-like vacuole. *The EMBO Journal*, *12*(5), 2195–2204.

Stefan, C. J., & Blumer, K. J. (1999). A syntaxin homolog encoded by VAM3 mediates down-regulation of a yeast G protein-coupled receptor. *The Journal of Biological Chemistry*, *274*(3), 1835–1841.

Stephan, J. S., Yeh, Y. Y., Ramachandran, V., Deminoff, S. J., & Herman, P. K. (2009). The tor and PKA signaling pathways independently target the Atg1/Atg13 protein kinase complex to control autophagy. *Proceedings of the National Academy of Sciences of the United States of America*, *106*(40), 17049–17054. http://dx.doi.org/10.1073/pnas.0903316106.

Stotz, A., & Linder, P. (1990). The ADE2 gene from Saccharomyces cerevisiae: sequence and new vectors. *Gene*, *95*(1), 91–98. http://dx.doi.org/10.1016/0378-1119(90)90418-Q.

Straub, M., Bredschneider, M., & Thumm, M. (1997). AUT3, a serine/threonine kinase gene, is essential for autophagocytosis in Saccharomyces cerevisiae. *Journal of Bacteriology*, *179*(12), 3875–3883.

Stromhaug, P. E., Reggiori, F., Guan, J., Wang, C. W., & Klionsky, D. J. (2004). Atg21 is a phosphoinositide binding protein required for efficient lipidation and localization of Atg8 during uptake of aminopeptidase I by selective autophagy. *Molecular Biology of the Cell, 15*(8), 3553–3566. http://dx.doi.org/10.1091/mbc.E04-02-0147.

Sung, M. K., & Huh, W. K. (2007). Bimolecular fluorescence complementation analysis system for in vivo detection of protein-protein interaction in Saccharomyces cerevisiae. *Yeast (Chichester, England), 24*(9), 767–775. http://dx.doi.org/10.1002/yea.1504.

Suzuki, K., Akioka, M., Kondo-Kakuta, C., Yamamoto, H., & Ohsumi, Y. (2013). Fine mapping of autophagy-related proteins during autophagosome formation in Saccharomyces cerevisiae. *Journal of Cell Science, 126*(Pt. 11), 2534–2544. http://dx.doi.org/10.1242/jcs.122960.

Suzuki, K., Kirisako, T., Kamada, Y., Mizushima, N., Noda, T., & Ohsumi, Y. (2001). The pre-autophagosomal structure organized by concerted functions of APG genes is essential for autophagosome formation. *The EMBO Journal, 20*(21), 5971–5981. http://dx.doi.org/10.1093/emboj/20.21.5971.

Suzuki, K., Kondo, C., Morimoto, M., & Ohsumi, Y. (2010). Selective transport of alpha-mannosidase by autophagic pathways: identification of a novel receptor, Atg34p. *The Journal of Biological Chemistry, 285*(39), 30019–30025. http://dx.doi.org/10.1074/jbc.M110.143511.

Suzuki, K., Kubota, Y., Sekito, T., & Ohsumi, Y. (2007). Hierarchy of atg proteins in pre-autophagosomal structure organization. *Genes to Cells: Devoted to Molecular & Cellular Mechanisms, 12*(2), 209–218. http://dx.doi.org/10.1111/j.1365-2443.2007.01050.x.

Suzuki, S. W., Yamamoto, H., Oikawa, Y., Kondo-Kakuta, C., Kimura, Y., Hirano, H., & Ohsumi, Y. (2015). Atg13 HORMA domain recruits Atg9 vesicles during autophagosome formation. *Proceedings of the National Academy of Sciences of the United States of America, 112*(11), 3350–3355. http://dx.doi.org/10.1073/pnas.1421092112.

Tanaka, C., Tan, L. J., Mochida, K., Kirisako, H., Koizumi, M., Asai, E., … Nakatogawa, H. (2014). Hrr25 triggers selective autophagy-related pathways by phosphorylating receptor proteins. *The Journal of Cell Biology, 207*(1), 91–105. http://dx.doi.org/10.1083/jcb.201402128.

Tanida, I., Mizushima, N., Kiyooka, M., Ohsumi, M., Ueno, T., Ohsumi, Y., & Kominami, E. (1999). Apg7p/Cvt2p: a novel protein-activating enzyme essential for autophagy. *Molecular Biology of the Cell, 10*(5), 1367–1379.

Tucker, K. A., Reggiori, F., Dunn, W. A., Jr., & Klionsky, D. J. (2003). Atg23 is essential for the cytoplasm to vacuole targeting pathway and efficient autophagy but not pexophagy. *The Journal of Biological Chemistry, 278*(48), 48445–48452. http://dx.doi.org/10.1074/jbc.M309238200.

Urbanowski, J. L., & Piper, R. C. (1999). The iron transporter Fth1p forms a complex with the Fet5 iron oxidase and resides on the vacuolar membrane. *The Journal of Biological Chemistry, 274*(53), 38061–38070.

van der Vaart, A., Griffith, J., & Reggiori, F. (2010). Exit from the golgi is required for the expansion of the autophagosomal phagophore in yeast Saccharomyces cerevisiae. *Molecular Biology of the Cell, 21*(13), 2270–2284. http://dx.doi.org/10.1091/mbc.E09-04-0345.

Vida, T. A., & Emr, S. D. (1995). A new vital stain for visualizing vacuolar membrane dynamics and endocytosis in yeast. *The Journal of Cell Biology, 128*(5), 779–792.

Wang, K., Jin, M., Liu, X., & Klionsky, D. J. (2013). Proteolytic processing of Atg32 by the mitochondrial i-AAA protease Yme1 regulates mitophagy. *Autophagy, 9*(11), 1828–1836. http://dx.doi.org/10.4161/auto.26281.

Wang, C. W., Kim, J., Huang, W. P., Abeliovich, H., Stromhaug, P. E., Dunn, W. A., Jr., & Klionsky, D. J. (2001). Apg2 is a novel protein required for the cytoplasm to vacuole targeting, autophagy, and pexophagy pathways. *The Journal of Biological Chemistry, 276*(32), 30442–30451. http://dx.doi.org/10.1074/jbc.M102342200.

Wang, J., Menon, S., Yamasaki, A., Chou, H. T., Walz, T., Jiang, Y., & Ferro-Novick, S. (2013). Ypt1 recruits the Atg1 kinase to the preautophagosomal structure. *Proceedings of the National Academy of Sciences of the United States of America*, *110*(24), 9800–9805. http://dx.doi.org/10.1073/pnas.1302337110.

Wang, K., Yang, Z., Liu, X., Mao, K., Nair, U., & Klionsky, D. J. (2012). Phosphatidylinositol 4-kinases are required for autophagic membrane trafficking. *The Journal of Biological Chemistry*, *287*(45), 37964–37972. http://dx.doi.org/10.1074/jbc.M112.371591.

Watanabe, Y., Kobayashi, T., Yamamoto, H., Hoshida, H., Akada, R., Inagaki, F., … Noda, N. N. (2012). Structure-based analyses reveal distinct binding sites for Atg2 and phosphoinositides in Atg18. *The Journal of Biological Chemistry*, *287*(38), 31681–31690. http://dx.doi.org/10.1074/jbc.M112.397570.

Wurmser, A. E., & Emr, S. D. (2002). Novel PtdIns(3)P-binding protein Etf1 functions as an effector of the Vps34 PtdIns 3-kinase in autophagy. *The Journal of Cell Biology*, *158*(4), 761–772. http://dx.doi.org/10.1083/jcb.200112050.

Yamamoto, H., Kakuta, S., Watanabe, T. M., Kitamura, A., Sekito, T., Kondo-Kakuta, C., … Ohsumi, Y. (2012). Atg9 vesicles are an important membrane source during early steps of autophagosome formation. *The Journal of Cell Biology*, *198*(2), 219–233. http://dx.doi.org/10.1083/jcb.201202061.

Yamazaki-Sato, H., Tanida, I., Ueno, T., & Kominami, E. (2003). The carboxyl terminal 17 amino acids within Apg7 are essential for Apg8 lipidation, but not for Apg12 conjugation. *FEBS Letters*, *551*(1–3), 71–77. http://dx.doi.org/10.1016/S0014-5793(03)00899-8.

Yeh, Y. Y., Shah, K. H., & Herman, P. K. (2011). An Atg13 protein-mediated self-association of the Atg1 protein kinase is important for the induction of autophagy. *The Journal of Biological Chemistry*, *286*(33), 28931–28939. http://dx.doi.org/10.1074/jbc.M111.250324.

Yeh, Y. Y., Wrasman, K., & Herman, P. K. (2010). Autophosphorylation within the Atg1 activation loop is required for both kinase activity and the induction of autophagy in Saccharomyces cerevisiae. *Genetics*, *185*(3), 871–882. http://dx.doi.org/10.1534/genetics.110.116566.

Yen, W. L., & Klionsky, D. J. (2012). Proteinase protection of prApe1 as a tool to monitor cvt vesicle/autophagosome biogenesis. *Autophagy*, *8*(8), 1245–1249. http://dx.doi.org/10.4161/auto.20916.

Yen, W. L., Legakis, J. E., Nair, U., & Klionsky, D. J. (2007). Atg27 is required for autophagy-dependent cycling of Atg9. *Molecular Biology of the Cell*, *18*(2), 581–593. http://dx.doi.org/10.1091/mbc.E06-07-0612.

Yorimitsu, T., & Klionsky, D. J. (2005). Atg11 links cargo to the vesicle-forming machinery in the cytoplasm to vacuole targeting pathway. *Molecular Biology of the Cell*, *16*(4), 1593–1605. http://dx.doi.org/10.1091/mbc.E04-11-1035.

CHAPTER NINETEEN

Methods to Assess Autophagy and Chronological Aging in Yeast

K. Kainz*[,1], J. Tadic*[,1], A. Zimmermann*, T. Pendl*, D. Carmona-Gutierrez*, C. Ruckenstuhl*, T. Eisenberg*[,†], F. Madeo*[,†,2]

*Institute of Molecular Biosciences, NAWI Graz, University of Graz, Graz, Austria
[†]BioTechMed-Graz, Graz, Austria
[2]Corresponding author: e-mail address: frank.madeo@uni-graz.at

Contents

Abstract

Autophagy is a catabolic process that is crucial for cellular homeostasis and adaptive response to changing environments. Importantly, autophagy has been shown to be induced in many longevity-associated scenarios and to be required to maintain lifespan extension. Notably, autophagy is a highly conserved cellular process among eukaryotes, and the yeast *Saccharomyces cerevisiae* has become a universal model system for unraveling the molecular machinery underlying autophagic mechanisms. Here, we discuss different protocols to monitor survival and autophagy of yeast cells upon chronological aging. These include the use of propidium iodide to assess the loss of cell membrane integrity, as well as clonogenic assays to directly determine survival rates. Additionally, we describe methods to quantify autophagic flux, including the alkaline phosphatase activity or the GFP liberation assays, which measure the delivery of autophagosomal cargo to the vacuole. In sum, we have recapped established protocols used to evaluate a link between lifespan extension and autophagy in yeast.

[1] These authors contributed equally to this work.

Methods in Enzymology, Volume 588
ISSN 0076-6879
http://dx.doi.org/10.1016/bs.mie.2016.09.086

1. INTRODUCTION

The term "autophagy" was first introduced by Christian de Duve at the CIBA Foundation Symposium on Lysosomes, which emerged following the first report on lysosomes in 1955 (de Duve, Pressman, Gianetto, Wattiaux, & Appelmans, 1955; Feng, He, Yao, & Klionsky, 2014). Since then, various autophagy-related degradation pathways have been characterized. Two main types of autophagy are known: macroautophagy and microautophagy. Both these processes can either occur in a nonselective or a selective manner (e.g., selective autophagy of mitochondria, peroxisomes, nucleus, etc.) (Reggiori & Klionsky, 2013). The cytoplasm to vacuole targeting pathway is considered to be a type of selective autophagy in yeast. This process is a vesicular trafficking pathway that is morphologically and mechanistically related to autophagy as it requires some of the autophagy-related (*ATG*) genes for the delivery of its primary cargo, precursor aminopeptidase 1 (Lynch-Day & Klionsky, 2010; Reggiori & Klionsky, 2013; Shintani & Klionsky, 2004). Here, we will focus on the best-studied form of autophagy, nonselective macroautophagy (hereafter referred to as autophagy), which recycles portions of the cytoplasm or complete (damaged) organelles, replenishing energy-rich macromolecules to maintain cellular homeostasis. This not only allows cells to adapt to various stress conditions and rigorous changes in their nutrient environment but also provides a means of intracellular quality control and cytoprotection (Kundu & Thompson, 2005; Puleston & Simon, 2015; Reggiori & Klionsky, 2013; Sampaio-Marques, Burhans, & Ludovico, 2014). Importantly, autophagy is evolutionary conserved. For instance, to date, more than 35 *ATG*s have been characterized in yeast (Jin & Klionsky, 2014). In fact, *Saccharomyces cerevisiae* has advanced to become a preferred model system to study the conserved regulators and mechanisms underlying autophagy. The autophagic core machinery and the functional and structural characterization of Atg proteins have been extensively described elsewhere recently (Feng et al., 2014; Reggiori & Klionsky, 2013).

S. cerevisiae is also a widely used model organism for aging research, particularly for two types of aging: chronological and replicative aging, which serve as models for the senescence of human postmitotic and proliferating cells, respectively. Replicative lifespan (RLS) is defined by the number of daughter cells produced by a single mother cell, while chronological lifespan (CLS) represents the time a nonproliferating culture remains viable in a nutrient-limited environment (Fabrizio & Longo, 2003;

Kaeberlein, Burtner, & Kennedy, 2007; Longo & Fabrizio, 2012; Mortimer & Johnston, 1959; Müller, Zimmermann, Becker, & Flömer, 1980; Sinclair & Guarente, 1997). Interestingly, yeast lifespan can be increased by reduced nutrient concentrations in the media. This process is reminiscent of caloric restriction (CR), which is known to increase lifespan in other model organisms such as worms, flies, or mammals (Alvers et al., 2009; Aris et al., 2013; Jiang, Jaruga, Repnevskaya, & Jazwinski, 2000; Powers, Kaeberlein, Caldwell, Kennedy, & Fields, 2006; Wu, Liu, & Huang, 2013). Thus, yeast appears to bear at least some conserved molecular pathways that underlie lifespan extension. Importantly, using yeast as a model, several conserved genes that increase lifespan were identified as genes that orchestrate autophagy (Klionsky et al., 2003; Thumm et al., 1994; Tsukada & Ohsumi, 1993). This mechanistic connection between autophagy and aging can thus be explored in *S. cerevisiae* and is the topic of the present review.

2. AUTOPHAGY AND LONGEVITY: AN INTIMATE CONNECTION

In yeast, chronological and replicative aging not only share some mechanistic similarities but also have striking differences. For instance, an important molecular pathway involved in extending RLS is linked to Sir2, an NAD-dependent histone deacetylase, which connects metabolism with aging through genomic silencing. If *SIR2* is deleted, RLS is reduced, whereas increased dosage of Sir2 promotes RLS extension (Imai, Armstrong, Kaeberlein, & Guarente, 2000; Kaeberlein, McVey, & Guarente, 1999). Nevertheless, deletion of *ATG15* can abolish RLS extension induced by *SIR2* overexpression (Tang et al., 2008). Notably, the first sirtuin gene (*SIR2*) was found in yeast, and ever since, many members of this family have been discovered in higher organisms (Dali-Youcef et al., 2007; Rogina & Helfand, 2004; Tissenbaum & Guarente, 2001). In general, sirtuins may play an important role in autophagy and aging regulation (Guedes, Ludovico, & Sampaio-Marques, 2016; Sampaio-Marques et al., 2012), since their activity is required for starvation-triggered autophagy induction, and may stimulate deacetylation of some Atg proteins involved in the autophagy machinery (Lee et al., 2008). However, Sir2 is not involved in CR-mediated CLS extension (Fabrizio et al., 2005; Smith, Daniel, McClure, Matecic, & Smith, 2007). On the other hand, both types of aging are similarly regulated by the nutrient signaling pathways controlled by serine–threonine protein kinase (Sch9) and the target of rapamycin (TOR) kinase (Kaeberlein et al., 2005; Matecic et al., 2010; Powers et al., 2006). Both TOR and Sch9 are negative regulators of CLS and RLS (Fabrizio et al., 2003; Kaeberlein et al., 2005; Powers et al., 2006). In association with cAMP-dependent protein kinase (PKA), TOR and Sch9 regulate several downstream processes, such as autophagy, protein synthesis, stress resistance, and mitochondrial function, all in response to nutrient availability. Moreover, the TOR kinase plays a central role in cellular nutrient sensing, possibly regulating CR-dependent lifespan extension (Bonawitz, Chatenay-Lapointe, Pan, & Shadel, 2007; Kapahi & Zid, 2004; Powers et al., 2006). In yeast, two TOR kinases (Tor1 and Tor2) coordinate the starvation response (Helliwell et al., 1994). Both are active in a nutrient-rich environment, whereas they become inactive under starvation conditions (Noda & Ohsumi, 1998). Generally, both of these proteins assemble into two TOR complexes, TOR complex 1 (TORC-1) and complex 2 (TORC-2), respectively. TORC-1 may contain either Tor1 or Tor2 and can be inhibited by rapamycin, whereas TORC-2 contains only Tor2 (Loewith et al., 2002). To keep in mind, TORC-1 and TORC-2 regulate distinct pathways; however, they can be partly overlapping such as in the regulation of sphingolipid and ceramide signaling (Teixeira, Medeiros, Vilaça, Moradas-Ferreira, & Costa, 2014). Intriguingly, CLS extension upon *TOR1* deletion requires Atg16 (Matecic et al., 2010), suggesting a direct dependence on the autophagic process. Upon nutrient availability, Atg13 is hyperphosphorylated by the TOR complex, resulting in reduced affinity to Atg1 and downregulation of autophagy (Kamada et al., 2000; Kundu & Thompson, 2005). Furthermore, inhibition of TOR signaling by rapamycin extends CLS in an autophagy-dependent manner (Alvers, Wood, et al., 2009). Importantly, induction of autophagy by rapamycin has been revealed to have a protective role in neurodegenerative-related disorders (Malagelada, Jin, Jackson-Lewis, Przedborski, & Greene, 2010; Sarkar, Ravikumar, Floto, & Rubinsztein, 2008). Further, it has been demonstrated that TOR signaling plays a central role in yeast cells modeling CLN3 disease (neuronal ceroid lipofuscinosis) (Bond, Brown, Ralli, Bähler, & Mole, 2015). With this in mind, promoting autophagy and selective targeting of the TOR pathway might yield novel strategies to treat age-dependent neurodegenerative diseases. It should be noted, though, that inhibition of the TOR pathway may affect protein translation, and thus promote lifespan extension through autophagy-independent changes in the proteome as well (Madeo, Tavernarakis, & Kroemer, 2010).

The progressive loss of cellular quality control is linked to aging, for instance due to accumulation of toxic protein aggregates or damaged organelles (Madeo et al., 2010; Rajawat, Hilioti, & Bossis, 2009; Yen & Klionsky, 2008). Since such toxic components can be eliminated by autophagy, this is believed to underlie at least some longevity phenotypes (Madeo, Zimmermann, Maiuri, & Kroemer, 2015; Schroeder et al., 2014). In line, many scenarios that lead to lifespan extension are directly linked to enhanced autophagy (Alvers, Fishwick, et al., 2009; Alvers, Wood, et al., 2009; Bergamini, Cavallini, Donati, & Gori, 2007; Eisenberg et al., 2009; Rubinsztein, Mariño, & Kroemer, 2011). Vice versa, accumulating evidence indicates that *ATG* genes are regulated by longevity pathways (Füllgrabe, Klionsky, & Joseph, 2014; Vellai, Takács-Vellai, Sass, & Klionsky, 2009).

The most common method to assess autophagy dependency of a given longevity phenotype is to examine whether the effect is abolished in strains lacking essential *ATG* genes. If that is the case, autophagy is a probable executor of longevity in the tested scenario. Although deletion of *ATG* genes is used as a common tool in the field of aging research, it has to be noted that *ATG* knockout strains exhibit a shortened lifespan both in nutrient-rich (Alvers, Fishwick, et al., 2009; Alvers, Wood, et al., 2009; Eisenberg et al., 2009) and under starvation conditions (Suzuki, Onodera, & Ohsumi, 2011; Tsukada & Ohsumi, 1993). Therefore, heterozygous *ATG* knockouts may be used as an alternative avenue. Such mutants may exhibit normal lifespan under standard conditions while specifically impairing the autophagy response in longevity-inducing regimens, a strategy that has been successfully applied to autophagy-dependent memory restoration during aging in flies (Gupta et al., 2013). Generally, these approaches target the canonical pathway of autophagy, which requires the hierarchical recruitment of the core Atg proteins. However, recent evidence in mammalian cells points toward the existence of a noncanonical type of autophagy, which only requires a small subset of the Atg proteins (Dupont & Codogno, 2013). The first study to raise such a possibility showed that autophagy is induced in neurons treated with 1-methyl-4-phenylpyridinium independently of Atg6/Beclin1 and Vps34 (Zhu et al., 2007). Ever since, many other forms of noncanonical autophagy have been described, including one that requires Atg8 and Uba1 (an E1 enzyme involved in ubiquitinylation) but not Atg3 or Atg7 for the formation of autophagosomes (Chang et al., 2013). Another study provided first evidence that besides the classical Atg5/Atg7-dependent autophagy process, an alternative Atg5/Atg7-independent pathway exists. This process seems to be regulated by ULK1/Atg1 and Beclin1/Atg6, and requires Rab9 and vesicles from the trans-Golgi and the late endosome for autophagosome formation (Nishida et al., 2009). Recently, it was reported that phospholipids like phosphatidylethanolamine (PE) and unsaturated fatty acids are able to induce autophagy (Niso-Santano et al., 2015; Rockenfeller et al., 2015). Niso-Santano and colleagues could show that oleate induces a noncanonical form of autophagy that operates independently of ULK1/Atg1, phosphatidylinositol 3-kinase, and Vps34 but involves the Golgi apparatus (Niso-Santano et al., 2015). Notably, these pathways are conserved in yeast, nematodes, human cells, and mice, suggesting that *S. cerevisiae* might serve as a model for at least some types of noncanonical autophagy, as well.

Yeast has been instrumental for elucidating the crucial role of autophagy in several lifespan-extending scenarios. The most consistent antiaging intervention that extends lifespan from yeast to nonhuman primates known to date is CR (Bergamini et al., 2007; Jia & Levine, 2007). In yeast, various protocols exist to apply CR: growth in low glucose (e.g., 0.5%) medium (Lin, Defossez, & Guarente, 2000), growth in standard glucose medium (2%) followed by transfer to water (Fabrizio & Longo, 2003), or amino acid restriction (e.g., asparagine/glutamate) (Powers et al., 2006). Using two of these methods (low glucose and water transfer) Aris et al. found that—although not yielding equivalent results—both of these interventions extend CLS in an autophagy-dependent manner (Aris et al., 2013). Notably, reduced supply of specific amino acids, e.g., methionine, leucine, isoleucine, or valine, increases CLS in yeast (Alvers, Fishwick, et al., 2009; Aris et al., 2013; Ruckenstuhl et al., 2014a, 2014b). Interestingly, methionine restriction enhanced vacuolar acidification, which—concomitant with lifespan extension—was abolished in autophagy-deficient strains (Ruckenstuhl et al., 2014a, 2014b).

Besides these nutrient-associated scenarios, pharmacological treatment with specific substances also suggests a connection between autophagy and lifespan extension. For instance, rapamycin, a small macrolide antibiotic, targets TOR signaling by specifically inhibiting TORC1 (Alvers, Wood, et al., 2009; Beck & Hall, 1999; Cardenas, Cutler, Lorenz, Di Como, & Heitman, 1999). Since TOR is a negative regulator of autophagy, rapamycin application induces autophagy in many organisms, including yeast (Kamada, Sekito, & Ohsumi, 2004; Li, Kim, & Blenis, 2014).

Alvers et al. could show that low doses of rapamycin are able to prolong CLS in yeast and that this lifespan extension requires functional autophagy (Alvers, Wood, et al., 2009). Nonetheless, rapamycin treatment also influences other metabolic pathways besides autophagy (De Virgilio & Loewith, 2006; Li et al., 2014). Treatment with spermidine, a natural polyamine, has been shown to increase lifespan of yeast, flies, and worms in an autophagy-dependent manner (Eisenberg et al., 2009; Morselli et al., 2011; Pietrocola et al., 2015; Speldewinde & Grant, 2015). Thereby, it possibly acts on acetyltransferases, resulting in global hypoacetylation of histone H3 and a specific hyperacetylation of the *ATG7* promoter region, which allows increased expression of *ATG7* in yeast (Eisenberg et al., 2009; Morselli et al., 2011; Pietrocola et al., 2015). Further studies are necessary to reveal the precise mechanisms behind the decline of autophagy with ongoing age (Cuervo, 2008; Rajawat et al., 2009; Yen & Klionsky, 2008) and to find new ways to activate cellular recycling. Due to the known crucial impact of acetylation on autophagy and aging regulation (Eisenberg et al., 2009, 2014; Fahrenkrog, 2015; Madeo et al., 2015; Yoshida, Lengronne, & Pasero, 2014) the impact of epigenetic and posttranslational modifications on autophagy will be of vital interest.

3. METHODS TO ASSESS CLS IN YEAST

Yeast CLS can be monitored by determining the survival rate in a yeast cell culture during the postdiauxic phase in minimal media. This phase follows the switch from fermentation to respiration upon depletion of the fermentable carbon source (diauxic shift). At this stage, yeast cells stop dividing and enter a stationary phase characterized by a high metabolic rate and mostly relying on mitochondrial oxidative phosphorylation for ATP production. This is the beginning of chronological aging (Fabrizio & Longo, 2003).

It should be noted that the composition of media is of high relevance when monitoring yeast aging, since autophagy and amino acid homeostasis are crucial factors for chronological longevity (Alvers, Fishwick, et al., 2009; Alvers, Wood, et al., 2009; Longo, Shadel, Kaeberlein, & Kennedy, 2012; Powers et al., 2006). Different laboratories use different growth media for yeast aging experiments, employ various kinds of culture flasks, and follow different inoculation schemes, which may influence CLS (Longo et al., 2012; Piper, 2011). For instance, it has been reported that using plastic caps or aluminum foil caps which are used alternatively in different laboratories can influence outcome of chronological aging experiments (Longo et al., 2012). Similarly, extracellular acetic acid, which can influence the pH of cultured media and induce apoptosis (Ludovico, Sousa, Silva, Leão, & Côrte-Real, 2001), can further modify the outcome of an aging experiment (Burtner, Murakami, Kennedy, & Kaeberlein, 2009). Equally important is oxygen availability in yeast culture, since it can influence CLS (Bisschops et al., 2015). Further, yeast growth temperature is a crucial aspect to be considered. Yeast optimal growth temperature is 28–32°C (Van Uden, 1985), which also depends on the yeast strain. Experiments in our laboratory and few others are performed at 28°C (Eisenberg et al., 2014; Lewinska, Macierzynska, Grzelak, & Bartosz, 2011), with the intention to grow yeast at the lowest basal stress. In addition, some laboratories are growing yeast at 26°C (Guedes et al., 2016), whereas some do at 30°C (Matecic et al., 2010). Finally, the genetic background of yeast strains should be considered (Aris et al., 2013). In fact, a single genetic difference between two widely used yeast wild-type strains (BY4741 and BY4742), which involves the ability to produce methionine, influences CLS: BY4741 shows a longer CLS due to *MET15* deletion, while BY4742, which is methionine prototroph, is shorter lived (Ruckenstuhl et al., 2014b).

Within the following sections, we describe the specific strategies and experimental details established in our laboratories that have proven eligibility to address diverse aspects of longevity and suit well for many longevity scenarios (Büttner et al., 2008, 2007; Carmona-Gutiérrez et al., 2011; Eisenberg et al., 2014; Ruckenstuhl et al., 2014b).

3.1 Yeast Survival Plating

Yeast survival plating is based on clonogenicity and determined by plating, a defined number of cells on YPD plates, and counting the resulting colony forming units (CFUs) that form after 2–3 days of incubation (Bonawitz et al., 2007; Büttner et al., 2008, 2007; Carmona-Gutiérrez et al., 2011; Longo & Fabrizio, 2012; Longo, Gralla, & Valentine, 1996; Ruckenstuhl et al., 2014b).

Of note, cells that are not able to divide and thus do not form colonies on a plate are not necessarily dead cells, but may have entered a senescent-like state that prevents reentry into cell proliferation (Werner-Washburne, Roy, & Davidson, 2011). Therefore, the estimation of yeast survival during chronological aging requires multiple methods besides clonogenicity assays;

for instance, the quantification of the dead cell population based on markers of programmed cell death (Carmona-Gutierrez et al., 2010; Eisenberg, Carmona-Gutierrez, Büttner, Tavernarakis, & Madeo, 2010) or the assessment of loss of cell membrane permeability.

3.1.1 Experimental Setup

3.1.1.1 Reagents and Materials

- Synthetic complete medium: 0.17% yeast nitrogen base (without ammonium sulfate and amino acids), 0.5% $(NH_4)_2SO_4$, 30 mg/L of all amino acids (except 80 mg/L histidine and 200 mg/L leucine), 30 mg/L adenine, and 320 mg/L uracil with 2% carbon source.

  Note: Growth media are differently composed in various laboratories and may influence reproducibility of data due to growing of yeast culture under different conditions.
- YPD agar plates: 2% peptone, 1% yeast extract, 2–4% glucose, and 2% agar.
- CASYton: 0.9% NaCl (w/v), 1 m*M* EDTA.

3.1.2 Experimental Procedure

3.1.2.1 Preparation of Cells

- For chronological aging experiments, yeast cells are incubated at 145 rpm and 28°C, unless other conditions, depending on the yeast strain, are required.
- Inoculate overnight yeast culture in suitable minimal/synthetic media.
- Inoculate corresponding media with fresh overnight yeast culture to an OD_{600} of 0.1 (1×10^6 cells/mL) in 100 mL baffled flasks (10% of flask volume).
- If media shifts are necessary (e.g., induction of gene expression), shift the cells to desired media in early to middle log phase ($OD_{600} \sim 0.3–0.5$).
- For high-throughput approaches, cells can be cultured in deep-well plates by incubating at 320 or 1000 rpm for 200 or 500 μL culture volume, respectively.

3.1.2.2 Survival Plating

- At selected time points (ideally every 2–3 days), cell survival is determined by clonogenicity.
- Take 10 μL of the yeast cell culture and dilute the samples 1:100 in sterile double-distilled water (ddH_2O) (Dilution 1, 1:100).
- Vortex the samples and dilute them again 1:100 (Dilution 2, 1:10,000).

- Dilute 100 µL of Dilution 1 in 10 mL CASYton (resulting in a dilution factor of 1:10,000) and count cells with a CASY cell counter.

  Note: If cells tend to aggregate or show incomplete budding, cell number estimation using automated cell counters can be misleading, resulting in unrealistic survival rates significantly above 100% when aggregates are separated during plating. To overcome this problem, sonication may be applied to the first dilution used for cell counting (Eisenberg et al., 2009).
- Using Dilution 2, plate 500 cells on YPD agar plates and incubate them at 28°C for 2 days, unless slow colony growth is observed (incubation up to 3–4 days).
- Count CFUs (either by hand or automatically i.e., Lemnatec colony-counter) and calculate the survival rate by dividing the number of observed colonies through the number of plated colonies.

  Note: Survival rates are frequently normalized to day 1 or day 3 of a chronological aging experiment. It is important, however, that the absolute survival rates at the time point used for normalization are comparable in all strains/conditions, since strong differences in early survival may otherwise artificially affect CLS.

3.2 Propidium Iodide Staining

Propidium iodide (PI) is an intercalating and fluorescent agent. PI staining has become a common method for assessing yeast cell viability in high-throughput studies that apply flow cytometry. PI stains dead cells as a result of porous membrane, since PI cannot diffuse into viable cells or cells which are dead but still have an intact membrane (i.e., primary apoptotic cells) (Braun et al., 2015; Deere et al., 1998; Eisenberg et al., 2010, 2014; Madeo, Fröhlich, & Fröhlich, 1997; Ocampo & Barrientos, 2011; Zheng et al., 2007). Even though this method is reliable and reproducible, it should be considered that some yeast strains do not retain membrane integrity under specific stress conditions and could be interpreted as dead cells, although they can regain membrane integrity after a short incubation time (Davey & Hexley, 2011).

3.2.1 Experimental Setup

3.2.1.1 Reagents and Materials

- PBS: 25 m*M* potassium phosphate buffer, pH 7.0, 0.9% NaCl (w/v).
- PI solution: 100 µg/mL PI in ddH_2O; protect from light and store at −20°C; dilute 1:1000 in 1 × PBS before use.

3.2.2 *Experimental Procedure*

3.2.2.1 Preparation of Cells

As already described in Section 3.1.

PI staining

- At selected time points (should be the same as used for survival plating), take 1–2 × 10^6 yeast cells and centrifuge samples in a 96-well microplate for 5 min at 4000 rpm and at room temperature.
- Remove supernatant and resuspend cell pellets in 250 μL PI solution.
- Incubate samples for 5 min in the dark.
- Centrifuge the samples for 5 min at 4000 rpm and at room temperature.
- Remove supernatant and resuspend the cell pellets in 250 μL PBS.
- Fluorescence should be measured using a flow cytometer (extinction at 488 nm/emission at 695 nm) preferably by using an automated high-throughput sampling system (e.g., BD HTS™ option for flow cytometers). For a nonautomated flow cytometer, transfer cells into appropriate sampling tubes.

  Note: It is also possible to measure fluorescence using a fluorescence plate reader (extinction 515 nm/emission 595 nm). For that purpose, use 2 × 10^6 yeast cells (equal cell numbers or densities are required for comparability of results) and measure in duplicates, employing a light-protected 96-well microplate suitable for fluorescence measurements. However, such measurement will quantify total fluorescence intensity of the cell population and not provide results at the single-cell level. Importantly, this can falsify results dramatically in case of strong differences in fluorescent intensities of PI positive cells that may be observed under the distinct conditions to be compared.

4. METHODS TO MEASURE NONSELECTIVE AUTOPHAGY IN YEAST

Besides the methods that exist to monitor nonselective autophagy (as described in this manuscript), there are several established methods to assess selective forms of autophagy. In particular, the process of mitophagy has been studied extensively in the past years regarding its importance in aging and disease (Mendl et al., 2011; Palikaras, Lionaki, & Tavernarakis, 2016; Sampaio-Marques et al., 2014). Hereafter, we describe two assays that are commonly used to monitor nonselective autophagy during yeast aging. Indeed, various studies connecting lifespan extension to autophagy have made use of them; for examples, see Aris et al. (2013), Eisenberg et al.

(2009, 2014), Guedes et al. (2016), Rockenfeller et al. (2015), Ruckenstuhl et al. (2014b), and Sampaio-Marques et al. (2012). For a more detailed recommendation on the interpretation and the use of different autophagy assays, see Klionsky et al. (2016).

4.1 Alkaline Phosphatase Activity Assay

4.1.1 Background

In yeast, *PHO8* encodes for the sole vacuolar alkaline phosphatase (ALP), a transmembrane glycoprotein that is synthesized as an inactive precursor at the ER and transported via the secretory pathway into the vacuole. Once delivered to the vacuole, the proregion at the C-terminus is cleaved by vacuolar proteases, resulting in an active form of the enzyme (Klionsky & Emr, 1989). Noda et al. described a method for monitoring autophagy using a truncated form of Pho8 (Pho8ΔN60), which lacks 60 amino acids at the N-terminus (Noda, Matsuura, Wada, & Ohsumi, 1995). This deletion includes the membrane spanning region necessary for the insertion of the protein into the ER membrane. Consequently, Pho8ΔN60 remains in the cytosol. Upon autophagy induction, nonselective portions of the cytosol (including Pho8ΔN60) are engulfed by autophagosomes and delivered to the vacuole. There, Pho8ΔN60 can be activated. Thus, the ALP activity of Pho8ΔN60 is directly proportional to the cytosolic cargo delivered for degradation and reflects the autophagic flux when analyzed in a *pho8Δ* background (Noda et al., 1995). Importantly, two key aspects allow the use of this assay: Pho8 is resistant to vacuolar degradation and the loss of *PHO8* shows no evident phenotype regarding the cell growth or the vacuolar degradation of autophagic bodies (Noda & Klionsky, 2008; Noda et al., 1995). This assay shows a high signal-to-noise ratio, providing a very sensitive method for monitoring autophagy in yeast.

4.1.2 Experimental Setup

4.1.2.1 Strains

The strains expressing Pho8ΔN60 can be either generated via a stable insertion of the mutant gene or by using a plasmid (e.g., pCC5; Campbell & Thorsness, 1998). For construction of the strains, see Janke et al. (2004), Longtine et al. (1998), and Yen et al. (2010). Importantly, yeast bears another ALP, Pho13p, which is located in the cytosol. Hence, in the original protocol, the *PHO13* gene was deleted in addition to *PHO8* mutation in order to avoid any autophagy-independent influence on ALP activity (Noda et al., 1995). However, another possibility is to employ a substrate that is primarily utilized by Pho8, α-napthyl phosphate (Kaneko, Tamai, Toh-e, & Oshima, 1985; Kaneko, Toh-e, & Oshima, 1982; Noda & Klionsky, 2008); this method requires a fluorometer and is described hereafter. To improve the specificity of this alternative, the ALP activity of the corresponding strains without Pho8ΔN60 can be assessed and subtracted to correct for the intrinsic (background) ALP activity (Eisenberg et al., 2009; Ruckenstuhl et al., 2014b).

4.1.2.2 Reagents and Materials

- Assay buffer (AB): 250 m*M* Tris/HCl, pH 9.5, 10 m*M* $MgSO_4$, 10 μ*M* $ZnSO_4$; can be stored at −20°C.
- α-Naphtyl phosphate solution (α-NP): 55 m*M* α-naphtyl phosphate in AB; can be stored at −20°C.
- Glycine/NaOH buffer (GB): 2 *M* glycine/NaOH, pH 11; can be stored at −20°C.
- Glass beads: 400–600 nm in diameter, acid washed.

4.1.3 Experimental Procedure

4.1.3.1 Preparation of Cell Extracts

- Harvest 2–4 OD_{600} equivalents of each sample, wash once with 500 μL ddH_2O, and resuspend the cells in 200 μL ice-cold AB. From that moment on, samples should be kept on ice.

  Note: The amount of harvested cells should be adapted to the time points that are used to monitor the ALP activity in order to get a proper protein concentration after cell disruption.
- Transfer the samples into precooled 1.5 mL tubes containing 100 μL acid-washed glass beads and disrupt the cells by using a vortex or a minibead-beater. Disruption should be performed in intervals; in between those intervals, the samples should be chilled on ice (e.g., 2 × 2 min with a 2-min interval on ice).

  Note: The amount of glass beads should not exceed half of the AB volume. Disruption time and cycles strongly depend on the sampling time point and the type of disrupter. For cells older than 1 day, the number of cycles should be increased (e.g., 3 × 2 min with 2 × 2 min intervals on ice for 2-day-old cells). When employing a vortexer, disruption should be performed in a refrigerated room using several cycles. If a minibead-beater is used, the sample tray (ideally metallic material) should be kept at −20°C before use and chilled in between the intervals at −20°C (without samples!). In our experience, this assay can be done until days 5–6 of chronological aging (5–6 × 2 min disruption, longer chill intervals in between) without difficulties. Nevertheless, it has to be noted that later time points are not recommended as more cycles do not yield better disruption but may negatively influence the phosphatase activity.

- After disruption, centrifuge the samples at 4°C and 10,000 × *g* for 10 min and transfer the supernatant into a fresh, precooled tube. Keep on ice.

  Note: Cell extracts can be kept at 4°C (or on ice) for a few hours without affecting ALP activity.
- Determine the protein concentration using a standard method (e.g., a Bradford assay).

4.1.3.2 ALP Activity Assay

Add 1–1.5 μg of total protein extract to 500 μL prewarmed AB; according to the extract volume used, add AB to reach a final volume of 550 μL. Reactions should be performed in doubles. Additionally, a blank should be prepared, using AB instead of cell extracts, in order to control for the proper function of the enzymatic assay.

Note: As with any enzyme assay, linearity of the assay should be verified. For most conditions, 1–1.5 μg total protein suits well for high autophagic activities observed in CLS longevity experiments, but should be confirmed under the given condition by applying different amounts of protein (see also Klionsky et al., 2016).

- Add 50 μL α-NP to start the reaction, mix thoroughly, and incubate for 20 min at 30°C.
- Stop the reaction by adding 200 μL GB and mix thoroughly.
- Transfer 100 μL of each sample into a black bottom 96-well plate and measure fluorescence intensity with a fluorescence reader (extinction at 340 nm/emission at 485 nm).
- The relative ALP activity is calculated as relative fluorescent units (RFU)/μg protein.

  Note: The earlier description uses a defined protein concentration (resulting in different volumes for each sample that are then adjusted with AB). Alternatively, it is also common to use equal volumes for all samples (resulting in different protein concentrations subjected to the assay that are then normalized in a subsequent calculation step). In any case, obtaining similar protein concentrations in all samples after disruption is critical for results comparability, since the ALP activity may artificially depend on the volumes used (or subjected protein concentrations, see also note above concerning linearity issues). Moreover, ALP activity is likely affected by the expression level of Pho8ΔN60 (Klionsky et al., 2016; Noda & Klionsky, 2008). Thus, a dot blot analysis using Pho8-specific antibody (e.g., Invitrogen, 459260) of the cell extracts is recommended to control for similar expression.

4.2 GFP-Liberation Assay

4.2.1 Background

The processing and/or degradation of several proteins involved in the autophagic pathway, e.g., precursor Ape1, Pex14-GFP, or GFP-Atg8, can also be monitored (Cheong, Klionsky, & Klionsky, 2008). For aging-related autophagy, the most commonly used assay is the determination of vacuolar Atg8 degradation, resulting in GFP liberation in strains carrying the GFP-Atg8 fusion construct. During the autophagic process, the ubiquitin-like protein Atg8 is conjugated to PE, thus associating with the phagophore/autophagosome at the internal and external membranes. Upon autophagy progression, Atg8 is cleaved off from the outer surface but is still associated with the inner membrane, and is thus trapped inside, causing Atg8 to remain associated with the completed autophagosome until it is delivered to the vacuole (Ichimura et al., 2000; Kirisako et al., 1999, 2000). Once in the vacuole, Atg8 is rapidly degraded. Coupling GFP to Atg8 allows to monitor this process and to quantify the actual delivery of autophagic vesicles to the vacuole, since free GFP, which is not sensitive to the acidic pH and resists vacuolar hydrolysis, accumulates in the vacuole after Atg8 degradation (Shintani & Klionsky, 2004). Notably, the position of the GFP tag is of great importance: as Atg8 is proteolytically processed at its C-terminus (followed by conjugation to PE), only N-terminally tagged Atg8 can be used as an autophagic flux marker, while Atg8-GFP has been employed to monitor Atg4-dependent Atg8 cleavage activity (Kim, Huang, & Klionsky, 2001; Klionsky, 2011).

Western blot analysis is one possible method to trace the fate of GFP-Atg8 during aging. For that purpose, the reduction of full-length GFP-Atg8 levels and simultaneous emergence of free GFP is followed over time (Fig. 1) (Guimaraes, Delorme-Axford, Klionsky, & Reggiori, 2015). The GFP signal can also be monitored via fluorescence microscopy: in that case, vacuoles show GFP fluorescence upon autophagy induction and delivery of Atg8/autophagosomes to and fusion with the vacuole (Fig. 2). Combining both methods, Western blot analysis and fluorescence microscopy, provides a powerful tool to monitor the autophagic flux during yeast chronological aging.

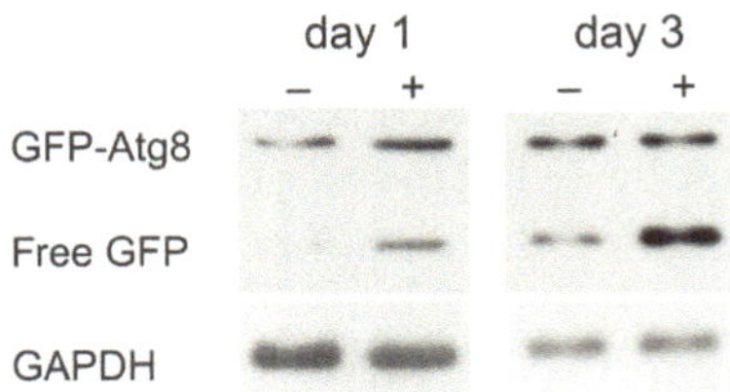

Fig. 1 Monitoring GFP-Atg8 liberation via Western blot analysis. Wild-type cells (BY4742) were grown in SD medium to mid-log phase and treated (+) or not (–) with rapamycin (200 ng/mL). Samples were taken at indicated time points and prepared for Western blot analysis. Induction of autophagy by rapamycin treatment can be detected through the increase of free GFP.

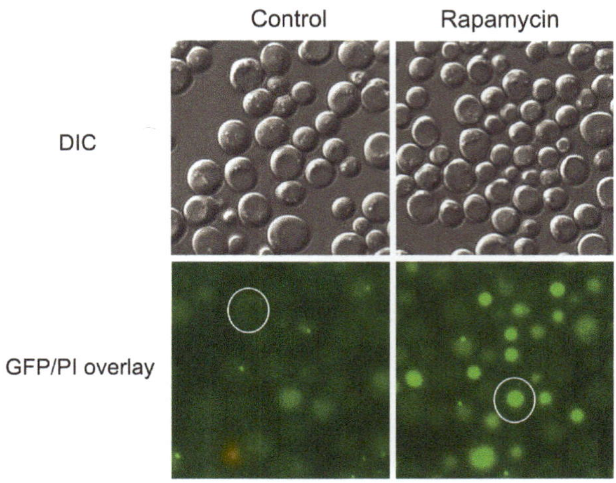

Fig. 2 Monitoring GFP-Atg8 degradation via fluorescence microscopy. Wild-type cells (BY4742) were treated or not with rapamycin (200 ng/mL). Cells were analyzed 20 h after intervention. The *left circle* shows a nonautophagic cell (GFP-Atg8 distributed in the cytosol, no GFP signal in the vacuole), whereas the *right circle* highlights an autophagic cell (GFP signal in the vacuole). Treatment with rapamycin causes an increased vacuolar GFP signal compared to control cells, indicating autophagy induction.

The expression of the GFP-Atg8 construct under the control of its endogenous promoter is strongly enhanced under starvation while it is only weak in rich media (Kirisako et al., 1999), suggesting that Atg8 expression is induced in at least some autophagy-activating conditions. Expressing GFP-Atg8 from its genomic locus and under control of its natural P_{ATG8} promoter, thus, enables monitoring Atg8 induction in the same assay.

The use of overexpression promoters (e.g., of *CUP1*) may overcome difficulties of GFP-Atg8 detection when this protein is only poorly expressed (e.g., in rich media or after short periods of autophagy induction). Notably, the levels of Atg8 were reported to regulate the size of autophagosomes (Jin & Klionsky, 2014b). Using a stronger promoter, though, clearly limits the reliability of estimating Atg8 expression levels (Shintani & Reggiori, 2008). The level of free GFP changes over time and strongly correlates with the delivery of autophagosomes to the vacuole, the degradation of Atg8, and the final breakdown of GFP. For a reasonable interpretation of results obtained from the GFP liberation assay, it has to be taken into account that the time, during which free GFP remains stable within the acidic vacuole, is yet unknown. Therefore, detectable free GFP levels at a given time point may reflect the integrated autophagic activity from a yet to be defined preceding time period. Thus, it is suggested to perform time course analyses (Mauvezin, Ayala, Braden, Kim, & Neufeld, 2014). Finally, while GFP is relatively stable in the vacuolar acidic environment, long incubation periods under starvation conditions might affect its stability (Cheong et al., 2008).

4.2.2 Monitoring GFP-Atg8 Degradation via Western Blot

The following protocol represents only one possible method to assess GFP-Atg8 degradation (GFP liberation) via Western blot. Other available protocols have proven to yield reproducible and reliable results, as well (Cheong et al., 2008; Guimaraes et al., 2015).

4.2.2.1 Experimental Setup

4.2.2.1.1 Strains Strains used to monitor GFP-Atg8 degradation must be transformed with and selected for a plasmid expressing the GFP-Atg8 construct (e.g., addgene, Plasmid #49425). Alternatively, GFP can be integrated at the 5′ end of the chromosomal *ATG8* locus using pYM-pATG8 plasmid, described in Eisenberg et al. (2014).

4.2.2.1.2 Reagents

- Lysis buffer (CP): 1.85 *M* NaOH, 7.5% β-mercaptoethanol; freshly prepared.
- Trichloroacetic acid (TCA): 55% TCA.
- Final sample buffer (FSB): 50 m*M* Tris/HCl, pH 6.8, 2% sodium dodecyl sulfate (SDS), 10% glycerol, 0.1% bromophenol blue, 100 m*M* dithiothreitol (DTT); can be stored at −20°C.

- Tris solution: 2 *M* Tris, untitrated.
- Blotting buffer (e.g., CAPS buffer): 10 m*M* 3-(cyclohexylamino)-1-propanesulfonic acid (CAPS), pH 11, 10% methanol.
- TBS (−T): 10 m*M* Tris/HCl, pH 7.4, 150 m*M* NaCl, (0.02% Triton-X 100).
- Blocking solution: e.g., 1% or 3% milk powder in TBS.
- α-GFP antibody (e.g., Roche, Cat. No. 11814460001): 1:5000 diluted in TBS + 1% milk powder.
- HRP-α-mouse antibody (e.g., Sigma, A9044): 1:10,000 diluted in TBS + 1% milk powder.

4.2.2.2 Experimental Procedure

4.2.2.2.1 Preparation of Cell Extracts

- Harvest 3 OD_{600} units of each sample and wash once with 500 μL ddH_2O. At this point, cell pellets can be stored at −20°C, if necessary. From now on, samples should be kept on ice until they are resuspended in FSB.
- Resuspend the pellets in 150 μL CP in order to lysate the cells, mix properly, and incubate on ice for 10 min.
- Add 150 μL TCA in order to precipitate the proteins, mix well, and incubate again on ice for 10 min.
- Centrifuge the samples at 4°C and 10,000 × *g* for 10 min and remove the supernatant.

  Note: In order to remove residual TCA, it is recommended to either centrifuge again for at least 3 min (completely remove supernatant) or to wash cells once with ice-cold acetone.
- Resuspend the samples in 150 μL FSB by pipetting carefully (foaming!) and, if necessary, add 1–5 μL Tris solution until the indicator shows blue color again. If required, samples can be stored at −20°C at this point.
- Heat the samples for 5 min at 95°C and centrifuge at full speed for 2 min in order to precipitate insoluble proteins/cell debris.

  Note: Cell extracts should be heated shortly before they are subjected to SDS-PAGE. If samples need to be reused it is recommended to heat smaller aliquots separately for each application.

4.2.2.2.2 Immunoblotting

- Use 10–15 μL of each cell extract to separate the proteins using SDS-PAGE (12.5%) at 12 mA for the stacking gel and 20 mA for the separating gel.

- Transfer the proteins on a membrane using a blotting buffer at 220–240 mA for 1 h.

  Note: For proper blotting, the buffer should be ice-cold and thus should be stored at 4°C. Additionally, blotting chambers can be cooled with a water-cooling-system during blotting, or blotting can be performed in a refrigerated room.
- After blotting, block unspecific binding sites (e.g., with 1% blocking solution at 4°C over night or 3% blocking solution at room temperature for 1 h).
- Wash the membranes three times for 5 min with TBS-T.
- Incubate with an α-GFP antibody for 1 h at room temperature or at 4°C over night under shaking.
- Wash the membranes three times for 5 min with TBS-T.
- Incubate with an α-mouse antibody for 1 h at room temperature under shaking.
- Wash the membranes three times for 5 min with TBS-T.
- For detection, incubate the membranes with ECL reaction mix according to the manufacturer's instructions.
- For quantification, calculate the ratio of each free GFP band (~25 kDa) to its corresponding GFP-Atg8 band (~40 kDa) or to the loading control, using a software program of choice.

  Note: Using the ratio of free GFP to GFP-Atg8 bears potential bias, as this ratio may be strongly affected by the overall expression levels of GFP-Atg8. If overall expression levels differ, free GFP normalized to a suitable loading control may be favored for interpretation of results.

4.2.3 Monitoring Vacuolar Localization of GFP-Atg8 via Microscopy

4.2.3.1 Experimental Setup

4.2.3.1.1 Strains The same strains as for the earlier described GFP liberation assay can be used.

4.2.3.1.2 Reagents

- PI: 100 μg/mL in ddH_2O; can be stored at −20°C; dilute 1:1000 in PBS before use.
- PBS: 25 m*M* potassium phosphate buffer, pH 7.2, 0.9% NaCl.

4.2.3.2 Experimental Procedure

- Harvest 50–100 μL cells (depending on the culture density) at various time points and resuspend in 250 μL PI/PBS solution.

Note: Cells can be washed once with water in order to overcome the influence of autofluorescent components of the medium used (especially for cells grown in YPD). Although in yeast cells, the vacuole can easily be recognized in light microscopy using differential interference contrast, additional staining of the vacuole (e.g., with FM4-64) can be performed.

- Incubate for 5–10 min at room temperature in the dark.
- Spin down for 2 min at full speed and take an aliquot (~2 μL) directly out of the cell pellet for microscopy.

  Note: Cell suspension can be directly transferred onto a slide and covered with a coverglass. If there are problems with floating of the cells, cells can be mounted on a pad of 1.2% agarose; for detailed description, see Rines, Thomann, Dorn, Goodwin, and Sorger (2011).
- Visualize the fluorescence image with a fluorescence microscope using appropriate filters (for detecting GFP: excitation 455–490 nm/emission at 500–540 nm, for detecting PI: excitation 535–585 nm/emission at 600–655 nm).
- Capture, color, and overlay images using a digital camera and a software program of choice.
- For quantification, count the cells that show a vacuolar localization of GFP-Atg8 and exclude PI-stained (= dead) cells.

4.3 Perspectives

Aging and autophagy are two evolutionary conserved and tightly connected pathways. Most of the known antiaging interventions were shown to be accompanied with enhanced autophagy and vice versa. Importantly, prominent longevity modulators like CR or spermidine require functional autophagy to gain lifespan extension. So far, aging research has mainly focused on macroautophagy. Thus, many different approaches to follow this unspecific process in the context of aging have been successfully implemented. Nevertheless, other autophagic networks like mitophagy, pexophagy, noncanonical autophagy, and others have become focus of interest regarding their role in aging- and age-related diseases. Further studies are needed to investigate the mechanism and regulation of these processes. Unfortunately, available methods to assess specific types of autophagy are often rare or do not allow an appropriate distinction from other pathways. Hence, improving classical methods and concomitantly establishing new approaches that allow to precisely distinguish between different types of autophagy is of great importance.

ACKNOWLEDGMENTS

F.M. is grateful to the Austrian Science Fund FWF (Austria) for Grants P23490-B12, P24381, P27893, I1000 and Grant "SFB Lipotox" and to BMWFW and the Karl-Franzens University for Grant "Unkonventionelle Forschung" and Grant DKplus Metabolic and Cardiovascular Diseases to K.K., J.T., and F.M. We acknowledge support from NAWI Graz.

REFERENCES

Alvers, A. L., Fishwick, L. K., Wood, M. S., Hu, D., Chung, H. S., Dunn, W. A., & Aris, J. P. (2009). Autophagy and amino acid homeostasis are required for chronological longevity in Saccharomyces cerevisiae. *Aging Cell*, *8*(4), 353–369. http://doi.org/10.1111/j.1474-9726.2009.00469.x.

Alvers, A. L., Wood, M. S., Hu, D., Kaywell, A. C., Dunn, W. A., & Aris, J. P. (2009). Autophagy is required for extension of yeast chronological life span by rapamycin. *Autophagy*, *5*(6), 847–849.

Aris, J. P., Alvers, A. L., Ferraiuolo, R. A., Fishwick, L. K., Hanvivatpong, A., Hu, D., … Dunn, W. A. (2013). Autophagy and leucine promote chronological longevity and respiration proficiency during calorie restriction in yeast. *Experimental Gerontology*, *48*(10), 1107–1119. http://doi.org/10.1016/j.exger.2013.01.006.

Beck, T., & Hall, M. N. (1999). The TOR signalling pathway controls nuclear localization of nutrient-regulated transcription factors. *Nature*, *402*(6762), 689–692. http://doi.org/10.1038/45287.

Bergamini, E., Cavallini, G., Donati, A., & Gori, Z. (2007). The role of autophagy in aging. *Annals of the New York Academy of Sciences*, *1114*(1), 69–78. http://doi.org/10.1196/annals.1396.020.

Bisschops, M., Vos, T., Martínez-Moreno, R., de la Torre Cortés, P., Pronk, J., & Daran-Lapujade, P. (2015). Oxygen availability strongly affects chronological lifespan and thermotolerance in batch cultures of Saccharomyces cerevisiae. *Microbial Cell*, *2*(11), 429–444. http://doi.org/10.15698/mic2015.11.238.

Bonawitz, N. D., Chatenay-Lapointe, M., Pan, Y., & Shadel, G. S. (2007). Reduced TOR signaling extends chronological life span via increased respiration and upregulation of mitochondrial gene expression. *Cell Metabolism*, *5*(4), 265–277. http://doi.org/10.1016/j.cmet.2007.02.009.

Bond, M., Brown, R., Ralli, C., Bähler, J., & Mole, S. (2015). A central role for TOR signalling in a yeast model for juvenile CLN3 disease. *Microbial Cell*, *2*(12), 466–480. http://doi.org/10.15698/mic2015.12.241.

Braun, R. J., Sommer, C., Leibiger, C., Gentier, R. J. G., Dumit, V. I., Paduch, K., … Madeo, F. (2015). Accumulation of basic amino acids at mitochondria dictates the cytotoxicity of aberrant ubiquitin. *Cell Reports*, *10*(9), 1557–1571. http://doi.org/10.1016/j.celrep.2015.02.009.

Burtner, C. R., Murakami, C. J., Kennedy, B. K., & Kaeberlein, M. (2009). A molecular mechanism of chronological aging in yeast. *Cell Cycle (Georgetown, Tex)*, *8*(8), 1256–1270.

Büttner, S., Bitto, A., Ring, J., Augsten, M., Zabrocki, P., Eisenberg, T., … Madeo, F. (2008). Functional mitochondria are required for α-synuclein toxicity in aging yeast.

Journal of Biological Chemistry, 283(12), 7554–7560. http://doi.org/10.1074/jbc.M708477200.

Büttner, S., Eisenberg, T., Carmona-Gutierrez, D., Ruli, D., Knauer, H., Ruckenstuhl, C., ... Madeo, F. (2007). Endonuclease G regulates budding yeast life and death. *Molecular Cell, 25*(2), 233–246. http://doi.org/10.1016/j.molcel.2006.12.021.

Campbell, C. L., & Thorsness, P. E. (1998). Escape of mitochondrial DNA to the nucleus in yme1 yeast is mediated by vacuolar-dependent turnover of abnormal mitochondrial compartments. *Journal of Cell Science, 111*(16), 2455–2464.

Cardenas, M. E., Cutler, N. S., Lorenz, M. C., Di Como, C. J., & Heitman, J. (1999). The TOR signaling cascade regulates gene expression in response to nutrients. *Genes & Development, 13*(24), 3271–3279.

Carmona-Gutiérrez, D., Bauer, M. A., Ring, J., Knauer, H., Eisenberg, T., Büttner, S., ... Madeo, F. (2011). The propeptide of yeast cathepsin D inhibits programmed necrosis. *Cell Death & Disease, 2*(5), e161. http://doi.org/10.1038/cddis.2011.43.

Carmona-Gutierrez, D., Eisenberg, T., Büttner, S., Meisinger, C., Kroemer, G., & Madeo, F. (2010). Apoptosis in yeast: Triggers, pathways, subroutines. *Cell Death & Differentiation, 17*(5), 763–773. http://doi.org/10.1038/cdd.2009.219.

Chang, T.-K., Shravage, B. V., Hayes, S. D., Powers, C. M., Simin, R. T., Harper, J. W., & Baehrecke, E. H. (2013). Uba1 functions in Atg7- and Atg3-independent autophagy. *Nature Cell Biology, 15*(9), 1067–1078. http://doi.org/10.1038/ncb2804.

Cheong, H., Klionsky, D. J., & Klionsky, D. J. (2008). Biochemical methods to monitor autophagy-related processes in yeast. In D. Klionsky (Ed.), *Methods in enzymology: Vol. 451.* (pp. 1–26). United States: Academic Press. Retrieved from http://www.sciencedirect.com/science/article/pii/S0076687908032011. Chapter 1.

Cuervo, A. M. (2008). Autophagy and aging. *Trends in Genetics: TIG, 24*(12), 604–612. http://doi.org/10.1016/j.tig.2008.10.002.

Dali-Youcef, N., Lagouge, M., Froelich, S., Koehl, C., Schoonjans, K., & Auwerx, J. (2007). Sirtuins: The "magnificent seven", function, metabolism and longevity. *Annals of Medicine, 39*(5), 335–345. http://doi.org/10.1080/07853890701408194.

Davey, H. M., & Hexley, P. (2011). Red but not dead? Membranes of stressed Saccharomyces cerevisiae are permeable to propidium iodide. *Environmental Microbiology, 13*(1), 163–171. http://doi.org/10.1111/j.1462-2920.2010.02317.x.

de Duve, C., Pressman, B. C., Gianetto, R., Wattiaux, R., & Appelmans, F. (1955). Tissue fractionation studies. 6. Intracellular distribution patterns of enzymes in rat-liver tissue. *Biochemical Journal, 60*(4), 604–617.

De Virgilio, C., & Loewith, R. (2006). Cell growth control: Little eukaryotes make big contributions. *Oncogene, 25*(48), 6392–6415. http://doi.org/10.1038/sj.onc.1209884.

Deere, D., Shen, J., Vesey, G., Bell, P., Bissinger, P., & Veal, D. (1998). Flow cytometry and cell sorting for yeast viability assessment and cell selection. *Yeast, 14*(2), 147–160. http://doi.org/10.1002/(SICI)1097-0061(19980130)14:2<147::AID-YEA207>3.0.CO;2-L.

Dupont, N., & Codogno, P. (2013). Non-canonical autophagy: Facts and prospects. *Current Pathobiology Reports, 1*(4), 263–271. http://doi.org/10.1007/s40139-013-0030-y.

Eisenberg, T., Carmona-Gutierrez, D., Büttner, S., Tavernarakis, N., & Madeo, F. (2010). Necrosis in yeast. *Apoptosis, 15*(3), 257–268. http://doi.org/10.1007/s10495-009-0453-4.

Eisenberg, T., Knauer, H., Schauer, A., Büttner, S., Ruckenstuhl, C., Carmona-Gutierrez, D., ... Madeo, F. (2009). Induction of autophagy by spermidine promotes longevity. *Nature Cell Biology, 11*(11), 1305–1314. http://doi.org/10.1038/ncb1975.

Eisenberg, T., Schroeder, S., Andryushkova, A., Pendl, T., Küttner, V., Bhukel, A., ... Madeo, F. (2014). Nucleocytosolic depletion of the energy metabolite acetyl-coenzyme a stimulates autophagy and prolongs lifespan. *Cell Metabolism, 19*(3), 431–444. http://doi.org/10.1016/j.cmet.2014.02.010.

Fabrizio, P., Gattazzo, C., Battistella, L., Wei, M., Cheng, C., McGrew, K., & Longo, V. D. (2005). Sir2 blocks extreme life-span extension. *Cell, 123*(4), 655–667. http://doi.org/10.1016/j.cell.2005.08.042.

Fabrizio, P., Liou, L.-L., Moy, V. N., Diaspro, A., SelverstoneValentine, J., Gralla, E. B., & Longo, V. D. (2003). SOD2 functions downstream of Sch9 to extend longevity in yeast. *Genetics, 163*(1), 35–46.

Fabrizio, P., & Longo, V. D. (2003). The chronological life span of Saccharomyces cerevisiae. *Aging Cell, 2*(2), 73–81. http://doi.org/10.1046/j.1474-9728.2003.00033.x.

Fahrenkrog, B. (2015). Histone modifications as regulators of life and death in Saccharomyces cerevisiae. *Microbial Cell, 2*(1), 1–13.

Feng, Y., He, D., Yao, Z., & Klionsky, D. J. (2014). The machinery of macroautophagy. *Cell Research, 24*(1), 24–41. http://doi.org/10.1038/cr.2013.168.

Füllgrabe, J., Klionsky, D. J., & Joseph, B. (2014). The return of the nucleus: Transcriptional and epigenetic control of autophagy. *Nature Reviews. Molecular Cell Biology, 15*(1), 65–74. http://doi.org/10.1038/nrm3716.

Guedes, A., Ludovico, P., & Sampaio-Marques, B. (2016). Caloric restriction alleviates alpha-synuclein toxicity in aged yeast cells by controlling the opposite roles of Tor1 and Sir2 on autophagy. *Mechanisms of Ageing and Development*, in press. http://doi.org/10.1016/j.mad.2016.04.006.

Guimaraes, R. S., Delorme-Axford, E., Klionsky, D. J., & Reggiori, F. (2015). Assays for the biochemical and ultrastructural measurement of selective and nonselective types of autophagy in the yeast Saccharomyces cerevisiae. *Methods, 75*, 141–150. http://doi.org/10.1016/j.ymeth.2014.11.023.

Gupta, V. K., Scheunemann, L., Eisenberg, T., Mertel, S., Bhukel, A., Koemans, T. S., … Sigrist, S. J. (2013). Restoring polyamines protects from age-induced memory impairment in an autophagy-dependent manner. *Nature Neuroscience, 16*(10), 1453–1460. http://doi.org/10.1038/nn.3512.

Helliwell, S. B., Wagner, P., Kunz, J., Deuter-Reinhard, M., Henriquez, R., & Hall, M. N. (1994). TOR1 and TOR2 are structurally and functionally similar but not identical phosphatidylinositol kinase homologues in yeast. *Molecular Biology of the Cell, 5*(1), 105–118.

Ichimura, Y., Kirisako, T., Takao, T., Satomi, Y., Shimonishi, Y., Ishihara, N., … Ohsumi, Y. (2000). A ubiquitin-like system mediates protein lipidation. *Nature, 408*(6811), 488–492. http://doi.org/10.1038/35044114.

Imai, S., Armstrong, C. M., Kaeberlein, M., & Guarente, L. (2000). Transcriptional silencing and longevity protein Sir2 is an NAD-dependent histone deacetylase. *Nature, 403*(6771), 795–800. http://doi.org/10.1038/35001622.

Janke, C., Magiera, M. M., Rathfelder, N., Taxis, C., Reber, S., Maekawa, H., … Knop, M. (2004). A versatile toolbox for PCR-based tagging of yeast genes: New fluorescent proteins, more markers and promoter substitution cassettes. *Yeast (Chichester, England), 21*(11), 947–962. http://doi.org/10.1002/yea.1142.

Jia, K., & Levine, B. (2007). Autophagy is required for dietary restriction-mediated life span extension in C. elegans. *Autophagy, 3*(6), 597–599. http://doi.org/10.4161/auto.4989.

Jiang, J. C., Jaruga, E., Repnevskaya, M. V., & Jazwinski, S. M. (2000). An intervention resembling caloric restriction prolongs life span and retards aging in yeast. *The FASEB Journal, 14*(14), 2135–2137. http://doi.org/10.1096/fj.00-0242fje.

Jin, M., & Klionsky, D. J. (2014). Regulation of autophagy: Modulation of the size and number of autophagosomes. *FEBS Letters, 588*(15), 2457–2463. http://doi.org/10.1016/j.febslet.2014.06.015.

Kaeberlein, M., Burtner, C. R., & Kennedy, B. K. (2007). Recent developments in yeast aging. *PLoS Genetics, 3*(5), e84. http://doi.org/10.1371/journal.pgen.0030084.

Kaeberlein, M., McVey, M., & Guarente, L. (1999). The SIR2/3/4 complex and SIR2 alone promote longevity in Saccharomyces cerevisiae by two different mechanisms. *Genes & Development, 13*(19), 2570–2580.

Kaeberlein, M., Powers, R. W., Steffen, K. K., Westman, E. A., Hu, D., Dang, N., … Kennedy, B. K. (2005). Regulation of yeast replicative life span by TOR and Sch9 in response to nutrients. *Science, 310*(5751), 1193–1196. http://doi.org/10.1126/science.1115535.

Kamada, Y., Funakoshi, T., Shintani, T., Nagano, K., Ohsumi, M., & Ohsumi, Y. (2000). Tor-mediated induction of autophagy via an Apg1 protein kinase complex. *The Journal of Cell Biology, 150*(6), 1507–1513.

Kamada, Y., Sekito, T., & Ohsumi, Y. (2004). *Autophagy in yeast: ATOR-mediated response to nutrient starvation. In G. Thomas, D. M. Sabatini, & M. N. Hall (Eds.), TOR* (pp. 73–84). Berlin Heidelberg: Springer. Retrieved from: http://link.springer.com/chapter/10.1007/978-3-642-18930-2_5.

Kaneko, Y., Tamai, Y., Toh-e, A., & Oshima, Y. (1985). Transcriptional and post-transcriptional control of PHO8 expression by PHO regulatory genes in Saccharomyces cerevisiae. *Molecular and Cellular Biology, 5*(1), 248–252.

Kaneko, Y., Toh-e, A., & Oshima, Y. (1982). Identification of the genetic locus for the structural gene and a new regulatory gene for the synthesis of repressible alkaline phosphatase in Saccharomyces cerevisiae. *Molecular and Cellular Biology, 2*(2), 127–137.

Kapahi, P., & Zid, B. (2004). TOR pathway: Linking nutrient sensing to life span. *Science of Aging Knowledge Environment : SAGE KE, 2004*(36), PE34. http://doi.org/10.1126/sageke.2004.36.pe34.

Kim, J., Huang, W.-P., & Klionsky, D. J. (2001). Membrane recruitment of Aut7p in the autophagy and cytoplasm to vacuole targeting pathways requires Aut1p, Aut2p, and the autophagy conjugation complex. *The Journal of Cell Biology, 152*(1), 51–64.

Kirisako, T., Baba, M., Ishihara, N., Miyazawa, K., Ohsumi, M., Yoshimori, T., … Ohsumi, Y. (1999). Formation process of autophagosome is traced with Apg8/Aut7p in yeast. *The Journal of Cell Biology, 147*(2), 435–446.

Kirisako, T., Ichimura, Y., Okada, H., Kabeya, Y., Mizushima, N., Yoshimori, T., … Ohsumi, Y. (2000). The reversible modification regulates the membrane-binding state of Apg8/Aut7 essential for autophagy and the cytoplasm to vacuole targeting pathway. *The Journal of Cell Biology, 151*(2), 263–276.

Klionsky, D. J. (2011). For the last time, it is GFP-Atg8, not Atg8-GFP (and the same goes for LC3). *Autophagy*, 7(10), 1093–1094. http://doi.org/10.4161/auto.7.10.15492.

Klionsky, D. J., Abdelmohsen, K., Abe, A., Abedin, M. J., Abeliovich, H., Arozena, A. A., … Zughaier, S. M. (2016). Guidelines for the use and interpretation of assays for monitoring autophagy (3rd edition). *Autophagy, 12*(1), 1–222. http://doi.org/10.1080/15548627.2015.1100356.

Klionsky, D. J., Cregg, J. M., Dunn, W. A., Jr., Emr, S. D., Sakai, Y., Sandoval, I. V., … Ohsumi, Y. (2003). A unified nomenclature for yeast autophagy-related genes. *Developmental Cell, 5*(4), 539–545. http://doi.org/10.1016/S1534-5807(03)00296-X.

Klionsky, D. J., & Emr, S. D. (1989). Membrane protein sorting: Biosynthesis, transport and processing of yeast vacuolar alkaline phosphatase. *The EMBO Journal, 8*(8), 2241–2250.

Kundu, M., & Thompson, C. B. (2005). Macroautophagy versus mitochondrial autophagy: A question of fate? *Cell Death & Differentiation, 12*(S2), 1484–1489. http://doi.org/10.1038/sj.cdd.4401780.

Lee, I. H., Cao, L., Mostoslavsky, R., Lombard, D. B., Liu, J., Bruns, N. E., … Finkel, T. (2008). A role for the NAD-dependent deacetylase Sirt1 in the regulation of autophagy. *Proceedings of the National Academy of Sciences, 105*(9), 3374–3379. http://doi.org/10.1073/pnas.0712145105.

Lewinska, A., Macierzynska, E., Grzelak, A., & Bartosz, G. (2011). A genetic analysis of nitric oxide-mediated signaling during chronological aging in the yeast. *Biogerontology*, *12*(4), 309–320. http://doi.org/10.1007/s10522-011-9329-4.

Li, J., Kim, S. G., & Blenis, J. (2014). Rapamycin: One drug, many effects. *Cell Metabolism*, *19*(3), 373–379. http://doi.org/10.1016/j.cmet.2014.01.001.

Lin, S.-J., Defossez, P.-A., & Guarente, L. (2000). Requirement of NAD and SIR2 for Life-Span Extension by Calorie Restriction in Saccharomyces cerevisiae. *Science*, *289*(5487), 2126–2128. http://doi.org/10.1126/science.289.5487.2126.

Loewith, R., Jacinto, E., Wullschleger, S., Lorberg, A., Crespo, J. L., Bonenfant, D., … Hall, M. N. (2002). Two TOR complexes, only One of which is rapamycin sensitive, have distinct roles in cell growth control. *Molecular Cell*, *10*(3), 457–468. http://doi.org/10.1016/S1097-2765(02)00636-6.

Longo, V. D., & Fabrizio, P. (2012). Chronological aging in Saccharomyces cerevisiae. *Sub-Cellular Biochemistry*, *57*, 101–121. http://doi.org/10.1007/978-94-007-2561-4_5.

Longo, V. D., Gralla, E. B., & Valentine, J. S. (1996). Superoxide dismutase activity is essential for stationary phase survival in Saccharomyces cerevisiae mitochondrial production of toxic oxygen species in vivo. *Journal of Biological Chemistry*, *271*(21), 12275–12280. http://doi.org/10.1074/jbc.271.21.12275.

Longo, V. D., Shadel, G. S., Kaeberlein, M., & Kennedy, B. (2012). Replicative and chronological aging in Saccharomyces cerevisiae. *Cell Metabolism*, *16*(1), 18–31. http://doi.org/10.1016/j.cmet.2012.06.002.

Longtine, M. S., McKenzie, A., Demarini, D. J., Shah, N. G., Wach, A., Brachat, A., … Pringle, J. R. (1998). Additional modules for versatile and economical PCR-based gene deletion and modification in Saccharomyces cerevisiae. *Yeast (Chichester, England)*, *14*(10), 953–961. http://doi.org/10.1002/(SICI)1097-0061(199807)14:10<953::AID-YEA293>3.0.CO;2-U.

Ludovico, P., Sousa, M. J., Silva, M. T., Leão, C., & Côrte-Real, M. (2001). Saccharomyces cerevisiae commits to a programmed cell death process in response to acetic acid. *Microbiology (Reading, England)*, *147*(Pt. 9), 2409–2415. http://doi.org/10.1099/00221287-147-9-2409.

Lynch-Day, M. A., & Klionsky, D. J. (2010). The Cvt pathway as a model for selective autophagy. *FEBS Letters*, *584*(7), 1359–1366. http://doi.org/10.1016/j.febslet.2010.02.013.

Madeo, F., Fröhlich, E., & Fröhlich, K. U. (1997). A yeast mutant showing diagnostic markers of early and late apoptosis. *The Journal of Cell Biology*, *139*(3), 729–734.

Madeo, F., Tavernarakis, N., & Kroemer, G. (2010). Can autophagy promote longevity? *Nature Cell Biology*, *12*(9), 842–846. http://doi.org/10.1038/ncb0910-842.

Madeo, F., Zimmermann, A., Maiuri, M. C., & Kroemer, G. (2015). Essential role for autophagy in life span extension. *The Journal of Clinical Investigation*, *125*(1), 85–93. http://doi.org/10.1172/JCI73946.

Malagelada, C., Jin, Z. H., Jackson-Lewis, V., Przedborski, S., & Greene, L. A. (2010). Rapamycin protects against neuron death in in vitro and in vivo models of Parkinson's disease. *The Journal of Neuroscience: The Official Journal of the Society for Neuroscience*, *30*(3), 1166–1175. http://doi.org/10.1523/JNEUROSCI.3944-09.2010.

Matecic, M., Smith, D. L., Pan, X., Maqani, N., Bekiranov, S., Boeke, J. D., & Smith, J. S. (2010). A microarray-based genetic screen for yeast chronological aging factors. *PLoS Genetics*, *6*(4), e1000921. http://doi.org/10.1371/journal.pgen.1000921.

Mauvezin, C., Ayala, C., Braden, C. R., Kim, J., & Neufeld, T. P. (2014). Assays to monitor autophagy in Drosophila. *Methods (San Diego, Calif.)*, *68*(1), 134–139. http://doi.org/10.1016/j.ymeth.2014.03.014.

Mendl, N., Occhipinti, A., Müller, M., Wild, P., Dikic, I., & Reichert, A. S. (2011). Mitophagy in yeast is independent of mitochondrial fission and requires the stress response gene WHI2. *Journal of Cell Science, 124*(8), 1339–1350. http://doi.org/10.1242/jcs.076406.

Morselli, E., Mariño, G., Bennetzen, M. V., Eisenberg, T., Megalou, E., Schroeder, S., … Kroemer, G. (2011). Spermidine and resveratrol induce autophagy by distinct pathways converging on the acetylproteome. *The Journal of Cell Biology, 192*(4), 615–629. http://doi.org/10.1083/jcb.201008167.

Mortimer, R. K., & Johnston, J. R. (1959). Life span of individual yeast cells. *Nature, 183*(4677), 1751–1752. http://doi.org/10.1038/1831751a0.

Müller, I., Zimmermann, M., Becker, D., & Flömer, M. (1980). Calendar life span versus budding lifespan of Saccharomyces cerevisiae. *Mechanisms of Ageing and Development, 12*(1), 47–52. http://doi.org/10.1016/0047-6374(80)90028-7.

Nishida, Y., Arakawa, S., Fujitani, K., Yamaguchi, H., Mizuta, T., Kanaseki, T., … Shimizu, S. (2009). Discovery of Atg5/Atg7-independent alternative macroautophagy. *Nature, 461*(7264), 654–658. http://doi.org/10.1038/nature08455.

Niso-Santano, M., Malik, S. A., Pietrocola, F., Bravo-San Pedro, J. M., Mariño, G., Cianfanelli, V., … Kroemer, G. (2015). Unsaturated fatty acids induce non-canonical autophagy. *The EMBO Journal, 34*(8), 1025–1041. http://doi.org/10.15252/embj.201489363.

Noda, T., & Klionsky, D. J. (2008). The quantitative Pho8Δ60 assay of nonspecific autophagy. In D. Klionsky (Ed.), *Methods in Enzymology: Vol. 451.* (pp. 33–42). United States: Academic Press. Retrieved from http://www.sciencedirect.com/science/article/pii/S0076687908032035. Chapter 3.

Noda, T., Matsuura, A., Wada, Y., & Ohsumi, Y. (1995). Novel system for monitoring autophagy in the yeast Saccharomyces cerevisiae. *Biochemical and Biophysical Research Communications, 210*(1), 126–132. http://doi.org/10.1006/bbrc.1995.1636.

Noda, T., & Ohsumi, Y. (1998). Tor, a phosphatidylinositol kinase homologue, controls autophagy in yeast. *Journal of Biological Chemistry, 273*(7), 3963–3966. http://doi.org/10.1074/jbc.273.7.3963.

Ocampo, A., & Barrientos, A. (2011). Quick and reliable assessment of chronological life span in yeast cell populations by flow cytometry. *Mechanisms of Ageing and Development, 132*(6–7), 315–323. http://doi.org/10.1016/j.mad.2011.06.007.

Palikaras, K., Lionaki, E., & Tavernarakis, N. (2016). Mitophagy: In sickness and in health. *Molecular & Cellular Oncology, 3*(1), e1056332. http://doi.org/10.1080/23723556.2015.1056332.

Pietrocola, F., Lachkar, S., Enot, D. P., Niso-Santano, M., Bravo-San Pedro, J. M., Sica, V., … Kroemer, G. (2015). Spermidine induces autophagy by inhibiting the acetyltransferase EP300. *Cell Death and Differentiation, 22*(3), 509–516. http://doi.org/10.1038/cdd.2014.215.

Piper, P. W. (2011). Maximising the yeast chronological lifespan. In M. Breitenbach, S. M. Jazwinski, & P. Laun (Eds.), *Aging research in yeast* (pp. 145–159). Netherland: Springer. Retrieved from http://link.springer.com/chapter/10.1007/978-94-007-2561-4_7.

Powers, R. W., Kaeberlein, M., Caldwell, S. D., Kennedy, B. K., & Fields, S. (2006). Extension of chronological life span in yeast by decreased TOR pathway signaling. *Genes & Development, 20*(2), 174–184. http://doi.org/10.1101/gad.1381406.

Puleston, D., & Simon, A. (2015). New roles for autophagy and spermidine in T cells. *Microbial Cell, 2*(3), 91–93. http://doi.org/10.15698/mic2015.03.195.

Rajawat, Y. S., Hilioti, Z., & Bossis, I. (2009). Aging: Central role for autophagy and the lysosomal degradative system. *Ageing Research Reviews, 8*(3), 199–213. http://doi.org/10.1016/j.arr.2009.05.001.

Reggiori, F., & Klionsky, D. J. (2013). Autophagic processes in yeast: Mechanism, machinery and regulation. *Genetics*, *194*(2), 341–361. http://doi.org/10.1534/genetics.112.149013.
Rines, D. R., Thomann, D., Dorn, J. F., Goodwin, P., & Sorger, P. K. (2011). Live cell imaging of yeast. *Cold Spring Harbor Protocols*, *2011*(9), 1026–1041. http://doi.org/10.1101/pdb.top065482. pdb.top065482.
Rockenfeller, P., Koska, M., Pietrocola, F., Minois, N., Knittelfelder, O., Sica, V., ... Madeo, F. (2015). Phosphatidylethanolamine positively regulates autophagy and longevity. *Cell Death and Differentiation*, *22*(3), 499–508. http://doi.org/10.1038/cdd.2014.219.
Rogina, B., & Helfand, S. L. (2004). Sir2 mediates longevity in the fly through a pathway related to calorie restriction. *Proceedings of the National Academy of Sciences of the United States of America*, *101*(45), 15998–16003. http://doi.org/10.1073/pnas.0404184101.
Rubinsztein, D. C., Mariño, G., & Kroemer, G. (2011). Autophagy and aging. *Cell*, *146*(5), 682–695. http://doi.org/10.1016/j.cell.2011.07.030.
Ruckenstuhl, C., Netzberger, C., Entfellner, I., Carmona-Gutierrez, D., Kickenweiz, T., Stekovic, S., ... Madeo, F. (2014a). Autophagy extends lifespan via vacuolar acidification. *Microbial Cell*, *1*(5), 160–162. http://doi.org/10.15698/mic2014.05.147.
Ruckenstuhl, C., Netzberger, C., Entfellner, I., Carmona-Gutierrez, D., Kickenweiz, T., Stekovic, S., ... Madeo, F. (2014b). Lifespan extension by methionine restriction requires autophagy-dependent vacuolar acidification. *PLoS Genetics*, *10*(5), e1004347. http://doi.org/10.1371/journal.pgen.1004347.
Sampaio-Marques, B., Burhans, W., & Ludovico, P. (2014). Longevity pathways and maintenance of the proteome: The role of autophagy and mitophagy during yeast ageing. *Microbial Cell*, *1*(4), 118–127. http://doi.org/10.15698/mic2014.04.136.
Sampaio-Marques, B., Felgueiras, C., Silva, A., Rodrigues, M., Tenreiro, S., Franssens, V., ... Ludovico, P. (2012). SNCA (α-synuclein)-induced toxicity in yeast cells is dependent on sirtuin 2 (Sir2)-mediated mitophagy. *Autophagy*, *8*(10), 1494–1509. http://doi.org/10.4161/auto.21275.
Sarkar, S., Ravikumar, B., Floto, R. A., & Rubinsztein, D. C. (2008). Rapamycin and mTOR-independent autophagy inducers ameliorate toxicity of polyglutamine-expanded huntingtin and related proteinopathies. *Cell Death & Differentiation*, *16*(1), 46–56. http://doi.org/10.1038/cdd.2008.110.
Schroeder, S., Zimmermann, A., Carmona-Gutierrez, D., Eisenberg, T., Ruckenstuhl, C., Andryushkova, A., ... Madeo, F. (2014). Metabolites in aging and autophagy. *Microbial Cell*, *1*(4), 110–114. http://doi.org/10.15698/mic2014.04.142.
Shintani, T., & Klionsky, D. J. (2004). Cargo proteins facilitate the formation of transport vesicles in the cytoplasm to vacuole targeting pathway. *The Journal of Biological Chemistry*, *279*(29), 29889–29894. http://doi.org/10.1074/jbc.M404399200.
Shintani, T., & Reggiori, F. (2008). Fluorescence microscopy-based assays for monitoring yeast Atg protein trafficking. In D. Klionsky (Ed.), *Methods in enzymology: Vol. 451*. (pp. 43–56). United States: Academic Press. Retrieved from *http://www.sciencedirect.com/science/article/pii/S0076687908032047*. Chapter 4.
Sinclair, D. A., & Guarente, L. (1997). Extrachromosomal rDNA circles—A cause of aging in yeast. *Cell*, *91*(7), 1033–1042. http://doi.org/10.1016/S0092-8674(00)80493-6.
Smith, J., Daniel, L., McClure, J. M., Matecic, M., & Smith, J. S. (2007). Calorie restriction extends the chronological lifespan of Saccharomyces cerevisiae independently of the Sirtuins. *Aging Cell*, *6*(5), 649–662. http://doi.org/10.1111/j.1474-9726.2007.00326.x.
Speldewinde, S., & Grant, C. (2015). Spermidine cures yeast of prions. *Microbial Cell*, *3*(1), 46–48. http://doi.org/10.15698/mic2016.01.474.
Suzuki, S. W., Onodera, J., & Ohsumi, Y. (2011). Starvation induced cell death in autophagy-defective yeast mutants is caused by mitochondria dysfunction. *PloS One*, *6*(2), e17412. http://doi.org/10.1371/journal.pone.0017412.

Tang, F., Watkins, J. W., Bermudez, M., Gray, R., Gaban, A., Portie, K., ... Craciun, G. (2008). A life-span extending form of autophagy employs the vacuole-vacuole fusion machinery. *Autophagy*, *4*(7), 874–886. http://dx.doi.org/10.4161/auto.6556.

Teixeira, V., Medeiros, T., Vilaça, R., Moradas-Ferreira, P., & Costa, V. (2014). Reduced TORC1 signaling abolishes mitochondrial dysfunctions and shortened chronological lifespan of Isc1p-deficient cells. *Microbial Cell*, *1*(1), 21–36. http://doi.org/10.15698/mic2014.01.121.

Thumm, M., Egner, R., Koch, B., Schlumpberger, M., Straub, M., Veenhuis, M., & Wolf, D. h. (1994). Isolation of autophagocytosis mutants of Saccharomyces cerevisiae. *FEBS Letters*, *349*(2), 275–280. http://doi.org/10.1016/0014-5793(94)00672-5.

Tissenbaum, H. A., & Guarente, L. (2001). Increased dosage of a sir-2 gene extends lifespan in Caenorhabditis elegans. *Nature*, *410*(6825), 227–230. http://doi.org/10.1038/35065638.

Tsukada, M., & Ohsumi, Y. (1993). Isolation and characterization of autophagy-defective mutants of Saccharomyces cerevisiae. *FEBS Letters*, *333*(1–2), 169–174. http://doi.org/10.1016/0014-5793(93)80398-E.

Van Uden, N. (1985). Temperature profiles of yeasts. In A. H. Rose & D. W. Tempest (Eds.), *Advances in microbial physiology: Vol. 25*. (pp. 195–251). United States: Academic Press. Retrieved from http://www.sciencedirect.com/science/article/pii/S0065291108602933.

Vellai, T., Takács-Vellai, K., Sass, M., & Klionsky, D. J. (2009). The regulation of aging: Does autophagy underlie longevity? *Trends in Cell Biology*, *19*(10), 487–494. http://doi.org/10.1016/j.tcb.2009.07.007.

Werner-Washburne, M., Roy, S., & Davidson, G. S. (2011). Aging and the survival of quiescent and non-quiescent cells in yeast stationary-phase cultures. In M. Breitenbach, S. M. Jazwinski, & P. Laun (Eds.), *Aging research in yeast* (pp. 123–143). Netherlands: Springer. Retrieved from http://link.springer.com/chapter/10.1007/978-94-007-2561-4_6.

Wu, Z., Liu, S. Q., & Huang, D. (2013). Dietary restriction depends on nutrient composition to extend chronological lifespan in budding yeast Saccharomyces cerevisiae. *PloS One*, *8*(5), e64448. http://doi.org/10.1371/journal.pone.0064448.

Yen, W.-L., & Klionsky, D. J. (2008). How to live long and prosper: Autophagy, mitochondria, and aging. *Physiology*, *23*(5), 248–262. http://doi.org/10.1152/physiol.00013.2008.

Yen, W.-L., Shintani, T., Nair, U., Cao, Y., Richardson, B. C., Li, Z., ... Klionsky, D. J. (2010). The conserved oligomeric Golgi complex is involved in double-membrane vesicle formation during autophagy. *The Journal of Cell Biology*, *188*(1), 101–114. http://doi.org/10.1083/jcb.200904075.

Yoshida, K., Lengronne, A., & Pasero, P. (2014). The replication timing program in the hands of two HDACs. *Microbial Cell*, *1*(8), 273–275. http://doi.org/10.15698/mic2014.08.163.

Zheng, K., Pan, J.-W., Ye, L., Fu, Y., Peng, H.-Z., Wan, B.-Y., ... Zhu, M.-Y. (2007). Programmed cell death-involved aluminum toxicity in yeast alleviated by antiapoptotic members with decreased calcium signals. *Plant Physiology*, *143*(1), 38–49. http://doi.org/10.1104/pp. 106.082495.

Zhu, J., Horbinski, C., Guo, F., Watkins, S., Uchiyama, Y., & Chu, C. T. (2007). Regulation of autophagy by extracellular signal-regulated protein kinases during 1-methyl-4-phenylpyridinium-induced cell death. *The American Journal of Pathology*, *170*(1), 75–86. http://doi.org/10.2353/ajpath.2007.060524.

CHAPTER TWENTY

Methods to Measure Lipophagy in Yeast

A. Cristobal-Sarramian, M. Radulovic, S.D. Kohlwein[1]
Institute of Molecular Biosciences, University of Graz and BioTechMed-Graz, Graz, Austria
[1]Corresponding author: e-mail address: sepp.kohlwein@uni-graz.at

Contents

Abstract

Maintenance of cellular and organismal lipid homeostasis is critical for life, and any deviation from a balanced equilibrium between fat uptake and degradation may have deleterious consequences, resulting in severe lipid-associated disorders. Excess fat is typically stored in cytoplasmic organelles termed "lipid droplets" (LDs); to adjust for a constantly fluctuating supply of and demand for cellular fat, these organelles are metabolically highly dynamic and subject to multiple levels of regulation. In addition to a well-described cytosolic lipid degradation pathway, recent evidence underscores the importance of "lipophagy" in cellular lipid homeostasis, i.e., the degradation of LD by autophagy in the lysosome/vacuole. Pioneering work in yeast mutant models has unveiled the requirement of key components of the autophagy machinery, providing

Methods in Enzymology, Volume 588
ISSN 0076-6879
http://dx.doi.org/10.1016/bs.mie.2016.09.087

evidence for a highly conserved process of lipophagy from yeast to man. However, further work is required to unveil the intricate metabolic interaction between LD metabolism and autophagy to sustain membrane homeostasis and cellular survival.

1. INTRODUCTION

Neutral lipids—triacylglycerols (TG), sterols, and steryl esters (SE)—are deposited in cytosolic lipid droplets (LDs), which serve as a reservoir to (i) buffer excess lipids and their precursors (e.g., fatty acids) in a rather inert form to protect cells from lipotoxic insults and (ii) mobilize these lipids in times of higher lipid demand, e.g., for energy production or membrane remodeling and biogenesis, during growth (Kohlwein, 2010; Kohlwein, Veenhuis, & van der Klei, 2013; Zechner et al., 2012). TG and SE breakdown is typically catalyzed by cytosolic, LD-resident lipases, and esterases, but recent evidence indicates that LD autophagy—termed "lipophagy"—also plays an important role in cellular (neutral) lipid homeostasis (Singh et al., 2009; van Zutphen et al., 2014; Wang, 2014, 2015). Numerous human pathologies have been associated with deficiencies in lipophagy, e.g., lipid storage disorders (Angelini, Nascimbeni, Cenacchi, & Tasca, 2016; Ward et al., 2016), aging (Rubinsztein, Mariño, & Kroemer, 2011), and cancer (Liu et al., 2013). The ease and versatility of yeast as an experimental system have significantly contributed to the discovery of some 38 autophagy-related proteins, many of which have orthologs in mammals (Klionsky, 2014; Klionsky et al., 2016; Reggiori & Klionsky, 2013) and are also required for lipophagy (van Zutphen et al., 2014; Wang, Miao, & Chang, 2014); a number of these genes not only plays a role in autophagy but also in alternative vacuolar targeting and degradation pathways. Lipophagy in yeast has been characterized as a selective microautophagic process that operates by direct interaction of LD with the vacuolar membrane prior to their internalization into the vacuole (Fig. 1) (Reggiori & Klionsky, 2013; van Zutphen et al., 2014; Wang et al., 2014); subsequent vacuolar TG breakdown is catalyzed by the Aut5/Cvt17/Atg15 lipase (Epple, Suriapranata, Eskelinen, & Thumm, 2001; Maeda, Oku, & Sakai, 2015; Teter et al., 2001; van Zutphen et al., 2014). In mammalian cells, LD internalization into the lysosome requires their prior engulfment into the autophagosome, which is a double membrane structure characteristic for a macroautophagic process (Singh et al., 2009); both degradation pathways involve a substantially overlapping set of proteins. Lipophagy in yeast may be induced by nitrogen starvation, by cultivating cells to stationary phase

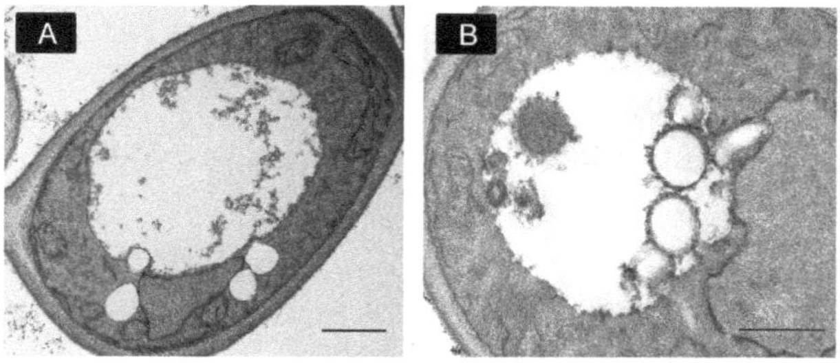

Fig. 1 LD may enter the vacuole/lysosome by autophagy. In growing cells or upon starvation, LDs are typically found in close vicinity of the vacuole (A). Cultivation in the presence of oleic acid leads to a massive proliferation of LDs and their subsequent internalization into the vacuole on starvation medium by a mechanism resembling "microautophagy" (B). Scale bar = 1 μm. *Reproduced from van Zutphen, T., Todde, V., de Boer, R., Kreim, M., Hofbauer, H. F., Wolinski, H., … Kohlwein, S. D. (2014). Lipid droplet autophagy in the yeast Saccharomyces cerevisiae.* Molecular Biology of the Cell, *25(2), 290–301. http://dx.doi.org/10.1091/mbc.E13-08-0448, with permission from The American Society of Cell Biology.*

(3–7 days of cultivation, which resembles starvation for carbon), or in the presence of oleic acid as the sole carbon source (van Zutphen et al., 2014; Wang, 2014; Wang et al., 2014). Autophagy may also be triggered by membrane stress as a consequence of impaired membrane phospholipid synthesis, a condition which also induces LD formation; adaptation to chronic membrane stress over time depends on LD degradation by lipophagy (Vevea et al., 2015). Notably, degradation of LD by lipophagy not only contributes to cellular lipid homeostasis but also LDs themselves play an important role in regulating the autophagic process (Li et al., 2015; Regnacq et al., 2016; Shpilka, Welter, Borovsky, Amar, Mari, et al., 2015; Velazquez, Tatsuta, Ghillebert, Drescher, & Graef, 2016). Sterols released from the SE pool of LD contribute to the formation of microdomains in the vacuolar membrane that are required for lipophagy in stationary phase cells (Wang et al., 2014). On the other hand, autophagy is impaired in the absence of LD, i.e., in mutants lacking TG and SE synthesizing enzymes, underscoring the tight regulatory and physiological interaction between LD metabolism and autophagy, to sustain membrane stress and support viability (Regnacq et al., 2016; Velazquez et al., 2016).

2. INDUCTION OF AUTOPHAGY BY STARVATION

Starvation for nitrogen or carbon is the most commonly used approach to induce autophagy, including lipophagy; alternatively, direct inhibition of the TOR signaling pathway by rapamycin (Delorme-Axford, Guimaraes,

Reggiori, & Klionsky, 2015; Klionsky et al., 2016) or extended maintenance of cells in stationary phase (3–7 days, Wang et al., 2014) may be used as well. Typically, LD autophagy (lipophagy) in yeast is induced by shifting cells for 4–8 h to starvation medium containing glucose but lacking a nitrogen source (van Zutphen et al., 2014). To induce LD formation prior to autophagy induction, cells are cultivated over night in the presence of 0.1% oleic acid. As reference strains, mutants lacking core components of the autophagy machinery should be used (e.g., *atg1*).

1. Streak out a strain of interest from a glycerol stock onto YPD agar plates (1% yeast extract, 2% peptone, 2% glucose, and 2% agar) and incubate for 2 days at 30°C.
2. Inoculate a single colony of the strain in 5 mL YPD liquid medium and cultivate cells overnight at 30°C on a rotary shaker (180 rpm).
3. Redilute cells to OD_{600} 0.1–0.3 in fresh YPD liquid medium and incubate to mid-log phase for 4–6 h at 30°C on a rotary shaker (180 rpm).
4. For additional induction of LDs, cells are inoculated into oleate medium containing 0.17% yeast nitrogen base without ammonium sulfate (Difco), 0.5% ammonium sulfate, 0.05% yeast extract, 0.1% oleic acid, and 0.05% Tween.
5. Harvest the desired amount of cells by centrifugation at 1000 × *g* for 2 min; snap freeze pellet in liquid nitrogen, and store at −80°C until further use; this sample serves as $t=0$ h reference point.
6. Wash the remaining cell culture three times with double-distilled water (DDW) (1000 × *g*, 5 min), transfer cells to starvation media (SD-N, containing 0.17% yeast nitrogen base without amino acids and ammonium sulfate (Difco) and 2% glucose).
7. Maintain cells for 4–8 h in SD-N media at 30°C on a rotary shaker (180 rpm).
8. Collect over time the desired amount of cells by centrifugation at 1000 × *g* for 5 min; snap freeze pellets in liquid nitrogen and store at −80°C until further use.
9. Cell pellets are boiled in PAGE sample buffer, centrifuged at 1000 × *g* for 5 min, protein extracts separated by SDS-PAGE and analyzed by Western blotting.

3. VISUALIZATION OF LDs AND VACUOLES IN VIVO

Subcellular localization and internalization of LDs into the vacuole are routinely visualized by fluorescence microscopy, using specific fluorophores that label the hydrophobic LD core (Nile Red, BODIPY 493/503, LD540;

Spandl, White, Peychl, & Thiele, 2009; Wolinski & Kohlwein, 2015) or the vacuolar membrane (FM4-64; Vida & Emr, 1995). In addition, green fluorescent protein (GFP)-tagged marker proteins that are associated with the LD surface, such as Faa4-GFP or Erg6-GFP, or integral vacuolar membrane proteins, such as Vph1-GFP, can be used to observe the interaction between these two organelles (Radulovic et al., 2013; van Zutphen et al., 2014; Wolinski, Bredies, & Kohlwein, 2012; Wolinski & Kohlwein, 2015). As an alternative to LD-specific fluorescence probes, and to overcome inherent problems in fluorescence crosstalk in multilabeling applications, the spectroscopic imaging technique, CARS (coherent anti-Stokes Raman scattering), has been successfully employed to monitor LD dynamics in the presence of fluorescent fusion proteins (van Zutphen et al., 2014; Wolinski & Kohlwein, 2015) (Fig. 2A).

3.1 Staining of LDs With Nile Red, BODIPY 493/503, or LD540 (Wolinski & Kohlwein, 2015) (Fig. 2A)

1. Grow cells in appropriate liquid growth media to the desired cell density.
2. Harvest an aliquot of cells by centrifugation in a 1.5 mL Eppendorf tube.
3. If YPD was used for cultivation, wash cells two times with DDW or SD medium to avoid background fluorescence from the growth medium.
4. Resuspend cells in 1000 μL DDW or SD medium.
5. Add 1 μL of Nile Red, BODIPY 493/503, or LD540 stock solution (typically 1 mg/mL in DMSO) to the cell suspension and stain for 15 min at room temperature (RT).
6. Centrifuge cells at 1000 × *g* for 2 min; remove 995 μL of supernatant; resuspend cells in remaining liquid.
7. Transfer 0.5 μL of cell suspension to an agarose-covered microscope slide and cover cells with a coverslip; make sure that air bubbles are excluded and that cells are evenly spread out underneath the coverslip.
8. Observe fluorescence on a confocal laser scanning microscope, preferably with adjustable emission bandwidth: excitation/emission ($\lambda_{ex}/\lambda_{em}$) for Nile Red: $\lambda_{ex}=488$ nm/$\lambda_{em}=550$–575 nm, BODIPY 493/503: $\lambda_{ex}=488$ nm/$\lambda_{em}=500$–550 nm, and LD540: $\lambda_{ex}=543$ nm/$\lambda_{em}=550$–600 nm.
9. Note: Nile Red fluorescence bleaches very rapidly; use very low laser intensity for excitation. The emission spectra of Nile Red and BODIPY 493/503 significantly overlap with GFP fluorescence. For simultaneous LD and GFP visualization, we recommend the use of LD540 (Spandl et al., 2009; Wolinski & Kohlwein, 2015).

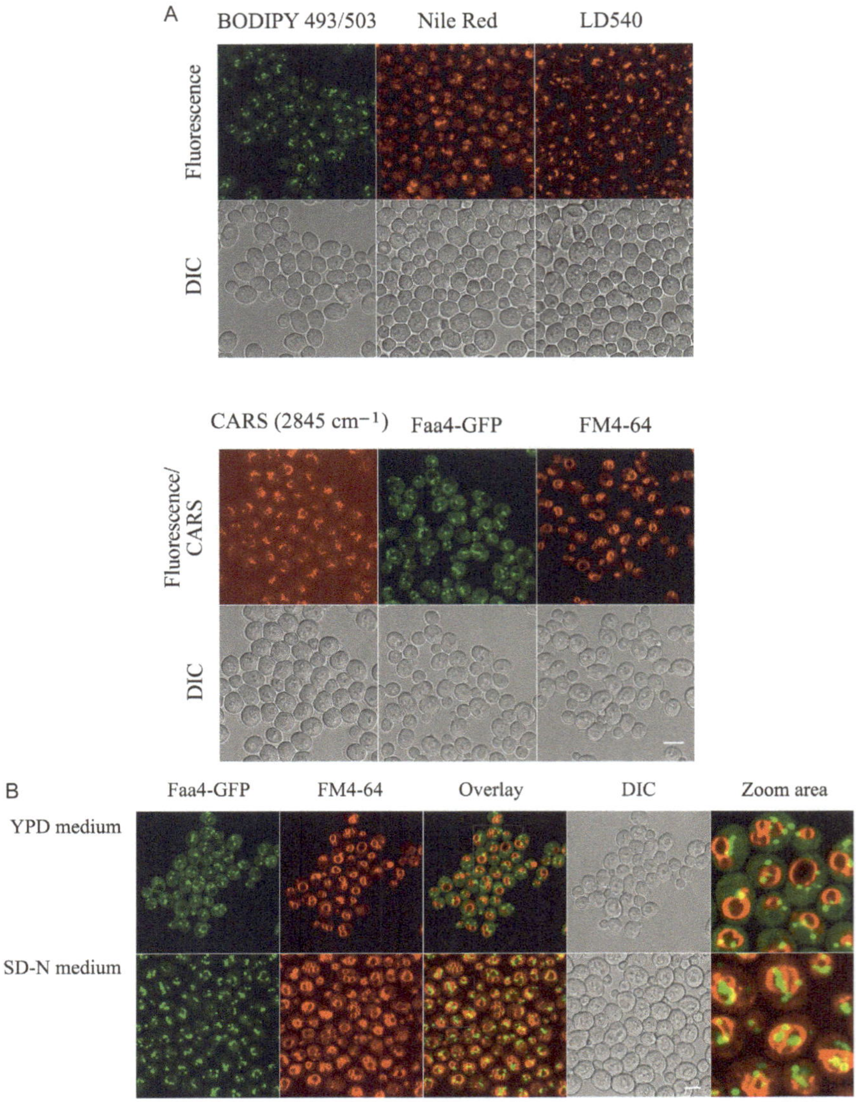

Fig. 2 Microscopic imaging techniques to visualize lipid droplet internalization into the vacuole by lipophagy. (A) LDs are visualized by the use of lipophilic fluorescent dyes (Nile Red, BODIPY 493/503, or LD540; see text for details) that enter the cell by diffusion and accumulate in the organelle; background fluorescence of these dyes in the aqueous cytosolic environment is typically very low. Alternatively, LD may be visualized by label-free CARS microscopy, making use of the high density of C–H bonds in lipids that are excited and visualized in a multiphoton process. LDs can also be visualized by expression of chromosomal fusion constructs with GFP; prototypic LD marker proteins are Faa4, Erg6, or Tgl3/4. The vacuolar membrane is typically labeled using FM4-64 which enters the cell via endocytosis. (B) Faa4-GFP and FM4-64 are visualized simultaneously by two-channel fluorescence excitation/detection. After induction of lipophagy by starvation (SD-N medium), LDs are frequently observed inside the vacuole. This becomes also apparent in transmission microscopy, by the much faster movement of LDs in the vacuolar lumen due to the reduced protein density/viscosity. Scale bar = 5 μm.

3.2 Visualization of LDs by CARS Microscopy (Wolinski & Kohlwein, 2015) (Fig. 2A)

CARS is a spectroscopic imaging methods that is particularly suited for the detection of neutral lipids that are packaged into LD. The method is based on a multiphoton excitation and scattering process to monitor vibrations of C–H bonds, which are present in high density in the hydrocarbon chains of lipids. This method does not require specific labeling for lipids and is compatible with GFP fluorescence detection (Fig. 2A).

1. Cells are cultivated and mounted on microscope slides as described earlier.
2. The CARS laser system (implemented in a confocal laser scanning microscope) is tuned to 2845 cm^{-1}, which excites CH_2 stretching vibrations in the aliphatic acyl-chains of the lipids.
3. The CARS signal is detected in F- (forward) or E- (epi) modes.
4. GFP-tagged proteins are preferentially imaged prior to CARS imaging to avoid potential bleaching.

3.3 Staining of Vacuolar Membranes With FM4-64 (Vida & Emr, 1995) (Fig. 2A)

1. Grow cells in appropriate liquid growth media to the desired cell density.
2. Harvest an aliquot of cells by centrifugation in a 1.5 mL Eppendorf tube.
3. If YPD was used, wash cells two times with DDW to avoid background fluorescence from the growth medium.
4. Resuspend cells in 1000 μL SD medium.
5. Add 1 μL of FM4-64 stock solution (typically 1 mg/mL in DMSO) to the cell suspension and stain for at least 15 min at RT. Note: FM4-64 first labels the plasma membrane and enters cells via endocytosis and ultimately ends up specifically in the vacuolar membrane. If plasma membrane or vesicular structures are still visible, extend the incubation time.
6. Centrifuge cells at 1000 × *g* for 2 min; remove 995 μL of supernatant; resuspend cells in remaining liquid.
7. Transfer 0.5 μL of cell suspension to an agarose-covered microscope slide and cover cells with a coverslip; make sure that air bubbles are excluded and that cells are evenly spread out underneath the coverslip.
8. Observe fluorescence on a confocal laser scanning microscope, preferably with adjustable emission bandwidth: excitation/emission ($\lambda_{ex}/\lambda_{em}$) for FM4-64: $\lambda_{ex} = 543$ nm/$\lambda_{em} = 565–615$ nm.

3.4 Staining of LDs and Vacuolar Membranes With GFP Fusion Proteins (Fig. 2A)

The procedure follows the same steps of cell preparation as described earlier, using cells that express chromosomally encoded C-terminal GFP, mCherry, or RFP fusions (Huh et al., 2003).

1. As a LD marker use Faa4-GFP (van Zutphen et al., 2014) or Erg6-GFP (Shpilka, Welter, Borovsky, Amar, Shimron, et al., 2015); as a vacuolar membrane marker use Vph1-GFP; alternatives described in the literature are: Fas1-GFP (Shpilka, Welter, Borovsky, Amar, Shimron, et al., 2015), Tgl3-mCherry, Tgl4-mCherry, and Atg15-GFP (Maeda et al., 2015).
2. After mounting of cells on agarose pads, microscopy is performed on a confocal laser scanning microscope, preferably with adjustable emission bandwidth: excitation/emission ($\lambda_{ex}/\lambda_{em}$) for GFP: $\lambda_{ex}=488$ nm/ $\lambda_{em}=500$–550 nm. If double labeling experiments are performed with LD540 (LD) use $\lambda_{ex}=488$ nm/$\lambda_{em}=500$–530 nm for GFP detection, and $\lambda_{ex}=543$ nm/$\lambda_{em}=550$–600 nm for LD540 detection.

3.5 Monitoring LD Internalization Into the Vacuole Upon Starvation (Fig. 2B)

1. Inoculate a single colony of the desired strain expressing a GFP-tagged LD marker protein (e.g., Faa4-GFP) in 5 mL YPD liquid media and cultivate cells over night at 30°C on a rotary shaker (180 rpm).
2. Redilute the cell culture and cultivate cells to the desired growth phase in YPD liquid media. If necessary, cells can be grown in the presence or absence of oleic acid, as described earlier, to induce the formation of LD.
3. Centrifuge 1 mL of cell culture in a 1.5 mL Eppendorf tube at 1000 × *g* for 2 min and wash cells twice with DDW. At this time point, FM4-64 staining can be performed to visualize the vacuoles (see Section 3.3).
4. Transfer 0.5 μL of cell suspension to an agar-coated microscope slide and image cells by transmission and fluorescence microscopy. At this point, LDs are typically localized in the cytoplasm, in close vicinity to the vacuole (Fig. 2B).
5. Transfer the remaining cells to SD-N media.
6. Cultivate cells in SD-N media for up to 8 h at 30°C on a rotary shaker (180 rpm).
7. Centrifuge 1 mL of cell culture in a 1.5 mL Eppendorf tube at 1000 × *g* for 2 min and wash cells twice with DDW. FM4-64 staining can be performed to visualize the vacuoles (see Section 3.3).

8. Transfer 0.5 μL of cell suspension to an agar-coated microscope slide and image cells by transmission and fluorescence microscopy, using GFP and FM4-64 excitation and emission channels simultaneously. At this point, a fraction of LDs is typically localized inside the vacuole (Fig. 2B).

4. BIOCHEMICAL ASSAYS TO MONITOR LIPOPHAGY IN YEAST

Lipophagy can be measured either by analyzing the internalization of LD-resident marker proteins or by monitoring the cellular TG content. Protein-based assays make use of the fact that GFP is rather stable against proteolytic attack inside the vacuole/lysosome; this partial degradation of a GFP-tagged LD marker protein gives an easily detectable readout for Western blot analysis by using GFP-specific antibodies. The measurement of total cellular TG content is also commonly used as a method to study lipophagy in several organisms, and incorporation of a control strain that lacks the vacuolar lipase (*atg15*) is recommended (Klionsky et al., 2016).

4.1 Monitoring Lipophagy Based on LD Protein Processing

1. Inoculate a single colony of the desired strain expressing a GFP-tagged LD marker protein (e.g., Faa4-GFP) in 5 mL YPD liquid media and cultivate cells overnight at 30°C on a rotary shaker (180 rpm).
2. Redilute the culture to OD_{600} 0.1–0.3 in fresh YPD and incubate to mid-log phase for 4–6 h at 30°C on a rotary shaker (180 rpm).
3. For Western blot analysis, harvest 3 OD_{600} units of cell material by centrifugation at 1000 × *g* for 2 min; nap freeze pellet in liquid nitrogen and store at −80°C until further use; this sample serves as $t=0$ h reference point.
4. Wash the remaining cell culture three times with DDW (1000 × *g*, 5 min), transfer cells to starvation media (SD-N, containing 0.17% yeast nitrogen base without amino acids and ammonium sulfate (Difco) and 2% glucose).
5. Maintain cells for 4–8 h in SD-N media at 30°C on a rotary shaker (180 rpm).
6. Collect over time the desired amount of cells by centrifugation at 1000 × *g* for 2 min; snap freeze pellets in liquid nitrogen and store at −80°C until further use.

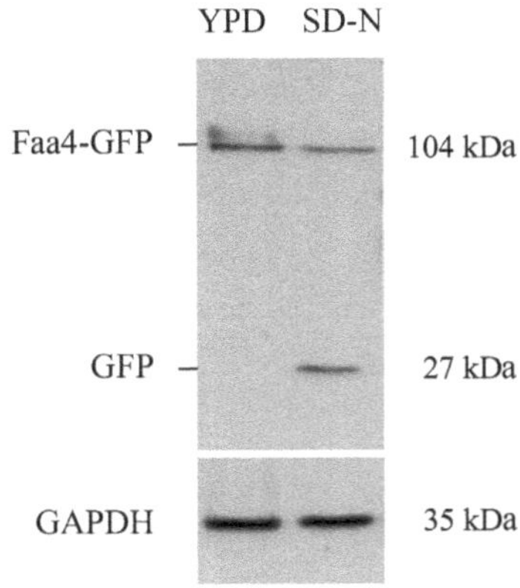

Fig. 3 Biochemical assay to monitor lipophagy after induction of starvation. The assay makes use of fusion constructs of the protein of interest to GFP and the increased stability of the barrel structure of the GFP protein against proteolysis after internalization into the vacuole. The appearance of the GFP fragment of the fusion protein indicates protein uptake and partial degradation, which can be monitored after 4–8 h after induction of starvation (SD-N medium); glyceraldehyde phosphate dehydrogenase (GAPDH) is used as a control.

7. Cell pellets are boiled in PAGE sample buffer, centrifuged at 1000 × *g* for 5 min, protein extracts separated by SDS-PAGE, and analyzed by Western blotting.
8. Proceed with Western blot analysis using anti-GFP antibodies as described (van Zutphen et al., 2014) (Fig. 3).

4.2 Monitoring Lipophagy Based on TG Analysis

1. Inoculate a single colony of the desired strain in 5 mL YPD liquid media and cultivate cells over night at 30°C on a rotary shaker (180 rpm).
2. Redilute the culture to OD_{600} 0.1–0.3 in fresh YPD and incubate to mid-log phase for 4–6 h at 30°C on a rotary shaker (180 rpm).
3. Harvest 20 OD_{600} units of cells by centrifugation at 1000 × *g* for 5 min for lipid extraction. Snap freeze in liquid nitrogen and store at −80°C until further use; this sample serves as $t=0$ h reference point.
4. Wash preculture three times with DDW; centrifuge at 1000 × *g* for 3–5 min and transfer cells to SD-N media.
5. Cultivate cells for 4–8 h in SD-N media at 30°C on a rotary shaker (180 rpm).

6. Harvest 20 OD_{600} units of cells by centrifugation at 1000 × *g* for 5 min for lipid extraction. Snap freeze in liquid nitrogen and store at −80°C until further use.
7. Lipids are extracted according to the Folch method and analyzed by UPLC-MS as previously described (Folch, Lees, & Stanley, 1957; Knittelfelder, Weberhofer, Eichmann, Kohlwein, & Rechberger, 2014).
8. Standardization and quantitative analysis of TG is done using Lipid Data Analyzer software (Hartler et al., 2011).
9. As an alternative to UPLC-MS, lipids can be separated by thin layer chromatography and quantified after charring and densitometric scanning of the TLC plates (Knittelfelder & Kohlwein, 2016).

5. ISOLATION OF VACUOLES FROM YEAST FOR ENZYMATIC ANALYSES

A more detailed analysis of enzymatic activities associated with lipophagy may require the isolation of intact vacuoles. The protocol is based on density gradient centrifugation/flotation of the vacuole-LD fractions and subsequent sedimentation of vacuolar membranes (Zinser & Daum, 1995).

5.1 Preparation of Spheroplasts

1. Cultivate cells in 500 mL YPD medium.
2. Harvest cells by centrifugation at 4000 × *g* for 5 min and wash cells twice with DDW. Determine the cell wet weight (CWW) of the resulting cell pellet in a previously weighed empty centrifuge tube.
3. Resuspend the cell pellet in 0.1 *M* Tris-buffer (pH 9.4) and 10 m*M* dithiothreitol (DTT) and incubate in the water bath at 30°C with shaking for 10 min.
4. Centrifuge at 4000 × *g* for 5 min, resuspend pellet in buffer containing 1.2 *M* sorbitol and 20 m*M* KH_2PO_4 (pH 7.4).
5. Add Zymolyase 20T to the cell suspension at a concentration of 5 mg/g CWW and incubate for 45 min at 30°C with gentle shaking. To assess the extent of cell wall digestion, dilute 10 μL of cell suspension in 200 μL sterile water, and mix gently. A clear solution after mixing indicates complete lysis of spheroplasts; this sample can also be inspected under the microscope to estimate the number of unbroken cells.

5.2 Isolation of Vacuoles

1. After Zymolyase treatment is completed, centrifuge cell suspension at 1000 × *g* for 5 min. Resuspend spheroplasts gently in buffer containing 1.2 *M* sorbitol, 20 m*M* K-PO_4 pH 7.4, and centrifuge again.
2. Resuspend the resulting pellet in breaking buffer (10 m*M* Tris–MES pH 6.9, 12% Ficoll 400, 0.2 m*M* EDTA) at 1 mL/0.3 g CWW.
3. Add 1 m*M* phenylmethylsulfonyl fluoride (PMSF) to the suspension and gently homogenize spheroplasts using a Dounce homogenizer with 20–30 strokes with a loose-fitting pestle (Wheaton, Millville, NJ) on ice.
4. Dilute the homogenized suspension with 1 vol. of breaking buffer and centrifuge at 5000 × *g* for 5 min at 4°C.
5. Collect the supernatant, overlay with 1 vol. breaking buffer that contains 1 m*M* PMSF, and centrifuge at 100,000 × *g* in a swing-out rotor (SW28 rotor, Beckman, Fullerton, CA) for 60 min at 4°C.
6. Collect the top-floating layer containing LD and attached vacuoles with a spatula and resuspend it gently in breaking buffer using a Dounce homogenizer.
7. Transfer into centrifuge tube and overlay the suspension first with ½ vol. of buffer containing 10 m*M* Tris–MES pH 6.9, 8% Ficoll 400, 0.2 m*M* EDTA with 1 m*M* PMSF; overlay with another ½ vol. of 10 m*M* Tris–MES pH 6.9, 4% Ficoll 400, 0.2 m*M* EDTA with 1 m*M* PMSF.
8. Centrifuge at 100,000 × *g* in a swing-out rotor (SW28 rotor, Beckman, Fullerton, CA) for 60 min at 4°C.
9. Remove and resuspend the top layer in 0.6 *M* sorbitol, 4% Ficoll 400, 0.2 m*M* EDTA, and 10 m*M* Tris–MES pH 6.9, and overlay with 1 vol. of 0.25 *M* sorbitol, 0.2 *M* EDTA and 10 m*M* Tris–MES pH 6.9.
10. Centrifuge at 100,000 × *g* in a swing-out rotor (SW28 rotor, Beckman, Fullerton, CA) for 30 min at 4°C.
11. Discard the top-floating layer that contains LD; vacuoles are in the pellet.
12. Resuspend the pellet in 500 μL of 0.6 *M* sorbitol, 4% Ficoll 400, 0.2 m*M* EDTA, and 10 m*M* Tris–MES pH 6.9; add 200 μg/mL trypsin; and incubate for 15 min on ice.
13. Overlay 14 mL of this preparation with 1 vol. of 0.25 *M* sorbitol, 0.2 *M* EDTA, and 10 m*M* Tris–MES pH 6.9.
14. Centrifuge at 100,000 × *g* in a swing-out rotor (SW28 rotor, Beckman, Fullerton, CA) for 30 min at 4°C; the resulting pellet contains purified

vacuoles which are suspended in buffer containing 0.25 *M* sucrose, 1 m*M* EDTA, and 1 m*M* DTT.

5.3 TG Lipase and Sterylester Hydrolase Assays

1. Resuspend the vacuole preparation in a final volume of 100 μL of buffer A (0.25 *M* sucrose, 1 m*M* EDTA, 1 m*M* DTT, pH 7.0) containing protease inhibitors (20 μg/mL leupeptin, 2 μg/mL antipain, 1 μg/mL pepstatin).
2. Prepare the TG substrate by emulsifying 0.33 μ*M* triolein (Sigma-Aldrich, St. Louis, MO), 0.45 μ*M* phosphatidylcholine/phosphatidylinositol (3:1, Sigma-Aldrich), 0.5% defatted bovine serum albumin (Carl Roth, Karlsruhe, Germany), and [9,10-^{3}H]triolein (10,000 cpm/μL; Perkin Elmer Life Sciences, Waltham, MA) as a radioactive tracer, as described (Holm, Olivecrona, & Ottosson, 2001). For the SE hydrolase assay, use [^{3}H]cholesteryl-oleate (43 Ci/mmol) as a radioactive tracer (Müllner et al., 2005; Rajakumari & Daum, 2010).
3. Incubate vacuole preparations with TG or SE substrates for 60 min in a water bath at 37°C.
4. Terminate the reaction by adding 3.25 mL of methanol/chloroform/heptane (10:9:7 per vol.), 1 mL of 0.1 *M* potassium carbonate, and 0.1 *M* boric acid; adjust pH 10.5.
5. Extract free fatty acids by rigorous vortexing, and separate organic phase by centrifugation at 800 × *g* for 15 min. Collect 1 mL of the upper phase to determine radioactivity by liquid scintillation counting (van Zutphen et al., 2014).

6. ELECTRON MICROSCOPY

Electron microscopy (EM) techniques have been crucial to depict morphological details of autophagic processes in yeast cells at the nanometer scale (Baba, 2008; Backues, Chen, Ruan, Xie, & Klionsky, 2014; Eskelinen, Reggiori, Baba, Kovacs, & Seglen, 2011; Griffith, Mari, De Maziere, & Reggiori, 2008; Mari & Reggiori, 2010; Osumi, 2012). EM requires extensive sample preparation including fixation, dehydration, and chemical substitution steps to increase contrast; since autophagy is a highly dynamic and multistep cellular process accompanied by rapid intracellular membrane rearrangements, good structural preservation for EM analysis is a challenging task (Baba, 2008; Backues et al., 2014). Induction of autophagy, for example, by nitrogen starvation reduces cell growth and affects cell wall morphology, which becomes thicker and less permeable to different chemical fixatives typically used in EM. To circumvent, the time consuming chemical fixation process, high-pressure freezing (HPF) of the specimen at greater than 2000 bar pressure followed by freeze substitution has become the method of choice (Dahl & Staehelin, 1989; Osumi, 2012; Vanhecke, Graber, & Studer, 2008). The advantage of HPF is an almost instantaneous fixation of the cell at ultralow temperature that prevents the formation of ice crystals and is not only faster but also more efficient than chemical fixation to preserve morphological features (Giddings et al., 2001). A recent addition to the microscopic methodology is the combination of light (fluorescence) microscopy and EM, termed correlative light/EM (CLEM) (Asakawa, Hiraoka, & Haraguchi, 2014; Razi & Tooze, 2009; Studer, Humbel, & Chiquet, 2008). In this approach, cells are first visualized under a light (fluorescence) microscope, and subsequently imaged by EM, allowing for molecular-specific imaging at the nanometer level.

For detailed protocols on EM, HPF, and CLEM, readers are referred to the following methods articles:

- Yeast EM (Baba, 2008; Wright, 2000);
- HPF, cryofixation (Dahl & Staehelin, 1989; Giddings et al., 2001; Osumi, 2012; Studer et al., 2008; Vanhecke et al., 2008); and
- CL/EM (Asakawa et al., 2014; Razi & Tooze, 2009; Studer et al., 2008).

7. SUMMARY

Lipophagy is a rather novel—yet undisputed—player in organismal lipid homeostasis. Despite some apparent mechanistic differences as to how LD (storage organelles for neutral fats) enter the vacuole in yeast (microautophagy) or the lysosome in mammalian cells (macroautophagy), yeast may offer significant advantages due to the highly conserved core components involved in these processes, to unravel some of the key questions that remain to be elucidated. These include the coordination of lipogenesis on the one hand and lipid degradation via cytosolic lipases vs lipophagy, on the other, and the contribution of LD to autophagy itself. Vacuolar/lysosomal LD breakdown yields fatty acids, glycerol, and sterols, and it remains to be uncovered by which mechanisms these compounds are exported from the vacuole and are trafficked to intracellular sites for further metabolism, e.g., activation of fatty acids by acyl-CoA synthetases in the endoplasmic reticulum (ER) for their recycling into lipids or peroxisomal membranes for degradation; sterols need to be reacylated in the ER and packaged into LD again or trafficked to the plasma membrane, where they serve essential functions as membrane components. Autophagy and LD metabolism are intimately linked: not only starvation but also membrane stress of the ER as a consequence of impaired phospholipid synthesis (leading to increased TG synthesis and LD formation) induces lipophagy (Vevea et al., 2015). On the other hand, total absence of neutral lipid synthesis also causes ER stress due to an overflow of de novo synthesized fatty acids that are now channeled into membrane lipids; in this lipotoxic scenario, autophagy is inhibited (Regnacq et al., 2016; Velazquez et al., 2016). Although TG and SE stored in LD may not directly provide lipid precursors for autophagosome formation, their synthesis is important to balance fatty acid homeostasis and to attenuate ER membrane stress. This finely tuned balance between fatty acid de novo synthesis to support phospholipid remodeling and membrane (autophagosome) biogenesis and fatty acid overflow may explain some of the controversy in the literature as to the requirement of LD for autophagy (Li et al., 2015; Shpilka, Welter, Borovsky, Amar, Mari, et al., 2015), and requires further elucidation.

ACKNOWLEDGMENTS

Heimo Wolinski is gratefully acknowledged for providing image data shown in Fig. 2. Work in the authors' laboratory is supported by grants from the Austrian Science Funds, FWF (PhD program W903 Molecular Enzymology), NAWI Graz, and BioTechMed-Graz. A.C.-S. is supported by a stipend from the Basque Government.

REFERENCES

Angelini, C., Nascimbeni, A. C., Cenacchi, G., & Tasca, E. (2016). Lipolysis and lipophagy in lipid storage myopathies. *Biochimica et Biophysica Acta (BBA)*, *1862*(7), 1367–1373. http://dx.doi.org/10.1016/j.bbadis.2016.04.008.

Asakawa, H., Hiraoka, Y., & Haraguchi, T. (2014). A method of correlative light and electron microscopy for yeast cells. *Micron*, *61*, 53–61. http://dx.doi.org/10.1016/j.micron.2014.02.007.

Baba, M. (2008). Electron microscopy in yeast. *Methods in Enzymology*, *451*, 133–149. http://dx.doi.org/10.1016/S0076-6879(08)03210-2.

Backues, S. K., Chen, D., Ruan, J., Xie, Z., & Klionsky, D. J. (2014). Estimating the size and number of autophagic bodies by electron microscopy. *Autophagy*, *10*(1), 155–164. http://dx.doi.org/10.4161/auto.26856.

Dahl, R., & Staehelin, L. A. (1989). High-pressure freezing for the preservation of biological structure: Theory and practice. *Journal of Electron Microscopy Technique*, *13*(3), 165–174. http://dx.doi.org/10.1002/jemt.1060130305.

Delorme-Axford, E., Guimaraes, R. S., Reggiori, F., & Klionsky, D. J. (2015). The yeast Saccharomyces cerevisiae: An overview of methods to study autophagy progression. *Methods*, *75*, 3–12. http://dx.doi.org/10.1016/j.ymeth.2014.12.008.

Epple, U. D., Suriapranata, I., Eskelinen, E. L., & Thumm, M. (2001). Aut5/Cvt17p, a putative lipase essential for disintegration of autophagic bodies inside the vacuole. *Journal of Bacteriology*, *183*, 5942–5955. http://dx.doi.org/10.1128/JB.183.20.5942-5955.2001.

Eskelinen, E. L., Reggiori, F., Baba, M., Kovacs, A. L., & Seglen, P. O. (2011). Seeing is believing: The impact of electron microscopy on autophagy research. *Autophagy*, 7(9), 935–956.

Folch, J., Lees, M., & Stanley, G. H. S. (1957). A simple method for the isolation and purification of total lipides from animal tissues. *Journal of Biological Chemistry*, *226*, 497–509.

Giddings, T. H., Jr., O'Toole, E. T., Morphew, M., Mastronarde, D. N., McIntosh, J. R., & Winey, M. (2001). Using rapid freeze and freeze-substitution for the preparation of yeast cells for electron microscopy and three-dimensional analysis. *Methods in Cell Biology*, *67*, 27–42.

Griffith, J., Mari, M., De Maziere, A., & Reggiori, F. (2008). A cryosectioning procedure for the ultrastructural analysis and the immunogold labelling of yeast Saccharomyces cerevisiae. *Traffic*, *9*(7), 1060–1072. http://dx.doi.org/10.1111/j.1600-0854.2008.00753.x.

Hartler, J., Trötzmüller, M., Chitraju, C., Spener, F., Köfeler, H. C., & Thallinger, G. G. (2011). Lipid Data Analyzer: Unattended identification and quantitation of lipids in LC-MS data. *Bioinformatics (Oxford, England)*, *27*, 572–577. http://dx.doi.org/10.1093/bioinformatics/btq699.

Holm, C., Olivecrona, G., & Ottosson, M. (2001). Assays of lipolytic enzymes. *Methods in Molecular Biology (Clifton, N.J.)*, *155*, 97–119. http://dx.doi.org/10.1385/1-59259-231-7:097.

Huh, W. K., Falvo, J. V., Gerke, L. C., Carroll, A. S., Howson, R. W., Weissman, J. S., & O'Shea, E. K. (2003). Global analysis of protein localization in budding yeast. *Nature*, *425*(6959), 686–691. http://dx.doi.org/10.1038/nature02026.

Klionsky, D. J. (2014). Citing recent declines in the discovery of new ATG genes, some scientists now suggest that the end of autophagy research may be within sight. *Autophagy*, *10*(5), 715–716. http://dx.doi.org/10.4161/auto.28813.

Klionsky, D. J., Abdelmohsen, K., Abe, A., Abedin, M. J., Abeliovich, H., Acevedo Arozena, A., … Zughaier, S. M. (2016). Guidelines for the use and interpretation of assays for monitoring autophagy (3rd edition). *Autophagy*, *12*(1), 1–222. http://dx.doi.org/10.1080/15548627.2015.1100356.

Knittelfelder, O. L., & Kohlwein, S. D. (2016). Thin-layer chromatography to separate phospholipids and neutral lipids from yeast. In B. Andrews, C. Boone, T. N. Davis, & S. Fields (Eds.), *Budding yeast: A laboratory manual* (pp. 217–238). New York: Cold Spring Harbor Laboratory Press.

Knittelfelder, O. L., Weberhofer, B. P., Eichmann, T. O., Kohlwein, S. D., & Rechberger, G. N. (2014). A versatile ultra-high performance LC-MS method for lipid profiling. *Journal of Chromatography. B, Analytical Technologies in the Biomedical and Life Sciences*, *951–952*, 119–128. http://dx.doi.org/10.1016/j.jchromb.2014.01.011.

Kohlwein, S. D. (2010). Triacylglycerol homeostasis: Insights from yeast. *Journal of Biological Chemistry*, *285*, 15663–15667. http://dx.doi.org/10.1074/jbc.R110.118356.

Kohlwein, S. D., Veenhuis, M., & van der Klei, I. J. (2013). Lipid droplets and peroxisomes: Key players in cellular lipid homeostasis or a matter of fat-store 'em up or burn 'em down. *Genetics*, *193*, 1–50. http://dx.doi.org/10.1534/genetics.112.143362.

Li, D., Song, J. Z., Li, H., Shan, M. H., Liang, Y., Zhu, J., & Xie, Z. (2015). Storage lipid synthesis is necessary for autophagy induced by nitrogen starvation. *FEBS Letters*, *589*(2), 269–276. http://dx.doi.org/10.1016/j.febslet.2014.11.050.

Liu, H., He, Z., von Rütte, T., Yousefi, S., Hunger, R. E., & Simon, H.-U. (2013). Down-regulation of autophagy-related protein 5 (ATG5) contributes to the pathogenesis of

early-stage cutaneous melanoma. *Science Translational Medicine*, *5*, 202ra123. http://dx.doi.org/10.1126/scitranslmed.3005864.
Maeda, Y., Oku, M., & Sakai, Y. (2015). A defect of the vacuolar putative lipase Atg15 accelerates degradation of lipid droplets through lipolysis. *Autophagy*, *276*, 1247–1258. http://dx.doi.org/10.1080/15548627.2015.1056969.
Mari, M., & Reggiori, F. (2010). Atg9 reservoirs, a new organelle of the yeast endomembrane system? *Autophagy*, *6*(8), 1221–1223. http://dx.doi.org/10.1083/jcb.200912089.
Müllner, H., Deutsch, G., Leitner, E., Ingolic, E., & Daum, G. (2005). YEH2/YLR020c encodes a novel steryl ester hydrolase of the yeast Saccharomyces cerevisiae. *The Journal of Biological Chemistry*, *280*, 13321–13328. http://dx.doi.org/10.1074/jbc.M409914200.
Osumi, M. (2012). Visualization of yeast cells by electron microscopy. *Journal of Electron Microscopy*, *61*(6), 343–365. http://dx.doi.org/10.1093/jmicro/dfs082.
Radulovic, M., Knittelfelder, O., Cristobal-Sarramian, A., Kolb, D., Wolinski, H., & Kohlwein, S. D. (2013). The emergence of lipid droplets in yeast: Current status and experimental approaches. *Current Genetics*, *59*, 231–242. http://dx.doi.org/10.1007/s00294-013-0407-9.
Rajakumari, S., & Daum, G. (2010). Janus-faced enzymes yeast Tgl3p and Tgl5p catalyze lipase and acyltransferase reactions. *Molecular Biology of the Cell*, *21*, 501–510. http://dx.doi.org/10.1091/mbc.E09.
Razi, M., & Tooze, S. A. (2009). Correlative light and electron microscopy. *Methods in Enzymology*, *452*, 261–275. http://dx.doi.org/10.1016/S0076-6879(08)03617-3.
Reggiori, F., & Klionsky, D. J. (2013). Autophagic processes in yeast: Mechanism, machinery and regulation. *Genetics*, *194*(2), 341–361. http://dx.doi.org/10.1534/genetics.112.149013.
Regnacq, M., Voisin, P., Sere, Y. Y., Wan, B., Soeroso, V. M., Bernard, M., ... Berges, T. (2016). Increased fatty acid synthesis inhibits nitrogen starvation-induced autophagy in lipid droplet-deficient yeast. *Biochemical and Biophysical Research Communications*, *477*(1), 33–39. http://dx.doi.org/10.1016/j.bbrc.2016.06.017.
Rubinsztein, D. C., Mariño, G., & Kroemer, G. (2011). Autophagy and aging. *Cell*, *146*, 682–695. http://dx.doi.org/10.1016/j.cell.2011.07.030.
Shpilka, T., Welter, E., Borovsky, N., Amar, N., Mari, M., Reggiori, F., & Elazar, Z. (2015a). Lipid droplets and their component triglycerides and steryl esters regulate autophagosome biogenesis. *The EMBO Journal*, *34*(16), 2117–2131. http://dx.doi.org/10.15252/embj.201490315.
Shpilka, T., Welter, E., Borovsky, N., Amar, N., Shimron, F., Peleg, Y., & Elazar, Z. (2015b). Fatty acid synthase is preferentially degraded by autophagy upon nitrogen starvation in yeast. *Proceedings of the National Academy of Sciences*, *112*, 1434–1439. http://dx.doi.org/10.1073/pnas.1409476112.
Singh, R., Kaushik, S., Wang, Y., Xiang, Y., Novak, I., Komatsu, M., ... Czaja, M. J. (2009). Autophagy regulates lipid metabolism. *Nature*, *458*, 1131–1135. http://dx.doi.org/10.1038/nature07976.
Spandl, J., White, D. J., Peychl, J., & Thiele, C. (2009). Live cell multicolor imaging of lipid droplets with a new dye, LD540. *Traffic*, *10*(11), 1579–1584. http://dx.doi.org/10.1111/j.1600-0854.2009.00980.x.
Studer, D., Humbel, B. M., & Chiquet, M. (2008). Electron microscopy of high pressure frozen samples: Bridging the gap between cellular ultrastructure and atomic resolution. *Histochemistry and Cell Biology*, *130*(5), 877–889. http://dx.doi.org/10.1007/s00418-008-0500-1.
Teter, S. A., Eggerton, K. P., Scott, S. V., Kim, J., Fischer, A. M., & Klionsky, D. J. (2001). Degradation of lipid vesicles in the yeast vacuole requires function of Cvt17, a putative

lipase. *Journal of Biological Chemistry*, *276*, 2083–2087. http://dx.doi.org/10.1074/jbc.C000739200.

van Zutphen, T., Todde, V., de Boer, R., Kreim, M., Hofbauer, H. F., Wolinski, H., … Kohlwein, S. D. (2014). Lipid droplet autophagy in the yeast Saccharomyces cerevisiae. *Molecular Biology of the Cell*, *25*(2), 290–301. http://dx.doi.org/10.1091/mbc.E13-08-0448.

Vanhecke, D., Graber, W., & Studer, D. (2008). Close-to-native ultrastructural preservation by high pressure freezing. *Methods in Cell Biology*, *88*, 151–164. http://dx.doi.org/10.1016/S0091-679X(08)00409-3.

Velazquez, A. P., Tatsuta, T., Ghillebert, R., Drescher, I., & Graef, M. (2016). Lipid droplet-mediated ER homeostasis regulates autophagy and cell survival during starvation. *The Journal of Cell Biology*, *212*(6), 621–631. http://dx.doi.org/10.1083/jcb.201508102.

Vevea, J. D., Garcia, E. J., Chan, R. B., Zhou, B., Schultz, M., Di Paolo, G., … Pon, L. A. (2015). Role for lipid droplet biogenesis and microlipophagy in adaptation to lipid imbalance in yeast. *Developmental Cell*, *35*(5), 584–599. http://dx.doi.org/10.1016/j.devcel.2015.11.010.

Vida, T. A., & Emr, S. D. (1995). A new vital stain for visualizing vacuolar membrane dynamics and endocytosis in yeast. *The Journal of Cell Biology*, *128*, 779–792.

Wang, C. W. (2014). Stationary phase lipophagy as a cellular mechanism to recycle sterols during quiescence. *Autophagy*, *10*(11), 2075–2076. http://dx.doi.org/10.4161/auto.36137.

Wang, C. W. (2015). Lipid droplets, lipophagy, and beyond. *Biochimica et Biophysica Acta*, *186*(8 Pt. B), 793–805. http://dx.doi.org/10.1016/j.bbalip.2015.12.010.

Wang, C. W., Miao, Y. H., & Chang, Y. S. (2014). A sterol-enriched vacuolar microdomain mediates stationary phase lipophagy in budding yeast. *The Journal of Cell Biology*, *206*(3), 357–366. http://dx.doi.org/10.1083/jcb.201404115.

Ward, C., Martinez-Lopez, N., Otten, E. G., Carroll, B., Maetzel, D., Singh, R., … Korolchuk, V. I. (2016). Autophagy, Lipophagy and lysosomal lipid storage disorders. *Biochimica et Biophysica Acta*, *1861*, 269–284. http://dx.doi.org/10.1016/j.bbalip.2016.01.006.

Wolinski, H., Bredies, K., & Kohlwein, S. D. (2012). Quantitative imaging of lipid metabolism in yeast: From 4D analysis to high content screens of mutant libraries. *Methods in Cell Biology*, *108*, 345–365. http://dx.doi.org/10.1016/B978-0-12-386487-1.00016-X.

Wolinski, H., & Kohlwein, S. D. (2015). Microscopic and spectroscopic techniques to investigate lipid droplet formation and turnover in yeast. *Methods in Molecular Biology*, *1270*, 289–305. http://dx.doi.org/10.1007/978-1-4939-2309-0_21.

Wright, R. (2000). Transmission electron microscopy of yeast. *Microscopy Research and Technique*, *51*(6), 496–510. http://dx.doi.org/10.1002/1097-0029(20001215)51:6<496::AID-JEMT2>3.0.CO;2-9.

Zechner, R., Zimmermann, R., Eichmann, T. O., Kohlwein, S. D., Haemmerle, G., Lass, A., & Madeo, F. (2012). FAT SIGNALS—Lipases and lipolysis in lipid metabolism and signaling. *Cell Metabolism*, *15*(3), 279–291. http://dx.doi.org/10.1016/j.cmet.2011.12.018.

Zinser, E., & Daum, G. (1995). Isolation and biochemical characterization of organelles from the yeast, Saccharomyces cerevisiae. *Yeast*, *11*, 493–536. http://dx.doi.org/10.1002/yea.320110602.

CHAPTER TWENTY-ONE

Assays to Monitor Pexophagy in Yeast

W. Wang, S. Subramani[1]
Section of Molecular Biology, Division of Biological Sciences, University of California, San Diego, CA, United States
[1]Corresponding author: e-mail address: ssubramani@ucsd.edu

Contents

Abstract

Pexophagy is a selective autophagy process that degrades damaged and/or superfluous peroxisomes in the yeast vacuole or in mammalian lysosomes. The molecular mechanisms of pexophagy are well studied in yeast. Peroxisomes can be rapidly induced by oleate in the budding yeast, *Saccharomyces cerevisiae*, and by oleate or methanol in the methylotrophic yeast, *Pichia pastoris*. A number of peroxisomal matrix enzymes, such as 3-ketoacyl CoA thiolase (thiolase) and alcohol oxidase (AOX), are upregulated correspondingly to meet metabolic demands of the cells. Removal of these peroxisome-inducing carbon sources creates conditions wherein peroxisomes are superfluous and results in pexophagy and the degradation of these peroxisomal matrix enzymes. In this chapter, we discuss different assays to monitor pexophagy in yeast. These assays rely on tracking the localization of the BFP–SKL protein (a peroxisomally targeted version of the blue fluorescent protein) by microscopy, biochemical analysis of the degradation of peroxisomal matrix proteins, thiolase and AOX, and/or measuring the reduction of AOX activity during pexophagy.

Methods in Enzymology, Volume 588
ISSN 0076-6879
http://dx.doi.org/10.1016/bs.mie.2016.09.088

1. INTRODUCTION

Peroxisomes are single-membrane-bound organelles present in most eukaryotic cells. They are involved in essential cellular metabolism, most notably the β-oxidation of fatty acids, as well as the production and degradation of toxic hydrogen peroxide and other reactive oxygen species. The human diseases caused by peroxisome dysfunction underscore the vital importance of this organelle (Ma, Agrawal, & Subramani, 2011).

The number of intracellular peroxisomes can be rapidly adjusted and tightly controlled in response to the changing environment and/or physiological conditions. This can be achieved through the coordination of peroxisome biogenesis and degradation. In rodents, the abundance of peroxisomes increases when the animals are administered with hypolipidemic drugs that are known peroxisome proliferators (Fahimi et al., 1982) or some other chemicals, such as phthalate esters (Yokota, 1986). The number of peroxisomes decreases rapidly and almost returns to the preproliferation state within a week if the drugs are discontinued (Yokota, 1993). Similar dynamic patterns of peroxisome homeostasis (balance between biogenesis and degradation) are also observed in yeast. Peroxisomes are induced when yeast cells are grown in the presence of the specific carbon sources like oleic acid or methanol, whose metabolism in that yeast requires peroxisomal enzymes and metabolism. However, when the favored peroxisome proliferation stimulus is removed, the superfluous and redundant peroxisomes are no longer required and are subjected instead to pexophagy, the selective degradation of peroxisome through autophagy.

Autophagy is a self-eating, dynamic, cellular process involving the sequestration of cytoplasmic constituents into a double-membrane structure called the autophagosome, followed by autophagosome/lysosome fusion and degradation of the cargos in the lysosome or vacuole (Levine & Klionsky, 2004). It is dramatically induced by various extra- and intracellular stresses, such as nutrient or growth factor depletion or mTOR kinase inhibition. Autophagy was initially described as a nonselective bulk degradation process. However, autophagy can also target selective cargos (Farré & Subramani, 2016). Pexophagy involves the general autophagy machinery, as well as specific elements such as the pexophagy receptor, Atg30 in *Pichia pastoris* (Farre, Manjithaya, Mathewson, & Subramani, 2008), and Atg36 in *Saccharomyces cerevisiae* (Motley, Nuttall, & Hettema, 2012). These receptors tag peroxisomes for degradation: Atg30 and Atg36 are associated with the peroxisomal membrane proteins (PMPs) and can recruit the core autophagy machinery to peroxisomes by binding the scaffold proteins (Atg11 and Atg17) and the ubiquitin-like protein Atg8 (Farre, Burkenroad, Burnett, & Subramani, 2013; Farre et al., 2008).

P. pastoris serves as an ideal cell model to study pexophagy due to (1) the large size and/or abundance of peroxisomes that can be induced by both methanol and oleate; (2) this process can be easily visualized by both biochemical and microscopy assays; and (3) convenient genetic manipulation of the cells is possible. In this chapter, we will discuss mainly the methods used to monitor pexophagy in *P. pastoris*. Similar methods can be also applied to *S. cerevisiae*.

2. ASSAYS TO MONITOR PEXOPHAGY

2.1 BFP–SKL Fluorescence Assay

2.1.1 Background

In peroxisome proliferation conditions, such as methanol or oleate medium, a large number of peroxisomal proteins are induced and imported into the peroxisomes utilizing one or more peroxisome-targeting signals (PTSs) present in the amino acid sequences of the proteins. The import of most peroxisomal matrix cargos depends on the PTS1 signal composed of a noncleavable, C-terminal tripeptide, Ser–Lys–Leu (SKL), or its conserved variants (Gould, Keller, Hosken, Wilkinson, & Subramani, 1989). Hence, peroxisomes can be labeled by fusing the BFP (blue fluorescent protein) tag to the SKL sequence (BFP–SKL). Alternatively, PMPs, which are targeted to peroxisomes via an mPTS (membrane PTS) sequence, or peroxisomal matrix proteins may also be fused to GFP or BFP to follow peroxisomes. The fluorescent dye FM4-64 [N-(3-triethylammoniumpropyl)-4-(*p*-diethylaminophenyl-hexatrienyl) pyridinium dibromide] stains the vacuole membrane through endocytosis and can be used as a marker for the vacuole (Vida & Emr, 1995). BFP–SKL and FM4-64 are used together to follow the localization of peroxisomes relative to the vacuole during pexophagy by fluorescence microscopy.

Peroxisomes are induced in oleate or methanol media, as evidenced by the increased number and size of punctuate BFP–SKL-positive dots compared to that in YPD medium (Fig. 1). Shifting cells from oleate to synthetic dextrose without nitrogen (SD-N) medium results in delivery of the BFP–SKL protein into the vacuole (Fig. 2). Atg30 acts as a selective receptor for

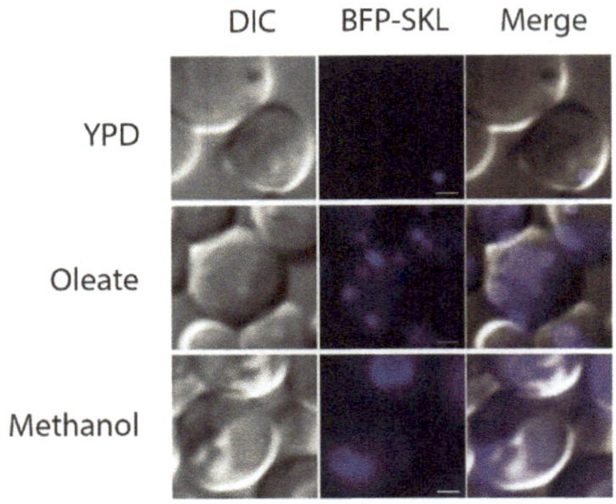

Fig. 1 Induction of peroxisomes in *P. pastoris*. The PPY12 strain overexpressing BFP–SKL was cultured in YPD medium to mid-log phase and shifted to oleate or methanol medium to induce peroxisomes (15 h). Scale bars, 1 μm.

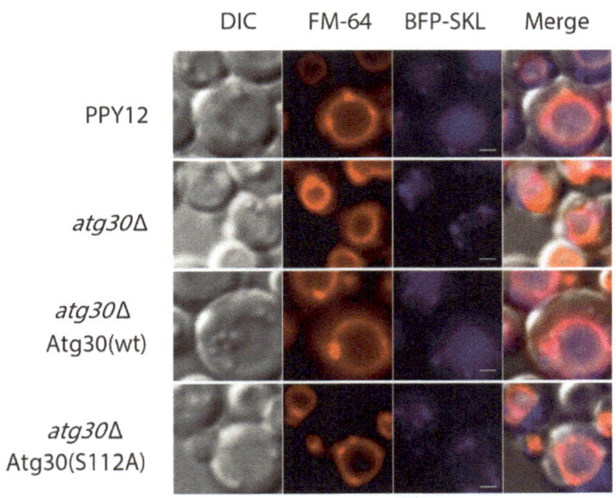

Fig. 2 BFP–SKL localization during pexophagy. Strain PPY12 (wt) and the mutants were cultured in oleate medium to induce peroxisome and then shifted to nitrogen starvation medium (SD-N) to induce pexophagy. The pictures are taken after 2 h of starvation. The vacuoles are labeled red with FM4-64 and peroxisomes blue with BFP–SKL. Scale bars, 1 μm.

pexophagy and its activity depends on the phosphorylation sites S71 and S112 (Farre et al., 2013, 2008). The *atg30* deletion mutant strain (*atg30Δ*) is not able to deliver BFP–SKL to the vacuole and serves as a good negative control; reexpression of Atg30 (wt), but not of Atg30 (S112A), rescues the defect of *atg30Δ* strain (Fig. 2).

2.1.2 Methods

2.1.2.1 Growth Medium Components

- YPD medium (2% (w/v) glucose, 2% (w/v) Bactopeptone, 1% (w/v) yeast extract).
- SD (defined synthetic dextrose medium) (0.17% (w/v) yeast nitrogen base without amino acids and ammonium sulfate, 0.5% (w/v) ammonium sulfate, 1 × complete supplement mixture (CSM) of amino acids (#1001-010, Sunrise Science), 2% (w/v) glucose).
- SD-N (nitrogen starvation medium) (0.17% (w/v) yeast nitrogen base without ammonium sulfate and amino acids, 2% (w/v) glucose).
- Ethanol medium: (0.17% (w/v) yeast nitrogen base without ammonium sulfate and amino acids, 0.5% (w/v) ammonium sulfate, 1 × CSM, 0.5% (v/v) ethanol).
- Methanol medium (0.17% (w/v) yeast nitrogen base without amino acids and ammonium sulfate, 0.5% (w/v) ammonium sulfate, 1 × CSM, 0.5% (v/v) methanol, 0.05% (w/v) yeast extract).
- Oleate medium for *P. pastoris* (0.17% (w/v) yeast nitrogen base without amino acids and ammonium sulfate, 0.5% (w/v) ammonium sulfate, 1 × CSM, 0.2% (v/v) oleate, 0.02% (v/v) Tween 40, 0.05% (w/v) yeast extract).
- Oleate medium for *S. cerevisiae* (0.17% (w/v) yeast nitrogen base without amino acids and ammonium sulfate, 0.5% (w/v) ammonium sulfate, 1 × CSM, 0.1% (v/v) oleate, 0.5% (v/v) Tween 40, 0.1% (w/v) yeast extract).

2.1.2.2 Strains

Strains Name	Description	Genotype	Source
PPY12	wt *P. pastoris* strain	*his4 arg4*	Gould, McCollum, Spong, Heyman, and Subramani (1992)
BFP–SKL	Overexpression of BFP–SKL in PPY12	PPY12 Pr_{GAPDH}-BFP–SKL::*ARG4 his4 arg4*	Farre et al. (2008)

Continued

Strains Name	Description	Genotype	Source
atg30Δ	*atg30* deletion in *BFP–SKL*-expressing strain	PPY12 *atg30Δ::KanMX*, Pr_{GAPDH}-BFP–SKL::*ARG4 his4 arg4*	Farre et al. (2008)
atg30Δ + Atg30 (wt)	Reexpress Atg30 (wt) in *atg30Δ* strain	PPY12 *atg30Δ::KanMX*, Pr_{ATG30}-Atg30(wt)-HA::*HIS4*, Pr_{GAPDH}-BFP–SKL::*ARG4 his4 arg4*	Farre et al. (2008)
atg30Δ + Atg30 (S112A)	Reexpress Atg30 (S112A) mutant in *atg30Δ* strain	PPY12 *atg30Δ::KanMX*, Pr_{ATG30}-Atg30(S112A)-HA::*HIS4*, Pr_{GAPDH}-BFP–SKL::*ARG4 his4 arg4*	Farre et al. (2013)

2.1.2.3 Protocol

1. Inoculate the *P. pastoris* strains in a 50 mL YPD culture overnight at 30°C.
   Note: (i) The flask for the culture should be large enough to provide good aeration. (ii) Use an autophagy- (e.g. *atg1Δ*) or pexophagy-defective mutant strain (e.g. *atg30Δ*) as a negative control.
2. Measure the cell density (OD_{600}) and make sure the cells are in log phase.
   Note: It is important to keep the cells in log phase (OD_{600} of 0.4–0.8); autophagy would occur if the cells enter starvation phase due to nutrient depletion in medium. If cells are overgrown next day, dilute them to 0.1–0.2 OD_{600}/mL and culture them till they reach the log phase.
3. Collect cell pellets and wash them two times with autoclaved H_2O by centrifugation at 3000 rpm, 3 min. The purpose is to wash away the carbon source that may be associated with the cells.
4. Resuspend the pellets in the methanol or oleate medium at starting OD_{600} of 0.1/mL in the presence of FM4-64 (Molecular Probes, 2 μg/mL). Grow cells overnight to 1–2 OD/mL at 30°C with shaking (250 rpm, ~15 h).
   Note: The medium would turn purple with addition of FM4-64. Adjust the pH to 6.0–6.8 with dilute KOH if it turns yellow.
5. Harvest 10 OD_{600} cells and wash twice with autoclaved H_2O, pelleting cells after each wash using centrifugation. Resuspend the cell pellets in 10 mL of SD-N medium and incubate in the shaker at 30°C to induce pexophagy.

Note: As is the case with general autophagy, there are two modes of pexophagy, i.e. macro- and micropexophagy which depend on the availability of the carbon source in the medium (Tuttle & Dunn, 1995). Moving cells from methanol to glucose medium or alternatively to ethanol medium induces micropexophagy and macropexophagy, respectively, in *P. pastoris*.

6. Take 1 mL of cells at 0, 0.5, 2, 4, and 6 h, respectively. Centrifuge at 3000 rpm for 3 min and wash once with autoclaved H_2O.
7. Discard the supernatant, and resuspend the cell pellets in 100 μL SD-N medium.
8. Mount the cells by spotting 5 μL of the cells on a dust-free slide and then drop the cover glass and move it horizontal until stop to have a single layer.
9. Observe the cells under a fluorescence microscope, using the Blue GFP (Ex: 402 nm; Em: 457 nm) and Rhodamine filter (Ex: 515 nm; Em: 640 nm) sets to view the BFP–SKL and FM4-64 signal, respectively.

   Notes: (i) Two negative controls can be used to avoid the "bleed through" of signals in the channels: BFP–SKL cells with no FM4-64 staining; cells (no BFP–SKL expression) stained with FM4-64. (ii) Pictures taken at different time points reflect the different stages of pexophagy, which can be observed by colocalization of BFP–SKL with FM4-64, as well as of other autophagy markers, such as the isolation membrane resident, Atg8 fused with GFP (GFP-Atg8). (iii) Peroxisomes can be induced by oleate or methanol in *P. pastoris* as shown earlier. However, peroxisomes can be induced only by oleate in *S. cerevisiae*. The components of the oleate medium also differ as shown in Section 2.1.2.1. In the case of oleate-induced peroxisomes, methods to monitor pexophagy in *P. pastoris* (including the BFP–SKL localization and the following thiolase degradation and thiolase-GFP processing assays) can be also applied to *S. cerevisiae*.

2.2 Thiolase and AOX Degradation Assay

2.2.1 Background

Peroxisomes are induced rapidly by shifting methylotrophic yeast from glucose to methanol or oleate medium. Correspondingly, a number of enzymes are dramatically upregulated to allow cells to metabolize these carbon sources. Examples are the alcohol oxidase (AOX) for methanol utilization and 3-ketoacyl CoA thiolase (thiolase) for β-oxidation of oleate. Replacing these carbon sources by glucose or ethanol results in the induction of

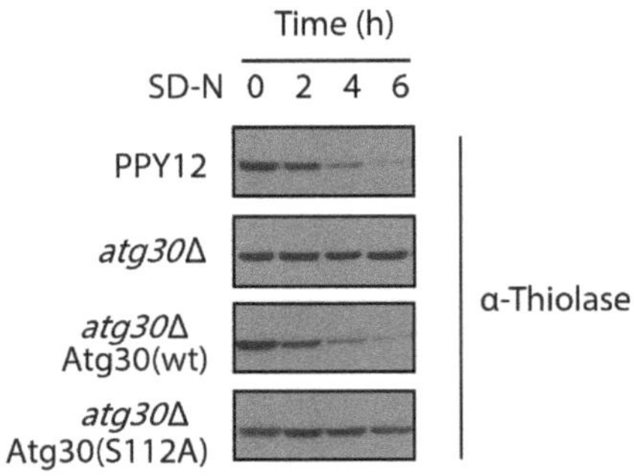

Fig. 3 Thiolase degradation assay. The *P. pastoris* strains were grown in oleate medium to induce peroxisomes and shifted to nitrogen starvation medium (SD-N) for the indicated times. Proteins were extracted by TCA precipitation, resolved by SDS-PAGE, and detected with antithiolase antibody.

pexophagy and degradation of these enzymes. Hence the protein levels of AOX or thiolase can be used to monitor pexophagy. Below is an example of thiolase detection during pexophagy. In the wt strain, thiolase protein is degraded upon moving cells from oleate to SD-N medium. Consistent with the results based on BFP–SKL localization (Fig. 2), thiolase degradation was blocked in *atg30Δ* cells and rescued by reexpressing Atg30 (wt) but not the mutant, Atg30 (S112A) (Fig. 3).

2.2.2 Method

2.2.2.1 Growth Media and Strains

The growth media and strains are the same as in Section 2.1.2.

2.2.2.2 Protocol

1. Induce peroxisomes in methanol or oleate media as in Section 2.1.2.
2. Harvest 10 OD_{600} cells from either methanol or oleate medium and wash twice with autoclaved H_2O, pelleting cells after each wash by centrifugation.
3. Resuspend the cell pellets in 10 mL of SD-N starvation medium and incubate in the shaker at 30°C to induce pexophagy.
4. For detection of AOX, take 1 mL of the methanol-induced culture at 0, 2, 4, and 6 h (starvation time), respectively, to a new tube containing 150 μL 100% trichloroacetic acid (TCA). For detection of thiolase, take 1 mL of oleate-induced cell culture at 0, 2, 4, and 6 h (starvation time), respectively, to a new tube containing 150 μL 100% TCA.
5. Mix and store at −80°C for at least 1 h before TCA precipitation.

6. Thaw the frozen TCA cell mixture at room temperature and pellet the protein lysates by centrifugation at 12,000 rpm, 10 min.
7. Aspirate off the supernatants and add 500 μL 80% ice cold acetone. The pellets are fully suspended using a vortex mixer or water bath sonication. Centrifuge contents at 12,000 rpm, 3 min. Discard the supernatants and wash the pellets again. Air-dry the pellets in a hood (~1 h).

   Note: Water bath sonication offers better disruption of the pellets. Remove as much of the supernatants as possible before air-drying the pellets. Be careful not to touch or disturb the pellets.
8. Resuspend the pellets in 100 μL of the following buffer (1% sodium dodecyl sulfate or SDS, 0.1 N NaOH) followed by water bath sonication.
9. Add 20 μL 6 × SDS loading buffer (10% SDS, 50% Glycerol, 0.5 *M* DTT; 0.25% bromophenol blue) and boil the samples for 5 min.
10. Centrifuge the samples at 12,000 rpm, 5 min. Load 10 μL of the samples in the 10% SDS-PAGE gel and detect AOX or thiolase proteins by standard Western blot assays.

    Note: We use nitrocellulose membranes and wet transfer (100 V, 1 h). The antibodies for AOX and thiolase detection are rabbit polyclonal antiserum made in our lab and incubated with the membrane overnight. An enhanced chemiluminescent (ECL) kit (RPN2106, GE Healthcare) is used to detect the signal.

2.3 Thiolase–GFP Processing Assay

2.3.1 Background

Another way to view thiolase degradation is to detect the cleavage of a thiolase-GFP fusion protein in the vacuole. The GFP tag is fused to the C-terminus of thiolase and its endogenous promoter drives the expression of the fusion protein. During pexophagy, thiolase-GFP protein is delivered to the vacuole, where the GFP moiety is cleaved off by vacuolar hydrolases. GFP is relatively stable in vacuoles compared to the GFP–fusion protein. This allows pexophagy to be monitored through the appearance of free GFP, which would appear and accumulate after shifting the strains from oleate to starvation medium (Fig. 4). The *atg8* deletion mutant has a defect in the processing of thiolase-GFP due to blockage of the general autophagy. Hence, the abundance of free GFP reflects the flux of pexophagy and alternatively, the ratio of full-length thiolase-GFP/GFP can give a semiquantitative measure of pexophagy. Our lab used this assay to test the essential phosphorylation sites of Atg30/Atg36 for pexophagy (Farre et al., 2013).

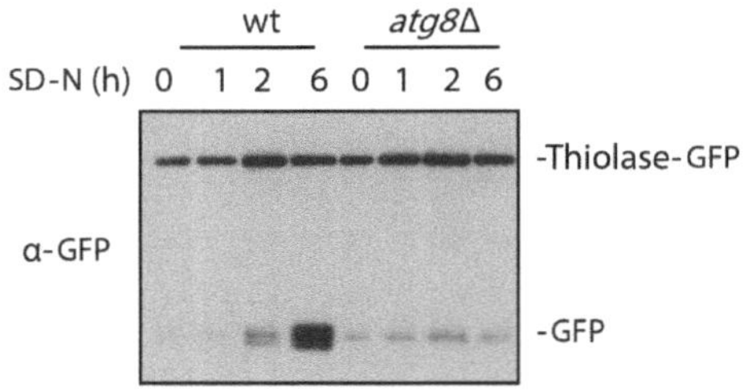

Fig. 4 Thiolase–GFP processing assay. The *P. pastoris* strains, PPY12 (wt or *atg8Δ* overexpressing thiolase–GFP), were grown in oleate medium to induce peroxisomes and shifted to starvation medium (SD-N) for the indicated times. Proteins were extracted by TCA precipitation, resolved by SDS-PAGE, and detected with anti-GFP antibody. *The figure was kindly provided by Dr. Jean-Claude Farré.*

In this section, we discuss the methods used to establish stable thiolase-GFP overexpressing strains, as well as detection of processing of thiolase-GFP during pexophagy.

2.3.2 Methods

2.3.2.1 Strains

Strains Name	Description	Genotype	Source
wt	PPY12 strain overexpressing thiolase–GFP	PPY12 $Pr_{THIOLASE}$-thiolase–GFP::*ARG4 his4 arg4*	Farre et al. (2013)
atg8Δ	*atg8* deletion in thiolase–GFP expressing strain	PPY12 *atg8Δ::KanMX*, $Pr_{THIOLASE}$-thiolase–GFP::*ARG4 his4 arg4*	Farre et al. (2013)

2.3.2.2 Preparation of Competent Cells

1. Inoculate PPY12 in a 50 mL YPD culture in a flask large enough to provide good aeration overnight.
2. Measure the OD_{600} of cells (the ideal OD should be in the range of log phase). Harvest 30 OD cells by centrifugation at 3000 rpm, 3 min.
3. Remove the supernatant, resuspend, and vortex in 5 mL fresh YPD medium.
4. Add 100 μL Hepes buffer (1 *M*, pH 8.0) and 125 μL DTT (1 *M*), vortex the cells.
5. Incubate the cells with shaking at 30°C for 15 min.

6. Wash with ice cold sterile H_2O (~45 mL) and centrifuge at 3000 rpm, 3 min (repeat the above wash step).
7. Discard the supernatant, vortex in 5 mL sorbitol (1 *M*), and centrifuge cells at 3000 rpm, 3 min.
8. Remove the supernatant, resupend the cells in 500 μL sorbitol (1 *M*), and incubate them on ice for 1 h. The cells are ready for transformation.

   Note: For highest efficiency, transform the cells directly without freezing. The cells can be also stored in small aliquots at −80°C freezer.

2.3.2.3 Electroporation and Screening the Positive Thiolase–GFP Colonies

1. Mix 50 μL of competent cells with 2 μg linearized thiolase-GFP plasmid on ice. Incubate electroporation cuvettes (2 mm) on ice for 10 min.

   Note: Plasmids need to be linearized in order to integrate into the genome of *P. pastoris*. The thiolase-GFP construct contains the *ARG4* selection marker and is integrated into the *ARG4* genomic locus of PPY12. Hence, SD-Arg plates are used as the selection plate as described later.
2. Pulse cells at (1500 V, 200 Ω, 25 μF) using electroporation instrument (BTX Electro Cell Manipulator 630).
3. Add 1 mL ice cold sorbitol immediately and place the cuvettes in 30°C incubator overnight.
4. Transfer cells from the cuvettes to a 1.5 mL tube and centrifuge at 3000 rpm, 3 min.
5. Wash cells with autoclaved H_2O and plate one-fourth of the cells on the SD-Arg selection plates. Place the plates upside down in 30°C incubator till visible colonies forms (~2 days).

   Note: Components for SD-Arg plate: 0.17% (w/v) yeast nitrogen base without amino acids and ammonium sulfate, 0.5% (w/v) ammonium sulfate, 1× CSM without Arg (#1031-010, Sunrise Science), 2% (w/v) glucose, 2% Agar.
6. Streak the colonies using sterile toothpicks to a new SD-Arg plate and grow them in a 30°C incubator. This provides another round of selection to the colonies and also makes sure that the colony is derived from a single one.
7. Inoculate a single colony to a 5 mL YPD media overnight and transfer cells to oleate medium to induce peroxisomes as in Section 2.1.2.
8. Extract proteins by TCA precipitation and detect thiolase-GFP expression by Western blot (anti-GFP antibody). They can be also confirmed

by observing the thiolase-GFP signal under the fluorescence microscope.

9. Make glycerol stocks of the positive colony by generating a 1:1 mix of the cell culture with 50% autoclaved glycerol. Store them at −80°C freezer for future use.

2.3.2.4 Thiolase–GFP Processing Detection

1. Induce peroxisomes by oleate medium as in Section 2.1.2.
2. Harvest 10 OD_{600} cells from the oleate medium and wash cells twice with autoclaved H_2O and collecting cells after each wash by centrifugation.
3. Resuspend the cell pellets in 10 mL of SD-N starvation (1 OD/mL) medium and incubate in the shaker at 30°C to induce pexophagy.
4. Collect 1 OD_{600} of cells at time 0, 1, 2, and 6 h, respectively, and extract the proteins by TCA precipitation (see more details in Section 2.2.2.2).
5. Detect the thiolase–GFP fusion and free GFP proteins by standard Western blot assay using anti-GFP antibody (#11814460001, Roche, dilution: 1:1000). The molecular weight of thiolase and GFP protein are 43 kDa and 26 kDa, respectively. Hence thiolase–GFP should be at 69 kDa.

2.4 AOX Plate Assay

2.4.1 Background

All methylotrophic yeasts possess a common methanol-utilizing pathway. Methanol is first oxidized by AOX in peroxisomes to form formaldehyde and hydrogen peroxide (H_2O_2), both of which are toxic to cells and are broken down by the enzymes located in the peroxisome (Yurimoto, Oku, & Sakai, 2011). Catalase is a peroxidase that catalyzes and decomposes H_2O_2 into H_2O and oxygen. We can use a similar chemical reaction in vitro to measure the activity of AOX. ABTS (2,2′-Azino-bis(3-ethylbenzothiazoline-6-sulfonic acid) can be used as a substrate of peroxidase: oxidized by peroxidase in the presence of the oxidizing agent, H_2O_2. The oxidized ABTS turns green, which is a measure of the activity of AOX and indirectly the abundance of peroxisomes in the AOX plate assay. When cells are shifted from methanol to glucose or ethanol, pexophagy occurs and results in the degradation of AOX protein and disappearance of the green color in the strain patches. Our lab used the assay to demonstrate that Atg37 is required for pexophagy as the *atg37* mutant still turns green even under pexophagy-inducing conditions (Nazarko et al., 2014).

2.4.2 Methods

2.4.2.1 Growth Plates

- YPD plates (2% (w/v) glucose, 2% (w/v) Bactopeptone, and 1% (w/v) yeast extract, 2% agar).
- Methanol plates (0.17% (w/v) yeast nitrogen base without amino acids and ammonium sulfate, 0.5% (w/v) ammonium sulfate, 0.5% (v/v) methanol, supplemented with 1 × CSM, 2% agar).
- SD plates (glucose plate) (0.17% (w/v) yeast nitrogen base without amino acids and ammonium sulfate, 0.5% (w/v) ammonium sulfate, 1 × CSM of amino acids (#1001-010, Sunrise Science), 2% (w/v) glucose, 2% agar).

2.4.2.2 Protocol

1. Plate the strains on YPD plates as small patches and place them in the 30°C incubator for 2 days.
2. Replicate the YPD plates onto the methanol medium plates and incubate at 30°C for another 2 days.
3. Replicate the cells from the methanol plates to the SD plates and incubate them at 30°C (~16 h).
4. Overlay the SD plates with 10 mL of the AOX activity mixture (50 m*M* Tris–HCl, pH 8.0, 0.5 mg/mL ABTS, 1 mg/mL digitonin, 0.3% low melt agarose, 6 IU/mL peroxidase, 1% methanol).

   Note: peroxidase and methanol are added in the last step after other components of the mixture are boiled and cooled to 40°C.
5. Keep the mixture plates at room temperature for 30 min and incubate them with the medium side up at 30°C till the strains are stained green.

3. CONCLUDING REMARKS

In summary, we have described different ways to monitor pexophagy in yeast. Each of them has its own advantages. Biochemical analysis of thiolase or AOX degradation and thiolase–GFP processing is less likely to be biased and can be easily handled for many samples. It can also be quantitative. Microscopic observation of BFP–SKL localization facilitates tracking pexophagy at different stages (Sakai, Koller, Rangell, Keller, & Subramani, 1998). The AOX plate assay allows high-throughput screening of genes required for pexophagy (Stasyk, Nazarko, & Sibirny, 2008). Additionally, however, it is possible that the gene of interest might affect general autophagy and hence pexophagy. Atg8 lipidation and GFP-Atg8 cleavage assays can be employed to exclude this possibility (Cheong & Klionsky, 2008). The above assays are based on different aspects of pexophagy and we recommend using at least two different assays to monitor pexophagy. Similar assays, except the AOX degradation and plate assays, can also be applied to *S. cerevisiae* because oleate can also induce peroxisomes in this organism.

Since autophagy is a highly conserved cellular process from lower to higher eukaryotes, we assume that this is also true for pexophagy. In fact, the earliest description of pexophagy came from mammalian cells where peroxisomes were detected in the lysosomes (De Duve & Baudhuin, 1966). Though the molecular mechanisms of pexophagy are well established in yeast, little is known in mammalian cells. This is largely due to the challenging detection of the pexophagy process in higher organisms. The pexophagy studies in yeast would definitely help uncover the mysteries of pexophagy in mammalian cells.

ACKNOWLEDGMENTS

We thank Dr. Jean-Claude Farré for careful and critical reading of the manuscript. This work was supported by NIH grant GM 069373 to S.S.

REFERENCES

Cheong, H., & Klionsky, D. J. (2008). Biochemical methods to monitor autophagy-related processes in yeast. *Methods in Enzymology*, *451*, 1–26.

De Duve, C., & Baudhuin, P. (1966). Peroxisomes (microbodies and related particles). *Physiological Reviews*, *46*, 323–357.

Fahimi, H. D., Reinicke, A., Sujatta, M., Yokota, S., Ozel, M., Hartig, F., et al. (1982). The short- and long-term effects of bezafibrate in the rat. *Annals of the New York Academy of Sciences*, *386*, 111–135.

Farre, J. C., Burkenroad, A., Burnett, S. F., & Subramani, S. (2013). Phosphorylation of mitophagy and pexophagy receptors coordinates their interaction with Atg8 and Atg11. *EMBO Reports*, *14*, 441–449.

Farre, J. C., Manjithaya, R., Mathewson, R. D., & Subramani, S. (2008). PpAtg30 tags peroxisomes for turnover by selective autophagy. *Developmental Cell*, *14*, 365–376.

Farré, J. C., & Subramani, S. (2016). Mechanistic insights into selective autophagy pathways: Lessons from yeast, Nature Reviews. *Molecular Cell Biology*, *17*(9), 537–552. http://dx.doi.org/10.1038/nrm.2016.74. Epub 2016 Jul 6.

Gould, S. J., Keller, G. A., Hosken, N., Wilkinson, J., & Subramani, S. (1989). A conserved tripeptide sorts proteins to peroxisomes. *The Journal of Cell Biology*, *108*, 1657–1664.

Gould, S. J., McCollum, D., Spong, A. P., Heyman, J. A., & Subramani, S. (1992). Development of the yeast *Pichia pastoris* as a model organism for a genetic and molecular analysis of peroxisome assembly. *Yeast*, *8*, 613–628.

Levine, B., & Klionsky, D. J. (2004). Development by self-digestion: Molecular mechanisms and biological functions of autophagy. *Developmental Cell*, *6*, 463–477.

Ma, C., Agrawal, G., & Subramani, S. (2011). Peroxisome assembly: Matrix and membrane protein biogenesis. *The Journal of Cell Biology*, *193*, 7–16.

Motley, A. M., Nuttall, J. M., & Hettema, E. H. (2012). Pex3-anchored Atg36 tags peroxisomes for degradation in *Saccharomyces cerevisiae*. *The EMBO Journal*, *31*, 2852–2868.

Nazarko, T. Y., Ozeki, K., Till, A., Ramakrishnan, G., Lotfi, P., Yan, M., et al. (2014). Peroxisomal Atg37 binds Atg30 or palmitoyl-CoA to regulate phagophore formation during pexophagy. *The Journal of Cell Biology*, *204*, 541–557.

Sakai, Y., Koller, A., Rangell, L. K., Keller, G. A., & Subramani, S. (1998). Peroxisome degradation by microautophagy in *Pichia pastoris*: Identification of specific steps and morphological intermediates. *The Journal of Cell Biology*, *141*, 625–636.

Stasyk, O. V., Nazarko, T. Y., & Sibirny, A. A. (2008). Methods of plate pexophagy monitoring and positive selection for ATG gene cloning in yeasts. *Methods in Enzymology*, *451*, 229–239.

Tuttle, D. L., & Dunn, W. A., Jr. (1995). Divergent modes of autophagy in the methylotrophic yeast *Pichia pastoris*. *Journal of Cell Science*, *108*(Pt. 1), 25–35.

Vida, T. A., & Emr, S. D. (1995). A new vital stain for visualizing vacuolar membrane dynamics and endocytosis in yeast. *The Journal of Cell Biology*, *128*, 779–792.

Yokota, S. (1986). Quantitative immunocytochemical studies on differential induction of serine: Pyruvate aminotransferase in mitochondria and peroxisomes of rat liver cells by administration of glucagon or di-(2-ethylhexyl)phthalate. *Histochemistry*, *85*, 145–155.

Yokota, S. (1993). Formation of autophagosomes during degradation of excess peroxisomes induced by administration of dioctyl phthalate. *European Journal of Cell Biology*, *61*, 67–80.

Yurimoto, H., Oku, M., & Sakai, Y. (2011). Yeast methylotrophy: Metabolism, gene regulation and peroxisome homeostasis. *International Journal of Microbiology*, *2011*, 101298.

CHAPTER TWENTY-TWO

Monitoring Autophagic Responses in *Caenorhabditis elegans*

M.E. Papandreou*,†, N. Tavernarakis*,†,1

*Institute of Molecular Biology and Biotechnology, Foundation for Research and Technology—Hellas, Heraklion, Greece
†Faculty of Medicine, University of Crete, Heraklion, Greece
[1]Corresponding author: e-mail address: tavernarakis@imbb.forth.gr

Contents

Abstract

Autophagy, from the Greek *auto* (self) and *phagy* (eating), is a self-degradative process critical for eukaryotic cell homeostasis. Its rapidly responsive, highly dynamic nature renders this process essential for adapting to and offsetting acute/harsh conditions such as starvation, organelle dysfunction, and deoxyribonucleic acid (DNA) damage. Autophagy involves an intricate network of interacting factors with multiple levels of control. Importantly, dysregulation of autophagy has been linked to numerous debilitating pathologies, including cancer and neurodegenerative conditions in humans. Methods to monitor and quantify autophagic activity reliably are essential both for studying the basic mechanisms of autophagy and for dissecting its involvement in disease. The nematode *Caenorhabditis elegans* is a particularly suitable model organism to effectively

Methods in Enzymology, Volume 588
ISSN 0076-6879
http://dx.doi.org/10.1016/bs.mie.2016.10.039

visualize and study autophagy, in vivo, in a physiological and pathological context due to its optical transparency, experimental malleability, and precise developmental and anatomical characterization. Here, we outline the main tools and approaches to monitor and measure autophagic responses in *C. elegans*.

1. INTRODUCTION

Caenorhabditis elegans is a simple invertebrate animal model that offers a particularly versatile platform to dissect fundamental cellular and molecular processes. Technological advances in genetic engineering and optical microscopy, coupled with the unique features of the nematode, including its transparency, unsurpassed developmental and anatomical characterization, and genetic malleability, have facilitated studies of physiological responses in vivo, at both the organismal and cellular level. Macroautophagy, hereafter referred to as *autophagy*, is part of such a physiological response to nutrient deprivation and other stressors. Autophagy entails the formation of a double-membrane-bound vesicle, the autophagosome, which ultimately fuses with the lysosome, a lytic enzyme-containing organelle, where autophagosomal contents are degraded. These include aggregated proteins (general autophagy) or damaged organelles (selective autophagy), which are destroyed via the autophagic pathway, in a highly controlled manner. Perturbed or aberrant autophagy can lead to severe pathology, such as neurodegeneration, cancer, and accelerated aging. The relevant basic mechanisms underlying both physiological and pathophysiological regulation of autophagy in various contexts can be readily studied in *C. elegans*.

While autophagy was initially considered to be a bulk degradation process, it is now clear that several highly selective types of autophagy are important for cell function and survival. Specific proteins and protein modifications mediate cargo recognition by the core autophagic machinery. For example, appropriate receptors/adaptors direct selective autophagy toward regulatory proteins or organelles such as mitochondria that are targeted for degradation. In vivo imaging and monitoring the levels and subcellular localization of autophagy components under both normal and stress conditions are essential for elucidating the involvement of autophagy in the relevant responses. In this chapter, we survey tools and resources that can be used to examine different types of autophagy (both general and selective) in *C. elegans*. In addition, we detail methodologies that have been developed to visualize autophagic processes in live animals under various conditions of interest.

2. REPORTERS FOR AUTOPHAGY

2.1 Core Autophagic Components

The nematode proteins LGG-1 and LGG-2 are homologous to Atg8/mammalian LC3B (cytosolic microtubule-associated protein 1 light chain 3-MAP1LC3) and are involved in autophagosome formation. Conversion of LC3B-I to its lipidated form, LC3B-II, occurs during its incorporation in the preautophagosomal and autophagosomal membrane. LC3B is used to visualize autophagosomes as cytoplasmic puncta under autophagy-inducing conditions (Wu et al., 2015). This ubiquitinlike modifier interacts with autophagic substrates containing an LC3-interacting-region (LIR) motif, the simplified version of which is F/W/Y-x-x-L/I/V. For efficient identification of proteins containing LIR motifs, two recent online tools, iLIR and SLiM, have been developed, which can be used to predict new LC3-interacting proteins that are potential substrates of autophagy (Kalvari et al., 2014; Popelka & Klionsky, 2015). The specific, differential roles of the closely related *C. elegans* LGG-1 and LGG-2 proteins remain largely elusive. Both proteins can be imaged in vivo by generating full-length green fluorescent protein (GFP) or DsRed translational reporter fusions. Upon autophagy induction, these proteins become lipidated and decorate the autophagosome. This conversion causes the diffuse cellular distribution pattern to become distinctively punctate. Autophagy induction can then be assessed by the number of LGG-1 and or LGG-2 GFP/DsRed positive puncta per cell (Keith et al., 2016). In principle, reporter fusions can be directed for expression in different cell types of interest where monitoring of autophagy is desired. Extrachromosomal transgenes can be utilized, although genomic integration of reporter constructs is recommended to minimize expression variability among animals and phenomena of mosaicism. Even when integrated, reporter constructs may show idiosyncratic expression. For example, the widely used *adIs2122* [$p_{lgg\text{-}1}$GFP::LGG-1; *rol-6(su1006)*] transgenic reporter strain shows expression mainly in seam cells, hypodermis, and intestinal cells (Klionsky et al., 2016).

Proteins of the WIPI family also colocalize and form distinct puncta, together with LC3 and ULK1 (Itakura & Mizushima, 2010). WIPI proteins are required for omegasome to autophagosome formation under autophagy-inducing, nutrient-deprived conditions. The absence of the *C. elegans* WIPI-4 homolog EPG-6 causes accumulation of LGG-1 and abortive early autophagic structures (Lu et al., 2011). The GFP::EPG-6 reporter has been used to monitor autophagy at embryonic stages and is widely expressed in pharyngeal, body wall muscle, and neuronal cells. This protein is not itself a substrate of autophagy; thus, it can be effectively used as an early step marker to visualize the autophagic response without its levels being affected/diminished.

2.2 Receptors/Adaptors for Selective Autophagy

Autophagic receptors mediate the initial recognition of substrates such as organelles for selective autophagic degradation. Contrary to the abovementioned example, the *C. elegans* homolog of mammalian SQSTM-1/p62, SQST-1 (SeQueSTosome-related protein) has a dual role as an autophagic receptor/adaptor, but also as an autophagic substrate being degraded via autophagy. Visualization of SQST-1 is quite challenging under control conditions at larval and adult stages, most probably due to the rate of basal autophagic flux. However, this is a suitable autophagy reporter under conditions where autophagy is suppressed and SQST-1 levels are elevated in tissues such as the epidermis, neurons, and intestine (Zhang et al., 2015). Other autophagic substrates used as translational reporters to observe autophagic responses include the nematode-specific W07G4.5::GFP fusion, which is expressed both in the cytosol and the nucleus of intestinal cells. Its expression is accentuated in animals with compromised autophagy (Lin et al., 2013).

Mitochondria are energy-producing organelles found in eukaryotic cells; fluctuations in their numbers have been reported in diverse pathologies (Palikaras, Lionaki, & Tavernarakis, 2015). The regulation of mitochondrial mass is highly dynamic due to the balance between mitochondrial biogenesis and mitochondria selective autophagy (mitophagy) that takes place either by direct interaction of LC3 with mitochondrial membrane proteins or via autophagic adaptors such as p62 or optineurin (OPTN). Mitophagy is implemented as a response to dysfunctional or unneeded mitochondria. Two main tools have been developed to monitor mitophagy in *C. elegans*. The first manipulates the core autophagic component LGG-1, tagged with either GFP or DsRed, and a mitochondria-tagged DsRed or GFP, respectively, which can be targeted to different tissues for instance body wall muscle cells, depending on the promoter that drives expression of the transgene (Palikaras et al., 2015). Using this double reporter, the extent of colocalization of LGG-1 puncta and mitochondria, referred to as *mitoautophagosomes*, can be measured under different autophagy-inducing/hindering conditions analyzed in this section and Section 3, and shown in Fig. 1. This can be applied to other types of selective autophagy as well, such as endoplasmic reticulum (ER)-phagy by using an ER-specific reporter or a

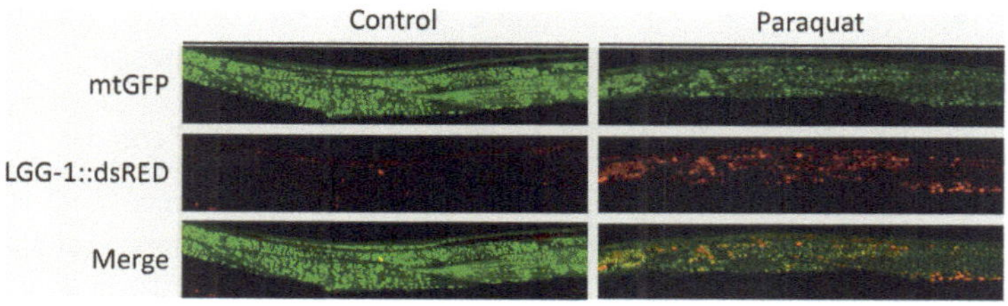

Fig. 1 LGG-1::DsReD colocalization with mtGFP in control and paraquat-treated worms. *Adapted from Palikaras, K., Lionaki, E., & Tavernarakis, N. (2015). Coordination of mitophagy and mitochondrial biogenesis during ageing in C. elegans.* Nature, *521(7553), 525–528. http://dx.doi.org/10.1038/nature14300.*

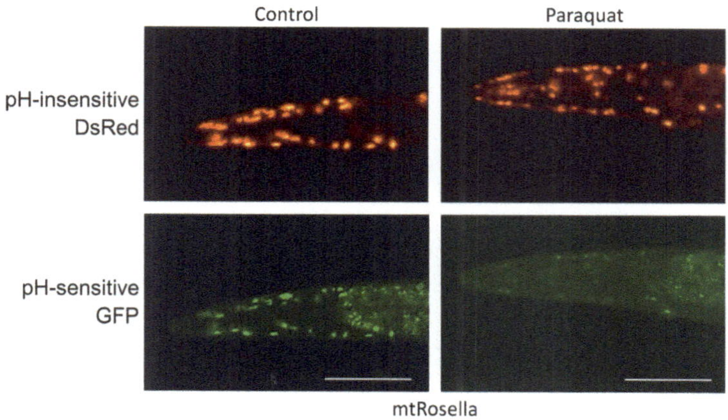

Fig. 2 mtRosella expression in control and paraquat-treated worms. *Adapted from Palikaras, K., Lionaki, E., & Tavernarakis, N. (2015). Coordination of mitophagy and mitochondrial biogenesis during ageing in C. elegans.* Nature, *521(7553), 525–528. http://dx.doi.org/10.1038/nature14300.*

reporter for the mammalian ER-phagy receptor ortholog in *C. elegans* (Khaminets et al., 2015).

The second tool takes advantage of the Rosella biosensor, a reporter initially used in the unicellular organism *Saccharomyces cerevisiae*. This construct, which was adapted for *C. elegans*, contains a fused, fast-maturing, and pH-insensitive variant of DsRed and a pH-sensitive variant of GFP (Rosado, Mijaljica, Hatzinisiriou, Prescott, & Devenish, 2008). Targeting to any organelle/cellular compartment (in this case, the mitochondrion, mtRosella) indicates the level of acidity of the environment/vacuole in which it is contained. Thus, under the autophagy-inducing conditions described here, the level of GFP/DsRed ratio is reduced when mitochondria are included in the autophagolysosome (Fig. 2). Furthermore, an even more refined approach to more accurately quantifying selective autophagy is

Table 1 Fluorescent Reporter Fusion Proteins, and *C. elegans* Strains Expressing the Respective Transgenes for Monitoring General and Selective Autophagy

Reporter	Nonselective Autophagy	Selective Autophagy	References
DsRed:: LGG-1 or GFP:: LGG-1	Ex011[p_{lgg-1} DsRed::LGG-1; *rol-6(su1006)*] *adIs2122*[p_{lgg-1} GFP::LGG-1; *rol-6(su1006)*]	R1284: N2;*Is* [p_{myo-3}mtGFP];Ex011 [p_{lgg-1}DsRed::LGG-1]	Kang, You, and Avery (2007); Palikaras et al. (2015)
Rosella biosensor	—	IR1631: N2;*Ex003* [p_{myo-3}TOMM-20:: Rosella]	Palikaras et al. (2015)
p_{hlh-30}HLH-30::GFP	JIN1679 *jinEx10* [p_{hlh-30}*hlh-30::gfp*; *rol-6 (su1006)*]	—	Lapierre et al. (2013)
W07G4.5:: GFP	*bpIs239*[*w07g4.5:: gfp, unc-76*]	—	Lin et al. (2013)

the construction of translational reporters for selective autophagy receptors or adaptors. Moreover, double-transgenic animals expressing either a core autophagic reporter, GFP::LGG-1, or the Rosella biosensor (as previously described), together with selective autophagy receptors, can also be generated (Table 1). In the case of mitophagy, the protein ortholog of NIX/BNIP3L and BNIP3 (the Nip3-like protein X/Bcl-2 and the adenovirus E1B-interacting protein) in *C. elegans* has been identified as DCT-1 (DAF-16/FOXO Controlled, germline Tumor affecting-1) (Zhang et al., 2012). This outer mitochondrial membrane protein is a critical regulator of the interplay between mitochondrial biogenesis and degradation through mitophagy. For instance, generation of a strain carrying a double reporter (LGG-1 fused with GFP as indicated previously, and DCT-1 fused with DsRed) will allow direct detection of colocalization between core autophagy components and a selective factor. Other selective factors include PDR-1/PARK2 and PINK-1/PINK1. Lipophagy, the degradation of lipid droplets by macroautophagy, has been a technically challenging task. More traditional techniques include monitoring by measuring lipid storage abundance by BODIPY and Nile Red staining, which have low specificity as they can also stain lysosomes. Coherent anti-Stokes Raman scattering microscopy is a more recent, noninvasive, label-free imaging method for monitoring lipid metabolism (Folick, Min, & Wang, 2011).

2.3 Transcriptional Activation

Upon autophagy-inducing conditions, besides the increased protein degradation, there are significant transcriptional alterations. The autophagic transcriptional program is mainly regulated by HLH-30, an ortholog of the mammalian TFEB, which is a helix-loop-helix transcription factor. Under autophagy-inducing stimuli, which cause the inhibition of LET-363/TOR, HLH-30 translocates to the nucleus to induce transcription of autophagy-related genes (Lapierre et al., 2013). Thus, evaluation of HLH-30 nuclear translocation can be done by confocal imaging of nematodes expressing the integrated translational fusion $p_{hlh\text{-}30}$HLH-30::GFP (Table 1). In this case, the percentage of animals with GFP localized in the nucleus is assessed. Moreover, another core transcription factor linking autophagy with aging is PHA-4/FOXA, which positively regulates *unc-51*/Ulk1 transcription. However, contrary to HLH-30::GFP translocating to the nucleus upon autophagy induction, PHA-4::GFP nuclear levels remain constant (Panowski, Wolff, Aguilaniu, Durieux, & Dillin, 2007).

3. AUTOPHAGY INDUCTION AND SUPPRESSION

3.1 Environmental Stress

Upon nutrient limitation or starvation, *C. elegans* upregulates autophagy to cope with this unfavorable condition, as do all other eukaryotes. Receptors send the signal to intracellular nutrient sensitive kinases such as LET-363/TOR, which is inhibited, causing the disinhibition of UNC-51/ULK1. Then vesicle nucleation begins and is mediated by the PI3K class III lipid kinase complex, including LET-512/VPS34 and BEC-1/BECLIN-1 (Fujioka et al., 2014). Autophagosome elongation and maturation involve two ubiquitin-like conjugation systems encompassing, the first, ATG5, LGG-3/ATG12, and ATG16, and the second, LGG-1(2)/MAP1LC3, ATG3, and ATG7. These steps are followed by autophagosome fusion with the lysosome. Nutrient deprivation is the most standard and evolutionarily conserved approach to upregulate the autophagic response.

Other assays that induce both general and selective macroautophagy are crowding, hypoxia, heat, and oxidative and DNA damage-induced stress. Thermotolerance has been used as an in vivo approach to activating autophagy in diverse organisms ranging from plants to nematodes, as it has been shown to induce transcription of autophagic genes, autophagosome accumulation, and increased mitochondrial degradation (Palikaras et al., 2015;

Zhou, Wang, Yu, & Chen, 2014). Similarly, DNA damage induced by ultraviolet irradiation (UV-B or UV-C) has also been shown to activate autophagy in *C. elegans* (Meyer & Bess, 2012). Moreover, oxidative stress induced by paraquat and mitochondrial stress by carbonyl cyanide *m*-chlorophenylhydrazone (CCCP), which is a chemical inhibitor of oxidative phosphorylation in mitochondria, both lead to mitochondrial defects, ultimately stimulating mitophagy.

3.2 Genetic Induction and Inhibition

Genetic manipulations, either knockouts or knockdowns, that influence longevity have been repeatedly associated with autophagy, and in some cases, autophagy is required for their effects on life span. For example, the *eat-2(ad1116)* mutant is a dietary restriction genetic model, and like other feeding-defective mutants, it has been shown to have naturally higher levels of autophagy in both larvae and adults, ultimately extending life span. Along similar lines, mutations in the gene encoding the insulin–IGF-1-like receptor abnormal Dauer Formation 2 (DAF-2) extend the *C. elegans* life span and activate autophagy. This life span extension requires both the activity of DAF-16/FOXO and autophagy. Notably, DAF-16 itself is dispensable for autophagy activation in *daf-2* mutants, and autophagy is not sufficient to promote longevity. Likewise, autophagy is required for life span extension by dietary restriction in *C. elegans*. Moreover, suppression of autophagy prevents life span extension by dietary restriction and TOR inhibition, which is known to mediate, at least in part, the beneficial effects of dietary restriction on longevity (Hansen et al., 2008).

Genetic inhibition of autophagy can also be performed with both mutants and knockdowns. Mutants of the autophagic machinery are usually nonviable, very sick, or have severe growth delay, causing technical difficulties in developmental synchronization. A list of mutants is provided in Table 2 (Cheng et al., 2013). Animals carrying mutations in *bec-1* (involved in vesicle nucleation) are not viable as homozygotes, while mutations in the *unc-51 gene* are viable but cause movement difficulties and defects in autophagy such as mislocalization of GFP::LGG-1. Alternatively, a mosaic analysis using the *bec-1(−)*; BEC-1::GFP transgenic strain could be applied (Takacs-Vellai et al., 2005). Further, *let-363* mutation inhibits the TOR pathway but cannot be effectively used, as it causes L3 developmental arrest (Vellai et al., 2003). On the other hand, the *lgg-2 (tm5755)* mutant is viable and reaches adulthood, although mutant phenotypes can be masked by

Table 2 *C. elegans* Genes That Are Routinely Targeted to Influence Autophagy, and the Corresponding Alleles or Genetic Manipulations Typically Used

Mammalian Homolog	*C. elegans* Gene	Genetic Manipulation	References
Beclin 1	*bec-1*	*RNAi*	Takacs-Vellai et al. (2005)
VPS-34	*let-512*	*RNAi*	Cheng et al. (2013)
WIPI3/4	*epg-6*	*bp242*	Lu et al. (2011)
ULK-1 (ATG1)	*unc-51*	*RNAi*	Hansen et al. (2008)
ATG2	*atg-2*	*bp576*	Cheng et al. (2013)
ATG4	*atg-4*	*bp501*	Cheng et al. (2013)
ATG5	*atg-5*	*RNAi*	Cheng et al. (2013)
ATG7	*atg-7*	*bp422*	Cheng et al. (2013)
LC3B (ATG8)	*lgg-1 & lgg-2*	*RNAi*	Cheng et al. (2013)
Sqstm1/p62	*sqst-1*	*ok2892*	Cheng et al. (2013)
HLH-30	*hlh-30*	*hlh-30(tm1978)*	Cheng et al. (2013)
IGF-1R	*daf-2*	*daf-2(e1370)*	Cheng et al. (2013)
TOR	*let-363*	*RNAi*	Vellai et al. (2003)
SH2D1B	*eat-2*	*eat-2(ad1116)*	Hansen et al. (2008)

functional redundancy or compensation by the LGG-1 protein. It is thus generally recommended that genetic inhibition of autophagy be performed by RNA interference (RNAi) knockdown of these genes after completion of development (larval stage 4) in order to examine the molecular and functional effects of autophagic ablation in itself. This is readily carried out in nematodes by feeding them with bacteria that have been transformed with double-stranded RNA (dsRNA)-expressing plasmid for specific knockdown of autophagy genes such as *bec-1*, *atg-7*, *vps-34*, *unc-51*, *lgg-1*, and *atg-18*.

3.3 Chemical Inhibition

Chemical suppression of autophagy has been widely used in higher eukaryotes. Cell culture experiments are usually conducted to investigate the effects of exogenous agents on autophagic responses, as there are no paracrine effects or boundaries that can affect the readout of the experiment. In vivo

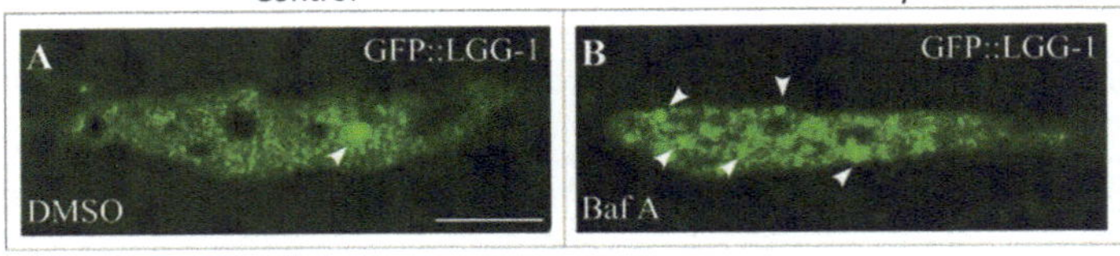

Fig. 3 LGG-1::GFP puncta (*arrowhead*) in seam cells of (A) control (DMSO) and (B) bafilomycin-treated worms. *Aadapted from Zhang, H., Chang, J. T., Guo, B., Hansen, M., Jia, K., Kovacs, A. L., ... Wu, F. (2015). Guidelines for monitoring autophagy in Caenorhabditis elegans.* Autophagy, 11*(1), 9–27. http://dx.doi.org/10.1080/15548627.2014.1003478.*

experiments need much higher concentrations for longer periods of time. Bafilomycin A1 is a V-ATPase inhibitor that blocks the autophagic flux acutely by inhibiting autolysosome acidification and autophagosome–lysosome fusion. Practically, this means that autophagic substrates cannot be degraded in the lysosome, and thus the total autophagic pool can be imaged or quantified. Two ways have been proposed to administer bafilomycin A1 to the animals, feeding and injection (Saha et al., 2015; Wilkinson et al., 2015; also see Fig. 3). Feeding requires the bleaching of adult worms carrying eggs, followed by egg hatching in liquid media containing a high concentration of bafilomycin A1; 24 h later, worms are transferred to nematode growth medium (NGM) plates containing bafilomycin until used for imaging. Alternatively, bafilomycin A1 can be injected to adult worms and after 2 h, there is a distinct increase in GFP::LGG-1 puncta (Klionsky et al., 2016). This is an interesting autophagic inhibitor, as it blocks late steps of autophagy after autophagosome formation, permitting visualization of autophagic substrates, which are otherwise rapidly degraded, preventing imaging.

4. EXPERIMENTAL CONSIDERATIONS

There are certain parameters during experimental design that should be considered. Measuring autophagic responses at different developmental stages can be critical as to the readout of the experiment. Synchronization of animals is an essential prerequisite, as aggregates degraded by autophagy vary with age. Note that when using certain mutants, the rate of growth is slower. Autophagy levels change with age, although whether it declines or increases during aging is still a matter of scientific debate (Chapin, Okada, Merz, & Miller, 2015). Moreover, most of the studies have been performed and perfected in embryos and early larval stages, where cells are more easily visualized. Although autophagy is high during early development, one should be cautious about the scientific question raised. For instance, if monitoring mitophagy, different parameters should be taken into consideration, as the mitochondrial number is low and mitochondria are still morphologically and consequently functionally immature during development. Mitophagy measurements are not ideal at that age. Also, after a certain age, it is important to discriminate functional effective autophagy. Changes in the number of autophagosomes and the abundance of autophagic substrates are merely indications. Monitoring different steps of the autophagic pathway, such as autophagy initiation, autophagosome formation/maturation, and degradation, is critical. Autophagosomes visualized as GFP::LGG-1 puncta may increase in number as a result of either increased autophagosome formation or decreased degradation, i.e., autophagic flux. This pitfall can be resolved by using bafilomycin A1, a specific V-ATPase inhibitor blocking lysosomal acidification, among others (hence autophagic turnover). In addition, autofluorescence increases with age, especially in the worm intestine, making it difficult to discriminate between an actual signal and background noise. Confocal microscopy reduces background noise. GFP::LGG-1 can also increase as a result of technical issues, such as unwanted starvation of the animals, contamination of the agar plate where they are maintained, and the anesthetic used after 5–10 min of treatment (Zhang et al., 2015).

5. EXPERIMENTAL PROCEDURES

5.1 Reagents/Equipment

- Worm pick
- Eppendorf tubes
- Tube rotator
- NGM agar plates: Weigh 3 g of NaCl (Merck, Nottingham, UK; cat. no. 1.06404.1000), 2.5 g of bactopeptone (BD Biosciences, San Jose, California, USA; cat. no. 211677), 0.2 g of streptomycin kept at 4°C (Sigma; cat. no. S-6501), 17 g of agar (Merck; cat. no. 1.01614) and add 0.8 L of distilled water. Autoclave with 500 mL of distilled water. While mixing, allow cooling for 1 h before adding 1 mL of 5 mg/mL cholesterol (SERVA Electrophoresis GmbH, Heidelberg, Germany; cat. no. 1701) in EtOH, 1 mL 1 *M* $MgSO_4$, 1 mL 1 *M* $CaCl_2$ (Sigma; cat. no. C-5080), (Sigma; cat. no. M-7506), 1 mL 10 mg/mL nystatin

(Sigma; cat. no. N-3503), 25 mL 1 *M* phosphate buffer with pH 6.0, and add up to 1 L of distilled water.

- M9 buffer for starvation: Weigh 3 g of KH_2PO_4 (Merck; cat. no. 1.04873.1000), 6 g of Na_2HPO_4 (Merck; cat. no. 1.06586.0500), 5 g of NaCl (Merck; cat. no. 1.06404.1000), and dissolve to 1 L of distilled water. Autoclave and add 1 mL of sterile 1 *M* $MgSO_4$ (sterile; Sigma; cat. no. M-7506). Store M9 buffer at 4°C.
- Levamisole: Weigh 1.2 g of levamisole (Sigma-Aldrich, St. Louis, Missouri, USA; cat. no. 196142) and add in 10 mL of distilled sterile water. Store levamisole solution at 4°C.
- Dimethyl sulfoxide cell culture grade BC (DMSO; Applichem, Maryland Heights, MO, USA; cat. no. A3672.0250).
- 50 μ*M* of bafilomycin A1 in 0.2% DMSO (Sigma-Aldrich; cat. no. B1793).
- *C. elegans* strains: R1284: N2;*Is*[$p_{myo\text{-}3}$mtGFP];Ex011[$p_{lgg\text{-}1}$DsRed::LGG-1] and IR1631: N2;Ex003[$p_{myo\text{-}3}$TOMM-20::Rosella].
- 20°C incubator.
- UV cross-linker (BIO-LINK–BLX-E365, Vilber Lourmat).
- Zeiss AxioObserver Z1 confocal microscope.
- Zeiss ZEN 2012 software.
- Volocity High-Performance 3D imaging software (PerkinElmer, Waltham, Massachusetts, USA).
- Microscope slides 75 × 25 × 1 mm (Marienfeld, Lauda-Koenigshofen, Germany; cat. no. 10 006 12).
- Microscope cover glass 18 × 18 mm (Marienfeld; cat. no. 01 010 30).
- Prism software package (GraphPad software).

5.2 Methodology

The following protocol outlines the basic steps to image selective autophagy of mitochondria in vivo in adult worms. First, synchronization of the strains N2;*Is*[$p_{myo\text{-}3}$mtGFP];Ex011[$p_{lgg\text{-}1}$DsRed::LGG-1] and IR1631: N2;Ex003 [$p_{myo\text{-}3}$TOMM-20::Rosella] is necessary to avoid variation due to developmental age differences. The first step is bleaching of gravid adult hermaphrodites, followed by egg hatching onto NGM plates seeded with *Escherichia coli* OP50 bacteria. In order to detect basal autophagy of mitochondria at different ages, injection of 50 μ*M* of bafilomycin A1 at d1, d2, d4, d6, d10, and d14 of adulthood should be done, as a control injection of DMSO should be performed. Feeding with bafilomycin A1 requires a concentration of 100 μg/mL from egg hatching until adulthood, which is both costly and increases variability, due to dependence on feeding conditions. Then, 2 h later, worms are collected with a worm pick, transferred to an empty NGM plate for 5 min to remove excess bacteria, and then transferred onto a levamisole droplet to immobilize worms on a microscope slide (approximately 13 μL). A coverslip is then gently placed on top and sealed around. Quick handling is essential at this point to avoid DsRed::LGG-1 puncta formed as a result of levamisole incubation.

A second alternative is to examine the autophagic flux under nutrient deprivation. In this case, all of these steps should be followed, but with two main alterations. First, egg laying is done on plates seeded with OP50 bacteria that have been UV-C irradiated for 15 min (0.5 J) using a UV cross-linker (BIO-LINK—BLX-E365, VilberLourmat). In this scenario, worms will have food to grow, but the bacteria will be dead; thus, when transferring to an empty NGM plate for acute starvation, no bacteria will be able to grow. In fact, even a small amount of OP50 can hinder the process, completely changing the results of the stress assay. Starvation can vary from 6 to 12 h, and bafilomycin can be administered 2 h before imaging. A more extreme approach to caloric restriction can be performed ideally for 24 h in an eppendorf tube containing M9 buffer on a rotator; however, this can cause bagging in gravid adults, and worms might not survive.

Image capturing using a confocal microscope (Zeiss AxioObserver Z1), for snapshots of single worms, as well as *Z*-stacks of single-body-wall muscle cells (1 μm step) is recommended. The mtRosella reporter fusion, expressed in body-wall muscle cells, is best imaged and quantified at the single-worm level, while colocalization between DsRed::LGG-1 and mitochondria-targeted GFP requires imaging at the cellular level. Lens, magnification, exposure time, resolution, laser intensity, and gain should be kept constant. Analysis of images can be performed directly using the Volocity software. Conversion is automatically done by transferring the Zeiss format images to a new folder of Volocity. Next, the pixel intensity of each fluorescent reporter/animal can be measured by encircling the worm, followed by choosing the Measurements panel, which contains the mean pixel intensity for each channel. Normalization can then be performed by dividing pixel intensity to the fluorescent area, and thereafter, the GFP to DsRed ratio can be calculated. For DsRed::GFP and mitoGFP colocalization assessment, in the Measurements panel, there is a "find spots" choice where, for each fluorescent spot on the image the pixel intensity is given. The list provided shows spots/puncta that contain a number representing pixel intensity over a certain manually selected threshold where it is apparent there is colocalization. After standardizing the thresholds, repetition for each body wall muscle cell should be done. All calculations can be performed using Microsoft Excel 2011 software package (Microsoft Corporation, Redmond, Washington, USA). Statistical analysis and graph design can be performed using the Prism software package (GraphPad Software). For comparison, use Student's *t*-test with a cutoff $p < 0.05$. For each experiment/biological replicate, examine 70 animals minimum (mtRosella) or 50 body-wall muscle cells (DsRed::LGG-1;mtGFP).

6. CONCLUDING REMARKS

The growing interest in selective forms of autophagy that are differentially regulated in various tissues has triggered the development of an arsenal of versatile tools that allow monitoring of both general and cargo-specific autophagy in *C. elegans*. The availability of reliable and quantitative methods for monitoring autophagy is a prerequisite for shedding light on the molecular mechanisms underlying autophagic responses in living cells and organisms. Delineating the role of autophagy at the organismal level is pivotal for the development of therapeutic interventions to treat several human pathologies, including neurodegenerative disorders. *C. elegans*-specific approaches and resources, combined with analogous methodologies developed in other organisms, including mice, will be instrumental toward facilitating progress in this direction.

ACKNOWLEDGMENTS

Work in the authors' laboratory is funded by grants from the European Research Council (ERC), the European Commission Framework Programmes, and the Greek Ministry of Education. Margarita Elena Papandreou is supported by the European Commission Marie Curie Initial Training Network (ITN) CODAGE.

REFERENCES

Chapin, H. C., Okada, M., Merz, A. J., & Miller, D. L. (2015). Tissue-specific autophagy responses to aging and stress in C. elegans. *Aging (Albany, NY)*, *7*(6), 419–434.

Cheng, S., Wu, Y., Lu, Q., Yan, J., Zhang, H., & Wang, X. (2013). Autophagy genes coordinate with the class II PI/PtdIns 3-kinase PIKI-1 to regulate apoptotic cell clearance in C. elegans. *Autophagy*, *9*(12), 2022–2032. http://dx.doi.org/10.4161/auto.26323.

Folick, A., Min, W., & Wang, M. C. (2011). Label-free imaging of lipid dynamics using coherent anti-stokes Raman scattering (CARS) and stimulated Raman scattering (SRS) microscopy. *Current Opinion in Genetics & Development*, *21*(5), 585–590. http://dx.doi.org/10.1016/j.gde.2011.09.003.

Fujioka, Y., Suzuki, S. W., Yamamoto, H., Kondo-Kakuta, C., Kimura, Y., Hirano, H., … Noda, N. N. (2014). Structural basis of starvation-induced assembly of the autophagy initiation complex. *Nature Structural & Molecular Biology, 21*(6), 513–521. http://dx.doi.org/10.1038/nsmb.2822.

Hansen, M., Chandra, A., Mitic, L. L., Onken, B., Driscoll, M., & Kenyon, C. (2008). A role for autophagy in the extension of lifespan by dietary restriction in C. elegans. *PLoS Genetics. 4*(2). e24. http://dx.doi.org/10.1371/journal.pgen.0040024.

Itakura, E., & Mizushima, N. (2010). Characterization of autophagosome formation site by a hierarchical analysis of mammalian Atg proteins. *Autophagy, 6*(6), 764–776.

Kalvari, I., Tsompanis, S., Mulakkal, N. C., Osgood, R., Johansen, T., Nezis, I. P., & Promponas, V. J. (2014). iLIR: A web resource for prediction of Atg8-family interacting proteins. *Autophagy, 10*(5), 913–925. http://dx.doi.org/10.4161/auto.28260.

Kang, C., You, Y. J., & Avery, L. (2007). Dual roles of autophagy in the survival of Caenorhabditis elegans during starvation. *Genes & Development, 21*(17), 2161–2171. http://dx.doi.org/10.1101/gad.1573107.

Keith, S. A., Maddux, S. K., Zhong, Y., Chinchankar, M. N., Ferguson, A. A., Ghazi, A., & Fisher, A. L. (2016). Graded proteasome dysfunction in Caenorhabditis elegans activates an adaptive response involving the conserved SKN-1 and ELT-2 transcription factors and the autophagy-lysosome pathway. *PLoS Genetics. 12*(2). e1005823. http://dx.doi.org/10.1371/journal.pgen.1005823.

Khaminets, A., Heinrich, T., Mari, M., Grumati, P., Huebner, A. K., Akutsu, M., … Dikic, I. (2015). Regulation of endoplasmic reticulum turnover by selective autophagy. *Nature, 522*(7556), 354–358. http://dx.doi.org/10.1038/nature14498.

Klionsky, D. J., Abdelmohsen, K., Abe, A., Abedin, M. J., Abeliovich, H., Acevedo Arozena, A., … Zughaier, S. M. (2016). Guidelines for the use and interpretation of assays for monitoring autophagy (3rd edition). *Autophagy, 12*(1), 1–222. http://dx.doi.org/10.1080/15548627.2015.1100356.

Lapierre, L. R., De Magalhaes Filho, C. D., McQuary, P. R., Chu, C. C., Visvikis, O., Chang, J. T., … Hansen, M. (2013). The TFEB orthologue HLH-30 regulates autophagy and modulates longevity in Caenorhabditis elegans. *Nature Communications, 4*, 2267. http://dx.doi.org/10.1038/ncomms3267.

Lin, L., Yang, P., Huang, X., Zhang, H., Lu, Q., & Zhang, H. (2013). The scaffold protein EPG-7 links cargo-receptor complexes with the autophagic assembly machinery. *The Journal of Cell Biology, 201*(1), 113–129. http://dx.doi.org/10.1083/jcb.201209098.

Lu, Q., Yang, P., Huang, X., Hu, W., Guo, B., Wu, F., … Zhang, H. (2011). The WD40 repeat PtdIns(3)P-binding protein EPG-6 regulates progression of omegasomes to autophagosomes. *Developmental Cell, 21*(2), 343–357. http://dx.doi.org/10.1016/j.devcel.2011.06.024.

Meyer, J. N., & Bess, A. S. (2012). Involvement of autophagy and mitochondrial dynamics in determining the fate and effects of irreparable mitochondrial DNA damage. *Autophagy, 8*(12), 1822–1823.

Palikaras, K., Lionaki, E., & Tavernarakis, N. (2015). Coordination of mitophagy and mitochondrial biogenesis during ageing in C. elegans. *Nature, 521*(7553), 525–528. http://dx.doi.org/10.1038/nature14300.

Panowski, S. H., Wolff, S., Aguilaniu, H., Durieux, J., & Dillin, A. (2007). PHA-4/Foxa mediates diet-restriction-induced longevity of C. elegans. *Nature, 447*(7144), 550–555. http://dx.doi.org/10.1038/nature05837.

Popelka, H., & Klionsky, D. J. (2015). Analysis of the native conformation of the LIR/AIM motif in the Atg8/LC3/GABARAP-binding proteins. *Autophagy, 11*(12), 2153–2159. http://dx.doi.org/10.1080/15548627.2015.1111503.

Rosado, C. J., Mijaljica, D., Hatzinisiriou, I., Prescott, M., & Devenish, R. J. (2008). Rosella: A fluorescent pH-biosensor for reporting vacuolar turnover of cytosol and organelles in yeast. *Autophagy, 4*(2), 205–213.

Saha, S., Ash, P. E., Gowda, V., Liu, L., Shirihai, O., & Wolozin, B. (2015). Mutations in LRRK2 potentiate age-related impairment of autophagic flux. *Molecular Neurodegeneration*, *10*, 26. http://dx.doi.org/10.1186/s13024-015-0022-y.

Takacs-Vellai, K., Vellai, T., Puoti, A., Passannante, M., Wicky, C., Streit, A., … Muller, F. (2005). Inactivation of the autophagy gene bec-1 triggers apoptotic cell death in C. elegans. *Current Biology*, *15*(16), 1513–1517. http://dx.doi.org/10.1016/j.cub.2005.07.035.

Vellai, T., Takacs-Vellai, K., Zhang, Y., Kovacs, A. L., Orosz, L., & Muller, F. (2003). Genetics: Influence of TOR kinase on lifespan in C. elegans. *Nature*, *426*(6967), 620. http://dx.doi.org/10.1038/426620a.

Wilkinson, D. S., Jariwala, J. S., Anderson, E., Mitra, K., Meisenhelder, J., Chang, J. T., … Hansen, M. (2015). Phosphorylation of LC3 by the Hippo kinases STK3/STK4 is essential for autophagy. *Molecular Cell*, *57*(1), 55–68. http://dx.doi.org/10.1016/j.molcel.2014.11.019.

Wu, F., Watanabe, Y., Guo, X. Y., Qi, X., Wang, P., Zhao, H. Y., … Zhang, H. (2015). Structural basis of the differential function of the two C. elegans Atg8 homologs, LGG-1 and LGG-2, in autophagy. *Molecular Cell*, *60*(6), 914–929. http://dx.doi.org/10.1016/j.molcel.2015.11.019.

Zhang, H., Chang, J. T., Guo, B., Hansen, M., Jia, K., Kovacs, A. L., … Wu, F. (2015). Guidelines for monitoring autophagy in Caenorhabditis elegans. *Autophagy*, *11*(1), 9–27. http://dx.doi.org/10.1080/15548627.2014.1003478.

Zhang, J., Loyd, M. R., Randall, M. S., Waddell, M. B., Kriwacki, R. W., & Ney, P. A. (2012). A short linear motif in BNIP3L (NIX) mediates mitochondrial clearance in reticulocytes. *Autophagy*, *8*(9), 1325–1332. http://dx.doi.org/10.4161/auto.20764.

Zhou, J., Wang, J., Yu, J. Q., & Chen, Z. (2014). Role and regulation of autophagy in heat stress responses of tomato plants. *Frontiers in Plant Science*, *5*, 174. http://dx.doi.org/10.3389/fpls.2014.00174.

CHAPTER TWENTY-THREE

Characterization of Autophagic Responses in *Drosophila melanogaster*

T. Xu, S. Kumar, D. Denton[1]
Centre for Cancer Biology, University of South Australia, Adelaide, SA, Australia
[1]Corresponding author: e-mail address: donna.denton@unisa.edu.au

Contents

Abstract

Drosophila is an excellent model system for studying autophagy during animal development due to the availability of genetic reagents and opportunity for in vivo cell biological analysis. The regulation and mechanism of autophagy are highly evolutionarily conserved and the role of autophagy has been characterized during various stages of *Drosophila* development as well as following starvation. Studies in *Drosophila* have revealed novel insights into the role of distinct components of the autophagy machinery.

Methods in Enzymology, Volume 588
ISSN 0076-6879
http://dx.doi.org/10.1016/bs.mie.2016.09.089

This chapter describes protocols for examining autophagy during *Drosophila* development. A crucial step in the induction of autophagy is the incorporation of Atg8a into the autophagosome. This can be measured as autophagic puncta using live fluorescent imaging, immunostaining, or immunoblot analysis of LC3/Atg8a processing. The level of autophagy can also be examined using other specific components of the autophagy pathway as markers detected by immunofluorescent imaging. Based on the distinct morphology of autophagy, it can also be examined by transmission electron microscopy. In addition, one of the advantages of using *Drosophila* as a model is the ability to undertake genetic analysis of individual components of the autophagy machinery. Current approaches that can be used to monitor autophagy, including the overall flux and individual steps in *Drosophila melanogaster*, will be discussed.

1. INTRODUCTION

Macroautophagy (hereafter referred to as autophagy) is a highly conserved catabolic process that degrades cytoplasmic components including organelles and large protein aggregates through the lysosome. Autophagy occurs at a low basal rate in most cells and is induced to high levels in response to cellular stress conditions such as nutrient limitation, organelle damage, and infection (Kroemer, Marino, & Levine, 2010). Autophagy as a cell survival mechanism has been well studied from yeast to mammals (Zhang & Baehrecke, 2015). Moreover, autophagy can also function as a cell death mechanism and has been best studied in model organisms (Denton, Xu, & Kumar, 2015). *Drosophila* is an exceptional in vivo model to study both starvation-induced autophagy and autophagy-dependent programmed cell death (Denton et al., 2015; Xu, Nicolson, Denton, & Kumar, 2015).

1.1 Autophagy Machinery

With the increased understanding of the molecular mechanisms of autophagy, a number of protein components, including a large group of distinct Autophagy-related (Atg) proteins that are required for each step of autophagy, have been identified (Zhang & Baehrecke, 2015). Most of these components are evolutionarily conserved from yeast to mammals, including in *Drosophila*. Autophagy is initiated by the formation of the phagophore that further expands to encapsulate part of the cytoplasm into a double membrane vesicle called the autophagosome, the organelle specifically required for autophagy (Seglen & Bohley, 1992). Once formed, the outer membrane of the autophagosome fuses with the lysosomal membrane where hydrolytic enzymes degrade the contents and the inner membrane. During this process,

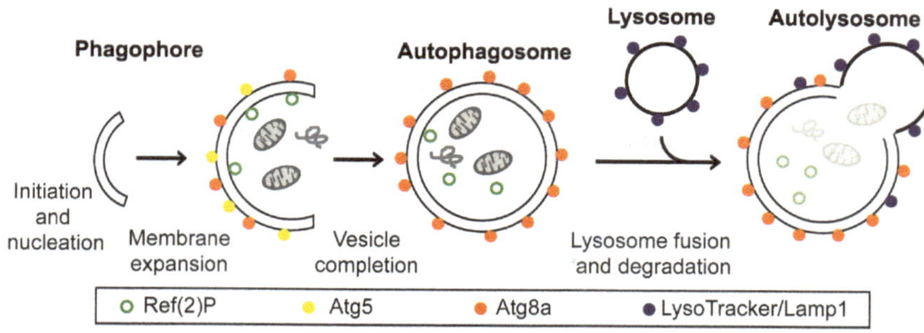

Fig. 1 Schematic diagram showing autophagy markers during the stages of autophagy. The multiple steps of autophagy can be identified due to the specific proteins that localize to the autophagic membrane during different stages. Following initiation Atg5–Atg12–Atg16 complex is localized to the membrane, as is Atg8a. Upon completion of the autophagosome, Atg5 is removed from the membrane with Atg8a maintaining its membrane localization. Lysosomes can be monitored by LysoTracker and Lamp1, which can also mark autolysosomes. The abundance of Ref(2)P can be monitored to assess its degradation by autophagy.

several Atg proteins are found in distinct localizations on autophagy structures and these can be used to monitor autophagy (Fig. 1). Transgenic *Drosophila* lines carrying these markers have been developed and have allowed detailed monitoring of autophagy in vivo.

1.2 *Drosophila* as a Model to Study Autophagy

In *Drosophila*, autophagy has been studied and visualized in many organs at different developmental stages. This includes during embryogenesis in the amnioserosa, at larval and pupal stages in fat body, midgut, salivary glands, and the imaginal discs, as well as in adults in neurons, wing epithelium, and during oogenesis (Butterworth & Forrest, 1984; Denton et al., 2009; Hou, Chittaranjan, Barbosa, McCall, & Gorski, 2008; Juhasz, Csikos, Sinka, Erdelyi, & Sass, 2003; Kimura, Kodama, Hayasaka, & Ohta, 2004; Lee & Baehrecke, 2001; Lee, Cooksey, & Baehrecke, 2002; Mohseni, McMillan, Chaudhary, Mok, & Reed, 2009; Nezis et al., 2009, 2010; Pandey, Batlevi, Baehrecke, & Taylor, 2007; Rusten et al., 2004, 2007; Scott, Juhasz, & Neufeld, 2007; Scott, Schuldiner, & Neufeld, 2004).

During *Drosophila* metamorphosis, larval tissues such as the midgut and salivary glands are degraded in response to increased levels of the steroid hormone 20-hydroxy-ecdysone (ecdysone) (Nicolson, Denton, & Kumar, 2015). There is a distinct requirement for autophagy in the removal of these tissues. At the larval–pupal transition a large ecdysone pulse triggers larval midgut degradation and the adult gut develops in its place. The larval midgut comprises a large tissue with anterior appendages called gastric caeca that undergo contraction together with condensation of the midgut occurring within 4 h relative to puparium formation (RPF). The degradation of the larval midgut is dependent on autophagy and can occur normally in the absence of caspase activity (Denton et al., 2009).

Approximately 12 h later another ecdysone pulse signals the salivary gland to undergo degradation. The expression of several apoptosis and autophagy genes is increased by ecdysone to execute salivary gland removal, partly due to the direct binding of the nuclear hormone heterodimer, ecdysone receptor/ultraspiracle, to their promoters (Cakouros, Daish, Martin, Baehrecke, & Kumar, 2002; Daish, Cakouros, & Kumar, 2003; Denton et al., 2013). Both caspase-dependent apoptosis and autophagy are required for removal of the salivary glands (Berry & Baehrecke, 2007).

A useful tissue to examine starvation-induced autophagy is the larval fat body, which functions as a nutrient-sensing organ. In early third instar larvae, starvation rapidly induces high levels of autophagy in the fat body (Scott et al., 2004). During the late third instar larval stage, the increase in ecdysone titer leads to an increase in basal autophagy in the fat body, even in the presence of nutrients (Kamoshida et al., 2012). Thus, *Drosophila* is a unique and useful model for studying autophagy in the context of both cell survival and cell death.

In the following sections, protocols using live fluorescent imaging and immunostaining with specific components of the autophagy pathway such as Atg8a, Atg5, lysosomal markers, and the autophagy receptor Ref(2)P will be described. In addition, protocols for the biochemical analysis of proteins for assaying autophagy and the genetic analysis of the autophagy pathway are also detailed. First, methods for the staging and preparation of *Drosophila* tissue will be described.

2. STAGING OF ANIMALS AND TISSUE DISSECTION

When studying autophagy in *Drosophila* animals need to be precisely staged and all crosses should be carried out at 25°C in noncrowded conditions on standard food. The methods described in this section focus on staging and dissection of the larval midgut, salivary glands, and fat body. A recent method for examination of autophagy in ovaries can be found in DeVorkin and Gorski (2014a).

2.1 Larval Midgut

During metamorphosis the larval midgut undergoes developmentally programmed degradation that is dependent on autophagy (Denton et al., 2009). Prior to metamorphosis third instar larvae begin wandering behavior where they cease feeding and begin emptying their gut contents; hence, larval age can be estimated based on gut clearance (Andres & Thummel, 1994; Maroni & Stamey, 1983). To stage animals to monitor midgut degradation, larvae can be propagated on food containing 0.05% bromophenol blue. Larvae with dark blue guts represent −24 to −12 h RPF, light blue guts −12 to −5 h RPF, and clear guts −6 to −1 h RPF. Immediately prior to pupariation, the larvae shorten, evert spiracles, and stop moving. In the first hour following pupariation the white cuticle subsequently tans. This coloring can be used to stage animals at 0 to +1 h RPF and is also referred to as white prepupae (WPP). These developmental stages can be used to monitor the progression of autophagy-dependent midgut degradation. Low levels of autophagy can be detected in the midgut of clear gut larvae (−6 to −1 h RPF) with high levels induced in WPP and in +2 h RPF.

To monitor for gut clearance, wandering third instar larvae are collected from the side of the vial with a moistened fine paintbrush and transferred to a Petri dish lined with moist Whatman paper. Larvae can be checked regularly (every 30–60 min) for clearance of the blue food from the gut or pupariation. Following pupariation, animals can be transferred to another dish (e.g., 96-well plate) with the time marked and aged for the desired time for experimental analysis.

Dissect larvae or pupae immersed in 1 × phosphate-buffered saline (PBS) [PBS: 137 m*M* NaCl, 2.7 m*M* KCl, 10 m*M* Na_2HPO_4, 1.8 m*M* KH_2PO_4 pH 7.4] on a Sylgard dissection dish under a dissecting microscope. Using fine forceps, hold down the mouth hooks while gently pulling the posterior end to reveal the midgut and then dissect away from other tissue (using a 26-G hypodermic needle).

2.2 Salivary Glands

For analysis of autophagy during salivary gland histolysis, animals are aged to +15 h RPF, which is equivalent to +3 h after head eversion (AHE). Head eversion marks the transition between prepupal and pupal stages, and can be monitored by the movement of an air bubble from the posterior to the anterior resulting in a space for the head to evert. This occurs rapidly in about

10 min. Collect animals in a Petri dish and examine every 30–60 min for head eversion, then transfer pupae into another dish (e.g., 96-well plate) with the time marked and allow to develop a further 3 h.

Salivary glands are dissected by submerging pupae in 1 × PBS on a Sylgard dissection dish under a dissecting microscope. Using fine forceps gently peel away the anterior cuticle to expose the head, and then gently push the head to release the salivary glands.

2.3 Larval Fat Body

The induction of autophagy in response to starvation or rapamycin (see below) can be examined in the fat body of third instar larvae. Larvae are aged for 96 h after egg deposition (AED) in normal food then gently collected with a fine paintbrush and transferred to a dish with moist Whatman paper for 2 h starvation. For control larvae, food is added to the dish. Starvation induces a robust autophagic response in the larval fat body after 2 h. As the level of autophagy in the larval fat body is sensitive to nutrient conditions, it is important to ensure that larvae are maintained in noncrowded conditions.

To dissect fat body tissue, transfer larvae to a Sylgard dissection dish containing 1 × PBS under a dissecting stereomicroscope. Use fine forceps, cut across the posterior, and invert the larvae by pushing the mouth hooks posteriorly through the body. The fat body can then be dissected away using a 26-G needle and the remaining carcass discarded. Fat body tissue is then transferred to a small drop of 1 × PBS on a microscope slide for further processing as needed.

2.4 Induction of Autophagy by Rapamycin

Rapamycin is widely used to induce autophagy in cultured cells and model organisms. In *Drosophila*, rapamycin can be fed to larvae to induce autophagy in the larval midgut and fat body (Xu et al., 2015). Larvae are aged in standard food for 96 h AED and then transferred to food containing the desired concentration of rapamycin (100 m*M*, diluted from a 1 m*M* stock dissolved in DMSO) for 2 h. Food containing the same volume of DMSO should be used as control treatment. To ensure the uptake of treated food, 0.05% bromophenol blue can also be added to the food. The fat body and midgut are then dissected as described above.

3. MORPHOLOGICAL MONITORING OF AUTOPHAGY BY TRANSMISSION ELECTRON MICROSCOPY

Electron microscopy was used in the earliest discovery of autophagy (Ashford & Porter, 1962), and is still an indispensable and reliable method for monitoring autophagy morphology. With the high resolution of transmission electron microscopy (TEM), autophagosomes can be observed as double membrane structures containing cargo such as organelles, and autolysosomes can be observed as single-membrane structure containing single-membrane vesicles and/or organelles that are to be or have been partially digested (Eskelinen, Reggiori, Baba, Kovacs, & Seglen, 2011).

1. Dissect the tissue of interest in 1 × PBS and fix in 2.5% glutaraldehyde in 0.1 *M* sodium cacodylate buffer, pH 7.4, for 30 min at room temperature.
2. Wash 3 × 5 min with 0.1 *M* sodium cacodylate buffer.
3. Postfix in 1% osmium tetroxide for 1 h.
4. Dehydrate in 70% ethanol 2 × 15 min, 90% ethanol 2 × 15 min, and 100% ethanol 3 × 15 min before propylene oxide treatment of 15 min.
5. Infiltrate resin in 1:1 mix of propylene oxide and epoxy resin for 1 h, 100% resin 2 × 1 h then embed in fresh resin and polymerize in an oven at 70°C for 24 h.
6. Cut ultrathin sections (80 nm) on grids, stain for 2–3 min with uranyl acetate (4% in 25% ethanol) and 2 min in Reynold's lead citrate.
7. Examine by TEM.

It is beneficial to obtain further advice from your local electron microscopy expert, as samples can be sensitive to the length of fixation and the fixation solution. Quantitative analysis can be performed by measuring and calculating the ratio of area or volume of autophagic vacuoles compared to the total cytoplasm (Lucocq & Hacker, 2013).

4. ANALYSIS OF AUTOPHAGY MARKERS IN VIVO

In the past two decades, the growing knowledge on autophagic structure and the specific localization of Atg proteins has aided in the identification and development of makers for monitoring autophagy through fluorescence microscopy. Membrane-localized proteins, such as Atg5 and Atg8a, can be used to monitor different stages of autophagy (Fig. 1). While there are a number of antibodies available for mammalian autophagy proteins, the number of antibodies for use in *Drosophila* is still limited. Rather, due to the ease of manipulating *Drosophila* genes, a number of transgenic lines with fluorescence-tagged autophagy markers are available and widely used. The number of available transgenic fluorescence lines and antibodies is still increasing.

Although multiple membrane markers are available to monitor autophagy, Atg8a is potentially the only one that marks autophagy from the earliest to the last stage of autophagy as it remains on the membrane (Fig. 1). In *Drosophila*, Atg8a is essential for autophagy required for both starvation-induced autophagy in the fat body and autophagy-dependent cell death in the midgut (Xu et al., 2015). Like mammalian LC3 and yeast Atg8, *Drosophila* Atg8a is widely used in autophagy studies. There are both GFP and mCherry-tagged Atg8a transgenic lines that enable autophagy induction to be visualized by the accumulation of fluorescent punctate structures in the cytoplasm (Fig. 2). Unlike GFP, which is sensitive to acidic conditions resulting in loss of fluorescence in the autolysosome, mCherry is more stable at lower pH (Kimura, Noda, & Yoshimori, 2007). The pmCherry-Atg8a and pGFP-Atg8a lines contain a region of the Atg8a promoter and result in expression of fluorescence-tagged Atg8a from its endogenous promoter (see Figs. 2 and 3 for pmCherry-Atg8a) (Denton et al., 2012, 2009).

One limitation of the Atg8a marker is that when the fusion between autophagosomes and lysosomes is blocked, a high level of Atg8a would be observed due to the accumulation of nonfused autophagosome (Klionsky et al., 2008). This can lead to false conclusions of increased autophagy. To more accurately determine autophagy flux a dual-tagged GFP-mCherry-Atg8a marker can be used to monitor autophagosome–lysosome fusion as GFP fluorescence becomes quenched in the acidic lysosomal environment while mCherry fluorescence remains (Nezis et al., 2010). Hence, it is important to monitor several stages of autophagy.

A marker for the early steps in autophagy is Atg5 (Fig. 1). The Atg5, Atg12, and Atg16 complex is localized on the phagophore and elongating autophagosome membrane, but is not present on the completed autophagosome (Fig. 1). A transgenic marker line, Atg5-GFP, and Atg5 antibodies are used to monitor autophagy (Fig. 2) (Barth, Glick, & Macleod, 2010; Rusten et al., 2004).

Similar to p62 in mammalian cells, the *Drosophila* ortholog, Ref(2)P, functions as an adapter protein that recruits ubiquitinated substrates to autophagosomes. Therefore, when autophagy activity is elevated cytoplasmic Ref(2)P decreases, whereas blocking autophagy results in an accumulation of

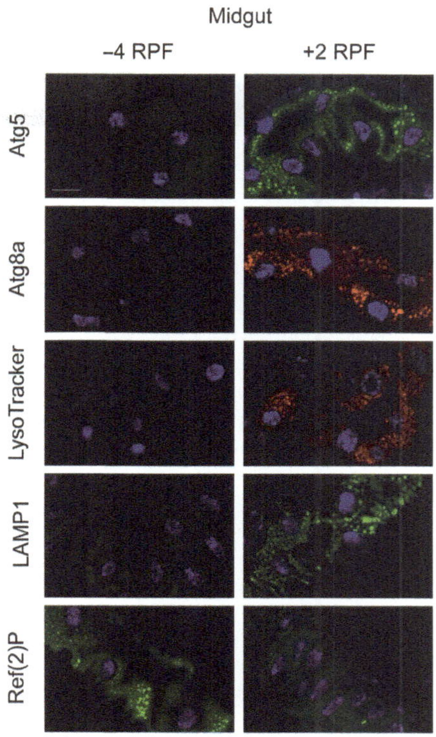

Fig. 2 Live fluorescent imaging of autophagy markers in the larval midgut. Live imaging of autophagy markers at this onset (−4 h RPF) and during (+2 h RPF) midgut degradation. The level of GFP-Atg5 puncta and pmCherry-Atg8a puncta increases during midgut removal with high levels detected at +2 h RPF compared to earlier stages (−4 h RPF). Similarly, lysosomal markers, LysoTracker Red and GFP-LAMP1, show low basal levels at −4 h RPF and increase during midgut removal (+2 h RPF). The high level of GFP-Ref(2)P in −4 h RPF midguts is rapidly reduced following induction of autophagy in the midgut +2 h RPF. In all panels, the nuclei are stained with Hoechst and scale bar represents 20 μm.

Ref(2)P (Bartlett et al., 2011; Cumming, Simonsen, & Finley, 2008; Nezis et al., 2008). This has been demonstrated using Ref(2)P antibodies as well as GFP-tagged Ref(2)P (Fig. 2) (Chang & Neufeld, 2009; Lindmo et al., 2008; Nezis et al., 2008). Assays that monitor the level of Ref(2)P can be used as a marker of autophagic activity in combination with other markers such as Atg8a puncta.

Assays that mark lysosome can also be used to monitor autophagy by specific staining of late-stage autophagic vesicles (Scott et al., 2004). Lysosomal markers, LysoTracker, and fluorescence-tagged lysosome-associated

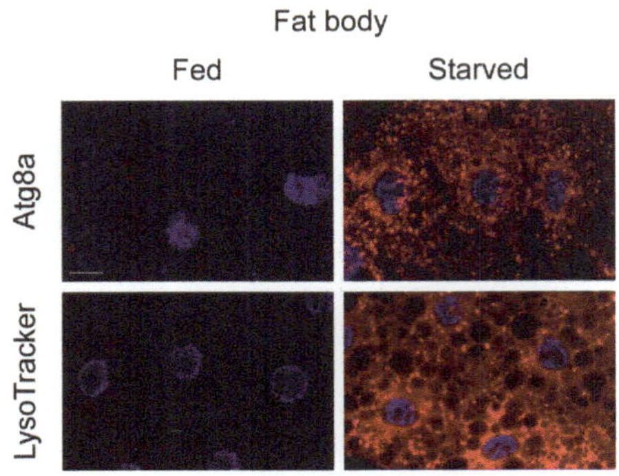

Fig. 3 Starvation-induced autophagy in the fat body. The fat body from larvae at 96 h AED either fed or starved for 2 h. The low levels of pmCherry-Atg8a and LysoTracker Red puncta in the fat body of fed larvae show a dramatic increase following starvation (*right*). The nuclei are stained with Hoechst and scale bar represents 20 μm.

membrane protein (LAMP1) transgenic lines are commonly used to monitor autophagy in *Drosophila*. LysoTracker is an acidotropic dye that stains cellular acidic compartments, including lysosomes and autolysosomes. While not a direct marker of autophagy, LysoTracker has been used to detect autophagy-associated lysosomal activity in several *Drosophila* tissues including the fat body (Fig. 3) (Scott et al., 2004), midgut (Fig. 2) (Xu et al., 2015), salivary gland (Berry & Baehrecke, 2007), and ovary (Barth, Szabad, Hafen, & Kohler, 2011; Hou et al., 2008), as well as in *Drosophila* cell culture (DeVorkin & Gorski, 2014b). There is also a GFP-tagged LAMP1 line available that marks lysosomes and autolysosomes similar as LysoTracker. Examining the colocalization between pmCherry-Atg8a and LysoTracker is useful to monitor fusion and has been observed in dying midguts (Xu et al., 2015).

Using the markers described above the following protocols will focus on examining the level autophagy in larval and pupal tissue including fat body, midgut, and salivary gland. These autophagy markers are most informative when they are used in combination, such as markers showing autophagic vesicles at different stages rather than on their own. Optimization of these methods may be required for use in other tissues.

4.1 Live Imaging of Fluorescently Tagged Markers of Autophagy

There are a number of transgenic lines available that express GFP- or mCherry-tagged autophagy markers. Lines available include pmCherry-Atg8a, pGFP-Atg8a (Denton et al., 2012, 2009), where expression is driven from the endogenous promoter negating the need for a GAL4 driver, and UAS-regulated lines including UAS-GFP-mCherry-Atg8a, UAS-GFP-Atg8a, UAS-mCherry-Atg8a, UAS-eGFP-Atg5, UAS-GFP-Ref(2)P, UAS-GFP-LAMP1, and UAS-GFP-LC3 (Chang & Neufeld, 2009; Juhasz et al., 2008; Nezis et al., 2010; Rusten et al., 2004).

1. Set up *Drosophila* crosses using the tissue-specific or ubiquitous driver of interest with UAS fluorescence-tagged autophagy marker line (see Section 6).
2. Dissect out the tissue of interest in 1 × PBS containing 1 μg/mL Hoechst 33342. (Optional fixation step: for fragile tissue this may be required but it will decrease the fluorescent signal. Fix briefly for 2 min in freshly prepared 4% paraformaldehyde in 1 × PBS.)
3. Transfer the tissue onto a slide with PBS (or mounting agent if fixed) that has double-sided adhesive tape as a bridge to prevent the tissue from being squashed when placing the coverslip over it.
4. Cover the sample with the coverslip positioned over the double-sided adhesive tape.
5. Image immediately by fluorescence microscopy (confocal or two-photon microscope).

In the midgut, red fluorescence is preferred over green florescence as it shows less background. It is recommended to examine the tissue live immediately after dissection; however, it may be necessary to briefly fix the tissue. Some *Drosophila* lines show stable fluorescence after fixing, while others quickly lose their fluorescence. Examples of staining in the midgut can be seen in Fig. 2, and in the fat body in Fig. 3. Image analysis software programs such as ImageJ can be used to quantitate the fluorescent puncta and/or fluorescence intensities as described in this chapter (see Section 4.4).

4.2 LysoTracker

Here, we provide a protocol for examining LysoTracker staining in *Drosophila* tissues using fluorescence microscopy. An example of staining in the midgut can be seen in Fig. 2, and in the fat body in Fig. 3.

1. Dissect appropriately staged animals in 1 × PBS (see Section 2).
2. Remove PBS from the tissue using a fine tip pipette (drawn out glass pipette) and replace with 200 μL of staining solution: 100 n*M* LysoTracker

Red DND-99 and 1 μg/mL Hoechst 33342 in 1 × PBS. Incubate in the dark for 2–5 min at room temperature.

3. Remove staining solution and wash with 200 μL of 1 × PBS protected from light for 5 min.
4. Mount tissue on a microscope slide between double-sided tape in 1 × PBS and add coverslip.
5. View immediately by fluorescence microscopy.

Compare the levels of LysoTracker puncta between control and experimental samples. In the midgut, we find that LysoTracker Red provides a strong signal with low background, but LysoTracker Green can be used for coimaging with mCherry-tagged marker. While LysoTracker staining is useful as an initial indicator of potential autophagy-associated lysosomal activity, additional autophagy assays must be undertaken to further substantiate the findings.

4.3 Immunofluorescence

There are a small number of commercially available antibodies that cross-react with *Drosophila* proteins including GABARAP1 (Kim et al., 2015), Atg9 (Bader, Shandala, Ng, Johnson, & Brooks, 2015), and Atg5 (Hindle et al., 2013) that have been shown to be useful in characterizing autophagy. In addition, Atg8a and Ref(2)P antibodies have been generated by several laboratories (Nezis et al., 2008; Shelly, Lukinova, Bambina, Berman, & Cherry, 2009; Wyers, Dru, Simonet, & Contamine, 1993).

1. Prepare midgut tissue (as described in Section 2) and collect at least 10 midguts for each sample. Larval midguts can be kept attached to the mouth hooks for ease of manipulation during the staining. Where this is not possible, the samples can be monitored under a dissection microscope. For the fat body, staining can also be done on a microscope slide.
2. Transfer the tissue to a 1.5-mL microfuge tube containing 1 mL of freshly prepared 4% paraformaldehyde in 1 × PBS and fix on a rotating platform for 20 min at room temperature.
3. Wash in 1 × PBTX (PBS + 0.1% Triton-X) for 3 × 5 min followed by 3 × 20 min at room temperature on rotating platform.
4. Block in 5% host species serum in 1 × PBTX (0.1%) at room temperature for 1 h. An alternative blocking solution is 1% BSA in 1 × PBTX (0.1%).

5. Add the primary antibody diluted in block and incubate overnight at 4°C.
6. Wash in 1 × PBTX (0.1%) 3 × 5 min followed by 3 × 20 min at room temperature on rotating platform.
7. Add the fluorescently tagged secondary antibody diluted in 1 × PBTX (0.1%). Alexa Fluor-conjugated secondary antibodies (Molecular Probes, OR, USA) are available in a wide range of fluorescence-emission wavelengths, and we routinely use Alexa Fluor 488 and Alexa Fluor 568, which possess good intensity and photostability. Cover the samples to keep in the dark in this and all subsequent steps and incubate at room temperature for 2 h on rotating platform.
8. Wash in 1 × PBTX (0.1%) for 3 × 5 min followed by 3 × 20 min at room temperature on a rotating platform.
9. Samples can also be costained to detect nuclei using Hoechst at 4 μg/mL in 1 × PBTX (0.1%) for 1 min prior to the last wash.
10. Mount the tissue of interest on a microscope slide with double-sided adhesive tape for thick tissue in 80% glycerol in 1 × PBS (or other mounting agent such as Vectashield). Place a coverslip across the tape and seal with clear nail varnish. Slides can be stored in the dark at 4°C before imaging.

4.4 Quantification of Autophagy Markers

Quantification of fluorescence signals can be done using software such as ImageJ (or Zeiss measurement software) that measures parameters such as fluorescence intensity.

1. Open the merged TIFF in ImageJ (single channel can also be done).
2. Select the option, "split channels."
3. Make sure that the channel that has the autophagy puncta (e.g., red channel) is selected and go to "threshold."
4. Set threshold so all punctate structures are selected (will be marked by a red dot). The lower threshold is 255 and the upper threshold will depend on how the image was captured (gain and laser power).
5. Analyze particles, change the size to 2-infinity, tick "display result" and "summarize" click OK.
6. The "count" is the number of particles.
7. Following this then count how many cells are present (i.e., number of nuclei).
8. Data can then be presented as puncta per cell.

5. BIOCHEMICAL METHODS

A biochemical method such as immunoblotting allows for the quantitation of the levels of autophagy protein markers and thus can be used to complement microscopy methods. A widely used marker of autophagy in mammals is the increased abundance of the lipidated form of LC3. Another established marker of autophagy induction in mammalian cells is monitoring the level of p62/SQSTM1 (Ref(2)P in *Drosophila*), which decreases following autophagy indication as it is directly degraded by autophagy. There are commercial antibodies targeting Atg5 (Hindle et al., 2013), Atg8a (GABARAP1, Kim et al., 2015), and Atg9 (Bader et al., 2015) that have been shown to be suitable for immunoblotting. While there are a limited number of commercially available antibodies, several laboratories have generated reagents for use in both immunostaining (see Section 4.3) and immunoblotting (Nezis et al., 2008; Shelly et al., 2009; Wyers et al., 1993). The protein levels can be determined by immunoblotting and quantified using standard techniques (Sambrook & Russell, 2001).

Preparation of protein lysate

1. Protein lysates can be prepared from appropriately staged and dissected tissues (e.g., 20 midguts per 100 μL) in lysis buffer.
2. The sample is homogenized in lysis buffer using pestle in a 1.5-mL microfuge tube.
3. Debris is removed by spinning the extracts at 13,000 rpm for 20 min at 4°C then transferring the supernatant to a fresh tube.
4. After determining protein concentration by Bradford or BCA assay, the lysates can be used for immunoblotting.

Samples can be snap frozen and stored at −70°C until use. The volume of lysis buffer used will vary depending on the protein yield of the sample.

Immunoblotting

1. Separate proteins by size using sodium dodecyl sulfate-polyacrylamide gel electrophoresis (load 30 μg of protein). If small molecules like Atg8a are to be visualized, use a 4–20% gradient gel.
2. Transfer to a polyvinylidene difluoride membrane (or nitrocellulose).
3. Block the membrane with 5% skim milk powder in 1 × PBST (PBS + 0.1% Tween 20) or in Tris-buffered saline containing 0.1% Tween 20 (TBST).
4. Incubate with a primary antibody in block overnight at 4°C.
5. Wash with 1 × PBST for 3 × 15 min.

6. Incubate the membrane with an appropriate secondary antibody conjugated to either horseradish peroxidase or alkaline phosphatase for 1 h at room temperature.
7. Visualize by enhanced chemiluminescence or enhanced chemifluorescence substrate.

6. GENETIC ANALYSIS

6.1 Expression of Fluorescently Tagged Autophagy Markers In Vivo

The GAL4/UAS system is commonly used in *Drosophila* to drive expression of a gene of interest (Brand & Perrimon, 1993). Tissue-specific or ubiquitous enhancers drive expression of GAL4 that results in transcriptional activation from the GAL4 binding sites (UAS) to drive expression of a gene of interest, and is widely used for expression of fluorescently tagged proteins and for gene knockdown by RNAi. To assay for autophagy using the fluorescently tagged markers as listed in the above section, a tissue-specific driver GAL4 line can be crossed to the fluorescently tagged UAS lines to generate stocks for further analysis. These lines can then be used to examine mutant alleles or RNAi-mediated gene knockdown.

6.2 Knockdown of *Atg* Genes in *Drosophila*

The most common method for RNAi-mediated tissue-specific knockdown of genes in vivo uses the GAL4/UAS system. The GAL4 driver can be selected for a specific tissue or developmental stage or during adulthood to drive the RNAi transgene to knockdown a given target gene. This can also be combined with a temperature-sensitive GAL80 line to achieve both spatial and temporal knockdown during development. Use of transgenic UAS-RNAi lines specific to autophagy genes has proven to be an excellent method to study the role of these genes in autophagy, as well as other cellular and developmental processes. There are a number of independent RNAi lines available for most *Atg* genes and other genes involved in autophagy regulation from Bloomington Stock Center USA; Vienna Drosophila Research Center, Austria; and National Institute of Genetics, Japan. The knockdown of numerous *Atg* genes has been verified in the midgut; a list of these lines can be found in Table 1 (Xu et al., 2015). Also a useful tool for inducing autophagy is knockdown of Tor (Xu et al., 2015).

Table 1 List of Knockdown Lines for *Atg* and Other Autophagy Genes

Gene	Line	Gene	Line
Atg1	BL26731	*Atg14*	BL55398
Atg2	BL27706	*Atg16*	BL34358
Atg3	BL34359	*Atg17*	BL36918
Atg4	BL35740	*Atg18*	BL28061
Atg5	BL27551	*Atg101*	BL34360
Atg6	BL35741 BL28060	*Vps15*	BL34092
Atg7	BL27707	*Vps34*	BL33384
Atg8a	BL34340	*Tor*	BL35578 BL33951
Atg9	BL28055	*raptor*	BL31529
Atg12	v29790	*rictor*	BL36584
Atg13	BL40861		

BL, Bloomington; *v*, VDRC.

6.2.1 Verifying RNAi-Mediated Atg Gene Knockdown by Quantitative Real-Time PCR

The level of knockdown of the RNAi target gene should be verified in the specific tissue by qRT-PCR. This protocol can also be used to examine alterations in *Atg* gene expression under different experimental conditions or during different developmental stages/tissue specific.

1. Collect appropriately staged animals and dissect tissue of interest (20 midguts in 1 mL TRIzol).
2. Isolate total RNA using TRIzol reagent protocol (Invitrogen, Carlsbad, CA, USA).
3. Synthesize cDNA from 1 μg of total RNA using High Capacity cDNA Reverse Transcription Kit (Applied Biosciences, Foster City, CA, USA) and oligo-dT primer.
4. Perform quantitative real-time PCR on a Rotor-Gene 3000 using RT2 Real-Time SYBR Green/ROX PCR MasterMix (Qiagen, Valencia, CA, USA) as per the manufacturer's instructions. A list of *Atg* primers can be found in Denton et al. (2009) and Xu et al. (2015).

5. Perform reactions in triplicate. Normalize mRNA expression levels against an internal control gene such as *rp49* using the ΔΔCT method. Alternatively the relative standard curve method can be used.

6.3 Analyzing Autophagy in Mosaic Clones

Another powerful genetic technique is the generation of mosaic clones that enables the direct comparison of homozygous mutant or knockdown cells to neighboring control cells within the same tissue. Clones are generated through flip recombinase (FLP) that mediates recombination between flip recombinase target (FRT) sites to generate homozygous mutant cell clones surrounded by control cells (Xu & Rubin, 1993). The clone cells can be identified by either the presence or absence of a marker, such as GFP. The *FLP* gene can be induced using either the heat shock promoter or the GAL4/UAS system. The main types of mosaic clones generate homozygous mutant cells from heterozygous cells (Xu & Rubin, 1993) or alternatively wild-type cells vs UAS transgene (e.g., RNAi) expressing cells (Brand & Perrimon, 1993).

1. Set up the cross between the clone line and mutant line with the same FRT sites: *hsFLP; ubi-GFP FRT40A; pmCherry-Atg8a/TM6B* can be crossed to mutant allele line on the same chromosome as the FRT (e.g., $Tor^{\Delta P}$ *FRT40A*).
2. To generate clones in the midgut or fat body allow the cross to lay for 6 h then heat shock embryos at 37°C for 60 min. The induction of mitotic recombination in the midgut and fat body cells needs to occur during early embryonic development before these tissues enter into endoreplicative cycles later in development.
3. Prepare and image the tissue of interest as described above (Section 2).
4. The clone will be distinguished by the absence of GFP.
5. The mCherry-Atg8a puncta can be compared between the homozygous mutant cells and the GFP-marked control cells.

The following technique results in the clonal activation of the GAL4-UAS system to induce GFP-marked cells that express the RNAi or overexpression transgene. There are a large number of UAS-RNAi lines available from the various stock centers. The *Actin 5C* is a constitutive promoter that is separated from the GAL4 transcription factor coding sequence by a cassette containing FRT sequence and *CD2* gene. Following heat shock-induced expression of FLP, the cassette is excised leading to the juxtaposition of *Actin 5c* with GAL4 resulting in stable expression of GAL4 that results in activation of UAS sites in the clone cell, UAS-GFP and UAS-RNAi.

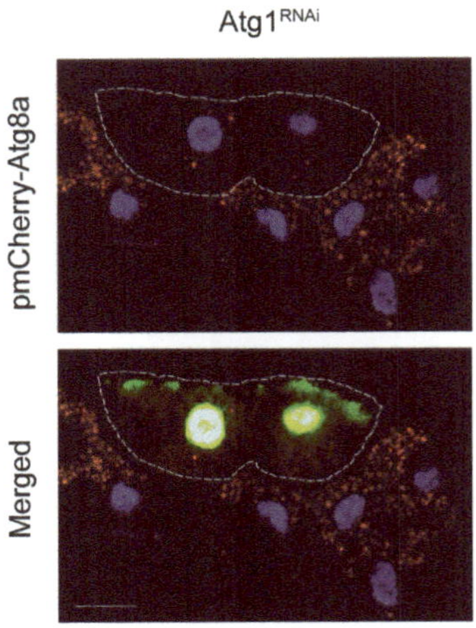

Fig. 4 Mosaic clones in the larval midgut. Mosaic clones, generated as in Section 6.3, showing an *Atg1* knockdown cell (*hsFLP; pmCherry-Atg8a/+; Act>CD2>GAL4, UAS-nlsGFP/UAS-Atg1 RNAi*) marked by GFP (*green, outlined*) with decreased levels of mCherry-Atg8a puncta (*red*) compared to the control cells (non-GFP) in +1 h RPF gastric caeca. Scale bar represents 20 μm.

The *hsFLP; pmCherry-Atg8a; Act>CD2>GAL4, UAS-nlsGFP/TM6B* line can be crossed to UAS line (either RNAi or overexpression) to enable the generation of mosaic clone of transgenic cells (Denton et al., 2012) (Fig. 4).

1. Set up the cross between the clone line and UAS transgenic line.
2. Heat shock 1-day-old embryo at 37°C for 5 min.
3. Prepare and image the tissue of interest as described above.
4. nlsGFP (nuclear localization tagged) will indicate the cells that express the UAS transgene.
5. The difference in the level of mCherry-Atg8a puncta between transgenic cells and neighboring control cells can then be visualized and analyzed.

ACKNOWLEDGMENTS

We thank the members of our laboratory for their comments on the protocols. The *Drosophila* research in our laboratory was supported by the National Health and Medical Research Council of Australia Project Grant (1041807) and a Senior Principal Research Fellowship (1103006), and funds from University of South Australia to S.K. T.X. was supported by a University of South Australia President's Scholarship.

REFERENCES

Andres, A. J., & Thummel, C. S. (1994). Methods for quantitative analysis of transcription in larvae and prepupae. *Methods in Cell Biology, 44*, 565–573.

Ashford, T. P., & Porter, K. R. (1962). Cytoplasmic components in hepatic cell lysosomes. *The Journal of Cell Biology, 12*, 198–202.

Bader, C. A., Shandala, T., Ng, Y. S., Johnson, I. R., & Brooks, D. A. (2015). Atg9 is required for intraluminal vesicles in amphisomes and autolysosomes. *Biology Open, 4*, 1345–1355.

Barth, S., Glick, D., & Macleod, K. F. (2010). Autophagy: Assays and artifacts. *The Journal of Pathology, 221*, 117–124.

Barth, J. M., Szabad, J., Hafen, E., & Kohler, K. (2011). Autophagy in *Drosophila* ovaries is induced by starvation and is required for oogenesis. *Cell Death and Differentiation, 18*, 915–924.

Bartlett, B. J., Isakson, P., Lewerenz, J., Sanchez, H., Kotzebue, R. W., Cumming, R. C., et al. (2011). p62, Ref(2)P and ubiquitinated proteins are conserved markers of neuronal aging, aggregate formation and progressive autophagic defects. *Autophagy, 7*, 572–583.

Berry, D. L., & Baehrecke, E. H. (2007). Growth arrest and autophagy are required for salivary gland cell degradation in *Drosophila*. *Cell, 131*, 1137–1148.

Brand, A. H., & Perrimon, N. (1993). Targeted gene expression as a means of altering cell fates and generating dominant phenotypes. *Development, 118*, 401–415.

Butterworth, F. M., & Forrest, E. C. (1984). Ultrastructure of the preparative phase of cell death in the larval fat body of *Drosophila melanogaster*. *Tissue & Cell, 16*, 237–250.

Cakouros, D., Daish, T., Martin, D., Baehrecke, E. H., & Kumar, S. (2002). Ecdysone-induced expression of the caspase DRONC during hormone-dependent programmed cell death in *Drosophila* is regulated by Broad-Complex. *The Journal of Cell Biology, 157*, 985–995.

Chang, Y. Y., & Neufeld, T. P. (2009). An Atg1/Atg13 complex with multiple roles in TOR-mediated autophagy regulation. *Molecular Biology of the Cell, 20*, 2004–2014.

Cumming, R. C., Simonsen, A., & Finley, K. D. (2008). Quantitative analysis of autophagic activity in Drosophila neural tissues by measuring the turnover rates of pathway substrates. *Methods in Enzymology, 451*, 639–651.

Daish, T. J., Cakouros, D., & Kumar, S. (2003). Distinct promoter regions regulate spatial and temporal expression of the *Drosophila* caspase dronc. *Cell Death and Differentiation, 10*, 1348–1356.

Denton, D., Aung-Htut, M. T., Lorensuhewa, N., Nicolson, S., Zhu, W., Mills, K., et al. (2013). UTX coordinates steroid hormone-mediated autophagy and cell death. *Nature Communications, 4*, 2916.

Denton, D., Chang, T. K., Nicolson, S., Shravage, B., Simin, R., Baehrecke, E. H., et al. (2012). Relationship between growth arrest and autophagy in midgut programmed cell death in *Drosophila*. *Cell Death and Differentiation, 19*, 1299–1307.

Denton, D., Shravage, B., Simin, R., Mills, K., Berry, D. L., Baehrecke, E. H., et al. (2009). Autophagy, not apoptosis, is essential for midgut cell death in *Drosophila*. *Current Biology, 19*, 1741–1746.

Denton, D., Xu, T., & Kumar, S. (2015). Autophagy as a pro-death pathway. *Immunology and Cell Biology, 93*, 35–42.

DeVorkin, L., & Gorski, S. M. (2014a). Genetic manipulation of autophagy in the *Drosophila* ovary. *Cold Spring Harbor Protocols, 2014*, 973–979.

DeVorkin, L., & Gorski, S. M. (2014b). LysoTracker staining to aid in monitoring autophagy in *Drosophila*. *Cold Spring Harbor Protocols, 2014*, 951–958.

Eskelinen, E. L., Reggiori, F., Baba, M., Kovacs, A. L., & Seglen, P. O. (2011). Seeing is believing: The impact of electron microscopy on autophagy research. *Autophagy*, *7*, 935–956.

Hindle, S., Afsari, F., Stark, M., Middleton, C. A., Evans, G. J., Sweeney, S. T., et al. (2013). Dopaminergic expression of the Parkinsonian gene LRRK2-G2019S leads to non-autonomous visual neurodegeneration, accelerated by increased neural demands for energy. *Human Molecular Genetics*, *22*, 2129–2140.

Hou, Y. C., Chittaranjan, S., Barbosa, S. G., McCall, K., & Gorski, S. M. (2008). Effector caspase Dcp-1 and IAP protein Bruce regulate starvation-induced autophagy during *Drosophila melanogaster* oogenesis. *The Journal of Cell Biology*, *182*, 1127–1139.

Juhasz, G., Csikos, G., Sinka, R., Erdelyi, M., & Sass, M. (2003). The *Drosophila* homolog of Aut1 is essential for autophagy and development. *FEBS Letters*, *543*, 154–158.

Juhasz, G., Hill, J. H., Yan, Y., Sass, M., Baehrecke, E. H., Backer, J. M., et al. (2008). The class III PI(3)K Vps34 promotes autophagy and endocytosis but not TOR signaling in Drosophila. *The Journal of Cell Biology*, *181*, 655–666.

Kamoshida, Y., Fujiyama-Nakamura, S., Kimura, S., Suzuki, E., Lim, J., Shiozaki-Sato, Y., et al. (2012). Ecdysone receptor (EcR) suppresses lipid accumulation in the Drosophila fat body via transcription control. *Biochemical and Biophysical Research Communications*, *421*, 203–207.

Kim, M., Semple, I., Kim, B., Kiers, A., Nam, S., Park, H. W., et al. (2015). *Drosophila* Gyf/GRB10 interacting GYF protein is an autophagy regulator that controls neuron and muscle homeostasis. *Autophagy*, *11*, 1358–1372.

Kimura, K., Kodama, A., Hayasaka, Y., & Ohta, T. (2004). Activation of the cAMP/PKA signaling pathway is required for post-ecdysial cell death in wing epidermal cells of *Drosophila melanogaster*. *Development*, *131*, 1597–1606.

Kimura, S., Noda, T., & Yoshimori, T. (2007). Dissection of the autophagosome maturation process by a novel reporter protein, tandem fluorescent-tagged LC3. *Autophagy*, *3*, 452–460.

Klionsky, D. J., Abeliovich, H., Agostinis, P., Agrawal, D. K., Aliev, G., Askew, D. S., et al. (2008). Guidelines for the use and interpretation of assays for monitoring autophagy in higher eukaryotes. *Autophagy*, *4*, 151–175.

Kroemer, G., Marino, G., & Levine, B. (2010). Autophagy and the integrated stress response. *Molecular Cell*, *40*, 280–293.

Lee, C. Y., & Baehrecke, E. H. (2001). Steroid regulation of autophagic programmed cell death during development. *Development*, *128*, 1443–1455.

Lee, C. Y., Cooksey, B. A., & Baehrecke, E. H. (2002). Steroid regulation of midgut cell death during Drosophila development. *Developmental Biology*, *250*, 101–111.

Lindmo, K., Brech, A., Finley, K. D., Gaumer, S., Contamine, D., Rusten, T. E., et al. (2008). The PI 3-kinase regulator Vps15 is required for autophagic clearance of protein aggregates. *Autophagy*, *4*, 500–506.

Lucocq, J. M., & Hacker, C. (2013). Cutting a fine figure: On the use of thin sections in electron microscopy to quantify autophagy. *Autophagy*, *9*, 1443–1448.

Maroni, G., & Stamey, S. C. (1983). Use of blue food to select synchronous, late third instar larvae. *Drosophila Information Service*, *59*, 142.

Mohseni, N., McMillan, S. C., Chaudhary, R., Mok, J., & Reed, B. H. (2009). Autophagy promotes caspase-dependent cell death during *Drosophila* development. *Autophagy*, *5*, 329–338.

Nezis, I. P., Lamark, T., Velentzas, A. D., Rusten, T. E., Bjorkoy, G., Johansen, T., et al. (2009). Cell death during *Drosophila melanogaster* early oogenesis is mediated through autophagy. *Autophagy*, *5*, 298–302.

Nezis, I. P., Shravage, B. V., Sagona, A. P., Lamark, T., Bjorkoy, G., Johansen, T., et al. (2010). Autophagic degradation of dBruce controls DNA fragmentation in nurse cells during late *Drosophila melanogaster* oogenesis. *The Journal of Cell Biology*, *190*, 523–531.

Nezis, I. P., Simonsen, A., Sagona, A. P., Finley, K., Gaumer, S., Contamine, D., et al. (2008). Ref(2)P, the *Drosophila melanogaster* homologue of mammalian p62, is required for the formation of protein aggregates in adult brain. *The Journal of Cell Biology*, *180*, 1065–1071.

Nicolson, S., Denton, D., & Kumar, S. (2015). Ecdysone-mediated programmed cell death in *Drosophila*. *The International Journal of Developmental Biology*, *59*, 23–32.

Pandey, U. B., Batlevi, Y., Baehrecke, E. H., & Taylor, J. P. (2007). HDAC6 at the intersection of autophagy, the ubiquitin-proteasome system and neurodegeneration. *Autophagy*, *3*, 643–645.

Rusten, T. E., Lindmo, K., Juhasz, G., Sass, M., Seglen, P. O., Brech, A., et al. (2004). Programmed autophagy in the *Drosophila* fat body is induced by ecdysone through regulation of the PI3K pathway. *Developmental Cell*, 7, 179–192.

Rusten, T. E., Vaccari, T., Lindmo, K., Rodahl, L. M., Nezis, I. P., Sem-Jacobsen, C., et al. (2007). ESCRTs and Fab1 regulate distinct steps of autophagy. *Current Biology*, *17*, 1817–1825.

Sambrook, J., & Russell, D. W. (2001). *Molecular cloning: A laboratory manual* (3rd ed.). Cold Spring Harbour, NY: Cold Spring Harbour Laboratory Press.

Scott, R. C., Juhasz, G., & Neufeld, T. P. (2007). Direct induction of autophagy by Atg1 inhibits cell growth and induces apoptotic cell death. *Current Biology*, *17*, 1–11.

Scott, R. C., Schuldiner, O., & Neufeld, T. P. (2004). Role and regulation of starvation-induced autophagy in the Drosophila fat body. *Developmental Cell*, 7, 167–178.

Seglen, P. O., & Bohley, P. (1992). Autophagy and other vacuolar protein degradation mechanisms. *Experientia*, *48*, 158–172.

Shelly, S., Lukinova, N., Bambina, S., Berman, A., & Cherry, S. (2009). Autophagy is an essential component of Drosophila immunity against vesicular stomatitis virus. *Immunity*, *30*, 588–598.

Wyers, F., Dru, P., Simonet, B., & Contamine, D. (1993). Immunological cross-reactions and interactions between the *Drosophila melanogaster* ref(2)P protein and sigma rhabdovirus proteins. *Journal of Virology*, *67*, 3208–3216.

Xu, T., Nicolson, S., Denton, D., & Kumar, S. (2015). Distinct requirements of Autophagy-related genes in programmed cell death. *Cell Death and Differentiation*, *22*, 1792–1802.

Xu, T., & Rubin, G. M. (1993). Analysis of genetic mosaics in developing and adult *Drosophila* tissues. *Development*, *117*, 1223–1237.

Zhang, H., & Baehrecke, E. H. (2015). Eaten alive: Novel insights into autophagy from multicellular model systems. *Trends in Cell Biology*, *25*, 376–387.

CHAPTER TWENTY-FOUR

Methods to Study Autophagy in Zebrafish

E. Fodor, T. Sigmond, E. Ari, K. Lengyel, K. Takács-Vellai, M. Varga[1], T. Vellai[1]

Eötvös Loránd University, Budapest, Hungary

[1]Corresponding authors: e-mail address: mvarga@ttk.elte.hu; vellai@falco.elte.hu

Contents

Abstract

Autophagy (cellular self-eating) is a highly regulated degradation process of the eukaryotic cell during which parts of the cytoplasm are delivered into, and broken down within, the lysosomal compartment. The process serves as a main route for the elimination of superfluous and damaged cellular constituents, thereby mediating macromolecular and organellar turnover. In addition to maintaining cellular homeostasis, autophagy is involved in various other cellular and developmental processes by degrading specific regulatory proteins, and contributing to the clearance of intracellular pathogens. The physiological roles and pathological involvement of autophagy can be effectively studied in divergent eukaryotic model systems ranging from yeast to mice. Such a tractable animal model—applied only recently for autophagy research—is the zebrafish *Danio rerio*, which also facilitates the analysis of more specific biological processes such as tissue regeneration. In this chapter, we overview the main methods and

Methods in Enzymology, Volume 588
ISSN 0076-6879
http://dx.doi.org/10.1016/bs.mie.2016.10.028

tools that are used to monitor autophagic structures and to assay autophagic responses in this vertebrate organism. We place emphasis on genetic (functional) approaches applied for exploring novel cellular and developmental roles of the autophagic process.

1. INTRODUCTION

Autophagy (cellular "self-eating," also called lysosome-mediated self-degradation) has become one of the most intensively studied cellular processes in the last decade; the origins of autophagy research include the discovery of autophagic structures by electron microscopy in 1957, the definition of "autophagy" as a lysosome-mediated process by Christian de Duve in 1967, the first genetic screens for identifying autophagy-related genes in the unicellular yeast in the 1990s, the initial implication of autophagy in human pathologies in the same decade, and the characterization of the first autophagy-deficient metazoan systems in 2003 (Clark, 1957; Deter & de Duve, 1967; Juhász, Csikós, Sinka, Erdélyi, & Sass, 2003; Liang et al., 1999; Meléndez et al., 2003; Schlumpberger et al., 1997; Scott et al., 1996; Tsukada & Ohsumi, 1993). Autophagy functions as a major catabolic process of eukaryotic cells, which delivers damaged (dysfunctional), superfluous, or regulatory components of the cytoplasm into lysosomes for enzymatic degradation (Mizushima, Levine, Cuervo, & Klionsky, 2008). Depending on the mechanism of delivery, three major classes of autophagy can be distinguished: microautophagy, chaperone-mediated autophagy, and macroautophagy. Quantitatively the most significant form of autophagy is macroautophagy (hereafter referred to as autophagy) during which a growing double-membrane (the so-called isolation membrane) sequesters the cytoplasmic materials destined for degradation. When membrane growth is completed, a vacuole-like structure called autophagosome is formed. The autophagosome then fuses with a lysosome to form an autolysosome in which the degradation by acidic hydrolases takes place. The end products of autophagic breakdown can be reused in the synthetic processes and provide energy for the survival of cell under prolonged starvation. Sequestration of autophagic substrates occurs either without or with target specificity (general or selective autophagy) (Kirkin et al., 2009; Klionsky, 2005). The former is considered as a bulk, unspecific degradation process while the latter requires specific adaptor/receptor proteins that recognize the target molecule, organelle, or infective agent.

The core molecular machinery of autophagy involves several distinct autophagy-related protein (ATG) complexes: (i) an initiation complex formed around the ATG1/UNC-51 protein kinase core; (ii) a membrane-growing complex containing the class III phosphatidylinositol-3 kinase (PtdIns-3K); (iii) a multiple protein conjugation complex that covalently links the ubiquitin-like protein ATG8-I/LC3B-I (soluble form) to the phosphatidylethanolamine (PE) component of the growing autophagosomal membrane (ATG8-II, also called ATG8-PE; insoluble form); and (iv) a protein recycling system that salvages several ATG proteins (Klionsky, 2005).

Both cellular and developmental functions of autophagy are rather diverse (Levine & Klionsky, 2004; Levine & Kroemer, 2008; Mizushima & Levine, 2010; Mizushima et al., 2008). It primarily maintains cellular homeostasis by ensuring macromolecular and organellar turnover under constantly changing environmental conditions, rejuvenates cellular constituents by degrading dysfunctional, damaged organelles, and macromolecules—in particular misfolded, oxidized, and aggregated proteins—(intracellular quality control), and specifically eliminates regulatory proteins (e.g., maternal effect factors or certain components of the miRNA pathway) and organelles (e.g., sperm-derived paternal mitochondria, thereby contributing to maternal inheritance of the organelle) at early stages of embryogenesis (Al Rawi et al., 2011; Mizushima & Levine, 2010; Sato & Sato, 2011; Zhang & Zhang, 2013). Due to its major function in the elimination of cellular damage, autophagy also plays a central role in the regulation of the aging process (Tóth et al., 2008; Vellai, 2009; Vellai, Takács-Vellai, Sass, & Klionsky, 2009). In humans, defects in autophagy are implicated in the development of various age-dependent degenerative diseases. Such pathologies include different types of cancer, neurodegenerative diseases, tissue atrophy (e.g., sarcopenia) and fibrosis, defective lipid metabolism (e.g., obesity), diabetes, immune deficiency, and infection by intracellular pathogens (Levine & Kroemer, 2008; Mizushima et al., 2008; Rubinsztein, 2006; Takacs-Vellai, Bayci, & Vellai, 2006). Not surprisingly, autophagy has emerged as a major cellular target of pharmacological interventions in order to treat such disorders (Billes et al., 2016; Fleming, Noda, Yoshimori, & Rubinsztein, 2011; Papp et al., 2016; Tan et al., 2014).

The function, mechanism, and regulation of autophagy, and its potential contribution to degenerative diseases, can be examined in tractable animal model systems such as the nematode *Caenorhabditis elegans*, the fruit fly *Drosophila melanogaster*, and the mouse *Mus musculus* (Kourtis & Tavernarakis, 2009; Neufeld & Baehrecke, 2008). In recent years, however, zebrafish (*Danio rerio*) has also become an effectively used vertebrate model to study different aspects of the autophagic process (reviewed in Varga, Fodor, & Vellai, 2015). This organism is particularly suitable to explore novel functions for autophagy in regeneration, differentiation, and development (for example, zebrafish is capable of regenerating its caudal fin after amputation and this process also involves the function of *atg* genes; Varga et al., 2014). In this chapter, we review the methods and tools that are frequently used for monitoring autophagy and assessing autophagic responses in *D. rerio*. This collection may be profitable for further research in this field.

2. MONITORING AUTOPHAGY IN ZEBRAFISH

The detection of autophagic activity in zebrafish can be achieved at multiple levels, including both molecular and cellular methods (Klionsky et al., 2016). The former is mainly represented by Western blotting and PCR-based transcription analyses while the latter generally involves fluorescence and electron microscopy.

2.1 Western Blotting

Among methods used in zebrafish autophagy research, Western blotting is most often applied to measure the levels of membrane-bound form of ATG8/LC3B (Atg8-PE/LC3B-II) (Klionsky et al., 2016). However, SQSTM1/P62 (a substrate of autophagy), BECLIN 1/ATG6, ATG5, and ATG7 (components of the core autophagic pathway) levels have also been estimated with this technique and used to characterize the autophagic process. These latter approaches are advised to be used rather as additional methods, as none of these markers can be considered specific for autophagy; therefore, changes in their expression level might not be related to the modulation of the process.

Compared with control samples, an increase in LC3B-II levels can be either a marker of increased autophagic activity, or it could be the result of decreased lysosomal turnover. To be able to differentiate between these two alternatives, when determining the autophagic flux by Western blot it is always advisable to repeat the experiment in the presence of lysosomal inhibitors (Mizushima & Yoshimori, 2007). Likewise, when performing an inhibitor assay one should carefully determine the optimal concentration for the compound, as not only too low, but too high concentrations/exposure times can mask the original effect on autophagy (Moreau et al., 2014). In some cases, LC3B-II is much more abundant than LC3B-I or the latter might seem to be even undetectable. As the molecular weights of the two forms are similar, it can be challenging to tell which one is missing when only one band is visible. A handful of ideas have been proposed to explain this phenomenon, which include: different binding specificity of the LC3B antibodies to the two forms of LC3B, different migration capacities on different membranes, or lower stability of LC3B-I than of LCB3-II (Klionsky et al., 2016).

Even when the same antibody has been used in separate experiments, sometimes it is difficult to compare the results due to the differential experimental circumstances. In general, we could not find evidence that using one or another commercially available antibody will increase the chances in properly detecting the two bands. However, there is evidence that optimizing detection methods can tremendously affect the outcome. For example, intensity of LC3B-I can be quite different even if the same antibodies and developmental stages are used probably at least partly due to the different antibody concentrations applied. LC3B-I might not be visible with shorter exposure times, but easily detectable with longer ones (Clancey, Beirl, Linbo, & Cooper, 2013). Membrane choice might also have a great impact on the final outcome of the procedure. Experimental evidence shows that PVDF membranes seem to bind LC3B-II more effectively, while nitrocellulose membranes have the ability to retain more LC3B-I. Thus, it might be challenging to ensure that both forms are in the linear range, making proper quantification almost impossible. Because of these aforementioned reasons comparing LC3B-II with LC3B-I levels might give misleading results; therefore, we recommend comparing LC3B-II levels with a housekeeping control, which is a more reliable approach (Barth, Glick, & Macleod, 2010). Antibodies that have been used for Western blotting to monitor autophagic activity in zebrafish were summarized in Varga et al. (2015).

2.2 PCR-Based Assays

Measuring changes in the transcript levels of *atg* genes does not provide a clear picture about the autophagic activity by itself. An elevated level of a specific mRNA might not only be due to the induction of autophagy, it could also reflect a compensatory mechanism (Hu, Zhang, & Zhang, 2011). When analyzing qPCR data, one should always keep in mind that autophagy is regulated through posttranslational modifications as well (Xie et al., 2015). That said, measuring transcriptional changes has been validated on several occasions and it can complement other methods.

PCR-based assays have been widely applied in various aspects of zebrafish autophagy research (Table 1). Real-time PCR has been used to

Table 1 List of Primers Used to Assess the Expression of Autophagy-Related (*atg*) Genes in Zebrafish

Primer Sequence (5′ to 3′)	References
ambra1a	
F: CTGCTGCTCATTGCCACC R: CGCATCTCCACACTGTCC	Benato et al. (2013)
F: CTGCTGCTCATTGCCACC R: CGCATCTCCACACTGTCC	Miccoli et al. (2015)
F: TCTTTCGAGAAATGGCACCT R: CTCTCTGCGTTAGGGACAGG	Santangeli et al. (2016)
ambra1b	
F: AGGTGACGGACAGTCAGC R: CCTACCATCACATAGCAGC	Benato et al. (2013)
F: GCATACCACGTCAGACTCG R: CCTACCATCACATAGCAGC	Miccoli et al. (2015)
atg3	
F: GGCTGTTTGGATATGATGAG R: AGCAGGTGGAGGGAGATTAG	Zheng et al. (2015)
atg4a	
F: CGCGGCTGTGGTTCACTTAT R: GATCCCACCTCCAATCTCGG	Huang, Zhang, Ye, and Wang (2016)
atg4d	
F: TTCATGTCGGCCTGGAACAA R: CATGACACAAACGTCTGCCG	Huang et al. (2016)
F: GCTCATGAGGACAAGGCTTC R: GCACCTCGAAATCCACATCT	Kyöstilä et al. (2015)
atg5	
F: GATTGCTGCCTGCTACTTCC R: CTCTGCTAAGGGACCGACTG	Mohanty et al. (2015)
F-ex2: TGACAAGGATGTGCTTCGAG R-ex3: ACCACATTTCCTCCACATCC R-in2: TTTAACAACCAAATGAACACTTATGTCT ATTCAACTG	Lee et al. (2014)
F: ATGATAATGGCAGATGACAAGG R: TCAGTCACTCGGTGCAGG	Hu et al. (2011)
F: AGAGAGGCAGAACCCTACTATC R: CCTCGTGTTCAAACCACATTTC	Elenbaas et al. (2016)

Table 1 List of Primers Used to Assess the Expression of Autophagy-Related (*atg*) Genes in Zebrafish—cont'd

Primer Sequence (5′ to 3′)	References
atg6vld	
F: TCAGGAGGGAGGAGACAGTTA R: TAGCCTCGTCCAGAGTCACA	Huang et al. (2016)
atg7	
F-ex1: GATTCTGGCATCAGCTCACA R-ex2: GCATCAAATGCGCTGAACT R2-ex2: TTTGTCGGTGGATTTGAAGG R-in1: AAGCGGGTAAGGTTAATATTGCT	Lee et al. (2014)
atg10	
F: ATGACCGGTGAGAGAAAGCCT R: AGCCTTCATCAGAGCCCTTGA	Tsai, Chen, Chen, and Wang (2013)
F: GCAGCTATCAGATCCCCGTC R: GAGGATGTTCCTGCTGTGTCA	Huang et al. (2016)
atg12	
F: ATGTCTGACAACGCAGAATC R: TCATCCCCAGGCCTGAGACTT	Hu et al. (2011)
F: TTCATCTCACGCTTCCTCAA R: CGTCACTTCCGAAACACTCA	Zheng et al. (2015)
F: TATGTTAATCAGTCGTTTGCGCC R: CTAGTTTGCCGTCACTTCCGA	Tsai et al. (2013)
atg16l	
F: AATTCGTTCAGCTCGTCTCC R: CAGCGTTCACTTCTCCATCA	Elenbaas et al. (2016)
beclin-1	
F: GGCAGTGAAGAGTCCAGGAG R: ACTTAGCAGTCAGGGGCAGA	Mohanty et al. (2015)
F-ex1: ACCCACTTTGTGAGGAGTGC R-ex2: GTCCCTCATCCAGCTCTTTG	Lee et al. (2014)
F: GGACCACTTGGAACAACT R: CCGAAGTTCTTCAGTGTCCATC	Benato et al. (2013)
F: GCATACCACGTCAGACTCG R: CCTACCATCACATAGCAGC	Miccoli et al. (2015)

Continued

Table 1 List of Primers Used to Assess the Expression of Autophagy-Related (*atg*) Genes in Zebrafish—cont'd

Primer Sequence (5′ to 3′)	References
F: GATCATGCAATGGTGGCTTTC R: CCTCCTGTGTCCTCAATCTTT	Elenbaas et al. (2016)
F1-ex3: CAAACAAGATGGCGTGGCTCGAAA F2-ex4: GTGGAACTATGGAGAACTTGAGTCGCA R-ex7: TCCAACTCCAGCTGCTGTCTCTT	Sasaki et al. (2014)
F: AGAGCATTGAGACAAAGCGTGAA R: TCTGCCAAGGCGGAAGTTATT	Zheng et al. (2015)
F: GGCTTTCCTTGACTGTGTCC R: CCTTTGTCCACATCCATTCTG	Jia et al. (2015)
F: GGACCACTTGGAACAACT R: CCGAAGTTCTTCAGTGTCCATC	Santangeli et al. (2016)
F: AGTCGCAGACTGAAAGTGACA R: TCTGGCACTCGTTCTCAGTG	Huang et al. (2016)
gabarapa	
F: GTCTGACCTCACAGTTGGGC R: TCCTGGTAGAGCAGTCCCAT	Huang et al. (2016)
map1lc3a	
F: CGAGTCGACCGACAATTTAGC R: TCCTTGCAACGATCAGCGAA	Ganesan, Moussavi Nik, Newman, and Lardelli (2014)
map1lc3b	
F: AATGTGACGATTGGACACGAGT R: AGTACAACAGCTCACGGTTATGC	Ganesan et al. (2014)
F: TCCAAACAAGATCCCGGTCA R: GACCAGCAGGAAGAAAGCCT	Huang et al. (2016)
mtor	
F: TTATCGTGCTGGTCCGAGCT R: AAGTGGGCCCTTATCGCTGT	Tsai et al. (2013)
p62	
F: CGATGTTTTTGTCGGTCTCA R: CAAGAGCCAAACCCATCATT	Jia et al. (2015)
ulk1b	
F: GGCAACTATGGGCAGTCTGT R: ACCTGTGGAGAGAGCTGGAA	Lee et al. (2014)

show that a handful of *atg* genes have a circadian expression pattern (Huang et al., 2016), that bacteria can effect expression of genes involved in autophagy and apoptosis (Miccoli et al., 2015), or that myobacterial phosphorybosiltransferase can inhibit autophagy in zebrafish macrophages (Mohanty et al., 2015). Similarly, among other techniques, real-time PCR helped to prove that in *kri1l* mutant fish, defective in the ribosomal small unit assembly, excessive autophagy contributes to hematopoietic stem cell/progenitor cell death (Jia et al., 2015).

Semiquantitative PCR has also been used to confirm that the most important *atg* genes are maternally deposited. During embryonic development, some *atg* genes (*becn1*, *ulk1b*, *atg12*, and *map1lc3a*) show stable activity, while the expression of others (*atg5*, *atg7*, *ambra1a1*, *ambra1a2*, *ambra1b*, and *map1lc3b*) is much more dynamic.

2.3 Fluorescence Microscopy (Reporters)

The body of the zebrafish embryo and larva is largely transparent, which enables the efficient studying of the autophagic process by fluorescence microscopy in this organism. The animal remains transparent even in later developmental stages if it is treated with 1-phenyl-2-thiourea (PTU), or carries a mutation causing defects in pigmentation. Basically two different approaches exist to express reporter transgenes in zebrafish. First, the microinjection of mRNAs coding for the reporter construct into embryos (Jia et al., 2015). When zygotes are injected, the transgene can usually be expressed in each somatic cell and remains detectable for an average of 3 days. Therefore, this method is not suitable for monitoring autophagic activity in later developmental stages and adulthood. Second, the generation of stable transgenic lines by using, most frequently, the Tol2 system. This allows both ubiquitous (e.g., CMV promoter-driven) and tissue-specific (e.g., liver- or photoreceptor-specific promoter-driven) expression activities at any life stages. Currently, the most frequently used reporter system in zebrafish autophagy research is Tg(*CMV:GFP-map1lc3*) (He, Bartholomew, Zhou, & Klionsky, 2009) (Fig. 1). When autophagy is induced, the reporter protein accumulates in distinct foci that label autophagic structures, mainly autophagosomes.

As GFP is sensitive to the acidic environment of lysosomal lumen, other reporters being largely insensitive to pH, such as mCherry and RFP, can be used to monitor autophagic flux unambiguously. For example, fish strains transgenic for an mCherry-GFP-LC3B tandem construct

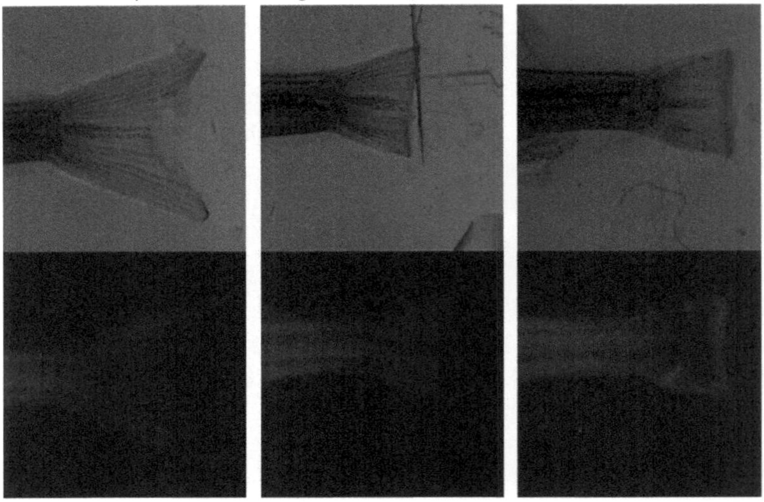

Fig. 1 MAP1LC3B accumulation in the regenerating caudal fin tissue, blastema. Expression of a GFP-LC3B (Tg(*CMV:GFP-map1lc3b*) reporter in caudal fin before (*left*), right after (*middle*), and 2 days later (*right*) of amputation. LC3B abundantly accumulates in the regenerating tissue called bastema (*intense green color*). Light microscopic images (*top*) and the corresponding fluorescence pictures (*bottom*) are shown (M. Varga, unpublished results).

(Tg(*TαCP:mCherry-GFP-map1lc3b*) were generated and assayed in wild-type vs mutant genetic backgrounds (George, Hayden, Stanton, & Brockerhoff, 2016) (Fig. 2).

Initial steps of the autophagic membrane formation involve the conversion of PtdIns(3)P to PtdInsI(3,5)P_2 by PtdIns-3K. Reporters that label the FYVE zinc-finger domain capable of binding PtdIns(3)P are also indicative for autophagic responses. Such reporters are represented by Tg(*TαCP:YFP-2XFYVE*), Tg(*TαCP:mCherry-ML1NX2*), and Tg(*TαCP:tRFP-t-2XFYVE*) (George et al., 2016). Reporter constructs that have been used to detect autophagic structures in zebrafish were summarized in Varga et al. (2015). In the near future, novel reporters targeting ATG proteins other than LC3B will likely to be constructed. In addition, expression of these reporters should be driven by endogenous promoters.

Selective autophagy is still a relatively unexplored area in zebrafish research and only few transgenic tools have been developed to date that could be reliably used to monitor this process. One possibility is to combine existing autophagy reporter transgenic lines such as Tg(*CMV:GFP-map1lc3b*) with transgenic tools that mark the cargo. For example,

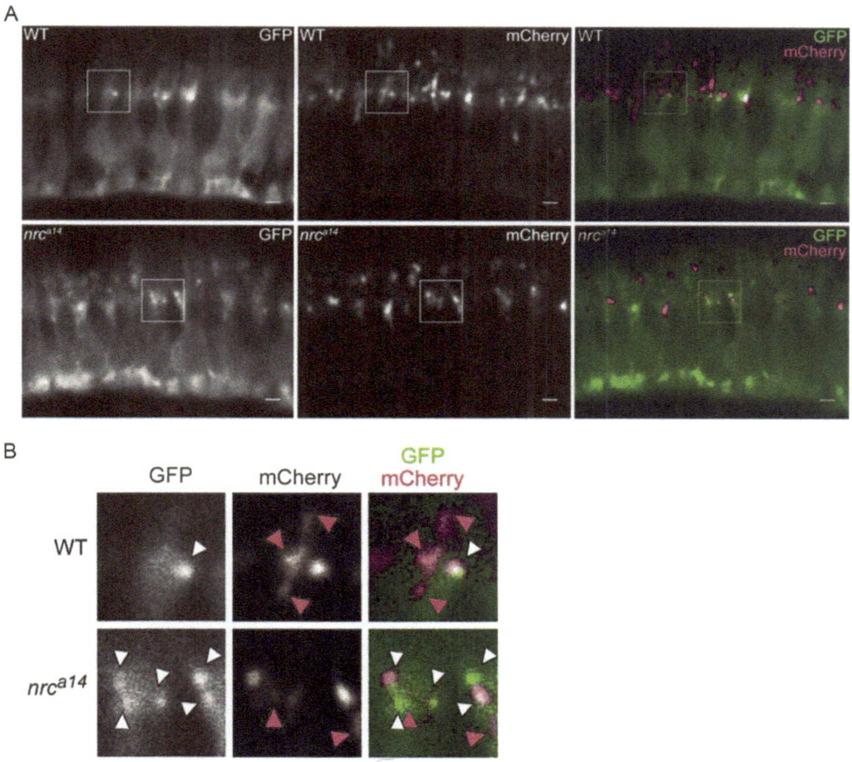

Fig. 2 Expression of an *mCherry-GFP-map1lc3b* reporter transgene in cone photoreceptors of wild-type and nrc^{a14} mutant zebrafish larvae at 5 dpf. (A) In wild-type (WT) cone photoreceptors, autoplysosomes (magenta-only foci) are visible. Fewer magenta-only punta are visible in nrc^{a14} mutant cones, indicating that autophagosomes had remained largely unfused with acidic compartments in the mutant cones. (B) Enlargement of areas shown in boxes in the panel (A). Magenta-only foci indicate autolysosomes. nrc^{a14} mutants are defective in the polyphosphoinositide phosphatase Synaptojanin 1 that is implicated in proper protein degradation. *dpf*, days postfertilization. *Adapted from George, A. A., Hayden, S., Stanton, G. R., & Brockerhoff, S. E. (2016). Arf6 and the 5′phosphatase of synaptojanin 1 regulate autophagy in cone photoreceptors.* Inside Cell, 1, *117–133.*

transgenic bacteria expressing a red fluorescent protein can be used to monitor the autophagy-dependent clearance of mycobacteria (van der Vaart et al., 2014) or *Shigella* (Mostowy et al., 2013) from zebrafish larvae. It has been also suggested that stable transgenic lines with fluorescently labeled mitochondria can be used to observe mitophagy in a living larvae (Wager & Russell, 2013). In all these cases, after the respective infection/cross protocol is carried out, zebrafish larvae have to be prepared for confocal microscopy as described later in the text.

2.4 Transmission Electron Microscopy

Transmission electron microscopy (TEM) is the classical and still the most obvious method to detect autophagic structures (Klionsky et al., 2016). Although direct visualization of autophagosomes and autolysosomes enables us to identify even subtle changes in autophagic responses, the method is quite laborious and requires a significant experience from the researcher. In zebrafish research, TEM is generally used as an additional technique to confirm results obtained by transgene reporters and protein detection-based molecular methods. TEM was applied, for example, in assessing the role of autophagy in zebrafish embryogenesis (Lee et al., 2014), in cellular response to bacterial infection (Hosseini et al., 2014; Mostowy et al., 2013), in Atrogin function (Bühler et al., 2016), and in caudal fin regeneration (Hosseini et al., 2014; Varga et al., 2014) (Fig. 3). Ultrastructural analysis of the autophagic process will certainly remain a preferred tool for high-quality future research in this field.

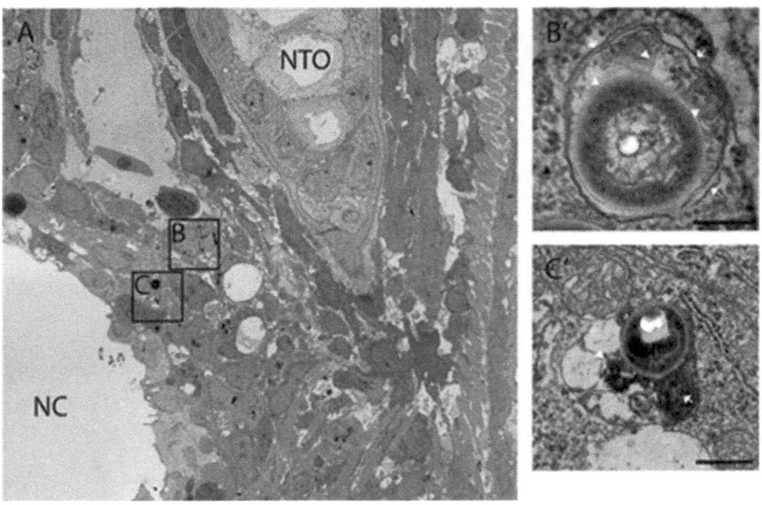

Fig. 3 TEM images of caudal fin of a zebrafish infected with a pathogen bacterium. (A) Overview image of a granuloma with necrotic center in the caudal fin at 5 dpi. (B) Higher magnification of a bacterium in an initial autophagic vacuole. (C) Higher magnification of region indicated in (A), showing a single bacterium in a degradative autophagic vacuole. *dpi*, days postinfection; *NC*, necrotic center; *NTO*, notochord; *TEM*, transmission electron microscopy. Scale bars indicate 1 μm. *Adapted from Hosseini, R., Lamers, G. E., Hodzic, Z., Meijer, A. H., Schaaf, M. J., & Spaink, H. P. (2014). Correlative light and electron microscopy imaging of autophagy in a zebrafish infection model.* Autophagy, 10, *1844–1857.*

3. MODULATION OF AUTOPHAGIC RESPONSES IN ZEBRAFISH

During the past couple of decades, efficient sequence-based technologies have been developed to identify and functionally analyze components of the core autophagy pathway. These include mainly bioinformatics (in silico analysis) and genetic approaches.

3.1 Zebrafish Autophagy Genes

In the era of reverse genetics, the functional characterization of ATG proteins requires first the identification of the corresponding coding (*atg*) sequences. Ortholog proteins in *D. rerio* were searched using the following protocol (Table 2). To find the RefSeq IDs (NCBI Reference Sequence; NCBI Resource Coordinators, 2016) of human proteins, we searched them on UniProt database (UniProt Consortium, 2015) by their protein names. Each UniProt entry contains the adherent RefSeq IDs of the protein among other cross-references. If there were multiple RefSeq IDs connected to a single UniProt protein, the one with the minimum NP number was chosen since that one was annotated earlier, therefore should be the canonical isoform. Then, we applied BLASTP (Altschul et al., 1997) sequence similarity search using the human protein ortholog as query sequence and *D. rerio*'s (taxid: 7955) reference proteins as search set. The first relevant hit with the smallest *E* value and the highest query coverage were considered as ortholog. If, among the best hits, there was a relevant annotated RefSeq hit (RefSeq ID starting with NP) as well as predicted proteins (RefSeq ID starting with XP) the annotated RefSeq protein was chosen as ortholog. The corresponding gene names were translated from the RefSeq IDs using the ID mapping function of the UniProt webpage. *D. rerio* genes in Table 2 without *E* value and identity percentage were described previously as othologs of autophagy genes in yeast or human.

3.2 Genetic Approaches

3.2.1 Gene Silencing With Morpholino Oligonucleotides

In the absence of *bona fide* autophagy mutants, early efforts to discern the role of autophagy in zebrafish biology were dominated by reverse genetic approaches based on the use of antisense morpholino oligonucleotides (MOs) (Table 3). These synthetic oligos can be easily injected into the developing embryos, where they interfere with gene expression, either by

Table 2 Zebarfish (*Danio rerio*) Orthologs of Yeast and Human Autophagy Genes

Function	Yeast	Human	*Danio rerio*	E Value	Identity
Adaptor/Receptor					
		Sqstm1/p62	*Sqstm1/p62*		
		Nbr1			
		Optn			
		Ndp52	*XP_005158014.1**	5.00E−44	42%
		Tax1bp1	*Tax1bp1a/b*	0	52%
	Atg32	*Fundc1*	*Fundc1*	2.00E−67	71%
	Atg32	*Bnip3*	*Bnip3La/b*	2.00E−76	62%
	Atg32	*Nix*			
	Atg34				
	Atg19b				
Induction					
	Atg1	*Ulk1, Ulk2*	*Ulk1a Ulk1b, Ulk2*		
	Atg13	*Kiaa0652*	*Atg13**	0	72%
	Atg101	*Atg101*	*Atg101*	5.00E−147	87%
	Atg11	*Htt*	*Huntingtin*	0	71%
	Atg17	*Flip200*	*Rb1cc1**	0	59%
Nucleation					
	Atg6	*Beclin 1*	*Beclin1*		
	Atg14	*Atg14*	*NP_001019983.1*	0	67%
	Vps34	*PtdIns-3K/Vps34*	*Pik3c3**	0	87%
	Vps15	*Vps15*	*Scpep1**	0	82%
		Ambra1	*Ambra1a Ambra1b*	0	53%
	Vps38	*Uvrag*	*Uvrag*	0	58%
	Mr1	*Mtmr14/Jumpy*	*Mtmr14**	0	72%
		Naf1	*Naf1**	2.00E−74	61%

Table 2 Zebarfish (*Danio rerio*) Orthologs of Yeast and Human Autophagy Genes—cont'd

Function	Yeast	Human	*Danio rerio*	E Value	Identity
Membrane formation					
	Atg18	*Wipi1/2*	*Wipi1*		
		Wipi3/4	*Wipi2*	0	96%
	Atg2	*Atg2*	*XP_009306033.1**	0	55%
			*XP_009306033.1**	0	42%
	Atg9	*Atg9*	*Atg9a Atg9b*	0	42%
		Vmp1	*Vmp1*	0	73%
		Ei24	*Ei24*	0	84%
	Pep12	*Stx7*	*Stx7**	8.00E−89	59%
		Stx8	*Stx8*	1.00E−92	58%
Membrane elonganation					
	Atg8	*Gabarap*	*Gabarap*		
	Atg8	*Lc3*	*Map1lc3a/b/c*	3.00E−82	96%
	Atg12	*Atg12*	*Atg12*		
	Atg3	*Atg3*	*Apg3l*		
	Atg4	*Atg4a/b*	*Atg4a Atg4b*		
	Atg4	*Atg4c/d*	*Atg4c*		
	Atg5	*Atg5*	*Atg5*		
	Atg7	*Atg7*	*Atg7*		
	Atg10	*Atg10*	*Atg10*		
	Atg16	*Atg16l1*	*Atg16l1*		
	Atg16	*Atg16l2*	*Atg16l2*		
		Flip	*Flip1**	0	65%
Autophag. maturation					
	Rab7	*Rab7*	*Rab7*	2.00E−150	98%
	Vam6	*Vps39*	*Vps39*		
	Vps41	*Vps41*	*Vps41*		
		Rubicon	*XP_009294779.1**	0	54%

Continued

Table 2 Zebarfish (*Danio rerio*) Orthologs of Yeast and Human Autophagy Genes—cont'd

Function	Yeast	Human	*Danio rerio*	E Value	Identity
		Vamp7	*XP_005157192.1**	9.00E−134	83%
		Vamp8	*Vamp8*	6.00E−30	69%
	Vti1	*Vti1B*	*Vti1B*	7.00E−97	63%
		Snap29	*Snap29*	7.00E−86	51%
		Stx17	*Syntaxin17*	4.00E−75	56%
		Mepg5	*Epg5*	0	61%

Where there is no BLAST value, see Fleming and Rubinsztein (2011). Where the gene name was ambiguous, we used the RefSeq ID of the gene. Predicted genes are marked with *.

masking the translational start site from the ribosome, or by binding to splice donor/acceptor sites, therefore, interfering with splicing. This relatively straightforward approach to knock down gene function has been used very efficiently, and several studies demonstrated that MOs can efficiently phenocopy known mutations (Nasevicius & Ekker, 2000).

In the past few years, MOs have been applied to the study of autophagy in zebrafish with vigor. Knockdowns of *ambra1a* and -*1b* (Benato et al., 2013; Skobo et al., 2014), *atg4da* (Kyöstilä et al., 2015), *atg5* (Hu et al., 2011; Lee et al., 2014), *atg7* (Lee et al., 2014), and *bcn1* (Lee et al., 2014) have all resulted in broadly similar, lethal, and severe phenotypic defects, suggesting a general role of autophagy in early embryogenesis. These embryos showed a shorter and bent trunk, often accompanied by pericardial edema and defects of cardiac morphogenesis and skeletal muscle development. In addition, they usually had small heads as the result of impaired neurogenesis.

The MO-driven depletion of gene products necessary for autophagosome maturation, such as Dram1, Spns1, or Snx14, have resulted in decreased survival upon bacterial infection (van der Vaart et al., 2014), embryonic senescence (Sasaki et al., 2014), and loss in cerebellar parenchyma (Akizu et al., 2015), respectively. Several efforts were also made to impair the function of autophagy-specific adaptor proteins, such as p62/Sqstm and optineurin, too. In these aforementioned studies, MO-based depletion resulted in increased susceptibility for bacterial infections (Chew et al., 2015; Mostowy et al., 2013). For optineurin knockdowns defects in axonal vesicle trafficking (Paulus & Link, 2014) and axonopathy (Korac et al., 2013) were also observed.

Table 3 Summary of MO Sequences Used to Silence Autophagy-Related Genes in Zebrafish

MO Sequence (5′ to 3′)	MO Type	References
ambra1a		
CTC CAA ACA CTC TTC CTC ACT CCC T	ATG	Benato et al. (2013) and Skobo et al. (2014)
TGT AAT CAA AGT GGT CTT ACC TGT C	Splice	
ambra1b		
TTT TCC TCT TTA GTG CTC CAC GGC C	ATG	Benato et al. (2013) and Skobo et al. (2014)
TGA AAT TGA TTG TTA CCT ATC TGG A	Splice	
atg4da		
CGG TCC AGC CTG AGA AAA TAA AAG A	Splice	Kyöstilä et al. (2015)
atg5		
CAT CCT TGT CAT CTG CCA TTA TCA T	ATG	Boglev et al. (2013)
CAT CCT TGT CAT CTG CCA TTA TCA T	ATG	Hu et al. (2011)
GTG CCC TTA AAA CCA AAA ATA ACA C	Splice	
CCT TGT CAT CTG CCA TTA TCA TCG T	ATG	Varga et al. (2014)
CAC ATC CTT GTC ATC TGC CAT TAT C	ATG	Lee et al. (2014)
ATT CCT TTA ACT TAC ATA GTA GGG T	Splice	
atg7		
AGC TTC AGA CTG GAT TCC GCC ATC G	ATG	Lee et al. (2014)
AGC TCG TTC TCC AAA CTC ACC GTT A	Splice	
beclin1		
ACC TCA AAG TCT CCA TGC TTC TTT C	ATG	Lee et al. (2014)
TGT TAT TGT GTG TTA CTG ACT TGT A	ATG	
CAT CCT GCA AAA CAC AAA TGG CTT A	Splice	Sasaki et al. (2014)
dram1		
AAG GCT GGA AAA CAA ACG TAC AGT A	Splice	Sasaki et al. (2014)
GTC GTC TCC TGT AAC AAA ACA TGC A	Splice	

Continued

Table 3 Summary of MO Sequences Used to Silence Autophagy-Related Genes in Zebrafish—cont'd

MO Sequence (5′ to 3′)	MO Type	References
optn		
CGA TGA TCC AGA TGC CAT GCT TTC T	ATG	Korac et al. (2013)
AAA TTT CTC TCA CCT CAG CTC CAC T	Splice	
TGT CCC CAT TCA TCA TCG ATG ATC C	ATG	Paulus and Link (2014)
TAA CCC GCA CCT TTC AGG TCT CGG T	Splice	
AGA GCC TCT GTG GGA TGC ATA TAA T	Splice	Chew et al. (2015)
p62/sqstm1		
CAC TGT CAT CGA CAT CGT AGC GGA A	ATG	Mostowy et al. (2013)
CTT CAT CTA GAG ACA AAG TTC AGG A	Splice	van der Vaart et al. (2014)
snx14		
GTC CGA CAT TAT TCC TCA CGG ATG A	ATG	Akizu et al. (2015)
spns1		
ATC TGC TTG TGA CAT CAC TGC TGG A	ATG	Sasaki et al. (2014)

More recently, a novel class of MOs has been developed, where special delivery moieties are tethered to the synthetic oligos, thus when introduced in the tissues, they can be absorbed by the surrounding cells. Such "vivo-MOs" can be used for targeted knockdown of genes in adults, too (Chablais & Jazwinska, 2010). Using this approach, a role for autophagy (more specifically Atg5) during adult caudal fin regeneration was recently demonstrated (Varga et al., 2014).

Although most studies performed with MOs have produced broadly similar phenotypes (which would suggest specificity), it is notable that they are also reminiscent of previously described off-target MO effects (Robu et al., 2007). This has become a major point of worry in the field recently. As the molar amount of MOs injected usually exceeds that of the target mRNAs by several orders of magnitude, it is possible that less-specific phenotypes are due to the (unspecific) effects of the unbound oligos (Schulte-Merker & Stainier, 2014). To complicate things further, even when the mutants do not show the same phenotype as the morphants, one cannot rule out the specificity of the MO. A recent study suggested that in some mutants the effect of genomic compensation might be able to make up for the loss of the mutated gene (Rossi et al., 2015). As this effect is not observed for MO-treated embryos, phenotypic differences will be evident between the two classes of functional knockdowns. To tackle this specificity problem, the field has adopted a more rigorous approach to MOs. A majority of zebrafish researchers now are in favor of accepting MO results only when there is either a mutant with an identical phenotype or the injection of the MO into the mutant background has no visible phenotypic effects.

3.2.2 Mutations

The power of zebrafish as a genetic model has been demonstrated decades ago in ENU-induced mutagenesis screens (Driever et al., 1996; Haffter, Granato, Brand, & Mullins, 1996). However, neither these original efforts nor the mutagenesis screens performed ever since have described in detail mutant alleles of core *atg* genes. One possible explanation for this, supported by the results of the past couple of years (Papp et al., 2016; Varga et al., 2014), could be the late onset of autophagy-related phenotypes in this organism. As most early screening efforts focused on easily observable phenotypes in the gross morphology of the embryo, it is possible that mutations in *atg* genes that, perhaps due to maternal effects, result in subtle later phenotypes were not picked up in these screens.

More recently, a systematic effort at the Sanger Institute (UK), the Zebrafish Mutation Project, has produced several putative hypomorphic and null-mutant alleles for core autophagy genes, including *ambra1a*, *atg5*, *atg7*, *beclin1*, and *ztor* (Kettleborough et al., 2013). However, a thorough characterization of these lines has not been performed yet.

3.2.3 Indirect Means to Interfere With Autophagy

In the past few years, a number of mutants have also been described, where an indirect attenuation of the autophagic process can be detected. For example, using two recently described models of disrupted ribosome biogenesis, *titania* (*tti*) and *rps7*, researchers have shown that autophagy has a role in cell survival (Boglev et al., 2013; Heijnen et al., 2014). Other examples include a TALEN-induced mutation in the nuclear receptor gene *nr1d1*, which results in the misexpression of *atg* genes and an increase in autophagosome density in larvae (an effect mediated through the regulation of the *ulk1a* promoter) (Huang et al., 2016), and the *nrc* mutant deficient in Synaptojanin 1

(SynJ1), where autophagy is blocked at the maturation step leading to an overabundance of autophagosomes in the cone photoreceptors of zebrafish larvae (George et al., 2016).

3.2.4 CRISPR/Cas9-Based Methods: A Sign of Things to Come?

Fortunately, just as the issues with the MO-based reverse genetic approach became evident, the novel genome-editing techniques TALEN and CRISPR/Cas9 also became available in zebrafish, and they are expected to revolutionize zebrafish genetics in the near future (Varshney, Sood, & Burgess, 2015). The most obvious use for precision genome-editing methods will be the creation of allelic loss-of-function series and floxed alleles for the core autophagy machinery genes. With these novel lines, we will be able to finally test the veracity of morphant phenotypes.

In the meantime, another experimental design, analogous to the RNAi approach, could be used to assess the tissue-specific roles of *atg* genes. For the genes, where very efficient gRNAs can be identified, stable transgenes could be created, with inducible (Gal4/UAS-dependent) Cas9 expression. In these lines, the tissue-specific induction of Cas9 could create biallelic edits of the respective genes in the surveyed tissues.

A further twist to this approach could come from the recently developed 2C-Cas9 method, combining the Cre/Lox and Cas9 systems to create tissue-specific mosaic knockdown of targeted genes (Di Donato et al., 2016). This method could be very powerful in identifying novel genes involved in autophagy. Some of these approaches could also be combined with nuclease-dead (dCas9)-based CRISPR interference (CRISPRi) methods, with the added benefit that potential genomic compensation effects could be overcome. All of these methods will require efficient sgRNAs. Therefore, the maintenance and expansion of existing gRNA databases is a priority for the field (Varshney et al., 2016).

3.3 Chemical/Pharmacological Applications

To date, several small molecules and pharmacological agents (drugs and drug candidates) have been identified that can potently enhance or inhibit the autophagic process. Since autophagy is implicated in various human pathologies and aging (Levine & Kroemer, 2008; Mizushima et al., 2008), these factors are potentially significant in respect of medical applicability. As an example, the immunosuppressant drug rapamycin impedes the autophagy inhibitor mammalian target of rapamycin (mTOR) kinase, thereby promoting autophagy. Rapamycin acts upstream of the autophagic pathway and targets TOR with multiple cellular functions. More specific modulators of autophagy act at different stages of the process. For example, Bafilomycin A1 impairs autophagosome—lysosome fusion. A nearly complete list of reagents/small molecules previously used to interfere with autophagy in zebrafish was presented in Varga et al. (2015).

Recently, testing a small molecule library, a screen was performed for more specific enhancers of the autophagic process. The screen targeted the myotubularin-like phosphatase Jumpy/MTMR14 that antagonizes PtdIns-3K and its autophagy membrane generating function (Vergne et al., 2009), and resulted few potent autophagy enhacer factors called AUTENs (Billes et al., 2016; Papp et al., 2016). At least two AUTEN molecules, AUTEN-67 and -99, are capable of strengthening basal autophagic activity in the fish model via inhibiting the zebrafish ortholog of mammalian MTMR14/Jumpy (Papp et al., 2016; Kovács et al., manuscript under revision). Together, these data imply that autophagy can be effectively modified in zebrafish by chemical substances, and that this organism is particularly suitable for small molecule screens to identify autophagy-inducing drug candidates.

4. EXPERIMENTAL PROCEDURES

4.1 Reagents/Equipments

- Sharp forceps (Dumont no. 5): forceps are used to remove the chorion before small molecular reagent treatments
- Glass Pasteur pipettes: wide bore pipettes with a flame-polished edge are used to collect embryos and larvae
- Microinjection needles: glass thin-walled microcapillaries with filament (TW100F-4) are purchased from World Precision Instrument, Inc., Sarasota, FL, USA. Needles are pulled with a micropipette puller device (Sutter Instruments Inc., CA, USA)
- Injector: a commercial injector (e.g., Tritech Instruments, Mumbai, India; Eppendorf, Hamburg, Germany) is necessary for injections
- Micrometer slide
- Mineral oil
- E3 embryo medium: for a 50× stock solution dilute 14.6 g NaCl, 0.65 g KCl, 2.20 g $CaCl_2$, and 4.05 g $MgSO_4$ per 1 L distilled water
- Dimethyl sulfoxide (DMSO): high purity DMSO (>99.5%) is purchased from a commercial seller
- Small molecular reagents: to create 1000× or 100× stock solutions, small molecular reagents are diluted at the appropriate concentrations in DMSO

- Morpholinos: synthetic antisense MOs of given sequence are ordered from GeneTools (OR, USA), diluted to a 100 × stock concentrations and stored at room temperature. (Preferentially, before use the stock solution can be heated for 5 min to 65°C, to increase to solubility of the morpholino.) For a detailed list of morpholinos used in experiments see Table 3
- Zebrafish strains: wild-type (AB, ekwill or tuebingen) and/or transgenic lines are used in for the experiments
- 28.5°C incubator
- Tricaine methanesulfonate (MS222): for a 25 × stock solution, dissolve 400 mg tricaine powder (Sigma-Aldrich, cat. no. MS222) in 97.9 mL distilled water and 2.1 mL 1 *M* Tris (pH 9). Adjust pH to 7 and store it in the freezer
- Phenylthiourea (PTU): for a 50 × stock solution 1-phenyl-2-thiourea (Sigma-Aldrich, cat. no. P7629) is diluted to a 10 m*M* stock concentration in sterile water (5 × 3 min sonication is used to completely dissolve the reagent)
- Methylcellulose: for a 4% solution weigh 4 g methylcellulose (Sigma-Aldrich, cat. no. M0387) and dissolve it in 100 mL dH_2O. As cold temperature helps the process, use a rotator, set to slow motion in a cold room. Once dissolved, methylcellulose can be stored at room temperature
- Low melting point agarose: a 0.8–2% solution of low melting point agarose (Sigma-Aldrich, cat. no. A9414) in E3 medium is prepared in advance and stored at 4°C. Before use, it is melted in a microwave oven and kept at 42°C (this temperature, while prevents the agarose to harden, is tolerated for the short time of embedding by zebrafish embryos)
- mRNA synthesis kit: for capped mRNA synthesis, a commercial kit (e.g., mMessage mMachine, ThermoFisher/Ambion) is used
- PBS/PMFS: freshly added 1 m*M* PMSF in 1 × PBS
- Sample buffer: 1:1 combination of E3 medium and 2 × Laemmli sample buffer (Bio-Rad, cat. no. 1610737)
- Acrylamide protein gel: 4–20% Mini-PROTEAN® TGX™ Precast Gels (Bio-Rad, cat. no. 4561095)
- Protein standard: Precision Plus Protein Dual Color Standard (Bio-Rad, cat. no. 1610374)
- 1 × Running buffer: dilute 10-fold a 10 × Tris/Glycine/SDS buffer stock (Bio-Rad, cat. no. 1610732)
- Mini-PROTEAN Electrophoresis system (Bio-Rad)
- PVDF membrane (Bio-Rad, cat. no. 1620177)

- 1 × Transfer buffer (Towbin): for 1 L add 200 mL methanol to 80 mL of 10 × Tris/Glycine buffer (Bio-Rad, cat. no. 1610734) and 720 mL distilled water
- Nonfat dry milk (any commercial brand can be used)
- ECL kit (ThermoFisher Scientific, cat. no. 32106)
- X-ray films (GE Healthcare LTD, Amersham Hyperfilm TM ECL, UK)
- Kodak—Carestream GBX developer/replenisher (Sigma-Aldrich, cat. no. P7042)
- Kodak—Carestream GBX fixer/replenisher (Sigma-Aldrich, cat. no. P7167)
- PBST: 0.5% Tween-20 in PBS
- Glutaraldehyde (TAAB, cat. no. G011)
- Sodium cacodylate (Sigma-Aldrich, cat. no. C0250)
- OsO_4 (Sigma-Aldrich, cat. no. 419494)
- Durcupan (Electron Microscopy Sciences, Hatfield, PA, USA, 14020)

4.2 Methodology

Zebrafish stocks have to be maintained according to standard protocols at 25–30°C, in an 14 h light/10 h dark cycle in a certified animal facility. Breeding pairs are set up the previous night (e.g., two males and four females) in an appropriate breeding tank, and embryos are collected in the morning and raised in a 28.5°C incubator till 4 dpf (days of post fertilization). (Larvae start to feed at 5 dpf, from which age they might be the subject of animal welfare laws, therefore, for longer experiments a specific license will be required, according to the local regulation.)

Embryos are injected with morpholinos or capped mRNAs at the 1–2 cell stage (approximately 20 min–1 h after fertilization). Microinjection needles are loaded with the solution and the tip of the needle is broken with a forceps. The micrometer slide is used for calibration (1–2 nL/embryo is a standard injection volume). Embryos are lined up along a microscope slide put on a Petri dish plate and are injected either with or without the help of a micromanipulator device (for experienced users, there is no difference between the two methods). After injections, embryos are washed into a Petri dish containing E3 medium mixed with small amounts of methylene blue and stored in the incubator.

Fish embryos can be subjected to small molecular treatments from the timepoint of the fertilizaion. However, in autophagy research treatments usually start at later stages. As the chorion is unpermeable for certain chemicals it is advisable to remove it prior the start of treatment. It is important to note that dechorionation before the end of the epiboly is tricky, as the yolk cell is cery sensitive and can burst. Treatments are usually performed on 24-well plates, 5–10 embryos are placed in single wells in 4–5 mL E3 embryo medium. To avoid pigment formation PTU can be also added into the medium. Aliquots from stock solutions of the respective small molecular compounds are pipetted into the wells and diluted to the desired concentration. As several compounds are light sensitive, the plate is covered in aluminum foil and stored in the incubator. During longer treatments, the plate is checked every 5 h to remove dead embryos. As some compounds degrade over time, solutions are refreshed every 24 h.

For imaging, embryos are anesthetized using tricaine methanesulfonate (MS 222) diluted to 1 × in E3 medium and embedded in 4% methylcellulose or 0.8–2% low melting point agarose, dependent on the observation method. Embedding in methylcellulose is quicker, therefore, preferred when using a fluorescent stereomicroscope, whereas agarose embedding provides greater stability, necessary for confocal microscopy. When using agarose, make sure to cover your sample in E3 medium supplemented with 1 × tricaine solution, otherwise your samples will dry and the embryos die.

Fluorescent pictures are analyzed using the ImageJ software package. Fluorescence intensities and/or punctae numbers are measured in a standardized way and the results are analyzed with a statistical program package (e.g., R).

When quantitative-PCR experiments are used to assess the experession level of *atg* genes, it is extremely important to use the respective primers with their standardized conditions. A list of qPCR primers can be found in Table 1. For the best PCR conditions, we suggest readers to consult the referenced papers.

For Western blot analysis, 15–20 embryos are collected. If younger than 4 dpf, fish need to be dechorionated and deyolked in order to avoid overloading artifacts on the SDS gel (Link, Shevchenko, & Heisenberg, 2006). Deyolking can be performed after sedation with tricaine in ice-cold PBS/PMFS by pressing the embryos up and down with a narrow glass pipet. Deyolked embryos are transferred to a new tube and washed with PBS/PMSF twice. After rapid centrifugation (3000 rpm, 4°C) excess liquid is removed. Samples are dissolved 50–100 μL of sample buffer and homogenized until solution has a uniform consistency. Lysates are boiled at 95°C for 5 min and centrifuged at top speed for 1 min. At this point, samples are either frozen at −70°C or run immediately on an acrylamide protein gel.

10–20 μL of the sample is loaded into a single lane and blotted to PVDF membrane. Blocking is done with 3% nonfat dry milk in PBST at room temperature. Membranes are incubated overnight at 4°C in blocking solution containing the primary antibody. After washing (4 × 10 min PBST) incubation is done for 1 h at room temperature with secondary antibody in blocking solution, and washed again (4 × 10 min) with PBST. Membranes are visualized using luminol-based ECL kit by exposure to X-ray films. (A list of antibodies previously used to observe with autophagy in zebrafish can be found in Varga et al. (2015). Corresponding horseradish peroxidase-conjugated antibodies can be purchased from all major suppliers.)

For TEM, depending on the size of the sample, embryos (or tissues) are fixed form 1 h to overnight in a mixture of paraform-glutaraldehyde in sodium cacodylate buffer (0.1 *M*, pH 7.4). The concentration of the fixatives can vary from 1.5% to 2% for glutaraldehyde and from 1% to 4% for paraformaldehyde. However, fixation even without paraformaldehyde does not seem to affect morphology at least in 5 dpf embryos. After incubating the samples in sodium cacodylate buffer (1 h) and rinsing twice, postfixation is performed with 1% osmium tetroxide for 1 h at room temperature followed by rinsing the samples. Dehydration is performed with graded series of acetone (30% for 15 min, 50%, 70%, 90%, dry acetone (2 ×) for 30 min). Afterward infiltration and embedding is done in Durcupan, following the supplier's protocol. Ultrathin sections are made, transferred to formvar coated grids, and are stained with lead citrate and uranyl acetate.

5. CONCLUDING REMARKS

The transparency and small size of zebrafish embryos make them an excellent subject for the in vivo observation of autophagy, whereas the availability of a high-quality genome sequence combined with easy to use genome-editing techniques hold the promise of creating a mutant collection that is unparalleled in any other vertebrate model organism. Indeed, the priority of the field should be in delivering on this latter objective, as at this point mutants are necessary to test the previous knockdown results obtained with MOs. We also need to understand if there are species-specific aspects of autophagy in zebrafish, so we can fully take them into account for future research. However, once the previous groundbreaking results related to development, regeneration, and disease can be reconfirmed or reassessed with new techniques, the power of the zebrafish model will become apparent to all researchers. The past two decades have seen the coming to age of zebrafish as a popular genetic model organism and a vigorous renaissance of autophagy research. These two fields are starting to intersect and based on the results that are already available, we can be confident that the future is promising for zebrafish autophagy research.

ACKNOWLEDGMENTS

This work was supported by the Hungarian Scientific Research Fund (OTKA K109349) to T.V., M.V., and K.T.-V., and MEDinPROT Protein Science Research Synergy Program to T.V., K.T.-V., and M.V.

The authors declare no competing interest.

REFERENCES

Akizu, N., Cantagrel, V., Zaki, M. S., Al-Gazali, L., Wang, X., Rosti, R. O., et al. (2015). Biallelic mutations in SNX14 cause a syndromic form of cerebellar atrophy and lysosome-autophagosome dysfunction. *Nature Genetics*, *47*, 528–534.

Al Rawi, S., Louvet-Vallée, S., Djedd, A., Sachse, M., Culetto, E., Hajjar, C., et al. (2011). Postfertilization autophagy of sperm organelles prevents paternal mitochondrial DNA transmission. *Science*, *334*, 1144–1147.

Altschul, S. F., Madden, T. L., Schäffer, A. A., Zhang, J., Zhang, Z., Miller, W., et al. (1997). Gapped BLAST and PSI-BLAST: A new generation of protein database search programs. *Nucleic Acids Research*, *25*, 3389–3402.

Barth, S., Glick, D., & Macleod, K. F. (2010). Autophagy: Assays and artifacts. *The Journal of Pathology*, *221*, 117–124.

Benato, F., Skobo, T., Gioacchini, G., Moro, I., Ciccosanti, F., Piacentini, M., et al. (2013). Ambra1 knockdown in zebrafish leads to incomplete development due to severe defects in organogenesis. *Autophagy*, *9*, 476–495.

Billes, V., Kovács, T., Hotzi, B., Manzéger, A., Tagscherer, K., Komlós, M., et al. (2016). AUTEN-67 (Autophagy Enhancer-67) hampers the progression of neurodegenerative symptoms in a Drosophila model of Huntington's disease. *Journal of Huntington's Disease*, *5*(2), 133–147. PMID:27163946.

Boglev, Y., Badrock, A. P., Trotter, A. J., Du, Q., Richardson, E. J., Parslow, A. C., et al. (2013). Autophagy induction is a tor- and tp53-independent cell survival response in a zebrafish model of disrupted ribosome biogenesis. *PLoS Genetics*, *9*, e1003279.

Bühler, A., Kustermann, M., Bummer, T., Rottbauer, W., Sandri, M., & Just, S. (2016). Atrogin-1 deficiency leads to myopathy and heart failure in Zebrafish. *International Journal of Molecular Sciences*, *17*, e187.

Chablais, F., & Jazwinska, A. (2010). IGF signaling between blastema and wound epidermis is required for fin regeneration. *Development*, *137*, 871–879.

Chew, T. S., O'Shea, N. R., Sewell, G. W., Oehlers, S. H., Mulvey, C. M., Crosier, P. S., et al. (2015). Optineurin deficiency in mice contributes to impaired cytokine secretion and neutrophil recruitment in bacteria-driven colitis. *Disease Models & Mechanisms*, *8*, 817–829.

Clancey, L. F., Beirl, A. J., Linbo, T. H., & Cooper, C. D. (2013). Maintenance of melanophore morphology and survival is cathepsin and vps11 dependent in zebrafish. *PLoS One*, *8*, e65096.

Clark, S. L., Jr. (1957). Cellular differentiation in the kidneys of newborn mice studies with the electron microscope. *The Journal of Biophysical and Biochemical Cytology*, *3*, 349–362.

Deter, R. L., & de Duve, C. (1967). Influence of glucagon, an inducer of cellular autophagy, on some physical properties of rat liver lysosomes. *The Journal of Cell Biology*, *33*, 437–449.

Di Donato, V., De Santis, F., Auer, T. O., Testa, N., Sánchez-Iranzo, H., Mercader, N., et al. (2016). 2C-Cas9: A versatile tool for clonal analysis of gene function. *Genome Research*, *26*, 681–692.

Driever, W., Solnica-Krezel, L., Schier, A. F., Neuhauss, S. C., Malicki, J., Stemple, D. L., et al. (1996). A genetic screen for mutations affecting embryogenesis in zebrafish. *Development*, *123*, 37–46.

Elenbaas, J. S., Maitra, D., Liu, Y., Lentz, S. I., Nelson, B., Hoenerhoff, M. J., et al. (2016). A precursor-inducible zebrafish model of acute protoporphyria with hepatic protein aggregation and multiorganelle stress. *The FASEB Journal*, *30*, 1798–1810.

Fleming, A., Noda, T., Yoshimori, T., & Rubinsztein, D. C. (2011). Chemical modulators of autophagy as biological probes and potential therapeutics. *Nature Chemical Biology*, *7*, 9–17.

Fleming, A., & Rubinsztein, D. C. (2011). Zebrafish as a model to understand autophagy and its role in neurological disease. *Biochimica et Biophysica Acta*, *1812*, 520–526.

Ganesan, S., Moussavi Nik, S. H., Newman, M., & Lardelli, M. (2014). Identification and expression analysis of the zebrafish orthologues of the mammalian MAP1LC3 gene family. *Experimental Cell Research*, *328*, 228–237.

George, A. A., Hayden, S., Stanton, G. R., & Brockerhoff, S. E. (2016). Arf6 and the 5′ phosphatase of synaptojanin 1 regulate autophagy in cone photoreceptors. *Inside the Cell*, *1*, 117–133.

Haffter, P., Granato, M., Brand, M., & Mullins, M. C. (1996). The identification of genes with unique and essential functions in the development of the zebrafish, Danio rerio. *Development*, *123*, 1–36.

He, C., Bartholomew, C. R., Zhou, W., & Klionsky, D. J. (2009). Assaying autophagic activity in transgenic GFP-Lc3 and GFP-Gabarap zebrafish embryos. *Autophagy*, *5*, 520–526.

Heijnen, H. F., van Wijk, R., Pereboom, T. C., Goos, Y. J., Seinen, C. W., van Oirschot, B. A., et al. (2014). Ribosomal protein mutations induce autophagy through S6 kinase inhibition of the insulin pathway. *PLoS Genetics*, *10*, e1004371.

Hosseini, R., Lamers, G. E., Hodzic, Z., Meijer, A. H., Schaaf, M. J., & Spaink, H. P. (2014). Correlative light and electron microscopy imaging of autophagy in a zebrafish infection model. *Autophagy*, *10*, 1844–1857.

Hu, Z., Zhang, J., & Zhang, Q. (2011). Expression pattern and functions of autophagy-related gene atg5 in zebrafish organogenesis. *Autophagy*, *7*, 1514–1527.

Huang, G., Zhang, F., Ye, Q., & Wang, H. (2016). The circadian clock regulates autophagy directly through the nuclear hormone receptor Nr1d1/Rev-erbα and indirectly via Cebpb/(C/ebpβ) in zebrafish. *Autophagy*, *12*(8), 1292–1309. PMID:27171500.

Jia, X.-E., Ma, K., Xu, T., Gao, L., Wu, S., Fu, C., et al. (2015). Mutation of kri11 causes definitive hematopoiesis failure via PERK-dependent excessive autophagy induction. *Cell Research*, *25*, 946–962.

Juhász, G., Csikós, G., Sinka, R., Erdélyi, M., & Sass, M. (2003). The Drosophila homolog of Aut1 is essential for autophagy and development. *FEBS Letters*, *543*, 154–158.

Kettleborough, R. N. W., Busch-Nentwich, E. M., Harvey, S. A., Dooley, C. M., de Bruijn, E., van Eeden, F., et al. (2013). A systematic genome-wide analysis of zebrafish protein-coding gene function. *Nature*, *496*, 494–497.

Kirkin, V., Lamark, T., Sou, Y. S., Bjørkøy, G., Nunn, J. L., Bruun, J. A., et al. (2009). A role for NBR1 in autophagosomal degradation of ubiquitinated substrates. *Molecular Cell*, *33*, 505–516.

Klionsky, D. J. (2005). The molecular machinery of autophagy: Unanswered questions. *Journal of Cell Science*, *118*, 7–18.

Klionsky, D. J., Abdelmohsen, K., Abe, A., Abedin, M. J., Abeliovich, H., et al. (2016). Guidelines for the use and interpretation of assays for monitoring autophagy (3rd edition). *Autophagy*, *12*, 1–222.

Korac, J., Schaeffer, V., Kovacevic, I., Clement, A. M., Jungblut, B., Behl, C., et al. (2013). Ubiquitin-independent function of optineurin in autophagic clearance of protein aggregates. *Journal of Cell Science*, *126*, 580–592.

Kourtis, N., & Tavernarakis, N. (2009). Autophagy and cell death in model organisms. *Cell Death and Differentiation*, *16*, 21–30.

Kyöstilä, K., Syrjä, P., Jagannathan, V., Chandrasekar, G., Jokinen, T. S., Seppälä, E. H., et al. (2015). A missense change in the ATG4D gene links aberrant autophagy to a neurodegenerative vacuolar storage disease. *PLoS Genetics*, *11*, e1005169.

Lee, E., Koo, Y., Ng, A., Wei, Y., Luby-Phelps, K., Juraszek, A., et al. (2014). Autophagy is essential for cardiac morphogenesis during vertebrate development. *Autophagy*, *10*, 572–587.

Levine, B., & Klionsky, D. J. (2004). Development by self-digestion: Molecular mechanisms and biological functions of autophagy. *Developmental Cell*, *6*, 463–477.

Levine, B., & Kroemer, G. (2008). Autophagy in the pathogenesis of disease. *Cell*, *132*, 27–42.

Liang, X. H., Jackson, S., Seaman, M., Brown, K., Kempkes, B., Hibshoosh, H., et al. (1999). Induction of autophagy and inhibition of tumorigenesis by beclin 1. *Nature*, *402*, 672–676.

Link, V., Shevchenko, A., & Heisenberg, C. P. (2006). Proteomics of early zebrafish embryos. *BMC Developmental Biology*, *6*, 1.

Meléndez, A., Tallóczy, Z., Seaman, M., Eskelinen, E. L., Hall, D. H., & Levine, B. (2003). Autophagy genes are essential for dauer development and life-span extension in C. elegans. *Science*, *301*, 1387–1391.

Miccoli, A., Gioacchini, G., Maradonna, F., Benato, F., Skobo, T., & Carnevali, O. (2015). Beneficial bacteria affect Danio rerio development by the modulation of maternal factors involved in autophagic, apoptotic and dorsalizing processes. *Cellular Physiology and Biochemistry*, *35*, 1706–1718.

Mizushima, N., & Levine, B. (2010). Autophagy in mammalian development and differentiation. *Nature Cell Biology*, *12*, 823–830.

Mizushima, N., Levine, B., Cuervo, A. M., & Klionsky, D. J. (2008). Autophagy fights disease through cellular self-digestion. *Nature*, *451*, 1069–1075.

Mizushima, N., & Yoshimori, T. (2007). How to interpret LC3 immunoblotting. *Autophagy*, *3*, 542–545.

Mohanty, S., Jagannathan, L., Ganguli, G., Padhi, A., Roy, D., Alaridah, N., et al. (2015). A mycobacterial phosphoribosyltransferase promotes bacillary survival by inhibiting oxidative stress and autophagy pathways in macrophages and zebrafish. *The Journal of Biological Chemistry*, *290*, 13321–13343.

Moreau, K., Fleming, A., Imarisio, S., Lopez Ramirez, A., Mercer, J. L., Jimenez-Sanchez, M., et al. (2014). PICALM modulates autophagy activity and tau accumulation. *Nature Communications*, *5*, 4998.

Mostowy, S., Boucontet, L., Mazon Moya, M. J., Sirianni, A., Boudinot, P., Hollinshead, M., et al. (2013). The zebrafish as a new model for the in vivo study of Shigella flexneri interaction with phagocytes and bacterial autophagy. *PLoS Pathology*, *9*, e1003588.

Nasevicius, A., & Ekker, S. C. (2000). Effective targeted gene "knockdown" in zebrafish. *Nature Genetics*, *26*, 216–220.

Neufeld, T. P., & Baehrecke, E. H. (2008). Eating on the fly: Function and regulation of autophagy during cell growth, survival and death in Drosophila. *Autophagy*, *4*, 557–562.

Papp, D., Kovács, T., Billes, V., Varga, M., Tarnóci, A., Hackler, L., Jr., et al. (2016). AUTEN-67, an autophagy-enhancing drug candidate with potent antiaging and neuroprotective effects. *Autophagy*, *12*, 273–286.

Paulus, J. D., & Link, B. A. (2014). Loss of optineurin in vivo results in elevated cell death and alters axonal trafficking dynamics. *PLoS One*, *9*, e109922.

NCBI Resource Coordinators. (2016). Database resources of the National Center for Biotechnology Information. *Nucleic Acids Research*, *44*, D7–D19.

Robu, M. E., Larson, J. D., Nasevicius, A., Beiraghi, S., Brenner, C., Farber, S. A., et al. (2007). p53 activation by knockdown technologies. *PLoS Genetics*, *3*, e78.

Rossi, A., Kontarakis, Z., Gerri, C., Nolte, H., Hölper, S., Krüger, M., et al. (2015). Genetic compensation induced by deleterious mutations but not gene knockdowns. *Nature*, *524*, 230–233.

Rubinsztein, D. (2006). The roles of intracellular protein-degradation pathways in neurdegradation. *Nature*, *443*, 780–786.

Santangeli, S., Maradonna, F., Gioacchini, G., Cobellis, G., Piccinetti, C. C., Dalla Valle, L., et al. (2016). BPA-induced deregulation of epigenetic patterns: Effects on female zebrafish reproduction. *Scientific Reports*, *6*, 21982.

Sasaki, T., Lian, S., Qi, J., Bayliss, P. E., Carr, C. E., Johnson, J. L., et al. (2014). Aberrant autolysosomal regulation is linked to the induction of embryonic senescence: Differential roles of Beclin 1 and p53 in vertebrate Spns1 deficiency. *PLoS Genetics*, *10*, e1004409.

Sato, M., & Sato, K. (2011). Degradation of paternal mitochondria by fertilization-triggered autophagy in C. elegans embryos. *Science*, *334*, 1141–1144.

Schlumpberger, M., Schaeffeler, E., Straub, M., Bredschneider, M., Wolf, D. H., & Thumm, M. (1997). AUT1, a gene essential for autophagocytosis in the yeast Saccharomyces cerevisiae. *Journal of Bacteriology*, *179*, 1068–1076.

Schulte-Merker, S., & Stainier, D. Y. R. (2014). Out with the old, in with the new: Reassessing morpholino knockdowns in light of genome editing technology. *Development*, *141*, 3103–3104.

Scott, S. V., Hefner-Gravink, A., Morano, K. A., Noda, T., Ohsumi, Y., & Klionsky, D. J. (1996). Cytoplasm-to-vacuole targeting and autophagy employ the same machinery to deliver proteins to the yeast vacuole. *Proceedings of the National Academy of Sciences of the United States of America*, *93*, 12304–12308.

Skobo, T., Benato, F., Grumati, P., Meneghetti, G., Cianfanelli, V., Castagnaro, S., et al. (2014). Zebrafish ambra1a and ambra1b knockdown impairs skeletal muscle development. *PLoS One*, *9*, e99210.

Takacs-Vellai, K., Bayci, A., & Vellai, T. (2006). Autophagy in neuronal cell loss: A road to death. *BioEssays*, *28*, 1126–1131.

Tan, C. C., Yu, J. T., Tan, M. S., Jiang, T., Zhu, X. C., & Tan, L. (2014). Autophagy in aging and neurodegenerative diseases: Implications for pathogenesis and therapy. *Neurobiology of Aging*, *35*, 941–957.

Tóth, M. L., Sigmond, T., Borsos, E., Barna, J., Erdélyi, P., Takács-Vellai, K., et al. (2008). Longevity pathways converge on autophagy genes to regulate life span in Caenorhabditis elegans. *Autophagy*, *4*, 330–338.

Tsai, I.-T., Chen, Y.-H., Chen, Y.-H., & Wang, Y.-H. (2013). Amikacin-induced fin reduction is mediated by autophagy. *Journal of Toxicologic Pathology*, *26*, 79–82.

Tsukada, M., & Ohsumi, Y. (1993). Isolation and characterization of autophagy-defective mutants of Saccharomyces cerevisiae. *FEBS Letters*, *333*, 169–174.

UniProt Consortium. (2015). UniProt: A hub for protein information. *Nucleic Acids Research*, *43*, D204–D212.

van der Vaart, M., Korbee, C. J., Lamers, G. E. M., Tengeler, A. C., Hosseini, R., Haks, M. C., et al. (2014). The DNA damage-regulated autophagy modulator DRAM1 links mycobacterial recognition via TLP-MYD88 to authophagic defense. *Cell Host & Microbe*, *15*, 753–767.

Varga, M., Fodor, E., & Vellai, T. (2015). Autophagy in zebrafish. *Methods*, *75*, 172–180.

Varga, M., Sass, M., Papp, D., Takács-Vellai, K., Kobolak, J., Dinnyés, A., et al. (2014). Autophagy is required for zebrafish caudal fin regeneration. *Cell Death and Differentiation*, *21*, 547–556.

Varshney, G. K., Sood, R., & Burgess, S. M. (2015). Understanding and editing the zebrafish genome. *Advances in Genetics*, *92*, 1–52.

Varshney, G. K., Zhang, S., Pei, W., Adomako-Ankomah, A., Fohtung, J., Schaffer, K., et al. (2016). CRISPRz: A database of zebrafish validated sgRNAs. *Nucleic Acids Research*, *44*, D822–D826.

Vellai, T. (2009). Autophagy genes and ageing. *Cell Death and Differentiation*, *16*, 94–102.

Vellai, T., Takács-Vellai, K., Sass, M., & Klionsky, D. J. (2009). The regulation of aging: Does autophagy underly longevity? *Trends in Cell Biology*, *19*, 487–494.

Vergne, I., Roberts, E., Elmaoued, R. A., Tosch, V., Delgado, M. A., Proikas-Cezanne, T., et al. (2009). Control of autophagy initiation by phosphoinositide 3-phosphatase Jumpy. *The EMBO Journal*, *28*, 2244–2258.

Wager, K., & Russell, C. (2013). Mitophagy and neurodegeneration. *Autophagy*, *9*, 1693–1709.

Xie, Y., Kang, R., Sun, X., Zhong, M., Huang, J., Klionsky, D. J., et al. (2015). Posttranslational modification of autophagy-related proteins in macroautophagy. *Autophagy*, *11*, 28–45.

Zhang, P., & Zhang, H. (2013). Autophagy modulates miRNA-mediated gene silencing and selectively degrades AIN-1/GW182 in C. elegans. *EMBO Reports*, *14*, 568–576.

Zheng, X., Dai, W., Chen, X., Wang, K., Zhang, W., Liu, L., et al. (2015). Caffeine reduces hepatic lipid accumulation through regulation of lipogenesis and ER stress in zebrafish larvae. *Journal of Biomedical Science*, *22*, 1–12.

CHAPTER TWENTY-FIVE

Biochemical Methods to Monitor Autophagic Responses in Plants

Y. Bao, Y. Mugume, D.C. Bassham[1]
Iowa State University, Ames, IA, United States
[1]Corresponding author: e-mail address: bassham@iastate.edu

Contents

Abstract

The study of autophagy in plants is rapidly increasing, due to its pivotal and fundamental roles in responding to stressful stimuli, recycling nutrients during senescence, and maintaining growth under normal conditions. Assays for detecting autophagy in plants have generally been based on microscopic observations, providing qualitative information on autophagy activity. Here, we discuss biochemical assays for detecting autophagy, which have the potential for providing more quantitative information, with a focus on immunoblotting with antibodies against ATG8, NBR1, or epitope tags fused to ATG proteins.

1. INTRODUCTION

Autophagy is a process for recycling of cellular components through the lysosome/vacuole and is conserved among eukaryotes (Kim & Klionsky, 2000; Li & Vierstra, 2012). It functions both in controlling development and in the response of organisms to environmental stimuli (Bassham et al., 2006; Liu & Bassham, 2012; Michaeli, Galili, Genschik, Fernie, & Avin-Wittenberg, 2016; Reggiori & Klionsky, 2002). In yeast and animals, at least three types of autophagy have been described to date: (1) microautophagy,

Methods in Enzymology, Volume 588
ISSN 0076-6879
http://dx.doi.org/10.1016/bs.mie.2016.09.090

in which a small portion of the lysosomal/vacuolar surface-associated cytoplasm is engulfed by membrane invagination (Mijaljica, Prescott, & Devenish, 2011; van Doorn & Papini, 2013); (2) macroautophagy, in which bulk cytoplasmic components are degraded via double-membrane vesicles termed autophagosomes (De Duve & Wattiaux, 1966; Yang & Klionsky, 2009); and (3) chaperone-mediated autophagy, which appears to occur only in animals, for selective lysosomal degradation of proteins containing a KFERQ-like recognition motif via a cytosolic chaperone complex (Cuervo, Gomes, Barnes, & Dice, 2000). Only microautophagy and macroautophagy have been observed in plants, and macroautophagy (hereafter called autophagy) is by far the best-studied type (Michaeli et al., 2016). Upon activation of autophagy, de novo-formed double-membrane autophagosomes initiate as cup-shaped phagophores and expand to enclose components in the cytoplasm for degradation (Klionsky & Emr, 2000; Li & Vierstra, 2012; Liu & Bassham, 2012). After fusion of the autophagosomal outer membrane with the tonoplast (the membrane surrounding the large central vacuole), the inner membrane and the contents inside the autophagic body are degraded in the vacuole (Michaeli et al., 2016; Yang & Bassham, 2015). Alternatively, plant autophagosomes may fuse with endosome- or lysosome-like structures and be transported into the central vacuole for degradation as a later event (Moriyasu & Ohsumi, 1996; Takatsuka, Inoue, Matsuoka, & Moriyasu, 2004).

Thus far, more than 40 autophagy-related (*ATG*) genes have been reported in yeast (Mochida et al., 2015; Ohsumi, 2001; Suzuki, Kubota, Sekito, & Ohsumi, 2007) and the core autophagy components can be divided into several groups (Behrends, Sowa, Gygi, & Harper, 2010; Suzuki & Ohsumi, 2010; Yang & Klionsky, 2010) according to their functions: (1) the ATG1–ATG13 kinase complex initiates the formation of autophagosomes; (2) ATG9 and associated proteins may function in acquiring lipids from the Golgi, endoplasmic reticulum (ER), mitochondria, plasma membrane, and endosomes to expand the phagophore membrane; (3) the phosphoinositide 3-kinase complex for vesicle nucleation; (4) ATG8–PE (ATG8 covalently linked to phosphatidylethanolamine (PE)) involved in phagophore membrane expansion and cargo selection; and (5) the ATG5–ATG12 and ATG16 complex that functions as an E3 ligase in ATG8–PE conjugation.

Based on the functional conservation with yeast *ATG* genes, many *ATG* genes have been identified in the model plant *Arabidopsis thaliana* (Avin-Wittenberg, Honig, & Galili, 2012). Facilitated by the identification of knockout mutants disrupted in *ATG* genes in *Arabidopsis*, plant autophagy has been extensively studied in the past two decades (Floyd, Morriss, Macintosh, & Bassham, 2012; Lv, Pu, Qin, Zhu, & Lin, 2014; Yang & Bassham, 2015). In addition to its critical role in plant responses to nutrient deprivation, the significance of autophagy in biotic and abiotic stress responses has been broadly demonstrated by studies of abiotic stress, such as drought, salt, and oxidative stress (Liu, Xiong, & Bassham, 2009; Xiong, Contento, Nguyen, & Bassham, 2007); innate immunity and programmed cell death (Hayward & Dinesh-Kumar, 2011; Hofius, Munch, Bressendorff, Mundy, & Petersen, 2011; Minina et al., 2013); nutrient remobilization in senescent leaves (Guiboileau et al., 2012); and even for removing unwanted organelles including mitochondria (Li, Chung, & Vierstra, 2014; Liu, Sakakibara, Chen, & Okamoto, 2014), chloroplasts (Chiba, Ishida, Nishizawa, Makino, & Mae, 2003; Ishida & Yoshimoto, 2008; Ono, Wada, Izumi, Makino, & Ishida, 2013; Wada & Ishida, 2009), peroxisomes (Kim et al., 2013; Voitsekhovskaja, Schiermeyer, & Reumann, 2014; Yoshimoto et al., 2014), and ER (Liu et al., 2012; Yang, Srivastava, Howell, & Bassham, 2016). Recently, the study of autophagy has expanded into other plant species, including rice and maize (Chung, Phillips, & Vierstra, 2010; Ghiglione et al., 2008; Kuzuoglu-Ozturk et al., 2012; Li et al., 2015; Rana, Dong, Ali, Huang, & Zhang, 2012; Shin, Yoshimoto, Ohsumi, Jeon, & An, 2009), which shows the potential of this pathway for crop yield optimization and stress resistance.

Autophagy was originally regarded as a nonselective process for bulk degradation of cellular components in response to a variety of stressful stimuli. However, accumulating evidence has shown that autophagy can also be a selective process for disposing of specific targets under certain circumstances (Johansen & Lamark, 2011; Yang & Klionsky, 2009). A number of receptors have been identified in animal and yeast cells that recognize autophagic cargo, interact with ATG8 family proteins through a motif known as an LIR (LC3-interacting region) or AIM (ATG8-interacting motif), and thus direct incorporation of the cargo into forming autophagosomes (Zaffagnini & Martens, 2016). For example, mammalian NBR1 (neighbor of BRCA1 gene 1) and p62 are two critical autophagic receptors for delivery of ubiquitinated proteins or protein aggregates to the lysosome for degradation (Kirkin et al., 2009; Waters, Marchbank, Solomon, Whitehouse, & Gautel, 2009). A single NBR1 gene was discovered in *Arabidopsis* (Joka2 in tobacco), which appears to be a chimera of mammalian NBR1 and p62 (Svenning, Lamark, Krause, & Johansen,

2011; Zientara-Rytter et al., 2011). *Arabidopsis* NBR1 plays a role in targeting protein aggregates for degradation under heat, salt, and oxidative stresses (Zhou et al., 2013, 2014; Zientara-Rytter & Sirko, 2014). Many more selective autophagy receptors have been reported in yeast and mammals; however, homologs of most receptors are absent from plants.

The study of autophagy in plants has been advanced by employing several microscopic assays, most commonly morphological observation by electron microscopy (Ghiglione et al., 2008; Liu et al., 2005), labeling of autophagosomes with the fluorescent protein fusion GFP (green fluorescent protein)–ATG8 (Yoshimoto et al., 2004), and the use of acidotropic fluorescent dyes (Contento, Xiong, & Bassham, 2005; Moriyasu, Hattori, Jauh, & Rogers, 2003) to label autophagosomes (Klionsky et al., 2016; Mitou, Budak, & Gozuacik, 2009). Given the limitation that morphological methods for autophagy detection are mainly qualitative, biochemical assays for assessing autophagy have value in providing more quantitative methods. Here, we describe several biochemical assays for monitoring plant autophagy, using sucrose and/or nitrogen starvation for autophagy induction unless otherwise specified.

2. BIOCHEMICAL ASSAYS FOR AUTOPHAGY MONITORING

The majority of biochemical assays for monitoring plant autophagy are based on immunoblotting, either using antibodies against endogenous ATG proteins, most commonly ATG8, or by generation of transgenic plants expressing epitope-tagged ATG proteins, followed by immunoblotting with antibodies against the tag. Comparison with microscopic assays to visualize autophagosomes or autophagic bodies is extremely useful to confirm that the assays reflect activity of the autophagy pathway.

2.1 Buffers and Reagents

Protein extraction buffer (100 m*M* Tris–HCl, 300 m*M* sucrose, 1 m*M* EDTA, pH 7.5);

1 × SDS-PAGE loading sample buffer (50 m*M* Tris–HCl, pH 6.8, 6% glycerol (v/v), 2% (w/v) SDS, 0.004% (w/v) Bromophenol blue, 1% 2-mercaptoethanol);

1 × PBS buffer (137 m*M* NaCl, 2.7 m*M* KCl, 10 m*M* Na_2HPO_4, 2 m*M* KH_2PO_4, pH 7.4);

15% protein gel with 6 *M* urea for ATG8 lipidation assay (10 mL): 5 mL 30% acrylamide, 2.5 mL 1.5 M Tris–HCl, pH 8.8, 100 μL 10% SDS; 75 μL 10% ammonium persulfate (APS), 12 μL TEMED, 3.6 g urea;
Nitrocellulose blotting membrane, NC (Millipore, Bedford, MA);
Concanamycin A (ConcA) (Sigma-Aldrich, St. Louis, MO): 1 μ*M* final concentration;
Protease inhibitor cocktail (Roche, Germany): 1 tablet per 50 mL buffer;
Antibodies: for detecting ATG8 (Abcam, http://www.abcam.com) and GFP (Miltenyi Biotec, Germany).

2.2 ATG8 Lipidation and ATG8 Accumulation Detected by Immunoblotting

2.2.1 Background

ATG8 is an ubiquitin-like protein which can be used as an autophagosome marker in yeast, metazoans, and plants (Fujioka et al., 2008; Kirisako et al., 1999), and *Arabidopsis* has nine genes (*ATG8a–ATG8i*) encoding ATG8 isoforms (Li & Vierstra, 2012). The majority of *Arabidopsis* ATG8 isoforms are synthesized as proproteins (ATG8a–ATG8g), which are processed through cleavage by ATG4 to expose a glycine residue at the C-terminus (Ohsumi, 2001; Woo, Park, & Dinesh-Kumar, 2014). The processed ATG8 proteins are then conjugated to PE through an amide bond between the glycine residue of ATG8 and the amino group of PE (Ichimura et al., 2000). This lipidation of ATG8 leads to its localization to the preautophagosomal structure where it is thought to play a role in autophagosome formation (Kirisako et al., 1999). Eventually ATG4 mediates the delipidation of ATG8 attached to the outer autophagosome membrane, thus recycling the protein, an important step required for normal autophagy (Slobodkin & Elazar, 2013). Therefore, ATG8–PE on the outer membrane is released, while the ATG8–PE on the inner membrane is trapped inside the autophagosome as it matures to form a complete vesicle and is consumed in the vacuole during autophagic cargo degradation (Li & Vierstra, 2012).

This cellular route makes ATG8 an excellent marker to follow autophagosomes during autophagy, first reported in yeast (Kirisako et al., 1999) and later shown to occur similarly in plants (Yoshimoto et al., 2004). A comparison of lipidated to nonlipidated ATG8 has been used as a measure of autophagic activity in animal cells (Kabeya et al., 2000), and the availability of ATG8 antibodies makes such assays now possible in plants. Using antibodies against *Arabidopsis* ATG8a (Thompson, Doelling,

Suttangkakul, & Vierstra, 2005), a comparison between lipidated ATG8 (ATG8–PE) and unmodified ATG8 gives a measure of autophagic activity (Chung et al., 2010; Thompson et al., 2005).

Wild-type (WT) seedlings can be compared with autophagy mutants (*atg7*, *atg12a-1*, and *atg5*) (Chung et al., 2010; Thompson et al., 2005) and grown in nutrient-rich vs nutrient-deficient conditions (nitrogen or fixed-carbon deficiency). These *atg* mutants block ATG8 lipidation (Chung et al., 2010) and so are used as negative controls. Protein is extracted from the seedlings, separated by SDS-PAGE in the presence of urea, followed by immunoblot analysis with antibodies against ATG8 (Chung et al., 2010). An unrelated antibody can be used as a loading control. ATG8–PE can be distinguished from unlipidated/free ATG8 by its faster migration in the presence of urea (Chung et al., 2010; Ichimura et al., 2000).

2.2.2 Plant Growth Conditions and Treatment

1. Surface-sterilize seeds of *Arabidopsis* accession Columbia-0 (Col-0) and derived mutants in diluted bleach (distilled water:bleach = 2:1) containing 0.1% (v/v) Triton X-100 for 20 min, and rinse briefly with sterile water at least five times.
2. Stratify sterilized seeds in water at 4°C in the dark for at least 2 days and sow on solid half-strength Murashige and Skoog (1/2 MS) medium plates containing 1% (w/v) sucrose and 0.6% (w/v) phytoblend agar (Caisson, http://www.caissonlabs.com); the pH is adjusted to 5.7 with MES–KOH before autoclaving.
3. Unless otherwise noted, incubate the plates at 22°C under a long day (16 h light/8 h dark) photoperiod.
4. To impose N limitation, grow seedlings as above on MS-N plates: MS micronutrient salts plus 3 m*M* $CaCl_2$, 1.5 m*M* $MgSO_4$, 1.25 m*M* KH_2PO_4, 5 m*M* KCl, and 2 m*M* MES–KOH, pH 5.7, supplemented with 1% sucrose.

2.2.3 Procedures for ATG8 Lipidation Assay

1. Transfer 7-day-old seedlings to fresh 1/2 MS plates or onto MS-N plates for starvation. Typical growth times on N limitation medium are 4 days to allow induction of autophagy.
2. Homogenize seedlings in ice-cold protein extraction buffer (volume: gram fresh weight = 2:1) supplemented with a protease inhibitor cocktail.

3. Centrifuge at 10,000 × *g* for 30 min to pellet debris, then transfer the clarified supernatant to a new microcentrifuge tube.
4. Separate equal amounts of protein per sample (25 μg or as appropriate) by SDS-PAGE; for analysis of ATG8 lipidation, 6 *M* urea is added to the resolving gel. Add solid urea directly to the gel solution prior to addition of APS and TEMED, and rock gently until dissolved (see Section 2.1, for gel composition). Electrophoretically transfer onto nitrocellulose membrane for immunoblot analysis with anti-ATG8 antibodies.
5. Block membranes for 2 h at room temperature in 6% low-fat milk, followed by primary antibody (1:1000 dilution) incubation in 1 × PBS plus 0.1% (v/v) Tween-20 with 1% (w/v) low-fat dried milk. Incubate membranes in primary antibodies at room temperature for 2 h or at 4°C overnight.
6. Wash membranes with 1 × PBS plus 0.1% Tween-20 at least five times (5 min per wash). Incubate membranes with peroxidase- or alkaline phosphatase-labeled secondary antibodies (1:15,000–1:20,000 dilution) in 1 × PBS plus 0.1% Tween-20 with 1% (w/v) low-fat dried milk for 2 h at room temperature, wash with 1 × PBS plus 0.1% Tween-20 at least five times (5 min per wash), and detect as appropriate.
7. NIH ImageJ (Schneider, Rasband, & Eliceiri, 2012) (http://rsb.info.nih.gov/ij/) can be used to quantify the signal strength by comparison with control antibodies.

2.2.3.1 Issues Arising From ATG8 Immunoblotting

There are challenges in the detection of lipidated ATG8 by standard immunoblot procedures in plants. Plants contain an *ATG8* gene family with nine members (*ATG8a–ATG8i*) (Doelling, Walker, Friedman, Thompson, & Vierstra, 2002), which encode closely related proteins such that immunoblotting using ATG8a antibody is not specific to one particular ATG8 isoform. This leads to multiple bands which can obscure the identity of lipidated vs unmodified ATG8 (Li & Vierstra, 2012). In addition, the ATG8a antibodies reported in Chung et al. (2010) and available from Abcam (www.abcam.com) cross-react with an additional protein of unknown identity (Fig. 1), but with a similar mobility on SDS-PAGE as the lipidated form (Yoshimoto et al., 2004). It also cannot be excluded that this band is a breakdown product of ATG8 isoform(s) that is not lipidated.

Enriching the ATG8–PE adducts by isolation of membranes and solubilization with Triton X-100 followed by immunoblotting of gels run in the presence of urea results in more obvious differences between lipidated and

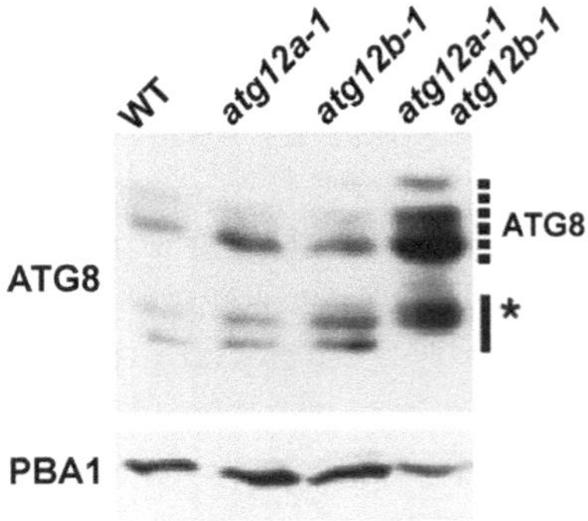

Fig. 1 Lipidation of ATG8 in wild-type (WT) plants and *atg12* mutants. Total protein was extracted directly into SDS-PAGE sample buffer from WT, *atg12a-1*, *atg12b-1*, and *atg12a-1 atg12b-1* double-mutant seedlings 3 days after germination, subjected to SDS-PAGE in the presence of 6 *M* urea, and analyzed by immunoblot analysis with antibodies against ATG8a and PBA1 as a loading control. *Dashed line*, free ATG8 proteins; *solid line*, ATG8-phosphatidylethanolamine (PE) adducts; *asterisk*, cross-reacting band. *Figure modified from Chung, T., Phillips, A. R., & Vierstra, R. D. 2010. ATG8 lipidation and ATG8-mediated autophagy in Arabidopsis require ATG12 expressed from the differentially controlled ATG12A AND ATG12B loci.* The Plant Journal, 62, *483–493, © 2010 The Authors; Journal compilation © 2010 Blackwell Publishing Ltd.*

nonlipidated species. Treatment of samples with phospholipase prior to electrophoresis removes the PE from the ATG8–PE adduct (Bassham, 2015; Li & Vierstra, 2012), facilitating identification of the lipidated form. Finally, comparison of protein from WT plants with that of autophagy mutants known to be defective in ATG8 lipidation demonstrates conclusively the identity of the observed bands.

2.2.4 ATG8 Accumulation After ConcA Treatment

ATG8 expression and the amount that localizes to autophagosomes increases following autophagy induction, as does the extent of autophagosome formation (Kirisako et al., 1999). The level of ATG8 can therefore be used as a measure of autophagy activity. Upon activation of autophagy, autophagosomes containing cargo that is to be degraded are trafficked to the vacuole, where both the cargo and the inner membrane are degraded by vacuolar hydrolases (Liu & Bassham, 2012). The amount of ATG8 as detected by immunoblotting is dependent on the balance between the rate of autophagosome formation and the rate of degradation; an increase in ATG8 could be due to increased formation or decreased degradation of autophagosomes. ConcA is a vacuolar ATPase inhibitor that leads to the neutralization of the vacuolar pH, blocking trafficking to the vacuole and inhibiting vacuolar degradation activity (Dettmer, Hong-Hermesdorf, Stierhof, & Schumacher, 2006; Dröse et al., 1993; Huss et al., 2002). Autophagic bodies containing ATG8–PE therefore accumulate inside the vacuole (Yoshimoto et al., 2004). ConcA can thus be used to distinguish between changes in ATG8 level due to autophagosome formation or autophagosome degradation. Samples analyzed by immunoblotting using antibodies against *Arabidopsis* ATG8 protein with or without ConcA treatment are compared; upon activation of autophagy, the level of both lipidated and nonlipidated ATG8 increases only or primarily in the presence of ConcA. In conditions in which autophagosome degradation is inhibited, for example, by chemicals or mutants affecting later steps in the autophagy pathway, high levels of ATG8 may be present but the level is insensitive to ConcA. The *Arabidopsis* ATG8 antibodies commonly in use react with ATG8 isoforms in additional plant species, and ATG8 lipidation and accumulation assays can therefore be adapted for other plants, including maize as shown in Fig. 2.

To detect the accumulation of ATG8 after ConcA treatment in *Arabidopsis*:

1. Grow WT and *atg* mutant seedlings for 7 days under nutrient-rich or starved/stressed conditions as described in Section 2.2.3.
2. Incubate seedlings plus or minus 1 μ*M* ConcA for 16 h in 1 × PBS buffer with gentle shaking at 50 rpm.
3. Extract protein directly into SDS-PAGE sample buffer and separate by SDS-PAGE as described in Section 2.2.3.
4. The amount of ATG8 protein is detected by immunoblotting and quantified as described in Section 2.2.3.

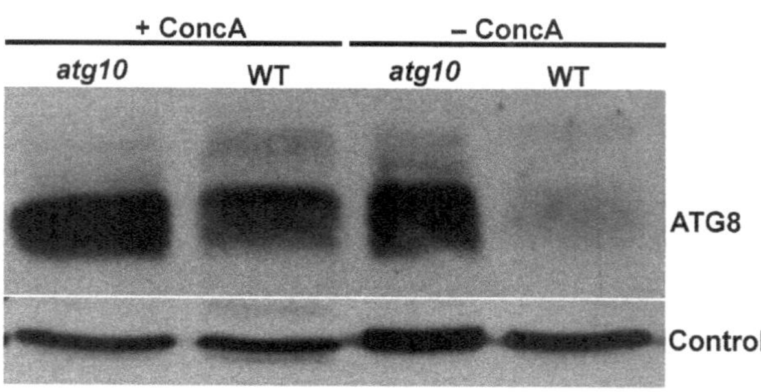

Fig. 2 Comparison of ATG8 accumulation in wild-type (WT) maize and an *atg10* mutant. 10-day-old WT and *atg10* maize seedlings were incubated with 1 μ*M* ConcA or DMSO for 6–8 h. Samples were analyzed by immunoblotting with anti-ATG8 antibodies. Control antibodies were used to demonstrate equal protein loading.

atg mutants that act downstream of ATG8 upregulation can be used as controls, as ATG8 will accumulate in these mutants independent of ConcA treatment. In contrast, upon autophagy activation, ATG8 accumulates to high levels in WT plants only upon 1 μM ConcA treatment, as in the absence of ConcA it is constantly degraded by vacuolar hydrolases upon delivery to the vacuole with autophagosomes.

2.3 GFP–ATG8 Cleavage Assay

ATG8 fused with GFP is a widely used marker for monitoring autophagy, predominantly via fluorescence microscopy in which GFP-labeled autophagosomes and autophagic bodies can be observed directly. This marker can also be used in immunoblot assays, based on the differential stability of GFP compared with ATG8 within the vacuole. Seeds of transgenic *Arabidopsis* plants expressing GFP–ATG8 can be obtained from the *Arabidopsis* Biological Resource Center (http://abrc.osu.edu/), stock numbers CS66943 (GFP–ATG8e) (Xiong et al., 2007), CS39996 (Thompson et al., 2005), or CS68819 (Shin, Lee, & Chung, 2014) (GFP–ATG8a). Upon activation of autophagy, the GFP–ATG8 is delivered to the vacuole via autophagosomes, and while the ATG8 is rapidly degraded, the released free GFP remains relatively stable, and can be detected by immunoblot using commercially available antibodies against GFP (Fig. 3). Thus, when GFP is fused to the N-terminus of ATG8, the abundance of free GFP is a reliable assay for detecting autophagy activity. C-terminal GFP fusions are inappropriate for this assay, as ATG8 is initially synthesized with a C-terminal extension and is proteolytically processed by ATG4 prior to lipidation (Ichimura et al., 2000;

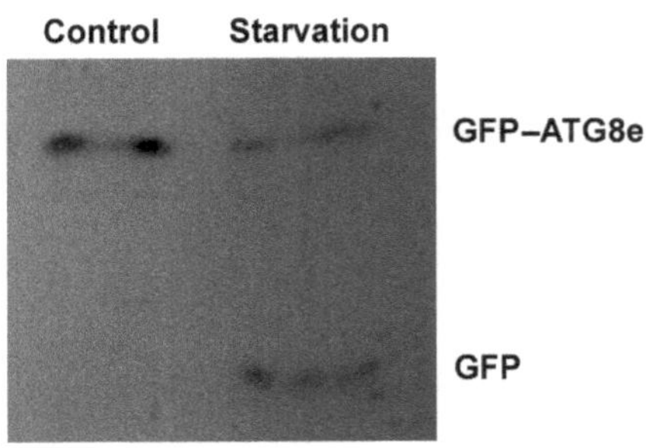

Fig. 3 Detection of free GFP released from GFP–ATG8e upon fixed-carbon starvation. 7-day-old transgenic seedlings expressing GFP–ATG8e were transferred to control or fixed-carbon starvation conditions and grown for an additional 4 days. Samples were analyzed by immunoblotting using anti-GFP antibodies.

Ohsumi, 2001). A C-terminal tag would therefore either prevent ATG8 function or would be cleaved.

2.4 ATG1–ATG13 Protein Degradation

TOR (target of rapamycin) kinase regulates many cellular processes by activation or inactivation of its downstream targets, both in development and in response to environmental stimuli. TOR is a negative regulator of autophagy; during nutrient deficiency, induction of autophagy is mediated by the inhibition of TOR (Liu & Bassham, 2010; Mizushima, 2010; Suzuki et al., 2007). A direct downstream target of TOR that mediates activation of autophagy is the ATG1–ATG13 protein kinase complex (Rabinowitz & White, 2010), which contains the core serine/threonine kinase ATG1 and several accessory partners including ATG11, 13, 17, 29, and 31 (ATG17, 29, and 31 are not present in plants). In yeast, TOR can hyperphosphorylate ATG13 under nutrient-rich conditions, decreasing its binding affinity for ATG1. In contrast, in nutrient deficiency, TOR activity is inhibited, and hypophosphorylated ATG13 has increased affinity for binding to ATG1, leading to autophagy induction (Mizushima, 2010).

Homologs of ATG1 and ATG13 are found in plants; four *ATG1* genes (*ATG1a*, *ATG1b*, *ATG1c*, and *ATG1t*) and two *ATG13* genes (*ATG13a* and *ATG13b*) have been identified in *Arabidopsis* (Suttangkakul, Li, Chung, & Vierstra, 2011). In an *Arabidopsis atg13a atg13b* double mutant, autophagosome formation is completely blocked, but ATG12–ATG5 complex formation and ATG8–PE conjugation still occur. Interestingly, the ATG1–ATG13 complex itself is rapidly turned over by autophagy (Suttangkakul et al., 2011). Thus, immunoblotting using ATG1a or ATG13a antibodies after starvation treatment can indicate autophagy induction; increased degradation of ATG1a or ATG13a denotes more autophagy induction.

Notes for performing this assay: Detailed procedures for detecting the protein levels of ATG1a and ATG13a can be performed as above in Section 2.2.3, but using antibodies against ATG1a and ATG13a. Specific bands detected byATG1a (70 kDa) or ATG13a antibodies (three isoforms of ATG13a at 80, 74, and 70 kDa) will be observed after immunoblotting, and all three ATG13a bands should be included when quantification is applied (Suttangkakul et al., 2011). Success of the assay can be confirmed by the absence of bands in extracts from *atg1a-1* single or *atg13a-1 atg13b-2* double-mutant plants.

2.5 NBR1 Accumulation in *atg* Mutants

Arabidopsis NBR1 (At4g24690) is most likely a selective autophagy cargo receptor that recognizes protein aggregates for degradation and, based on its domain organization, appears to be a functional hybrid of mNBR1 and p62 found in animal cells (Svenning et al., 2011). p62 and NBR1 mediate the autophagic clearance of cytosolic ubiquitinated substrates by binding ubiquitinated proteins via their C-terminal ubiquitin-associated UBA domains, and binding LC3, a mammalian ATG8 homolog, via LIR motifs (Johansen & Lamark, 2011; Svenning et al., 2011). Similarly, *Arabidopsis* NBR1 recognizes ubiquitinated protein substrates and binds ATG8 through its conserved LIR motif (Svenning et al., 2011; Waters et al., 2009), thus selecting cargo for degradation during stressful conditions (Zhou et al., 2013). In mammalian cells, NBR1 is degraded along with its cargo upon activation of autophagy and accumulates in autophagy mutants bound to insoluble protein aggregates (Kirkin et al., 2009). Thus, the level of AtNBR1, determined by immunoblotting using AtNBR1 antibodies, can be used as a measure of autophagy induction in *Arabidopsis*, at least under certain conditions such as heat stress.

Procedures for detecting the level of AtNBR1 protein are as described in Section 2.2.3.

1. Antibodies for detecting endogenous AtNBR1 have been described (Svenning et al., 2011), or transgenic plants expressing epitope-tagged AtNBR1 can be used.
2. Heat-stress 2-week-old soil-grown *Arabidopsis* Col-0 WT, autophagy mutants (*atg7* and *atg5*), and *nbr1* knockout mutants (*nbr1-1* and *nbr1-2*) by exposing them to 45°C for 10 h, with controls maintained at 22°C (Zhou et al., 2013).
3. Treat seedlings with 1 μ*M* ConcA for 16 h to block vacuolar degradation, or with DMSO as a control.
4. Extract total proteins from both stressed and unstressed seedlings, separate by SDS-PAGE, and immunoblot using antibodies against AtNBR1 (Svenning et al., 2011) or the epitope tag as appropriate.

Arabidopsis AtNBR1 is both an autophagy cargo receptor and an autophagy substrate, and is degraded in the vacuole (Zhou et al., 2013); its protein level is dramatically increased by blocking vacuolar acidification (Svenning et al., 2011). Under heat stress, AtNBR1 accumulation in WT plants will increase upon ConcA treatment due to blocked vacuole degradation, whereas in autophagy mutants, it will accumulate to the same level independently of

ConcA treatment (Svenning et al., 2011; Zhou et al., 2013). A comparison between ConcA-treated heat-stressed or unstressed WT plants can therefore be used as a measure of autophagy activation.

Note for performing this assay: The protein level of AtNBR1 is very low in WT plants under normal growth conditions, making identification of the band on immunoblot difficult. The use of conditions in which AtNBR1 accumulates, for example, in *atg* mutants or in heat-stressed WT plants, can be helpful as positive controls in identifying the protein band.

ACKNOWLEDGMENTS

We thank Dr. Richard D. Vierstra (Washington University in St. Louis) for permission to reuse Fig. 1, and Xiaochen Yang and Jie Tang for providing the images used in Figs. 2 and 3. This work was supported by Grants No. DE-SC0014038 from the United States Department of Energy and No. IOS-1353867 from the National Science Foundation to D.C.B.

REFERENCES

Avin-Wittenberg, T., Honig, A., & Galili, G. (2012). Variations on a theme: Plant autophagy in comparison to yeast and mammals. *Protoplasma*, *249*, 285–299.

Bassham, D. C. (2015). Methods for analysis of autophagy in plants. *Methods (San Diego, Calif)*, *75*, 181–188.

Bassham, D. C., Laporte, M., Marty, F., Moriyasu, Y., Ohsumi, Y., Olsen, L. J., et al. (2006). Autophagy in development and stress responses of plants. *Autophagy*, *2*, 2–11.

Behrends, C., Sowa, M. E., Gygi, S. P., & Harper, J. W. (2010). Network organization of the human autophagy system. *Nature*, *466*, 68–76.

Chiba, A., Ishida, H., Nishizawa, N. K., Makino, A., & Mae, T. (2003). Exclusion of ribulose-1,5-bisphosphate carboxylase/oxygenase from chloroplasts by specific bodies in naturally senescing leaves of wheat. *Plant & Cell Physiology*, *44*, 914–921.

Chung, T., Phillips, A. R., & Vierstra, R. D. (2010). ATG8 lipidation and ATG8-mediated autophagy in Arabidopsis require ATG12 expressed from the differentially controlled ATG12A AND ATG12B loci. *The Plant Journal*, *62*, 483–493.

Contento, A. L., Xiong, Y., & Bassham, D. C. (2005). Visualization of autophagy in Arabidopsis using the fluorescent dye monodansylcadaverine and a GFP-AtATG8e fusion protein. *The Plant Journal*, *42*, 598–608.

Cuervo, A. M., Gomes, A. V., Barnes, J. A., & Dice, J. F. (2000). Selective degradation of annexins by chaperone-mediated autophagy. *The Journal of Biological Chemistry*, *75*, 33329–33335.

De Duve, C., & Wattiaux, R. (1966). Functions of lysosomes. *Annual Review of Physiology*, *28*, 435–492.

Dettmer, J., Hong-Hermesdorf, A., Stierhof, Y. D., & Schumacher, K. (2006). Vacuolar H^+-ATPase activity is required for endocytic and secretory trafficking in Arabidopsis. *Plant Cell*, *18*, 715–730.

Doelling, J. H., Walker, J. M., Friedman, E. M., Thompson, A. R., & Vierstra, R. D. (2002). The APG8/12-activating enzyme APG7 is required for proper nutrient recycling and senescence in Arabidopsis thaliana. *The Journal of Biological Chemistry*, *277*, 33105–33114.

Dröse, S., Bindseil, K. U., Bowman, E. J., Siebers, A., Zeeck, A., & Altendorf, K. (1993). Inhibitory effect of modified bafilomycins and concanamycins on P- and V-type adenosinetriphosphatases. *Biochemistry*, *32*, 3902–3906.

Floyd, B. E., Morriss, S. C., Macintosh, G. C., & Bassham, D. C. (2012). What to eat: Evidence for selective autophagy in plants. *Journal of Integrative Plant Biology*, *54*, 907–920.

Fujioka, Y., Noda, N. N., Fujii, K., Yoshimoto, K., Ohsumi, Y., & Inagaki, F. (2008). In vitro reconstitution of plant Atg8 and Atg12 conjugation systems essential for autophagy. *The Journal of Biological Chemistry*, *283*, 1921–1928.

Ghiglione, H. O., Gonzalez, F. G., Serrago, R., Maldonado, S. B., Chilcott, C., Curá, J. A., et al. (2008). Autophagy regulated by day length determines the number of fertile florets in wheat. *The Plant Journal*, *55*, 1010–1024.

Guiboileau, A., Yoshimoto, K., Soulay, F., Bataillé, M. P., Avice, J. C., & Masclaux-Daubresse, C. (2012). Autophagy machinery controls nitrogen remobilization at the whole-plant level under both limiting and ample nitrate conditions in Arabidopsis. *The New Phytologist*, *194*, 732–740.

Hayward, A. P., & Dinesh-Kumar, S. P. (2011). What can plant autophagy do for an innate immune response? *Annual Review of Phytopathology*, *49*, 557–576.

Hofius, D., Munch, D., Bressendorff, S., Mundy, J., & Petersen, M. (2011). Role of autophagy in disease resistance and hypersensitive response-associated cell death. *Cell Death and Differentiation*, *18*, 1257–1262.

Huss, M., Ingenhorst, G., König, S., Gassel, M., Dröse, S., Zeeck, A., et al. (2002). Concanamycin A, the specific inhibitor of V-ATPases, binds to the V(o) subunit c. *The Journal of Biological Chemistry*, *277*, 40544–40548.

Ichimura, Y., Kirisako, T., Takao, T., Satomi, Y., Shimonishi, Y., Ishihara, N., et al. (2000). A ubiquitin-like system mediates protein lipidation. *Nature*, *408*, 488–492.

Ishida, H., & Yoshimoto, K. (2008). Chloroplasts are partially mobilized to the vacuole by autophagy. *Autophagy*, *4*, 961–962.

Johansen, T., & Lamark, T. (2011). Selective autophagy mediated by autophagic adapter proteins. *Autophagy*, *7*, 279–296.

Kabeya, Y., Mizushima, N., Ueno, T., Yamamoto, A., Kirisako, T., Noda, T., et al. (2000). LC3, a mammalian homologue of yeast Apg8p, is localized in autophagosome membranes after processing. *The EMBO Journal*, *19*, 5720–5728.

Kim, J., & Klionsky, D. J. (2000). Autophagy, cytoplasm-to-vacuole targeting pathway, and pexophagy in yeast and mammalian cells. *Annual Review of Biochemistry*, *69*, 303–342.

Kim, J., Lee, H., Lee, H. N., Kim, S. H., Shin, K. D., & Chung, T. (2013). Autophagy-related proteins are required for degradation of peroxisomes in Arabidopsis hypocotyls during seedling growth. *Plant Cell*, *25*, 4956–4966.

Kirisako, T., Baba, M., Ishihara, N., Miyazawa, K., Ohsumi, M., Yoshimori, T., et al. (1999). Formation process of autophagosome is traced with Apg8/Aut7p in yeast. *The Journal of Cell Biology*, *147*, 435–446.

Kirkin, V., Lamark, T., Sou, Y. S., Bjørkøy, G., Nunn, J. L., Bruun, J. A., et al. (2009). A role for NBR1 in autophagosomal degradation of ubiquitinated substrates. *Molecular Cell*, *33*, 505–516.

Klionsky, D. J., Abdelmohsen, K., Abe, A., Abedin, M. J., Abeliovich, H., Acevedo Arozena, A., et al. (2016). Guidelines for the use and interpretation of assays for monitoring autophagy (3rd edition). *Autophagy*, *12*, 1–222.

Klionsky, D. J., & Emr, S. D. (2000). Autophagy as a regulated pathway of cellular degradation. *Science*, *290*, 1717–1721.

Kuzuoglu-Ozturk, D., Cebeci Yalcinkaya, O., Akpinar, B. A., Mitou, G., Korkmaz, G., Gozuacik, D., et al. (2012). Autophagy-related gene, TdAtg8, in wild emmer wheat plays a role in drought and osmotic stress response. *Planta*, *236*, 1081–1092.

Li, F., Chung, T., Pennington, J. G., Federico, M. L., Kaeppler, H. F., Kaeppler, S. M., et al. (2015). Autophagic recycling plays a central role in maize nitrogen remobilization. *Plant Cell, 27*, 1389–1408.
Li, F., Chung, T., & Vierstra, R. D. (2014). AUTOPHAGY-RELATED11 plays a critical role in general autophagy- and senescence-induced mitophagy in Arabidopsis. *Plant Cell, 26*, 788–807.
Li, F., & Vierstra, R. D. (2012). Autophagy: A multifaceted intracellular system for bulk and selective recycling. *Trends in Plant Science, 17*, 526–537.
Liu, Y., & Bassham, D. C. (2010). TOR is a negative regulator of autophagy in Arabidopsis thaliana. *PloS One, 5*, e11883.
Liu, Y., & Bassham, D. C. (2012). Autophagy: Pathways for self-eating in plant cells. *Annual Review of Plant Biology, 63*, 215–237.
Liu, Y., Burgos, J. S., Deng, Y., Srivastava, R., Howell, S. H., & Bassham, D. C. (2012). Degradation of the endoplasmic reticulum by autophagy during endoplasmic reticulum stress in Arabidopsis. *Plant Cell, 24*, 4635–4651.
Liu, L., Sakakibara, K., Chen, Q., & Okamoto, K. (2014). Receptor-mediated mitophagy in yeast and mammalian systems. *Cell Research, 24*, 787–795.
Liu, Y., Schiff, M., Czymmek, K., Tallóczy, Z., Levine, B., & Dinesh-Kumar, S. P. (2005). Autophagy regulates programmed cell death during the plant innate immune response. *Cell, 121*, 567–577.
Liu, Y., Xiong, Y., & Bassham, D. C. (2009). Autophagy is required for tolerance of drought and salt stress in plants. *Autophagy, 5*, 954–963.
Lv, X., Pu, X., Qin, G., Zhu, T., & Lin, H. (2014). The roles of autophagy in development and stress responses in Arabidopsis thaliana. *Apoptosis, 19*, 905–921.
Michaeli, S., Galili, G., Genschik, P., Fernie, A. R., & Avin-Wittenberg, T. (2016). Autophagy in plants—What's new on the menu? *Trends in Plant Science, 21*, 134–144.
Mijaljica, D., Prescott, M., & Devenish, R. J. (2011). Microautophagy in mammalian cells: Revisiting a 40-year-old conundrum. *Autophagy, 7*, 673–682.
Minina, E. A., Filonova, L. H., Fukada, K., Savenkov, E. I., Gogvadze, V., Clapham, D., et al. (2013). Autophagy and metacaspase determine the mode of cell death in plants. *The Journal of Cell Biology, 203*, 917–927.
Mitou, G., Budak, H., & Gozuacik, D. (2009). Techniques to study autophagy in plants. *International Journal of Plant Genomics, 2009*, 451357.
Mizushima, N. (2010). The role of the Atg1/ULK1 complex in autophagy regulation. *Current Opinion in Cell Biology, 22*, 132–139.
Mochida, K., Oikawa, Y., Kimura, Y., Kirisako, H., Hirano, H., Ohsumi, Y., et al. (2015). Receptor-mediated selective autophagy degrades the endoplasmic reticulum and the nucleus. *Nature, 522*, 359–362.
Moriyasu, Y., Hattori, M., Jauh, G. Y., & Rogers, J. C. (2003). Alpha tonoplast intrinsic protein is specifically associated with vacuole membrane involved in an autophagic process. *Plant & Cell Physiology, 44*, 795–802.
Moriyasu, Y., & Ohsumi, Y. (1996). Autophagy in tobacco suspension-cultured cells in response to sucrose starvation. *Plant Physiology, 111*, 1233–1241.
Ohsumi, Y. (2001). Molecular dissection of autophagy: Two ubiquitin-like systems. *Nature Reviews. Molecular Cell Biology, 2*, 211–216.
Ono, Y., Wada, S., Izumi, M., Makino, A., & Ishida, H. (2013). Evidence for contribution of autophagy to rubisco degradation during leaf senescence in Arabidopsis thaliana. *Plant, Cell & Environment, 36*, 1147–1159.
Rabinowitz, J. D., & White, E. (2010). Autophagy and metabolism. *Science, 330*, 1344–1348.
Rana, R. M., Dong, S., Ali, Z., Huang, J., & Zhang, H. S. (2012). Regulation of ATG6/ Beclin-1 homologs by abiotic stresses and hormones in rice (Oryza sativa L.). *Genetics and Molecular Research, 11*, 3676–3687.

Reggiori, F., & Klionsky, D. J. (2002). Autophagy in the eukaryotic cell. *Eukaryotic Cell, 1*, 11–21.

Schneider, C. A., Rasband, W. S., & Eliceiri, K. W. (2012). NIH Image to ImageJ: 25 Years of image analysis. *Nature Methods, 9*, 671–675.

Shin, K. D., Lee, H. N., & Chung, T. (2014). A revised assay for monitoring autophagic flux in *Arabidopsis thaliana* reveals involvement of AUTOPHAGY-RELATED9 in autophagy. *Molecules and Cells, 37*, 399–405.

Shin, J. H., Yoshimoto, K., Ohsumi, Y., Jeon, J. S., & An, G. (2009). OsATG10b, an autophagosome component, is needed for cell survival against oxidative stresses in rice. *Molecules and Cells, 27*, 67–74.

Slobodkin, M. R., & Elazar, Z. (2013). The Atg8 family: Multifunctional ubiquitin-like key regulators of autophagy. *Essays in Biochemistry, 55*, 51–64.

Suttangkakul, A., Li, F., Chung, T., & Vierstra, R. D. (2011). The ATG1/ATG13 protein kinase complex is both a regulator and a target of autophagic recycling in Arabidopsis. *Plant Cell, 23*, 3761–3779.

Suzuki, K., Kubota, Y., Sekito, T., & Ohsumi, Y. (2007). Hierarchy of Atg proteins in pre-autophagosomal structure organization. *Genes to Cells, 12*, 209–218.

Suzuki, K., & Ohsumi, Y. (2010). Current knowledge of the pre-autophagosomal structure (PAS). *FEBS Letters, 584*, 1280–1286.

Svenning, S., Lamark, T., Krause, K., & Johansen, T. (2011). Plant NBR1 is a selective autophagy substrate and a functional hybrid of the mammalian autophagic adapters NBR1 and p62/SQSTM1. *Autophagy*, 7, 993–1010.

Takatsuka, C., Inoue, Y., Matsuoka, K., & Moriyasu, Y. (2004). 3-Methyladenine inhibits autophagy in tobacco culture cells under sucrose starvation conditions. *Plant & Cell Physiology, 45*, 265–274.

Thompson, A. R., Doelling, J. H., Suttangkakul, A., & Vierstra, R. D. (2005). Autophagic nutrient recycling in Arabidopsis directed by the ATG8 and ATG12 conjugation pathways. *Plant Physiology, 138*, 2097–2110.

van Doorn, W. G., & Papini, A. (2013). Ultrastructure of autophagy in plant cells. *Autophagy, 9*, 1922–1936.

Voitsekhovskaja, O. V., Schiermeyer, A., & Reumann, S. (2014). Plant peroxisomes are degraded by starvation-induced and constitutive autophagy in tobacco BY-2 suspension-cultured cells. *Frontiers in Plant Science, 5*, 629.

Wada, S., & Ishida, H. (2009). Chloroplasts autophagy during senescence of individually darkened leaves. *Plant Signaling & Behavior, 4*, 565–567.

Waters, S., Marchbank, K., Solomon, E., Whitehouse, C., & Gautel, M. (2009). Interactions with LC3 and polyubiquitin chains link nbr1 to autophagic protein turnover. *FEBS Letters, 583*, 1846–1852.

Woo, J., Park, E., & Dinesh-Kumar, S. P. (2014). Differential processing of Arabidopsis ubiquitin-like Atg8 autophagy proteins by Atg4 cysteine proteases. *Proceedings of the National Academy of Sciences of the United States of America, 111*, 863–868.

Xiong, Y., Contento, A. L., Nguyen, P. Q., & Bassham, D. C. (2007). Degradation of oxidized proteins by autophagy during oxidative stress in Arabidopsis. *Plant Physiology, 143*, 291–299.

Yang, X., & Bassham, D. C. (2015). New insight into the mechanism and function of autophagy in plant cells. *International Review of Cell and Molecular Biology, 320*, 1–40.

Yang, Z., & Klionsky, D. J. (2009). An overview of the molecular mechanism of autophagy. *Current Topics in Microbiology and Immunology, 335*, 1–32.

Yang, Z., & Klionsky, D. J. (2010). Mammalian autophagy: Core molecular machinery and signaling regulation. *Current Opinion in Cell Biology, 22*, 124–131.

Yang, X., Srivastava, R., Howell, S. H., & Bassham, D. C. (2016). Activation of autophagy by unfolded proteins during endoplasmic reticulum stress. *The Plant Journal, 85*, 83–95.

Yoshimoto, K., Hanaoka, H., Sato, S., Kato, T., Tabata, S., Noda, T., et al. (2004). Processing of ATG8s, ubiquitin-like proteins, and their deconjugation by ATG4s are essential for plant autophagy. *Plant Cell*, *16*, 2967–2983.

Yoshimoto, K., Shibata, M., Kondo, M., Oikawa, K., Sato, M., Toyooka, K., et al. (2014). Organ-specific quality control of plant peroxisomes is mediated by autophagy. *Journal of Cell Science*, *127*, 1161–1168.

Zaffagnini, G., & Martens, S. (2016). Mechanisms of selective autophagy. *Journal of Molecular Biology*, *428*, 1714–1724.

Zhou, J., Wang, J., Cheng, Y., Chi, Y. J., Fan, B., Yu, J. Q., et al. (2013). NBR1-mediated selective autophagy targets insoluble ubiquitinated protein aggregates in plant stress responses. *PLoS Genetics*, *9*, e1003196.

Zhou, J., Zhang, Y., Qi, J., Chi, Y., Fan, B., Yu, J. Q., et al. (2014). E3 ubiquitin ligase CHIP and NBR1-mediated selective autophagy protect additively against proteotoxicity in plant stress responses. *PLoS Genetics*, *10*, e1004116.

Zientara-Rytter, K., Lukomska, J., Moniuszko, G., Gwozdecki, R., Surowiecki, P., Lewandowska, M., et al. (2011). Identification and functional analysis of Joka2, a tobacco member of the family of selective autophagy cargo receptors. *Autophagy*, *7*, 1145–1158.

Zientara-Rytter, K., & Sirko, A. (2014). Significant role of PB1 and UBA domains in multimerization of Joka2, a selective autophagy cargo receptor from tobacco. *Frontiers in Plant Science*, *5*, 13.

CHAPTER TWENTY-SIX

Using Photoconvertible and Extractable Fluorescent Proteins to Study Autophagy in Plants

M.O. Abiodun*, K. Matsuoka*,†,‡,1

*Laboratory of Plant Nutrition, Faculty of Agriculture, Kyushu University, Fukuoka, Japan
†Biotron Application Center, Kyushu University, Fukuoka, Japan
‡Research Center for Organelle Homeostasis, Kyushu University, Fukuoka, Japan
[1]Corresponding author: e-mail address: kenmat@agr.kyushu-u.ac.jp

Contents

Abstract

Several methodologies have been employed to understand the kinetics of induced autophagic degradation in plants, but most of them are not capable of distinguishing the autophagic cargo proteins before and after induction of autophagy in cells. Here, we

Methods in Enzymology, Volume 588
ISSN 0076-6879
http://dx.doi.org/10.1016/bs.mie.2016.10.040

designed a mass photoconverter that allowed us to simultaneously monitor protein synthesis and degradation in tobacco BY-2 cells using a photoconvertible fluorescence marker protein, Kikume Green Red (KikGR). An example of a new protocol for the analysis of autophagy progression using a fusion protein of cytochrome *b*5 and KikGR under phosphate starvation is described. The other example described is the analysis of the proliferation of Golgi apparatus in tobacco BY-2 cells using the fusion protein of a prolyl 4-hydroxylase NtP4H1.1 and monomeric KikGR. A detailed protocol on key analysis, as well as tips and notes for experiments using KikGR proteins, are described.

1. INTRODUCTION

When there is a need for plants to seek out alternative nutrients as a result of deficiencies, self-digestion of intracellular contents is triggered. In cultured cells, this important cellular process, called *autophagy,* can be induced by replacing the regular nutrient medium with a nutrient-deficient one. A limited supply of nutrients such as sugars, nitrogen, and phosphate (Pi) has been reported to induce autophagy in cultured tobacco BY-cells (Moriyasu & Ohsumi, 1996; Toyooka, Takeuchi, Moriyasu, Fukuda, & Matsuoka, 2006). Apart from BY-2 cells, starvation-induced autophagy has been reported in cultured rice cells (Chen, Liu, Chen, Wu, & Yu, 1994), cultured sycamore maple cells (Aubert et al., 1996), and *Arabidopsis* plants and cells (Contento, Kim, & Bassham, 2004; Yoshimoto et al., 2004). In their methodologies, some of these reports employed single, non-photoconvertible fluorescent proteins (Yoshimoto et al., 2004; Yuasa, Toyooka, Fukuda, & Matsuoka, 2005) like RFP and GFP, which has helped them to reach great experimental conclusions. Toyooka et al. (2006) showed that fusion proteins of cytochrome *b*5 (Cyt *b*5) and RFP (Cyt *b*5-RFP) were good substrates for autophagy in tobacco BY-2 cells because the intact and processed forms of Cyt *b*5-RFP can be distinguished easily after the separation of proteins by sodium dodecyl sulfate-polyacrylamide gel electrophoresis (SDS-PAGE) and that quantification of the fluorescent intensities of RFP-related polypeptides can be used to calculate the efficiency of autophagy. The challenge with this system of quantification is the inability to differentiate between the proteins synthesized before and after the induction of autophagy. To overcome this problem, a new method capable of monitoring the dynamics of proteins before and after induced autophagy will be a great breakthrough in biology.

In order to visualize the dynamics of proteins and organelle in vivo, Hawes, Schoberer, Hummel, and Osterrieder (2010) suggested the analysis of Golgi apparatus by utilizing photoconvertible fluorescent proteins, e.g., "Kaede" tagged with Golgi-resident proteins. When irradiated with ultraviolet (UV) or violet light, the fluorescence emitted by populations of Golgi stacks was converted from green to red. Upon incubation of the tissue, the new Golgi marker protein emitted green fluorescence, and when transported to existing Golgi stacks, the protein gave a mix of red, green, and yellow signals (Brown et al., 2010). However, the use of microscope for photoconversion has limitations in application as the condition is usually stressful to cells. In addition, the Kaede protein is not easy to use, as the intensity of fluorescence before and after conversion is quite different. We thus chose the Kikume green red protein KikGR as a photoconvertible reporter protein because it shows similar intensities of fluorescence before and after conversion (Tsutsui, Karasawa, Shimizu, Nukina, & Miyawaki, 2005). Our previously reported method (Toyooka et al., 2006) was improved by replacing the RFP with photoconvertible KikGR fluorescent marker protein. Also, being conscious of the effect of converting the photoconvertible proteins under the microscope, we used a mass photoconverter developed in our laboratory (Abiodun & Matsuoka, 2013a, 2013b), as this allows time-course analysis, reduces stress on cells, and can convert a large number of transformed cells. Using these methodologies, we analyzed Pi-starvation-induced autophagy using this new reporter protein.

Also, in a separate experiment using a monomeric form of the photoconvertible fluorescence marker protein (mKikGR) fused with a Golgi-localizing tobacco prolyl 4-hydroxylase (NtP4H1.1), but in a normal medium without starvation, we report that exchange and dilution of fluorescence protein in the Golgi takes place. A 3-h, time-course image acquisition and analysis after photoconversion and incubation in the dark revealed that even without starvation, newly synthesized Golgi proteins are capable of diluting preexisting ones. It would be interesting to use the same method to investigate the movement and exchange of proteins between organelles.

2. EXPRESSION OF KikGR FUSION PROTEINS IN TOBACCO BY-2 CELLS

2.1 Plasmid Construction and Transformation of Tobacco BY-2 Cells

The expression of tetrameric or monomeric KikGR fusion proteins as examples described in this work was as follows. For the tetrameric KikGR fusion, pMAT137–Cyt *b*5–cKikGR, an expression plasmid of Cyt *b*5–KikGR fusion protein under the enhancer-duplicated CaMV35S promoter, was constructed as described by Tasaki, Asatsuma, and Matsuoka (2014). Monomeric KikGR fusion with NtP4H1.1 construct was as described by Abiodun and Matsuoka (2013a). These were used to transform tobacco BY-2 cells using *Agrobacterium* via the standard procedure as described previously (Tasaki et al., 2014).

2.2 Cell Culture and Culture Media

Wild-type tobacco BY-2 cell lines were used as the control, while tobacco BY-2 cell lines expressing Cyt *b*5–KikGR and NtP4H1.1–mKikGR were used for the tetrameric and monomeric experimental lines in this study. The cells were cultured in a 100-mL aliquot in a 300-mL Erlenmeyer flask placed in a mechanical shaker at a temperature of 26°C as described previously (Matsuoka & Nakamura, 1991). For the induction of autophagy, we routinely used rapidly growing cells—cells grown in medium for 3 or 4 days. These cells have been starved of nutrients when they reach the stationary phase of growth. Thus, cells in the stationary phase of growth are not useful for the analysis of autophagy. For the nutrient-starvation experiments with either sucrose, nitrogen, or phosphate, the medium were prepared as described by Matsuoka (2009).

3. PHOTOCONVERSION OF KikGR AND mKikGR FUSION PROTEINS IN TRANSFORMED TOBACCO CELLS

3.1 The Mass Converter

The mass converter (Fig. 1) consisted of a lamp for color conversion and a rotary shaker with a light reflection chamber. A 100-W black light bulb (H100BL-L; Toshiba, Tokyo, Japan) was used in our converter. The lamp was fixed into the base of an outdoor light projector (CAD mirror HT-4051X, Toshiba) to concentrate the light beam onto the flask on the shaker. The front of the projector was covered with a plane glass that reduced the transfer of lamp heat to the target. The projector unit was attached to electrical wire, which was plugged directly into a power source. The lamp-reflector system was kept on a plank base and held together using a G-clamp to centralize it to the flask chamber. The converter setup was placed in a normally air-conditioned room.

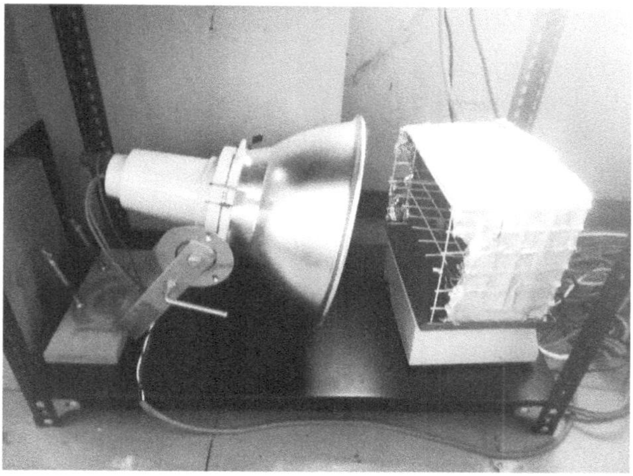

Fig. 1 The mass converter used to convert the fluorescence of KikGR-related proteins expressed in tobacco BY-2 cells. Details on how to assemble this set are given in the text.

The light spectrum of the black light bulb, which was composed of three mercury emission lines of 336, 365, and 405 nm (http://image.rakuten.co.jp/alllight/cabinet/hid/blackhid_bunko.jpg), allowed the efficient conversion of KikGR fluorescence. Unfortunately, the production of the bulb was terminated several years ago, although we could still purchase them from the market. Thus, in the future, similar bulbs from other companies or light-emitting diodes with a purple color may become inevitable in the construction of this converter.

The flask chamber was an Al foil-covered steel wire box on a small rotary shaker (BC-740; Bio Craft, Tokyo, Japan), with only one side opened so that the light from the lamp could be concentrated on the flask. The box had an internal flask holder cut and was hung to hold the necks of two flasks at the same time to avoid tilting off during shaking. We ensured that the size of the net box fit exactly into the square or rectangular size of the flat board on top of the shaker. This was necessary for firmness, as an undersized or oversized net box would not be able to stay firm on the shaker during photoconversion. As a result of wear and tear, the aluminum foil cover can be changed from time to time to ensure the necessary darkness during conversion. The flask chamber was placed in front of the lamp house with a reasonable distance of about 50 cm to allow for minimizing the excess heating of the flask in the chamber from the lamp.

3.2 Photoconversion

1. Switch on the lamp 20 min before placing the flask containing the cells into the flask chamber.
2. Set a 95–100-mL aliquot of the culture of log-phase-transformed cells, in a 300-mL Erlenmeyer flask in the flask chamber.
3. Switch on the shaker and shake the flask at 130 rpm.
4. Expose for an appropriate period of time.

The time for the conversion depends on construct and cell lines. For example, 1 h is sufficient for the conversion of Cyt *b*5–KikGR aggregate in transformed tobacco BY-2 cells, but 3 h is required to achieve a near-complete photoconversion for the NtP4H1.1–mKikGR fusion. For other constructs that are in use in our laboratory, different times are required for their conversion. Thus, the minimum time for the conversion will be defined experimentally for different fusion proteins.

3.3 Postconversion Handling

After the completion of the photoconversion, the cells were washed 3 × in centrifuge tubes using either a normal or a starvation medium. The resulting cells are then transferred into a flask containing the respective medium and incubated in the dark at 26°C in a mechanical shaker, where cell samples would be taken for microscopy and protein extraction at the time intervals indicated. Note that regular fluorescent lamps that are commonly used in the laboratory emit purple light that can cause the conversion of KikGR fluorescence. Thus, if the mechanical shaker for the culture cells has windows, the culture flask will require an Al foil cover to achieve complete darkness during the culture.

4. DETECTION OF PHOTOCONVERTED AND NONCONVERTED KikGR AND mKikGR FUSION PROTEINS AFTER SEPARATION OF PROTEINS BY SDS-PAGE

4.1 Protein Extraction From Transformed Tobacco BY-2 Cells

Gel electrophoresis is one of the easiest and common methods for the separation of proteins. As a fluorophore of red fluorescent protein is stable under partially denatured conditions with SDS (Baird, Zacharias, & Tsien, 2000), we tested the same conditions to detect other fluorescent proteins and found that the same conditions can be used to detect and to quantify fusion proteins with EGFP, YFP, KikGR, mKikGR, and DsRed and DsRed monomer (Liu, Hayashi, & Matsuoka, 2015; Tasaki et al., 2014; Toyooka et al., 2006; Yamauchi et al., 2013; Yuasa et al., 2005; Y. Oda & K. Matsuoka, unpublished result).

The following is the protocol for the extraction of proteins from tobacco cells to analyze the Cyt *b*5–KikGR fusion protein and its degradation products after induction of autophagy, as well as the decreased and novel appearance of YFP–NtAtg8a expressed in transformed tobacco BY-2 cells. In most cases, the handling of samples was carried out during the daytime in a laboratory with glass windows covered with UV protecting sheets, and without using fluorescent lamps as a source of light. In other cases, protein extraction was carried out during cloudy days or at night using fluorescent lamps as a light source in the room. In these cases, only the ceiling lamps were lit, while the sample tubes in an ice bath were covered with Al foil to minimize the exposure to light from the fluorescent lamps.

1. Precipitate cells from cell suspension taken from the incubated aliquots by centrifuging at $360 \times g$ for 1 min in a 1.5-mL microfuge tube.
2. Remove the supernatant and suspend the precipitated cells with phosphate-buffered saline (PBS; 4.3 mM of Na_2HPO_4; 1.4 mM of KH_2PO_4; 2.7 mM of KCl; 137 mM of NaCl, pH 7.4) using about 10 volumes of precipitated cells.
3. Centrifuge as described in step 1 and remove the supernatant.
4. Suspend the sedimented cells with equal volumes of PBS.
5. Subject the cell–PBS mix to sonication using a Bioruptor UCD-200TM sonicator (Cosmo Bio. Co. Ltd., Tokyo, Japan) in an ice-cold water bath with the following settings: M power setting for 1 min at 30-s intervals $10\times$.
6. Remove the microfuge tube from the sonicator. Centrifuge the tube at $360 \times g$ for 5 min, and then collect the supernatant and use as a total cell lysate.
7. An aliquot is kept on ice for protein quantification.
8. Freeze the total cell lysate with liquid nitrogen and store at –80°C until it is time for use.

This protocol can be use to extract many fluorescent protein fusions from transformed tobacco BY-2 cells, as well as from transformed *Arabidopsis* leaves and seedlings. Occasionally, however, this protocol may not be applied to some of the fluorescent proteins. For example, red-converted NtP4H1.1–mKikGR could not be recovered quantitatively (M.O. Abiodun and K. Matsuoka, unpublished observation) and HEPES-based buffer is required for the quantitative extraction of Bor1–GFP fusion proteins (Yamauchi et al., 2013).

4.2 Separation of Proteins by SDS-PAGE and Detection of Fluorescence Proteins in the Gel

1. Mix aliquots of the total proteins extracted with 0.25 volume of 5 × SDS sample buffer [250 m*M* of Tris–HCl, pH 6.8, 50% (w/v) of glycerol, 10% (w/v) of SDS, 0.2 *M* of dithiothreitol (DTT), 1% (w/v) of bromophenol blue].
2. Apply the mixed protein sample into the well of SDS-PAGE gel that is set in an electrophoresis chamber with a cooling system.
3. Separate proteins by electrophoresis in dark conditions. We usually cover the electrophoresis unit with a paper box.
4. At the completion of electrophoresis, remove the gel from the electrophoresis chamber and temporarily store it in distilled water.
5. Place the gel on a clean scanning area of a Typhoon 9400 image analyzer (GE Healthcare, Little Chalfont, UK).
6. Scan the gel in the following conditions: a 580 BP30 Cy3 filter set with a 532-nm laser at 650 V for recording the red fluorescence, and then a 526 SP Cy2 filter set with a 488-nm laser at 650 V for recording the green fluorescence for the KikGR fusion proteins.
7. Quantify fluorescence intensity of fluorescence bands using ImageQuant software attached to the Typhoon image analyzer.

   Note: We use a Hoefer Mighty Small mini-gel system and apply 12.5 mA per gel for separation. Excess application of current increases the temperature of the gel. This heat denatures the fluorophore in fluorescent proteins. Under the recording conditions given here, YFP gave signals in both scanning conditions (Fig. 2); thus, nonconverted and photoconverted KikGR can be easily distinguished. Sometimes the environmental contamination of fluorescent compounds may cause the irregular background (e.g., Fig. 2, red fluorescence). Thus, cleaning the electrophoresis apparatus is very important for the quantification of the fluorescence intensities of protein bands.

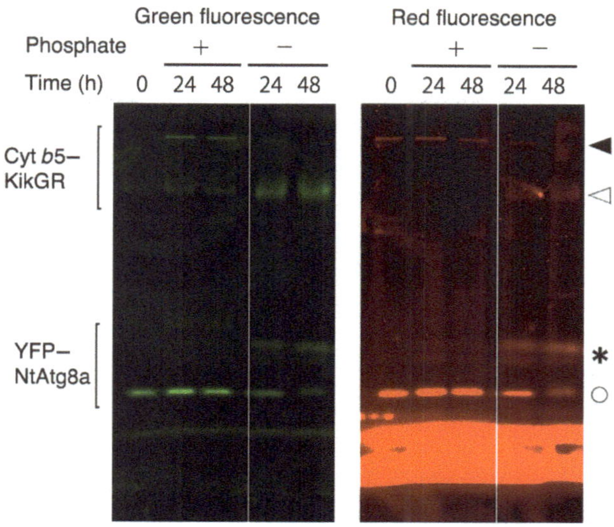

Fig. 2 Analysis of the phosphate-depletion dependent autophagy induction. Cells expressing both Cyt *b*5–KikGR and YFP–NtAtg8a (Toyooka et al., 2006) were exposed to purple light, incubated in a Pi-free medium, and cultured for the indicated times. Equal amounts of proteins (5 μg) from the cells were separated by SDS-PAGE, and fluorescence in the gel was recorded as described in the text. *Black* and *white arrowheads* indicate the migration position of intact and processed forms of Cyt *b*5–KikGR, respectively. *Open circle* and *asterisk* indicate the usual and slow-migrating forms of YFP–NtAtg8a, respectively. Note the high red fluorescence of compounds at the front of electrophoresis, as well as the uneven background of fluorescence in the image of red fluorescence. The uneven fluorescence, which comes from environmental contamination, prevented the quantification of the bands from the image. The uneven background can be reduced by repeated cleaning of glass plates for the casting of SDS-polyacrylamide gel.

5. DETECTION OF PHOTOCONVERTED AND NONCONVERTED KikGR AND mKikGR FUSION PROTEINS USING A MICROSCOPE

5.1 Using an Epifluorescence Microscope to Confirm the Expression of KikGR and mKikGR Fusion Proteins

The routine analysis of the expression of KikGR and mKikGR fusion proteins can be done using an epifluorescence microscope. We use Olympus IX50 microscope equipped with DP70 color CCD camera (Olympus, Tokyo, Japan). Images are collected with the aid of a 20× LC PLAN FL lens (Olympus) employing WIB filter/dichroic mirror cube

(Olympus) to detect the green fluorescence and RFP filter/dichroic mirror cube (Olympus) to detect the red fluorescence. For the confirmation that the green fluorescence is derived from KikGR or mKikGR fusion proteins, light from WBV filter/dichroic mirror cube is illuminated to the specimen with green fluorescence for 5 min, and then the fluorescent red color is investigated using an RFP cube. The conversion of fluorescence from green to red is an indication that the fluorescence comes from a KikGR or mKikGR protein, not from the autofluorescence of the cells.

5.2 Confocal Imaging Using Laser-Scanning Microscope

Using a Leica TCS SP8 confocal microscope system (Leica Microsystems, Mannheim, Germany) which was equipped with a white light laser and HyD detectors, we routinely captured images using an HCPL APO CS2 40 × 1.30 oil lens with a pinhole of 44.1 μm of the confocal unit at an image resolution of 1024 × 1024 pixels at 100 Hz. An excitation wavelength of 505 nm and emission between 514 and 546 nm is used to detect the green fluorescence, while an excitation wavelength of 555 nm and emission between 610 and 653 nm is used to detect the red fluorescence. Only a single scan of each color using line scan mode is used to collect each image. An example of the record of Cyt *b*5–KikGR aggregate 24 h after color conversion is shown in Fig. 3.

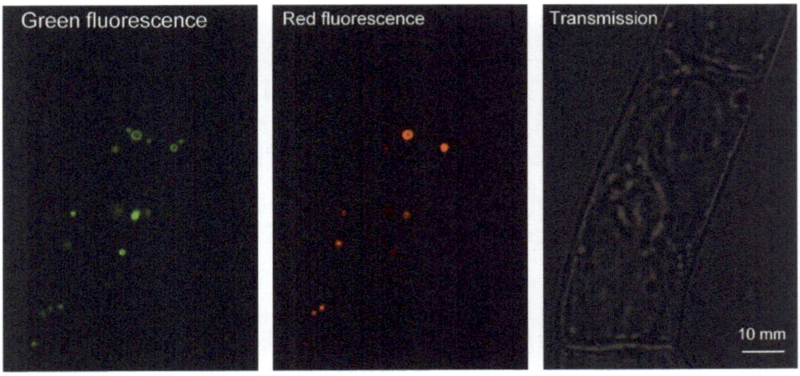

Fig. 3 Analysis of the proliferation and de novo generation of the aggregates. Confocal fluorescence microscopic images and the transmission image of Cyt *b*5–KikGR-expressing cells 24 h after color conversion and incubation in a normal medium are shown.

5.3 Confocal Imaging Using an Epifluorescence Microscope With a Spinning Disk Confocal Unit and an EMCCD Camera

Although the condition using a laser-scanning confocal microscope is sufficient to detect the aggregate generated from Cyt *b*5-RFP, weak fluorescence signals, such as signals from NtP4H1.1–mKikGR, cannot be quantified. In this case, we used an epifluorescence microscope with a spinning disk confocal unit and an electron-multiplying charge-coupled device (EMCCD) camera (Abiodun & Matsuoka, 2013a), as the excitation power is stronger using a band-pass filter than a single wavelength light beam from the white light laser and the sensitivity of an EMCCD camera is higher than that of a HyD system.

As we could not set up a condition to extract red-converted NtP4H1.1–mKikGR proteins for electrophoresis analysis, we quantified the green and red fluorescence intensities of the Golgi apparatus from the collected microscopic images. If a fusion protein cannot be extracted from the cells in this particular experiment, such quantification may be possible in any other experiment, including autophagic degradation of any intracellular structure.

Tobacco cells were mounted on a glass slide, and then their green and red fluorescence were observed using an Olympus IX80 inverted fluorescence microscope equipped with a DSU spinning disk confocal unit (Olympus) using an Uplan SApo 40 × /0.90 lens and GFP and RFP excitation/emission filter sets (Olympus). Images were collected using an iXON+-DU888E-CS0-BV camera (Andor) under the control of Meta Imaging Series 7.6.3 software (Molecular Devices Inc., Sunnyvale, California). Stored multicolor TIFF files were used for the quantification of red and green signals. After collecting image data of two colors, and as a 256-digit TIFF file, the intensities of dots are measured using ImageJ software (Abramoff, Magelhaes, & Ram, 2004) or using software in the Meta Image Series.

ACKNOWLEDGMENTS

This work was supported in part by MEXT KAKENHI Grants 19380045 and 26292194 to K.M.

REFERENCES

Abiodun, M. O., & Matsuoka, K. (2013a). Evidence that proliferation of Golgi apparatus depends on both de novo generation from the endoplasmic reticulum and formation from pre-existing stacks during the growth of tobacco BY-2 cells. *Plant & Cell Physiology*, *54*, 541–554. http://dx.doi.org/10.1093/pcp/pct014.

Abiodun, M. O., & Matsuoka, K. (2013b). Proliferation of the Golgi apparatus in tobacco BY-2 cells during cell proliferation after release from the stationary phase of growth. *Plant Signaling & Behavior*, *8*. e25027, http://dx.doi.org/10.4161/psb.25027.

Abramoff, M. D., Magelhaes, P. J., & Ram, S. J. (2004). Image processing with ImageJ. *Biophotonics International, 11*, 36–42.

Aubert, S., Gout, E., Bligny, R., Marty-Mazars, D., Barrieu, F., Alabouvette, J., et al. (1996). Ultrastructural and biochemical characterization of autophagy in higher plant cells subjected to carbon deprivation: Control by the supply of mitochondria with respiratory substrates. *The Journal of Cell Biology, 133*, 1251–1263.

Baird, G. S., Zacharias, D. A., & Tsien, R. Y. (2000). Biochemistry, mutagenesis, and oligomerization of DsRed, a red fluorescent protein from coral. *Proceedings of the National Academy of Sciences of the United States of America, 97*, 11984–11989.

Brown, S. C., Bolte, S., Gaudin, M., Pereira, C., Marion, J., Soler, M., et al. (2010). Exploring plant endomembrane dynamics using the photo-convertible protein kaede. *The Plant Journal, 63*, 696–711.

Chen, M. H., Liu, L. F., Chen, Y. R., Wu, H. K., & Yu, S. M. (1994). Expression of α-amylases, carbohydrate metabolism, and autophagy in cultured rice cells is coordinately regulated by sugar nutrient. *The Plant Journal, 6*, 625–636.

Contento, A. L., Kim, S. J., & Bassham, D. C. (2004). Transcriptome profiling of the response of Arabidopsis suspension culture cells to Suc starvation. *Plant Physiology, 135*, 2330–2347.

Hawes, C., Schoberer, J., Hummel, E., & Osterrieder, A. (2010). Biogenesis of the plant Golgi apparatus. *Biochemical Society Transactions, 38*, 761–767.

Liu, J., Hayashi, K., & Matsuoka, K. (2015). Membrane topology of Golgi-localized probable S-adenosylmethionine-dependent methyltransferase in tobacco (Nicotiana tabacum) BY-2 cells. *Bioscience, Biotechnology, and Biochemistry, 79*, 2007–2013.

Matsuoka, K. (2009). Chimeric fluorescent fusion proteins to monitor autophagy in plants. In D. Klionsky (Ed.), *Methods in enzymology: Vol. 451. Autophagy: Lower eukaryotes and non-mammalian systems* (pp. 541–555). Oxford: Elsevier Scientific Publishers.

Matsuoka, K., & Nakamura, K. (1991). Propeptide of a precursor to a plant vacuolar protein required for vacuolar targeting. *Proceedings of the National Academy of Sciences of the United States of America, 88*, 834–838.

Moriyasu, Y., & Ohsumi, Y. (1996). Autophagy in tobacco suspension-cultured cells in response to sucrose starvation. *Plant Physiology, 111*, 1233–1241.

Tasaki, M., Asatsuma, S., & Matsuoka, K. (2014). Monitoring protein turnover during phosphate starvation-dependent autophagic degradation using a photoconvertible fluorescent protein aggregate in tobacco BY-2 cells. *Frontiers in Plant Science, 5*, 172. http://dx.doi.org/10.3389/fpls.2014.00172.

Toyooka, K., Takeuchi, M., Moriyasu, Y., Fukuda, H., & Matsuoka, K. (2006). Protein aggregates are transported to vacuoles by macroautophagic mechanism in nutrient-starved plant cells. *Autophagy, 2*, 91–106.

Tsutsui, H., Karasawa, S., Shimizu, H., Nukina, N., & Miyawaki, A. (2005). Semi-rational engineering of a coral fluorescent protein into an efficient highlighter. *EMBO Reports, 6*, 233–238. http://dx.doi.org/10.1038/sj.embor.7400361.

Yamauchi, N., Gosho, T., Asatsuma, S., Toyooka, K., Fujiwara, T., & Matsuoka, K. (2013). Polarized localization and borate-dependent degradation of the Arabidopsis borate transporter BOR1 in tobacco BY-2 cells. *F1000Research, 2*, 185. http://dx.doi.org/10.12688/f1000research.2-185.v1.

Yoshimoto, K., Hanaoka, H., Sato, S., Kato, T., Tabata, S., Noda, T., et al. (2004). Processing of ATG8s, ubiquitin-like proteins, and their deconjugation by ATG4s are essential for plant autophagy. *The Plant Cell, 16*, 2967–2983.

Yuasa, K., Toyooka, K., Fukuda, H., & Matsuoka, K. (2005). Membrane-anchored prolyl hydroxylase with an export signal from the endoplasmic reticulum. *The Plant Journal, 41*, 81–94.

AUTHOR INDEX

Note: Page numbers followed by "*f*" indicate figures, and "*t*" indicate tables.

B

C

D

E

G

H

L

N

O

P

T

U

V

W

X

Y

SUBJECT INDEX

Note: Page numbers followed by "*f*" indicate figures, and "*t*" indicate tables.

B

N

O

P

Q

R